РОСТ КРИСТАЛЛОВ

ROST KRISTALLOV

GROWTH OF CRYSTALS

VOLUME 11

# *Growth of Crystals*

## *Volume 11*

**Edited by**

**A. A. Chernov**

**Institute of Crystallography**
**Academy of Sciences of the USSR, Moscow**

**Translated by**

**J. E. S. Bradley**

**Senior Lecturer in Physics**
**University of London**

CONSULTANTS BUREAU · NEW YORK AND LONDON

The Library of Congress cataloged the first volume of this title as follows:

Growth of crystals. v. [1]
New York, Consultants Bureau, 1958–

v. illus., diagrs. 28 cm.

Vols. 1, 3– constitute reports of 1st– Conference on Crystal Growth, 1956– v. 2 contains interim reports between the 1st and 2d Conference on Crystal Growth, Institute of Crystallography, Academy of Sciences, USSR.

"Authorized translation from the Russian" (varies slightly)
Editors: 1958– A. V. Shubnikov and N. N. Sheftal'.

1. Crystals – Growth. I. Shubnikov, Aleksei Vasil'evich, ed. II. Sheftal', N. N., ed. III. Consultants Bureau Enterprises, inc., New York. IV. Soveshchanie po rostu kristallov. V. Akademiia nauk SSSR. Institut kristallografii.
QD921.R633 548.5 58-1212

Library of Congress Catalog Card Number 58-1212
ISBN 978-1-4615-7115-5 ISBN 978-1-4615-7113-1 (eBook)
DOI 10.1007/978-1-4615-7113-1

The original Russian text, published for the Institute of Crystallography of the Academy of Sciences of the USSR by Nauka Press in Moscow in 1975, has been corrected by the editor for this edition.


Softcover reprint of the hardcover 1st edition 1979
A Division of Plenum Publishing Corporation
227 West 17th Street, New York, N.Y. 10011

# PREFACE

The Growth of Crystals series was begun in 1957 by A. V. Shubnikov and N. N. Sheftal' with the publication of the first volume, which contained the proceedings of the First All-Union Conference on Crystal Growth. The initiative and considerable efforts of the principal editor of the entire series, N. N. Sheftal', and his assistants led over the next 15 years to the publication of ten volumes which have assumed a leading position among the numerous books on crystal growth. It has become traditional in this series to adopt a broad approach to crystal growth problems, and this approach is continued in Volumes 11 and 12, which are composed mainly of papers presented at the Fourth All-Union Conference on Crystal Growth in Tsakhkadzor, September 17-22, 1972. These papers, presented by both Soviet and foreign workers, deal with crystal growth processes, growth methods, and crystal perfection.

Many of the papers reflect the tendency for our knowledge of crystallization processes to become increasingly more fundamental, with emphasis on quantitative treatments. There are some extremely difficult problems in this approach, especially when the requirements of practical uses are envisaged, and many of these are discussed in various ways in these two volumes. These topics include detailed theoretical and experimental analysis of cooperative phenomena in crystallization, with emphasis not only on statistical thermodynamics but also statistical kinetics. This approach involves research on the structure and properties of phase boundaries, including the composition and structure of surface layers in liquids, the frequencies and types of elementary acts at phase boundaries, impurity trapping, and the formation of metastable phases. Any ultimate solution in this field requires an enormously large body of information on liquid structure, particularly for multicomponent liquids, and the topics are of interest not only to physicists, but also to chemists and researchers in physical chemistry. Another aspect, equally important but less researched, is the kinetics of electronic states in elementary acts of crystallization and adsorption, together with the effects of external electrical and magnetic fields. In addition, such fields affect the formation of crystals generally. There are also major mathematical difficulties in analyzing macroscopic diffusion, convective and radiation processes, and the stresses in growing crystals.

Quantitative experimental studies in these areas involve major difficulties, which arise on account of the exceptional sensitivity of surface processes to the external conditions. Nevertheless, such researches are in hand, although they are concentrated mainly on macroscopic processes. Quantitative research on elementary steps is at present possible only for crystallization from the vapor state, although in recent years it has proved possible to measure exchange ion currents at elementary stages in solutions. The prospects for research on surface states in condensed phases are good, particularly when resonance methods and certain related techniques are employed.

Systematic fundamental researches on crystallization will lead to further progress in this area, but they would be inconceivable without a reasonably broad approach to the problems. It is becoming steadily more difficult to maintain this broad approach with the ever-increasing flow of information. The solution to this must be to take care not to reject what is of value in the older knowledge but to formulate fresh hypotheses providing deeper insights. In this way we should be able to facilitate fresh discoveries. It has repeatedly been observed that considerable assistance in maintaining a proper balanced viewpoint under conditions of superabundance of detailed information is derived from frequent reviews. Therefore, about one-third of the papers in Volumes 11 and 12 of Growth of Crystals are devoted to reviews on various aspects of crystal growth.

The reviews and the original papers are distributed by topic between the volumes. Following tradition, the first papers in Volume 11 deal with nucleation at surfaces and in the bulk; here much attention is given to molecular kinetics and the role of defects and inhomogeneities in the production of centers and critical nuclei.

The next section deals with layered growth, which presents researches on elementary parameters and kinetic coefficients in crystallization, step interaction, and growth-surface morphology.

The next section is a logical continuation and deals with the macroscopic consequences of elementary surface processes such as crystal growth forms, stability in growth forms, and heat and mass transport.

A considerable body of papers also deals with impurity trapping, with particular emphasis on the kinetic aspect of this problem.

The emphasis in Volume 12 will be on techniques of crystal growth and aspects of crystal perfection in relation to growth conditions. These topics have become particularly complicated and important on account of the enormous demand for large perfect single crystals for use in solid state physics and chemistry, and especially in novel technologies. The production of large single crystals requires not only proper use of all the latest advances in research on crystallization mechanisms but also a very detailed knowledge of the causes of defects such as inclusions, nonuniform impurity distributions, dislocations, blocks, internal stresses, etc. Solutions here must come largely from new technological principles and equipment designs, which allow one to provide closely controlled crystallization conditions. The very rigid current specifications for crystal perfection have raised various complicated but important problems in equipment design. This applies particularly to the precision maintenance and measurement of temperatures, including high temperatures, and the provision of appropriate temperature patterns, physicochemical problems in crucible choice, difficulties over reagent purity, etc. The single-crystal industry is at present growing very rapidly, and this has made some of these problems extremely acute. The specifications imposed in the production of large single crystals have a close relationship also to the production of single-crystal films.

Volume 12 will contain papers on the growth of crystals by various methods from the vapor state, from low-temperature solutions, from high-temperature solutions (including hydrothermal ones), and from melts. The section on defect formation is dominated by papers dealing with the formation of dislocations, the origin of internal stresses, and the causes of impurity distribution.

The editorial board is indebted to many collaborators in the Institute of Crystallography at the Academy of Sciences of the USSR and at Erevan State University, especially L. A. Solomentsev, T. A. Lebedev, L. N. Obolenskaya, S. A. Grinberg, A. M. Mel'nikov, L. V. Prikhod'ko, N. A. Mekhed, V. I. Muratov, A. G. Nalbandyan, K. B. Seiranyan, and A. Kh. Eritsyan, who have provided a great deal of assistance in preparing the manuscipts of both volumes. We are also very much indebted for collaboration from numerous Soviet and foreign specialists on crystal growth in the early publication of these volumes.

A. A. Chernov
Kh. S. Bagdasarov
E. I. Givargizov
R. O. Sharkhatunyan

# CONTENTS

## IV. IMPURITY TRAPPING

# REACTION FEATURES OF SILICA

**N. V. Belov and E. N. Belova**

*Institute of Crystallography, Academy of Sciences of the USSR, Moscow*

The history of the production of large synthetic quartz crystals goes back nearly 100 years; the problem involves considerable difficulties on account of the exceptionally low solubulity of $SiO_2$ and of silicates generally in ordinary solvents, together with the exceptional difficulty of handling $SiO_2$ in the form of silica in classical chemical analysis. A series of studies about 15 years ago, including some of our own, repeatedly confirmed the view that silicon is an extremely inert or immobile element, which as its oxide $SiO_2$ is almost completely incapable of participating in reactions, particularly those utilized in classical analytical chemistry. Some have sought to argue, in spite of the inertness, that silicon–oxygen patterns are readily adapted to other details of silicate architecture, but in that case one is really speaking of a form of chemical reaction. No conflict arises if we do not seek to equate reactions in dilute fluids with those in condensed phases, especially solids, since in the latter we find that silica can readily adapt not only to the cation framework but also to ephemeral rings of water molecules and even large clusters of other structures such as zeolites, A and X molecular sieves, etc. [1].

Undoubtedly, one of the major advances in mineralogy and silicate technology (e.g., cement and glass) has come from x-ray researches, which have shown that the basic unit in nearly all silicates is the $SiO_4$ tetrahedron, which is either individualized, when the four O atoms do not participate in the environment of adjacent Si atoms, or else the $SiO_4$ forms part of some larger silicate radical, while still remaining a reasonably regular tetrahedron, although some of the vertices, or perhaps all, participate in the environment of adjacent Si atoms, or rather in adjacent equivalent $SiO_4$ tetrahedra. It is generally accepted to say that a diortho group or pyro group $Si_2O_7$ is the result of condensation of two tetrahedra: $2SiO_4 = [Si_2O_7] + O$, with six $SiO_4$ condensing to a $[Si_6O_{18}]$ ring: $[Si_6O_{18}] = 6SiO_4 - 6O$. Similar equations are readily written for the unbounded radicals in pyroxenes, amphiboles, networks, and even frameworks, but in every case they involve release of free oxygen, whose subsequent external fate is of no particular interest. No search has been made for the reducing agent that would absorb this oxygen, and there is none. An exception may perhaps be made for those few statements by petrographers on the environment of amphibole formed after pyroxene, which contains grains of magnetite, and which is due to oxidation of the FeO from the primary pyroxene.

A $[SiO_4]^{4-}$ tetrahedron is itself a large cluster composed of four O anions linked to the highly charged $Si^{4+}$, and during the last 40 years this has become accepted in mineralogy and in earth sciences generally as capable as transferring as a whole from one compound or phase to another. In particular, the present authors have discussed the position of femic components from a differentiating magma and have written that the originally depositing MgO takes up individual $SiO_4$ tetrahedra to produce olivine: $2MgO + SiO_4 = Mg_2[SiO_4] + O_2$ (?), although no question of the released oxygen was raised. A similar problem arises over $Si_2O_7$ groups and even over the change required to produce pyroxene and biotite from MgO.

In recent years it has repeatedly been stated [2] that the $SiO_4$ tetrahedron is undoubtedly the basic static unit in any silicon–oxygen structure, i.e., is the silicate brick, but one which is produced directly at the point of use and introduced into preexisting structures as single mobile neutral $SiO_2$ molecules. Conversely, an entire $SiO_4$ brick cannot be released from a complex silicate framework, since all four vertices are closely linked to four equivalent $SiO_4$ bricks, and the latter would thus be destroyed by the release of one $SiO_4$. Such a yield of only 20% would be too small for a solid-state reaction. Therefore, only $SiO_2$ molecules may be considered as dynamic units, and they transport silica in living organisms (silicosis), in the walls of furnaces, and in other such circumstances, e.g., in the formation of olivine, where the neutral or basic MgO extracts neutral $SiO_2$ from the magma, although the mineralogist considers such molecules as acid (electronegative), and although when we write the equation $2MgO + SiO_2 = 2MgO \cdot SiO_2 = Mg_2SiO_4$ we take oxygen from the first member and attach it on paper to the Si, which leads to speak of the $[SiO_4]^{4-}$ tetrahedron as a basic silicate brick in the olivine building. As the magma cools, successive neutral batches of $SiO_2$ are detached, and the $SiO_4$ tetrahedra are bound into pyroxene chains: $Mg_2[SiO_4] + SiO_2 = Mg_2[SiO_6]$, the scheme being

$$
\begin{array}{cccccccc}
 & & O_2 & & & & & \\
 & & | & & & & & \\
O\diagdown\ \diagup O & \ldots\ldots & Si & \ldots\ldots & O\diagdown\ \diagup O & = & \ldots\ Si-O-Si-O-Si\ \ldots & = [Si_2O_6]_\infty \\
Si & & & & Si & & |\qquad |\qquad | & \\
\cdot & & & & | & & O_2\quad\ O_2\quad\ O_2 & \\
O_2 & & & & O_2 & & &
\end{array}
$$

As the silicification proceeds, i.e., as the pyroxenes are formed, new batches of $SiO_2$ are taken up and result in silicon–oxygen networks of talc–biotite type: $[Si_2O_6]_\infty + 2SiO_2 = [Si_4O_{10}]_{\infty\infty}$ [2].

We have already noted above the special position taken by amphibole in this particular sequence.

It is readily seen that if there are cations bearing their own oxygen (MgO, FeO, or $ZrO_2$), one can get silicification in a pure $SiO_2$ framework (quartz, tridymite, cristobalite), i.e., in a three-dimensional network of Si tetrahedra, in which all vertices are shared with neighbors. The pale (feldspar) constituents of rocks contain three-dimensional frameworks composed of tetrahedra, and these are formed in situ, i.e., within the initial magmatic glass around large cations such as Ca, Na, and K [2].

In the original studies on the formation of complex silicon–oxygen radicals in femic minerals, nothing was said on the state of aggregation of the silicifying $SiO_2$ molecule on the path to the final position, namely from the acid phase to the basic one; this was imagined as being gaseous, although it is difficult to imagine a gaseous layer between the phases in a condensed system at very high pressures and therefore any long life for the $SiO_2$ molecule.

Ten years ago it was supposed [3] that, if the true basis of a silicate structure is to be seen in the large cation polyhedra, with small Si atoms placed between the oxygen vertices (with all the O attached to these by tradition), then the small Si should readily migrate, and that without any oxygen burden, thereby being able to jump from oxygen tetrahedron to another or to an adjacent hole. The present writers for long viewed this possiblity at the very least skeptically since in our laboratory we obtained no evidence for such jumps from x-ray analysis of hydrated Ca–Al sodalite [4].

To find an explanation we need to go back to the 1890s, when coordination concepts were introduced into chemistry and mineralogy (by Werner and Vernadskii), when the coordination principle was received with considerable skepticism, and therefore the state, for instance, of aluminum in the coordination octahedron was still described by means of three valency bonds (solid lines) and three additional bonds (broken lines):

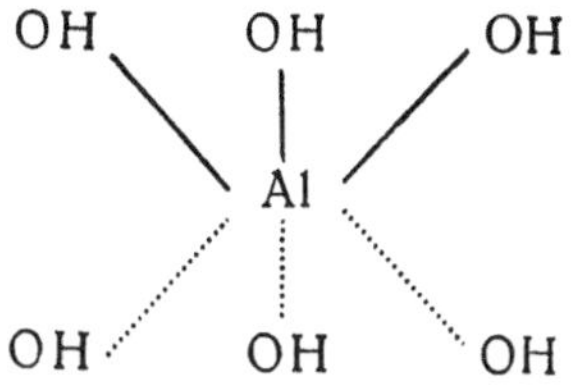

```
OH     OH     OH
  \     |     /
   \    |    /
       Al
   .    :    .
  .     :     .
OH     OH     OH
```

Similarly, the $SiO_4$ tetrahedron was represented by means of two solid lines and two broken lines,

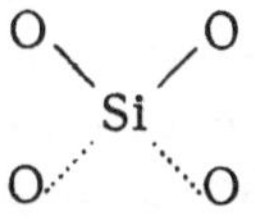

```
O     O
 \   /
  Si
 .   .
O     O
```

i.e., a distinction was drawn between two valency bonds and two auxiliary O.* Correspondingly, only one of the six edges of the tetrahedron was taken as a true or inherent edge. Today we can return to this concept, but with a modern cofrection for mesomerism (resonance): at any given instant, the true valency bonds are only two out of the four, but these readily exchange roles, and correspondingly they meet the requirements of any particular moment in that any two bonds can become the true ones, with only one of the six edges being the real edge, much as in the aldehyde – ketone resonance in organic chemistry. Therefore, the Si jumps through such an edge into an adjacent empty tetrahedron and reorganizes its bonds. We have observed the detailed result of such a jump in the hydration of Ca–Al sodalite [4]. Chemical crystallographers long ago established that an Si tetrahedron could not have edges in common with an adjacent Si tetrahedron, i.e., six tetrahedra having edges in common with some Si tetrahedron (for instance, in a close-packed oxygen environment) would result in a cavity, since potentially the central Si atom is already present.

Returning to the migration of silicon in the form of silica, we assume that the gas-molecular state of $SiO_2$ between two condensed phases persists for only a very short time, i.e., until the $SiO_4$ tetrahedron approaches along one edge (in the more acid phase) a perpendicular O–O edge in an Mg(Fe) octahedron in the more basic phase under high pressure (the basic phase is electronegative), and the resulting purely geometrical cavity is filled by an Si atom jumping through its own edge, thus leading to detachment from the previous phase, while the atom then appears in the adjacent phase as silicifying $SiO_2$ within a new $SiO_4$ tetrahedron (Fig. 1).

This "conjurer" mechanism for Si jumping from tetrahedron to tetrahedron (with shared edges) readily explains the fluidity of glass at elevated temperatures, the retention of continuity under plastic deformation, many effects observed in glass blowing, and so on,† but particularly the adaptability of $SiO_2$ to any basic pattern in a crystal structure, which is the basis of the dis-

* In order then (and now also) to avoid discussion of the $SiO_2$ molecule; for some while, until quantum chemistry arose, such bonds were called covalent, but of course not in the sense used nowadays.

† The formula for ordinary (window) glass $Na_2O \cdot CaO \cdot 6SiO_2$ has the ratio $Si : O = 6 : 14 \approx 2 : 5$, which was formerly considered characteristic of platy (network) silicates, and therefore silicon–oxygen networks frequently considered as the basis of the vitreous state. We have elucidated various silicate structures having $Si : O = 2 : 5$, but represented by double strips or chains: $[Si_6O_{15}]$, $[Si_4O_{10}]$, and so on, i.e., $[Si_2O_5]$ describes equally well silicates in macroscopic glass bodies (and also others represented by two subscript infinities in the formulas), as well as the glass tubes (one subscript infinity) commonly found in glass blowing.

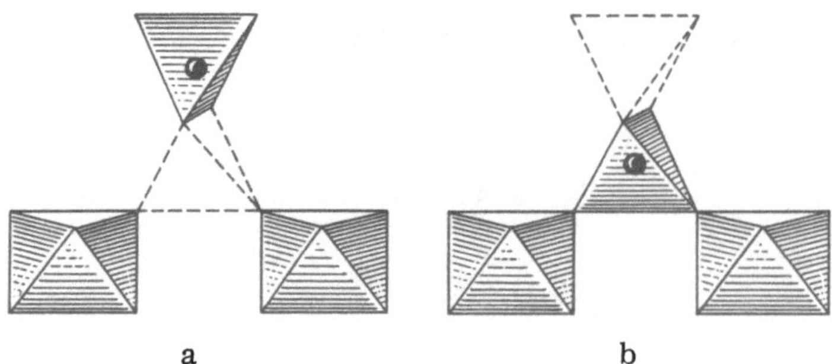

Fig. 1. Two stages in the transition of an $SiO_2$ molecule from one condensed phase to another: (a) a perpendicularly oriented O–O edge from the lower phase approaches a filled tetrahedron in the upper phase, and a cavity is formed between the phases (shown by broken lines); (b) this empty tetrahedron maintains its edges but is filled by an Si atom jumping from the other tetrahedron, and the new tetrahedron is then entirely in the lower phase when the secondary bonds have been formed. The upper tetrahedron becomes empty (broken lines).

tinction between the first and second chapters of silicate chemical crystallography, which involve intermediate and large cations respectively. The rarer occurrence of tetrahedra in larger components such as large nuclei (which arise in stable Debye-wave systems) is responsible for the devitrification in many forms of silica, e.g., cristobalite.

An unexpected solution (or approach to solution) was obtained for the position of aluminum in glass; there is no doubt that here all the aluminum is present in tetrahedra, but these tetrahedra are 10-12% larger than Si tetrahedra, and therefore there was a tendency to say that around each Si tetrahedron there should be six empty tetrahedra at least potentially available, but the edges would not be those that would link to the Si atom, which was seen as a basic characteristic of the vitreous state, since the silicon could then readily return to its original tetrahedron. This was also seen as the reason for the ready deposition of Al-rich phases as ordierite and so on, and attempts were made to rationalize this a priori undesirable phenomenon in the production of various types of sintered glasses. The plagioclase feldspars have the aluminum-rich anorthite crystallizing more readily, but then more readily weathering (decaying) than the silicon-rich and therefore more vitreous and stable albite (deanorthitization-albitization in feldspars).

Returning to the initial difficulties encountered in growing large quartz crystals, we see that the $SiO_4$ tetrahedron was the cause of the difficulties in analytical chemistry, since $SiO_4$ tetrahedra are normally associated, and the consequent jumps are therefore readily performed only within a body of associated tetrahedra ($[Si_2O_5]_{\infty\infty} \rightleftharpoons [Si_2O_5]_{\infty}$). The difficulties in analytical chemistry very largely arise because the latter envisages distinct particles, whereas silica acts in mass without adaptation to the cation background, with simple enclosure in a film of linked tetrahedra. A similar mechanism is impossible for free water molecules, since one requires a carrier, which for many reasons that cannot be considered here is the readily soluble Na, whose most common form of coordination polyhedron is a trigonal prism bearing an attached half-octahedron, whose short edges allow of ready adaptation to two $SiO_2$ molecules, which thus forms a kind of seed, which deposits by the same mechanism, but with the Na tending to fivefold coordination rather than sevenfold, as is characteristic of many crystalline Na silicates.

## Literature Cited


1. N. V. Belov, Chemical Crystallography of Silicates Containing Large Cations [in Russian], Izd. AN SSSR, Moscow (1961).
2. N. V. Belov, E. N. Belova, G. P. Litvinskaya, and Yu. A. Kharitonov, Vestn. Mosk. Univ., Ser. Geol., 25(4):8 (1970).
3. H. Taylor, ed., Chemistry of Cement [Russian translation], Mir, Moscow (1969).
4. V. I. Ponomarev, L. M. Kheiker, and N. V. Belov, Kristallografiya, 15:918 (1970); Neorgan. Mater., 7:1783 (1970).

Part I

# NUCLEATION AND INITIAL GROWTH STAGES

# HOMOGENEOUS NUCLEATION IN A LIQUID METAL

D. E. Ovsienko

*Institute of Metal Physics, Academy of Sciences of the Ukrainian SSR, Kiev*

There are many papers on nucleation kinetics, and various important laws have been established, which give an insight into the physical essence of solidification; however, some aspects remain unclear even now, especially as regards the observation of homogeneous (spontaneous) nucleation. This is due in part to the lack of a rigorous quantitative theory, but the main reason is the difficulty in eliminating the effects of insoluble components, which play a decisive part in nucleation and which often lead to conflicting results. This is also the main reason why there have been several changes of view on the subject over the last few decades. For instance, at one time it was assumed that Tamman curves characterize spontaneous nucleation, but it was later shown that these curves are certainly related to nucleation at impurities, while certain substances do not crystallize at all after removal of insoluble components, which meant that doubt was cast on spontaneous crystallization generally, and many came to believe that nuclei arise only at solid impurity particules. From about 1950 onwards, it gradually became clear again that homogeneous crystallization actually can be observed, particularly after microvolumes became generally used, which enable one to attain very high degrees of supercooling. It seemed that this method eliminated the effects of insoluble components and could give access to maximum supercoolings of about $0.18T_{mp}$ for metals, which corresponds to homogeneous nucleation. However, in recent years there has been much evidence that cast doubt on whether spontaneous nucleation can be observed at all for metals.

In the present paper we consider in more detail the state of the subject, with considerable given to nucleation at solid surfaces generally.

## 1. Nucleation Theory

### (a) Homogeneous Nucleation

The classical theory [1, 2] envisages that nuclei of a new phase are formed within the volume of an old one, which is in that case a metastable phase, via a fluctuation process, which gives the following expression for the nucleation rate:

$$J = Ke^{-W_0/kT}, \tag{1}$$

where $W_0$ is the work of formation for a critical nucleus, which for the spherical case takes the following form for a liquid-to-crystal transformation:

$$W_0 = \frac{16\pi}{kT}\left(\frac{M}{\rho}\right)^2\left(\frac{T_0}{q}\right)^2\frac{\sigma^3}{(\Delta T)}. \tag{2}$$

Here k is Boltzmann's constant, M molecular weight, $\rho$ density of the nucleus, q latent heat of fusion per molecule, $\sigma$ surface tension at the crystal–melt boundary, $T_0$ the equilibrium temperature for the solid and liquid phases, T the absolute temperature of the supercooled liquid, and $\Delta T = T_0 - T$ the supercooling. The preexponential factor K in (1) is governed by the rate of exchange of molecules between the nucleus and the initial phase, and it can [2] be put in the following form for crystallization:

$$K = K_0 e^{-U/kT}, \tag{3}$$

where U is the activation energy, which is close to the activation energy for self-diffusion in the liquid; the kinetic coefficient $K_0$ incorporates the nonstationary factor z ($z \approx 10^{-2}$) [3] and takes [4] the form

$$K_0 = z i_c (a\sigma/9\pi kT)^{\frac{1}{2}} n(kT/h), \tag{4}$$

where $i_c$ is the number of atoms at the surface of a critical nucleus, n is the number of atoms per $cm^3$, and h is Planck's constant. The numerical values for metals always give roughly the value $K_0 \approx 10^{33}$, although it is considered in some quarters [5] that the factor should be increased by roughly a factor $10^{10}$-$10^{12}$ on account of the contribution from rotational degrees of freedom to the free energy of a nucleus. On that view, there should be a contribution also from the translational degrees of freedom for vapor–liquid conversion, in addition to a contribution from the energy involved in separating the atoms in a nucleus from a large ensemble, which altogether multiply the preexponential constant by a factor of $10^{17}$. However, it has been shown [6-8] that these corrections have no real basis in the classical theory. The arguments on this topic still continue [10], and as yet we do not know the true value for $K_0$, which hinders quantitative evaluation of the theory.

We substitute (2) and (3) into (1) to get

$$J = K_0 e^{-U/kT} \cdot e^{-B\sigma^3/T(\Delta T)^2}, \tag{5}$$

where

$$B = \frac{16\pi}{kT}\left(\frac{M}{\rho}\right)^2\left(\frac{T_0}{q}\right)^2. \tag{6}$$

Figure 1 shows this relationship between J and $\Delta T$, which implies that the two increase together on account of the increase in the nucleation probability; on the other hand, the fall in J at large $\Delta T$ is due to the $e^{-U/kT}$ factor, which reflects the reduced molecular mobility at low temperatures, and hence the reduced rate of exchange between the nucleus and the liquid.

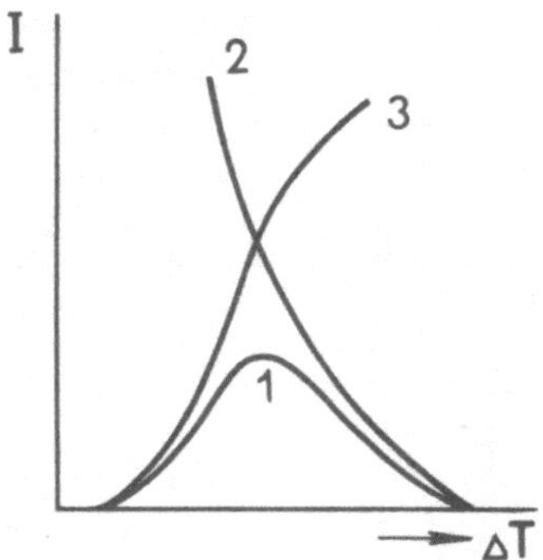


Fig. 1. Temperature dependence of: 1) I; 2) $e^{-U/kT}$; 3) $e^{-B\sigma^3 T(\Delta T)^2}$.

Therefore, U, B, and $\sigma$ should define the metastability boundary, as well as the height and position of the peak in the J($\Delta$T) curve; if we take the molecular mobility as very small, i.e., U/kT very large, the substance cannot crystallize and instead goes over to the vitreous state (salol, glycerol, etc). The atoms in a metal are highly mobile, so the retarding factor is unimportant and the nucleation rate is determined by the probability factor. Therefore, only the rising branch of the curve in Fig. 1 should be observed at normal cooling rates for metals.

These qualitative predictions are in general agreement with experiment; a quantitative check on (5) is difficult on account of the lack of an exact value for $K_0$ and the presence of the parameter $\sigma$, which is difficult to determine and also is a macroscopic quantity not applicable to the small groups of atoms involved in a nucleus. An attempt has been made [10] to incorporate this feature by expressing $\sigma_r$ as a function of the radius r as

$$\sigma_r = \sigma_\infty (1-2\delta/r), \tag{7}$$

where $\delta$ is the thickness of the transition layer, which is taken as 1-2 molecular diameters. Then $\sigma_r = 0.8\sigma_\infty$ for a nucleus. However, Buff [11] has shown that (7) is not applicable at the very high curvatures found in nuclei, and therefore the problem remains unresolved.

## (b) Heterogeneous Nucleation

In the condensation of a vapor, the work of nucleation (to produce a spherical segment) for a nucleus forming on a solid surface is given by the classical theory [1] as

$$W_s = W_0(\tfrac{1}{2} - \tfrac{3}{4}\cos\theta + \tfrac{1}{4}\cos^3\theta) \equiv W_0 f(\theta), \tag{8}$$

where $W_0$ is the work of formation of a droplet within the volume of the vapor and $\theta$ is the contact angle, which is expressed in terms of the surface energies: solid–vapor $\sigma_{sv}$, solid–liquid $\sigma_{sl}$, and liquid–vapor $\sigma_{lv}$:

$$\cos\theta = \frac{\sigma_{sv} - \sigma_{sl}}{\sigma_{lv}}. \tag{9}$$

It follows from (8) that if the surface is not wetted at all ($\theta$ = 180°), then $W_s = W_0$, i.e., the substrate has no effect. On the other hand, $W_s$ falls as the substrate becomes wetted, and it becomes zero for $\theta$ = 0°. Consequently, the nucleation energy for a wetted surface may be much less than the bulk value. Formally speaking, this theory can [4] be extended also to crystalline nuclei formed in a liquid simply by substituting other surface energies $\sigma_{sl}$, $\sigma_{sn}$, and $\sigma_{nl}$, where the subscripts s, l, and n now relate respectively to the substrate, liquid, and nucleus. Then instead of (9) we have

$$\cos\theta = \frac{\sigma_{sl} - \sigma_{sn}}{\sigma_{nl}}. \tag{9'}$$

Kozachkovskii [12] and Kaishev [13] envisaged a nucleus as a parallelepiped of height h and length $2l$ bounded by cube planes, and they derived the left side of (9) as $1 - h/2l$ instead of $\cos\theta$, and correspondingly $W_s = W_0 h/l$.

If we take a nucleus as a spherical segment, then the nucleation rate at a solid surface is

$$J_s = K_s e^{-U/kT} e^{-W_0 \cdot f(\theta)/kT} \tag{10}$$

or with (2) and (6),

$$J_s = K_s e^{-U/kT} e^{-B\sigma^3_\bullet/T(\Delta T)^2}, \tag{10'}$$

where

$$\sigma_{*}=\sigma_{nl}[f(\theta)]^{\frac{1}{3}}\equiv\sigma[f(\theta)]^{\frac{1}{3}}, \tag{11}$$

and $K_s$ is given by

$$K_s=K_0\frac{n_s}{n}[f(\theta)]^{\frac{1}{6}}. \tag{12}$$

Here $K_0$ is the kinetic coefficient of (4) for bulk nucleation, n is the number of atoms per $cm^3$, and $n_s$ is the number of atoms in the nucleus in contact with 1 $cm^2$ of solid surface. An estimate [4] gives $K_s \approx 10^{26}$.

Comparison of (10) and (5) shows that for $W_s \ll W_0$ the nuclei should arise on a substrate at much lower supercoolings, i.e., the metastability limit should be very much reduced. The physical meaning of these specifications for $\sigma_{sl}$ and $\sigma_{ns}$ amounts to specification of crystallochemical correspondence between the lattices. A dislocation model has been used [4] to derive an expression for $\sigma_{ns}$ as the sum of two terms: a chemical one $\gamma$, which is dependent on the bond type and strength, and a structural one $\alpha(\delta-\varepsilon)$, i.e.,

$$\sigma_{ns}=\gamma+\alpha(\delta-\varepsilon). \tag{13}$$

Then (10) gave the supercooling as dependent on the relative discrepancy in lattice constant for the coherent nucleus and substrate as ($\delta = \Delta a / a$).

$$\Delta T=\frac{C}{\Delta S_v}\cdot\delta^2, \tag{14}$$

where C is a constant dependent on the elastic coefficients and $\Delta S$ is the specific entropy of melting. If the nucleus is not coherent, $\Delta T$ becomes a linear function of $\delta$.

## 2. Experiments on the Effects of Insoluble Impurities

It has long been known that impurities facilitate crystallization, but the true mechanism has been established only relatively recently; Danilov and his coworkers [14] have made a considerable contribution here, since they have shown that a liquid always crystallizes on particles under ordinary conditions, and these particles are responsible for the relationship between the supercooling and the superheating, on account of activation and deactivation processes. Danilov and Kozachkovskii [14] have shown for salol that completely inactive solid particles become activated by contact with the solid phase, and the more so the higher the temperature; the activation rate is maximal near the melting point of salol, whereas it is virtually zero at the boiling point of nitrogen. Salol nuclei are readily formed on particles activated in this way, and these show all the features of the normal Tamman curve. The number of nuclei decreases as the superheating is increased, and it is possible to deactivate the particles completely, whereupon the salol will no longer crystallize but becomes vitreous, i.e., the effect is as if the particles had been removed by filtration.

Metals show only the rising branch in $J(\Delta T)$, and here impurity deactivation is accompanied by a gradual increase in the supercooling as the superheating increases (Fig. 2a). Danilov found that the activation arises by the formation of a molecular contact layer of the substance, which has an elevated melting point, while he considered deactivation as consisting of destruction of this layer above its melting point. The relationship between the supercooling and superheating may also be dependent on the presence of micropores and other nonunifor-

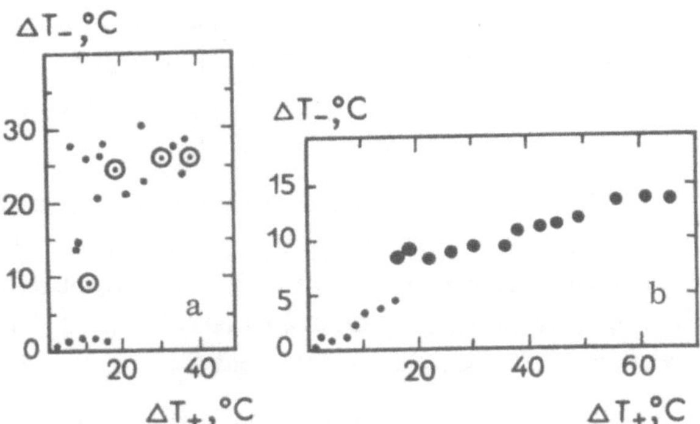


Fig. 2. Supercooling ($\Delta T_-$) as a function of superheating ($\Delta T_+$) for (a) tin; b) hydroquinone–$FeCO_3$ (after the specimen had been left in the crystalline state for 25 days).

mities, which Kozachkovskii [15] and Turnbull [16] have shown can persist above the melting point on account of the special form of phase equilibrium, which is determined by the relationship between the components of the surface tension.

Therefore, the relationship between the supercooling and superheating is due to this induced activity, which can sometimes be eliminated, as for salol, and thus can result in complete loss of impurity effects. However, there are solid particles that are naturally active and which are not deactivated by any superheating. Examples appear to be particularly numerous for metals, and they relate particularly to isomorphous components having some crystallochemical affinity with the main substance and thus exerting an orienting effect in phase transitions.

Detailed studies have been made of nucleation kinetics for naturally active surfaces [17-26], which have revealed various trends; for instance, a crystal of PbS introduced into an aqueous solution of NaCl or NaBr reduces the critical supersaturation from 16-20% to 6% for NaCl or from 5 to 0.6% for NaBr. These new metastability limits are due to nucleation on the PbS, and the values are very stable, i.e., are independent of the superheating and the number of repeated crystallizations, so it is possible to measure the nucleation rate $J_s = 1/S\tau$ (S is the area of the solid surface and $\tau$ is the nucleation delay) as a function of supersaturation $c/c_\infty$. The resulting curve is in general agreement with the predictions of the fluctuation theory. A major feature of this process is oriented nucleation on the PbS (Fig. 3), and the degree of orientation for the NaBr crystals is higher (97%) than that for NaCl (80%). This difference resembles the difference in the limiting supersaturations in being governed by the discrepancy between the lattice constants $\Delta a/a$, which is 6% for NaCl–PbS, as against 0.2% for NaBr–PbS. Another important point is that the most active nucleation sites on PbS are corners, edges, steps, and the like, as well as defective parts of faces generally. Figure 3a indicates

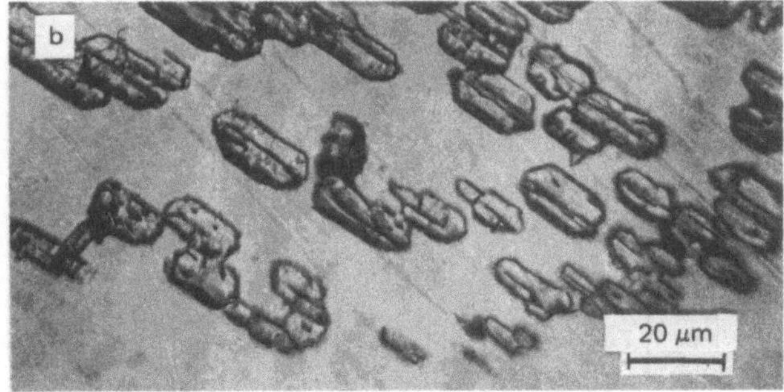


Fig. 3. Oriented nucleation: (a) NaBr from aqueous solution on a (001) face of a PbS single crystal; (b) molten hydroquinone on a $(10\bar{1}1)$ cleavage face of a $CaCO_3$ single crystal.

that the concentration of such nucleation sites at supersaturation of 0.6% is about $10^2$ $cm^{-2}$. However, if the supersaturation is higher, as is possible if a drop of solution is placed on the PbS, one can detect less active parts, and then the number of nuclei becomes as large as $10^6$ $cm^{-2}$.

Similar trends have been demonstrated [19, 20] for the crystallization of supercooled hydroquinone on calcite $CaCO_3$, zinc spar $ZnCO_3$, and siderite $FeCO_3$. All three minerals orient the hydroquinone crystals on deposition from solution and from the melt (Fig. 3b). The discrepancies $\Delta a/a$ and $\Delta b/b$ for the $(10\bar{1}1)$ planes are 13 and 9% respectively for $CaCO_3$, 20 and 0.5% for $ZnCO_3$, and 90 and 7% for $FeCO_3$.

This dimensional discrepancy implies that the minerals differ in catalytic activity (Table 1); calcite is the most active, and this produces nucleation at a supercooling of 4°C, which implies a displacement of the metastability limit for pure hydroquinone (36-37°) by almost a factor 10. $ZnCO_3$ and $FeCO_3$ crystals displace the limit to 15 and 18°C respectively. The width of the metastable range for heterogeneous nucleation, i.e., the range in which the crystallization delay $\tau$ varies from some hours down to 1-2 sec (essentially from infinity to zero) is relatively narrow and independent of the superheating, but the value varies from one mineral to another (Table 1). The observed relationship of $\ln J_s$ to $1/T(\Delta T)^2$ is linear for all the systems (Fig. 4), which agrees with (10) and indicates that nucleation is of fluctuation origin. Table 1 gives the $K_s$ derived from the intercept and the $\sigma_* = \sigma[f(\theta)]^{1/3}$ derived from the slope of the straight line.

TABLE 1

| No. | System | $\Delta T_m$, °C | Range $[\Delta T]$, °C | $K_s$ | $\sigma_*$, erg/cm² | Specimen volume, cm³ |
|---|---|---|---|---|---|---|
| 1 | Pure hydroquinone | 37 | 1.6 | $10^{29}$ | 23,5 | 0.06 |
| 2 | Hydroquinone–$FeCO_3$ | 18 | 2.5 | — | 10 | 0.06 |
| 3 | Hydroquinone–$ZnCO_3$ | 15 | 1.5 | $10^{11}$ | 9 | 0.06 |
| 4 | Hydroquinone–$CaCO_3$ | 4.4 | 1.0 | $10^5$ | 8,6 | 0.05 |
| 5 | Clean bismuth | 46 | 6 | $10^{11}$ | 25 | 0.05 |
| 6 | Oxidized bismuth | 13 | 1.5 | $10^{11}$ | 10 | 0.05 |
| 7 | Clean lead | 8 | 1 | — | 6 | 0.04 |
| 8 | Oxidized lead | 3.5 | 0.5 | — | 4 | 0.04 |
| 9 | Oxidized potassium | 1.5 | 0.25 | $10^4$ | 0,4 | 0.03 |
| 10 | Clean sodium | 3.5 | 0.15 | $10^{22}$ | 2 | 0.03 |
| 11 | Oxidized sodium | 2.5 | 0.15 | $10^6$ | 1 | 0.03 |
| 12 | Clean iron | 280 | — | — | — | 10 |
| 13 | Clean iron | 340 | 30 | $10^{16}$ | 160 | $10^{-6}$ |
| 14 | Oxidized iron | 100—200 | 35 | $10^{15}$ | 140 | $10^{-6}$ |
| 15 | Fe+MgO | 30 | — | — | — | 10 |
| 16 | Fe+BeO | 130 | — | — | — | 10 |
| 17 | Fe+$ZrO_2$ | 140 | — | — | — | 10 |
| 18 | Fe+$Al_2O_3$ | 280 | — | — | — | 10 |

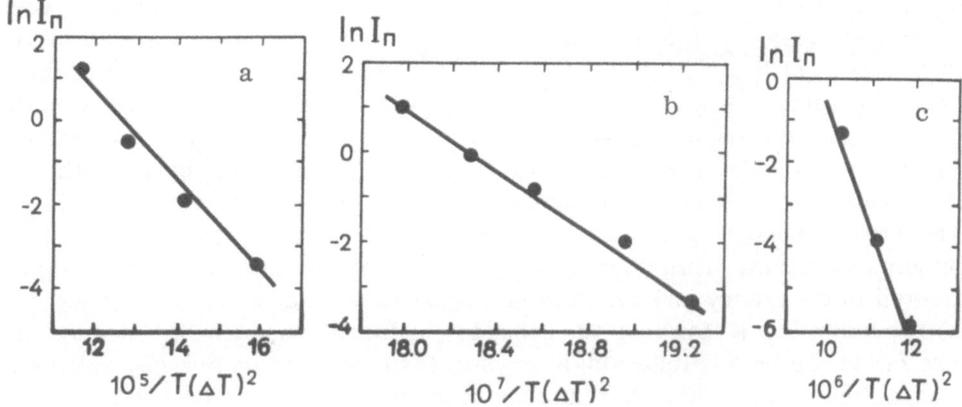


Fig. 4. Relationship of ln I to $1/T(\Delta T)^2$ for (a) hydroquinone – $CaCO_3$; (b) pure hydroquinone; (c) oxidized bismuth.

Microscopic observation shows that the nucleation centers on calcite arise primarily at the corners, edges, reentrant angles, and other defective points, whose surface density is around $10^3$ $cm^{-2}$. These crystallization centers arise by fluctuation and do not have very different effects on the width of the metastable region. If polycrystalline $ZnCO_3$ or $FeCO_3$ is used, the active-site concentration is far larger than that found on calcite single crystals.

A characteristic feature of the crystallization on such surfaces is the ideal reproducibility, which is independent of the superheating and of the number of recrystallizations. Further, if such a mineral is inserted in the liquid for the first time, there is no correlation between the supercooling and the superheating, but the solid surfaces become activated after prolonged contact with the crystalline material and nuclei begin to arise at supercoolings of about 1°. The activation rate is the greater the higher the temperature and the closer the crystallochemical affinity. The induced activity is readily eliminated by heating the material to 20-30° above the melting point of hydroquinone (Fig. 2b), whereas the natural activity persists for superheatings of about 200°C.

Detailed studies have been made of the effects of metal oxides [19, 21, 24] and of certain other oxides [23, 25] on metals. Both types of oxide are naturally active (Table 1). For instance, if the oxides of bismuth and lead introduced into the pure metals, which will supercool respectively by 46 and 8°C after vacuum distillation (in specimen volumes of 0.06 $cm^3$), they will reduce the supercooling to 13°C for Be or 3.5°C for Pb. Further, the oxides are additionally activated by contact with the crystalline metals, with the result that the supercooling is dependent on the superheating, but this induced activity is lost at superheatings of 80-100°C. When the induced activity has been eliminated, the nucleation kinetics may be described in terms of a sharp metastability limit and a linear relationship between $\ln J_s$ and $1/T(\Delta T)^2$ (Fig. 4c). The same occurs for K and Na [27] and also for Fe [21, 24]. It has been shown [23] that analogous features occur for other oxides. Table 1 shows for iron (100 g specimens) that the performance of an oxide is dependent on the composition; the most active particles are those of MgO, which nucleate iron at supercoolings of 30°, while the least active are $Al_2O_3$, which allow a supercooling of 280°C. However, in experiments on crystallization of droplets of iron of size 100-200 $\mu$m on substrates of these materials [25] there was no appreciable difference in the performance, and even MgO gave a supercooling of about 300°. Although the performance of the solid surfaces for droplets was much less than that for massive specimens, some activity was still present. In particular, droplets of iron gave the maximum supercooling of 340° as corresponding not to homogeneous nucleation but to heterogeneous nucleation on the $Al_2O_3$, as was clear from the small value of $K_s$ and the relationship between the supercooling and the superheating. The same applies to other clean metals (Table 1). Therefore, the results for $\sigma$, although higher than those for nucleation on the oxides of the metals themselves, do have some meaning. As $\sigma_*$ increases with the discrepancy between the lattice parameters, i.e., with the work of formation, the theory is confirmed in a qualitative sense; however, there are substantial quantitative discrepancies, particularly between the observed and theoretical values for $K_s$. Theory indicates that $K_s$ should be constant and fairly large, about $10^{26}$, whereas the observed $K_s$ (Table 1) are smaller by many orders of magnitude and vary from system to system. This discrepancy is due to the local character of nucleation, which is neglected in the theory and which is due to surface defects. For instance, the hydroquinone - $CaCO_3$ system has $K_s$ far less than that for hydroquinone — $ZnCO_3$ because the initiating defects are far fewer on a calcite single crystal than they are on polycrystalline $ZnCO_3$. Further, the experiments imply that the number of nuclei increases with the supercooling not simply on account of increased probability of formation on smooth surfaces (as assumed in the theory) but also because of increasing activation of less active parts. One expects that at very high supercoolings (supersaturations) nuclei would arise even at point defects [28], which are much less active, for example, than reentrant angles of steps. It is clear that these aspects should be incorporated in future developments of the theory.

This analysis implies that a feature common to any insoluble impurity is that it may be activated by molecular contact with the solid; the resulting induced activity causes the supercooling to be dependent on the superheating, and this is one of the main features indicating impurity effects. The induced activity can be eliminated completely by superheating, so some activatable impurities can be completely deactivated. On the other hand, naturally active impurities retain their natural activity even when the induced activity has been suppressed. A characteristic feature of crystallization on a naturally active surface is that there is a sharp metastability limit, which is independent of the superheating, while the nucleation is of fluctuation type, with K and $\sigma$ relatively small.

## 3. Homogeneous Crystallization in Metals

Homogeneous nucleation requires complete elimination of active solid particles, since even very minute amounts can radically alter the process; a check can be made only by reference to the features due to crystallization on solid surfaces. The laws of nucleation on impurities provide means of formulating [14] and refining [26] experimental criteria for spontaneous crystallization, which amount in essence to the requirement of complete absence of activation and deactivation, as well as of any relationship between supercooling and superheating, together with all the associated effects. Here, for instance, one needs to perform special tests to define the conditions that ensure activation in order to test for lack of correlation between supercooling and superheating. If the criteria are met, it can be said that impurities are absent and the temperature dependence of the nucleation rate may be considered as an unambiguous physical characteristic for the homogeneous crystallization of that substance.

It has been suggested [43] that homogeneous crystallization could be defined from the relationship between the lifetime $\tau_i$ of the supercooled state, the most probable crystallization temperature $T_m$, and the temperature coefficient of $\sigma$, namely $d\sigma/dT$, since the fluctuation theory implies certain relationships, although we shall see below that heterogeneous nucleation follows the same laws, and therefore such criteria can hardly serve to distinguish spontaneous crystallization from crystallization on impurities.

Experiments involving careful elimination of insoluble impurities have shown that liquids differ in their tendency to crystallize; for instance, salol, benzophenone, and other viscous substances do not crystallize spontaneously (without seeds) and go over a vitreous state, whereas metals can never be converted to amorphous form, even at very high cooling rates, and they always crystallize when an appropriate supercooling is attained.

Danilov and Neimark [14] examined the crystallization of various metals in volumes of about 0.5 $cm^3$ and found that metals such as Pb, Zn, and Al allow only very small supercoolings of about 3° when purified from insoluble materials, whereas metals such as Sn, Bi, and Sb are readily supercooled by 25-30°. They assumed that the observed supercoolings correspond to homogeneous crystallization. This was subsequently supported by the observation that in certain cases [14, 27] the observed kinetic coefficients were of the order of $10^{23}$, which was the value originally predicted by the theory [1]. These results were taken with X-ray data to conclude that the tendency to supercooling is dependent on the degree of destruction of the short-range order on melting. Metals with open lattices (Bi, Sn, or Sb) take up a closer atomic packing on melting, and therefore they should show greater supercooling than metals with close-packed lattices (Pb, Zn, or Al), which show no such major changes on melting. This explanation was generally accepted, since conversion of a liquid to a crystal with a compact structure would require less displacement of the atoms. Although this is a likely explanation, it does not have a firm experimental basis, since the observed supercoolings were later shown [29, 30] not to correspond to homogeneous crystallization.

It is difficult to provide the conditions for homogeneous nucleation in metals, since there are no reliable methods of eliminating insoluble impurities. The most promising technique would appear to be the use of microscopic volumes [29, 31], in which droplets of size about 50 $\mu$m are used, which are too small to contain active particles and therefore provide conditions for homogeneous nucleation.

The best example of this method appears to be Turnbull's [31] dilatometric study of solidification of mercury droplets covered with protective films. He found that films of mercury acetate and stearate caused droplets of mercury to begin to crystallize at appreciable rates at supercoolings of 40 and 60° respectively. Also, the nucleation frequency was proportional to the surface area of the droplet, while $K_s$ was around $10^{27}$, as envisaged in the theory of heterogeneous nucleation. In the case of droplets coated with mercury laurate, supercoolings of about 79° were attained, and then the nucleation frequency was proportional to the volume of the droplet, while the kinetic coefficient was $10^{42}$, i.e., substantially exceeded the $K_0$ calculated for homogeneous nucleation. Turnbull considered that homogeneous nucleation occurred in this case, while the discrepancy by a factor $10^7$ over $K_0$ was ascribed to the nuclei arising from a structure melting at a lower temperature than the stable microscopic structure. The temperature difference was estimated as about 11-12°, which would bring the observed K into agreement with the theoretical value. However, another explanation [5] was based on a correction factor of $10^{12}$ neglected in the theory, which arises from the contribution from the rotational degrees of freedom.

Although these explanations fail to provide a final decision, the very occurrence of high kinetic coefficients provides some basis for assuming homogeneous nucleation in mercury under these conditions.

Turnbull and Cech [29] used the microvolume method to determine $\Delta T_m$ (the maximum supercooling for droplets of diameter 50 $\mu$m for various metals, Table 2). On the assumption that homogeneous nucleation occurs at an appreciable rate at such supercoolings, they used (5) with $K = 10^{34}$ to calculate $\sigma$, and Table 2 gives these $\sigma_{exp}$, as well as the theoretical $\sigma_{theor}$ [33].

Table 2 and Fig. 5 show that the relative supercoolings $\Delta T/T_{mp}$ vary little from one metal to another (from 0.14 to 0.25), apart from mercury, the average value being $0.18T_{mp}$. Many workers have assumed that these supercoolings represent characteristics of homogeneous nucleation for liquid metals, which tends to be confirmed by the similarity in the $\sigma$

TABLE 2. Maximal Supercooling Quoted in Various Sources

| Metal | $\Delta T_m$ °C, [29, 31] | $\sigma_{exp}$, ergs/cm² [20] | $\sigma_{theor}$, ergs/cm² [33] | $\Delta T_m$ °C, from new data: drops 100 μm | Ref. | Spec. 1–10 g | Ref. | Spec. 100-500g | Ref. |
|---|---|---|---|---|---|---|---|---|---|
| Mercury | 79 | 31.2 | 28.6 | — | | — | | — | |
| Gallium | 76 | 56 | 62.8 | 106 | [41, 42] | — | | — | |
| Antimony | 135 | 101 | 89 | — | | — | | — | |
| Bismuth | 90 | 54.4 | 59.3 | 115 | [41] | — | | — | |
| Lead | 80 | 33.3 | — | — | | — | | — | |
| Tin | 118 | 59 | 56.4 | 132 | [41, 48] | — | | — | |
| Germanium | 227 | 181 | 168 | 316 | [45] | 200 | [34] | — | |
| Iron | 295 | 204 | 198 | 500 | [30] | 420 | [39] | 280 | [23] |
| Aluminum | 130 | 98 | 95 | — | | — | | — | |
| Silver | 227 | 126 | 119 | 252 | [45] | — | | 250 | [36] |
| Gold | 230 | 132 | 145 | — | | 190 | [34] | — | |
| Copper | 236 | 177 | 170 | 277 | [45] | 180 | [34] | 280 | [38] |
| Manganese | 308 | 206 | 187 | — | | — | | — | |
| Nickel | 319 | 265 | 223 | 480 | [40] | 480 | [39] | 290 | [35–37] |
| | | | | | | | | 305 | [40] |
| Cobalt | 330 | 234 | 227 | 470 | [40] | 310 | [34] | — | |
| Platinum | 370 | 208 | — | — | | — | | — | |
| Palladium | 332 | 240 | 288 | — | | 310 | [34] | — | |

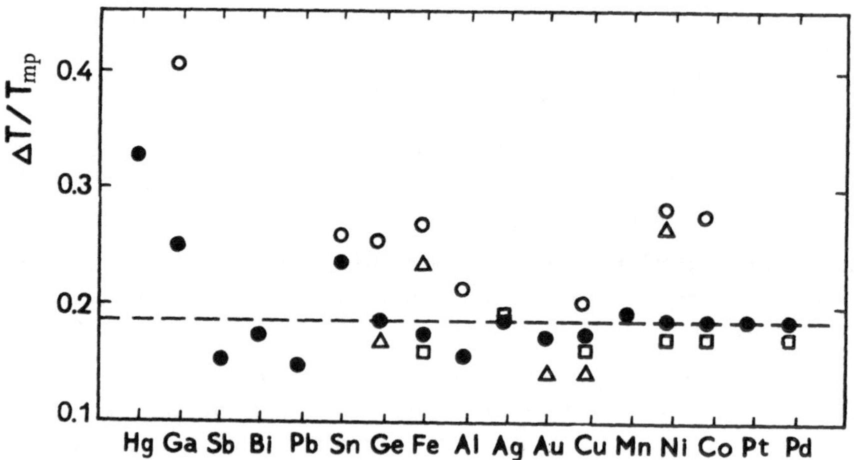


Fig. 5. Relative supercooling ($\Delta T/T_{mp}$) for various metals; ● droplets of diameter 10–20 μm given by Turnbull; ○ droplets of diameter 50–100 μm from other sources; △ specimens of mass 1–10 g, □ specimens of mass 100–500g.

(Table 2) as estimated by experiment and calculated from the theory [4, 31, 32]. However, this agreement is not a convincing argument for the assumption, because $\sigma$ is only slightly dependent on $\Delta T$, while the theory itself [31, 33] involves various simplifying assumptions. In particular, no allowance is made for the curvature, the structure of the boundary layer of the liquid [34], etc.

For a long time, the limiting supercoolings of $0.18T_{mp}$ obtained for droplets [29] were unattainable with massive specimens; however, in later studies [23, 34-38], in which melting under a glass slag was used, which eliminated oxides and other impurities, it proved possible to obtain nearly the same supercoolings on specimens of mass 10-500 g for Ag, Au, Cu, Co, Fe, Ni, Pd (see the latter columns of Table 2 and also Fig. 5). Melting in the suspended state, where the liquid metal was not in contact with the solid crucible, was used with 2-g specimens to obtain [39] much larger supercoolings: 420° for Fe and 480° for Ni. Similar supercoolings for these metals and Co were obtained [40] on droplets of about 100 μm melted in highly purified argon ($\Delta T_m$ was 470-480°C for Ni and Co, and over 500° for Fe [30]). Unfortunately, no study was made of the temperature dependence of the nucleation rate, so it was not possible to draw any conclusion on the type of nucleation at these supercoolings. However, the results represent indisputable proof that the supercoolings $0.18T_{mp}$ previously observed on droplets [4, 29] in most instances correspond to heterogeneous crystallization, not homogeneous. Clearly, the $\sigma$ derived from such values do not correspond to the true liquid–melt surface tension. Such estimates are of significance only in demonstrating whether homogeneous crystallization occurs, which can be established from appropriate criteria, including the temperature dependence of the nucleation rate. However, such experiments are very difficult, and therefore the number of papers is very restricted.

In this respect, apart from the experiments with mercury [4], we may note the data of [42-45] on the crystallization kinetics of droplets of 50-100 μm for Ga, Ge, Sn, Hg, Ag, and Cu, for which the critical supercoolings corresponded to $J = 10^5\text{-}10^7\ cm^{-3} \cdot sec^{-1}$, some of which are given in Table 2. The observed $J(\Delta T)$ gave the $K_{exp}$ in all cases as around $10^{32}$, i.e., essentially the theoretical values, which led the workers to assume that homogeneous nucleation occurred. However, the papers do not give a clear definition of the criteria for selecting the data from the set of droplets, since the droplets differed in tendency to supercool, and the selected data were used to plot $J(\Delta T)$, so not much confidence can be placed on the resulting $K_{exp}$. Further, data from other sources cast doubt on the supercoolings claimed for certain metals. For instance, one cannot accept the assertion [44, 45] that the silver droplets were supercooled by 252° and then showed homogeneous crystallization, since such supercoolings have been obtained with specimens of mass 500 g at much lower cooling rates [36]. The same applies for mercury, where the critical supercooling for droplets (for $J = 10^3\ cm^{-3} \cdot sec^{-1}$) is less by 20° than that in [31] and corresponds [31] to heterogeneous nucleation. The kinetic coefficient for mercury found in [31] was larger by a factor of $10^{10}$, while the value given for gallium [41] is $10^{65}$. The reasons for these very large differences in K are as yet uncertain, but it is completely clear that they lie outside the limits of the ordinary errors of experiment and should not occur for homogeneous nucleation, where the characteristics should be unambiguous.

Therefore, the observation of homogeneous crystallization, which previously has been considered as demonstrated, remains an open question and requires further research. In spite of this, the very large supercoolings obtained with metals do demand a change in approach to certain aspects. In particular, one is not justified in utilizing the macroscopic interfacial energy $\sigma$ for very small nuclei, which, for instance, for Ni may be of size about $10^{-7}$ cm for $\Delta T_m = 480°$ [40], i.e., comparable with the size of the ordered regions in the liquid [47]. However, such regions are not nuclei, while on the other hand, crystals of this size (nuclei) are stated [48] to have features of the amorphous state, so the very concept of surface energy becomes undefined. It would seem that one needs a radical reconsideration of all existing ideas on the relevant mechanisms, particularly as regards soluble impurities, in the sense that small amounts of soluble impurities cannot mask the effects of insoluble impurities and cause spontaneous crystallization, as the latter does not occur in pure metals even at the very large supercoolings of $0.2T_{mp}$. This applies to an even larger extent as regards the role of homogeneous nucleation in the solidification of castings, particularly under industrial conditions.

## Literature Cited


1. M. Volmer, Kinetik der Phasenbildung, Leipzig (1939).
2. Ya. M. Frenkel', Kinetic Theory of Liquids, Collected Works, Vol. 3 [in Russian], (1969).
3. Ya. B. Zel'dovich, Zh. Eksp. Teor. Fiz., 12:525 (1942).
4. D. Hollomon and D. Turnbull, Advances in Metal Physics, Vol. 1 [Russian translation], Metallurgizdat (1959), p. 304.
5. J. Lothe and G. M. Pound, J. Chem. Phys., 36(15):2080 (1962).
6. J. Reiss and J. L. Katz, Chem. Phys., 46:2496 (1967).
7. H. Reiss, J. L. Katz, and E. R. Cohen, J. Chem Phys., 48:5553 (1968).
8. A. G. Bashkirov, Phys. Letters, 28A:23 (1969).
9. K. Hishioka and G. M. Pound, Amer. J. Phys., 33:1211 (1970).
10. R. C. Tolman, J. Chem. Phys., 17:333 (1949).
11. F. R. Buff, J. Chem. Phys., 23:419 (1955).
12. O. D. Kozachkovskii, In: Aspects of Metal Physics and Metallography [in Russian], Izd. AN UkrSSR, Kiev (1948).
13. R. Kaishev, Izv. Bolg. Akad. Nauk, Ser. Fiz., 1:100 (1950).
14. V. I. Danilov, Structure and Crystallization of Liquids [in Russian], Naukova Dumka, Kiev (1956).
15. O. D. Kozachkovskii, In: Aspects of Metal Physics and Metallography [in Russian], Izd. AN UkrSSR, Kiev (1948).
16. D. Turnbull, J. Chem. Phys., 18:198 (1950).
17. D. E. Ovsienko and V. I. Danilov, In: Aspects of Metal Physics and Metallography [in Russian], Izd. AN UkrSSR, Kiev (1952), p. 89.
18. D. E. Ovsienko, In: Aspects of Metal Physics and Metallography [in Russian], Izd. AN UkrSSR, Kiev (1953), p. 153.
19. V. I. Danilov and D. E. Ovsienko, Zh. Eksp. Teor. Fiz., 8:879 (1951).
20. D. E. Ovsienko and E. I. Sosnina, In: Aspects of Metal Physics and Metallography [in Russian], No. 3, Izd. AN UkrSSR, Kiev (1952), p. 106.
21. D. E. Ovsienko and V. P. Kostyuchenko, In: Aspects of Metal Physics and Metallography [in Russian], No. 10, Naukova Dumka, Kiev (1960), p. 130; In: Growth of Crystals, Vol. 3, Consultants Bureau, New York (1962), p. 76.
22. D. E. Ovsienko and V. P. Kostyuchenko, In: Aspects of Metal Physics and Metallography [in Russian], No. 13, Naukova Dumka, Kiev (1962), p. 167.
23. D. E. Ovsienko, V. P. Kostyuchenko, In: Mechanisms and Kinetics of Phase Transitions [in Russian], Izd. AN BSSR, Minsk (1964).
24. V. P. Kostyuchenko and D. E. Ovsienko, Fiz. Met. Metalloved., 30:77 (1970).
25. D. E. Ovsienko, V. P. Kostyuchenko, and M. V. Leputskii, Metal Physics [in Russian], No. 3, Naukova Dumka, Kiev (1971), p. 62.
26. D. E. Ovsienko, In: Aspects of Metal Physics and Metallography [in Russian], No. 12, Izd. AN UkrSSR, Kiev (1961), p. 3.
27. V. I. Danilov and A. G. Pomegaibo, Dokl. Akad. Nauk SSSR, 68:843 (1949).
28. G. I. Distler, In: Growth of Crystals, Vol. 8, Consultants Bureau, New York (1969), p. 91; see also this volume, p. 44.
29. D. Turnbull and R. E. Cech, J. Appl. Phys., 20:411 (1952).

30. A. I. Dukhin, In: Aspects of Metal Physics and Metallography [in Russian], Metallurgizdat, Moscow (1959), p. 60.
31. D. Turnbull, J. Chem. Phys., 20:411 (1952).
32. A. S. Skapski, Acta Met., 4:576 (1956).
33. S. N. Zadumkin, Izv. Akad. Nauk SSSR, Metallurgiya i Toplivo, 6:119 (1960).
34. J. Fehling and E. Sceil, Z. Metallkunde, 54:592 (1962).
35. G. Colligan, V. A. Suprentent, and F. D. Leukey, J. Metals, 9:691 (1961).
36. G. L. F. Powell, J. Austral. Inst. Metals, 10:223 (1965).
37. L. Wolker, Physical Chemistry. of Process Metallurgy, Part 2, Interscience, New York (1961), p. 845.
38. G. L. Powell and L. M. Hagen, Trans. Met. Soc. AIME, 242:2133 (1968).
39. D. W. Gonnersal, S. U. Shirashi, and R. C. Ward, J. Austral. Inst. Metals, 10:220 (1965).
40. D. E. Ovsienko, V. V. Maslov, and V. P. Kostyuchenko, Kristallografiya, 16:405 (1971).
41. L. Bosio, Metaux, 40:425, 451 (1965).
42. V. P. Skripov, V. P. Koverda, and G. T. Butorin, Fiz. Met. Metalloved., 31:790 (1971).
43. V. P. Skripov, G. T. Butorin, and V. P. Koverda, Kristallografiya, 15:1219 (1970).
44. G. T. Butorin, Candidate's Dissertation, Sverdlovsk UPI (1971).
45. V. P. Skripov, V. P. Koverda, and G. T. Butorin, This volume, p. 22.
46. D. E. Ovsienko, G. A. Alfnitsev, and V. V. Maslov, Mechanism and Kinetics of Crystallization [in Russian], Minsk (in press).
47. A. V. Romanova and B. A. Mel'nik, Ukr. Fiz. Zh., 15:101 (1970).
48. V. K. Yatsimirskii, Teoret. Eksp. Khim., 6:704 (1970).

# CRYSTAL NUCLEATION KINETICS IN SMALL VOLUMES

**V. P. Skripov, V. P. Koverda, and G. T. Butorin**

*Urals Polytechnical Institute and Urals Scientific Center, Academy of Sciences of the USSR, Sverdlovsk*

A first order phase transition in a one-component system begins with nucleation; the most interesting case from the physical viewpoint is when the nuclei appear by a fluctuation mechanism (homogeneous nucleation). There is a monotonic fall in the minimal size of a nucleus as the supercooling increases (the number of molecules in such a nucleus is $n_*$). This means an increase in the probability that nuclei containing $n \simeq n_*$ molecules will arise. The transition states are usually neglected in the theory of homogeneous nucleation, and the frequency $J_1$ with which nuclei are formed per $cm^3$ is taken as an exponential function of the work of nucleation $W_*$ [1]:

$$J_1 = N_1 B \exp[-W_*/kT]. \tag{1}$$

Here $N_1$ is the number of molecules per $cm^3$ in the initial phase, while B is a kinetic factor. If the activation energy for self-diffusion in the liquid is comparatively small, then $B \simeq 10^{11}$-$10^{12}$ $sec^{-1}$, and $K_V = V_1 B \simeq 10^{33}$-$10^{34}$ $cm^{-3} \cdot sec^{-1}$, and the temperature dependence of $J_1$ is determined by the exponential factor, where $W_* \sim \sigma^3 \sim T^2$ (on the assumption of a constant entropy of melting), $\sigma$ is the surface tension at the crystal–melt boundary, $\sigma$ is the supercooling, and $T_0$ is the temperature of phase equilibrium for a planar boundary and a given external pressure. As one often lacks experimental evidence on $\sigma$, it is difficult to distinguish homogeneous nucleation from the heterogeneous form and to check the fluctuation theory.

Most published studies ignore the random character of nucleation and give no statistical analysis of the results; in most of them, data were obtained on nucleation kinetics at high degrees of supercooling by measuring the crystallization temperature on continuous cooling. The smooth curve of Fig. 1 represents the distribution of the random crystallization temperatures; and near the peak it takes the following form:

$$\frac{\delta N}{N} = -\frac{VJ_1(T)}{\dot{T}(d\ln J_1/dT)_{T=T_m}} \exp\left[\frac{VJ_1(T)}{\dot{T}(d\ln J_1/dT)_{T=T_m}}\right]\delta T, \tag{2}$$

where $\delta N$ is the number of crystallization occurring in the range $(T, T+\delta T)$, N is the total number of experiments, $\dot{T}$ is the cooling rate, and V is the volume of a drop.

The frequency corresponding to the temperature at which (2) has its maximum is [2] given by

$$J_1(T_m) = -\frac{\dot{T}}{V}(d\ln J_1/dT)_{T=T_m}. \tag{3}$$

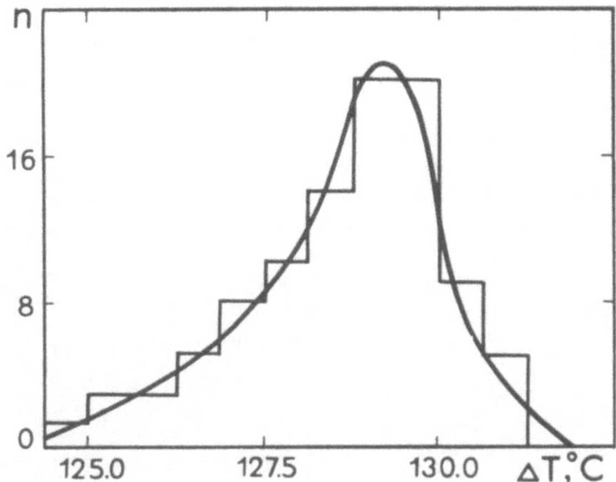


Fig. 1. Histogram of results from crystallization of tin under continuous cooling; N = 97, d = 20 μm, $\dot{T}$ = 1 deg/sec.

The temperature dependence of the nucleation frequency may be estimated from the half-width ($\Delta T_{1/2}$) of the distribution:

$$-\left(d\log J_1/dT\right)_{T=T_m}\simeq 0.95/\Delta T_{1/2}. \quad (4)$$

The temperature distribution is skewed, and on the low-temperature side it falls to zero much more rapidly than on the other side. Therefore, the mean crystallization temperature $\overline{T} = \Sigma T_i/N$ does not coincide with $T_m$. The difference $\overline{T} - T_m$ is dependent on the shape of the curve in a wide range around $T_m$. It is found that $\overline{T} - T_m$ varies from 3 to 10°C for different substances if the cooling rate is kept constant. Usually, an experiment is performed by reducing the temperature rapidly to a point close to $T_m$, after which the rate is reduced. In that case, $\overline{T} - T_m$ is smaller, but can still be several degrees. Therefore, a systematic error in $J_1$ can arise if one averages the random crystallization temperatures without taking account of these features of the distribution, and this error may exceed the errors in measuring the temperature, volume, and cooling rate very substantially. Also, the temperature dependence of ($d\log J_1/dT$) and the value for $K_V$ found from the isokinetic curve derived by such averaging may differ considerably from the real values.

The maximum supercooling attained in a given series of experiments is dependent on the number of experiments, and errors in $J_1$ by factors of 10-100 can arise if the maximum value is used instead of the most probable value.

In lifetime measurements, it is necessary to check that a Poisson distribution applies for the crystallization events and to exclude any range in which there is a thermally nonstationary state.

The above factors explain why there are often differences between published values, which can sometimes be eliminated if one has detailed information on the methods used and on the data processing.

We have obtained information on the nucleation kinetics of droplets of diameter between 5 and 500 μm. These small volumes allowed us to produce clearcut working conditions, and also to observe crystallization at centers of purely fluctuation origin. We employed two methods: (a) repeated measurement of droplet lifetime for a given supercooling and (b) measurement of the distribution of the crystallization temperatures for one or several droplets in a series of runs with continuous cooling at a constant rate.

In the first case, we measured the mean lifetime $\bar{\tau}$, and this gave the nucleation frequency $J_1 = (\bar{\tau} V)^{-1}$; the temperature of the environment was used as parameter, which enabled us to

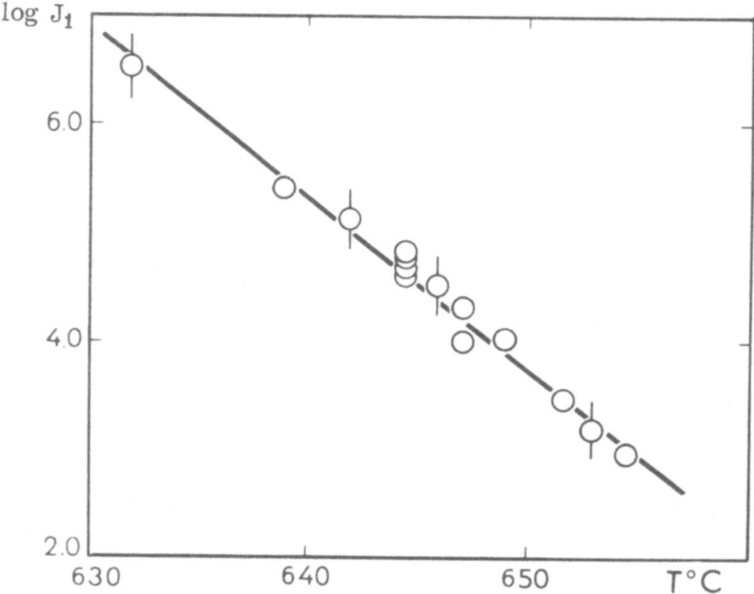


Fig. 2. Temperature dependence of the nucleation frequency for supercooled germanium ($T_0$ = 958°C).

vary $J_1$ by three orders of magnitude in the range $10^3$–$10^8$ $cm^{-3} \cdot sec^{-1}$. We checked that the distribution was of Poisson type (an exponential distribution for the intervals $\tau$) for a given $\Delta T$ and also that $\bar{\tau}$ and $1/V$ were proportional. In the second case, we used histograms (Fig. 1) to determine $T_m$ and $\Delta T_{1/2}$; from (3) and (4) we derive $J_1(T_m)$ and $(d \log J_1/dT)_{T=T_m}$. The two methods gave results in agreement for small droplets. Figure 2 shows $\log J_1$ as a function of T for germanium (first method). The variation in the nucleation frequency is described by the classical theory of homogeneous nucleation if $\sigma$ = 251 ergs · $cm^{-2}$. Similar results were obtained for other metals and for water.

Table 1 gives crystallization temperatures for various substances corresponding to $J_1$ = $10^5$ $cm^{-3} \cdot sec^{-1}$; the preexponential factor was of the order indicated by a rough calculation. The results over the range examined corresponded to an essentially constant surface tension,

TABLE 1

| Substance | $T_0$, °C | T, °C ($J_1 = 10^5$ $cm^{-3} \cdot sec^{-1}$) | $\sigma$, ergs · $cm^{-2}$ | d log $J_1$/dT, $deg^{-1}$ ($J_1 = 10^5$ $cm^{-3} \cdot sec^{-1}$) exp. | ther. | log $K_V$ | $n_*$ |
|---|---|---|---|---|---|---|---|
| $H_2O$ | 0 | −36.4 | 28.3 | 1.33 | 1.36 | 31.4 | 120 |
| Hg | −39.7 | −91.5 | 23.0 | 0.88 | 0.96 | 32.0 | 400 |
| Ga(β) | −16.3 | −115.5 | 40,4 | 0.39 | 0.41 | 32.0 | 170 |
| In | 156 | 75.0 | 30.8 | 0.62 | 0.64 | 32.2 | 570 |
| Sn | 232 | 110.2 | 59,0 | 0.39 | 0.40 | 32.1 | 270 |
| Bi | 271 | 170.6 | 61.2 | 0.53 | 0.54 | 31.8 | 215 |
| Ge | 958 | 642 | 251 | 0.153 | 0.162 | 31.0 | 150 |
| Ag | 960 | 708 | 143 | 0.201 | 0.206 | 32.5 | 470 |
| Cu | 1083 | 806 | 200 | 0.179 | 0.184 | 32.0 | 480 |

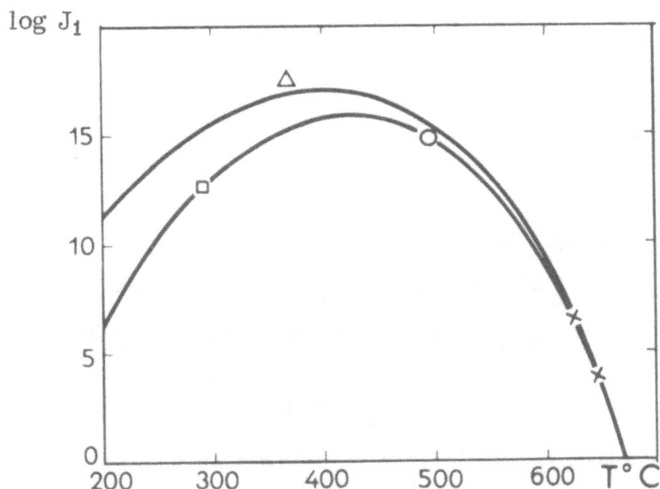


Fig. 3. Nucleation frequency for supercooled germanium extrapolated to very high degrees of supercooling: ×-× our data, ○ data of [4], Δ data of [5], □ data of [6].

which agrees with the view that $\sigma$ is only slightly dependent on temperature and that $(1/\sigma) \times (d\sigma/dT) \sim 10^{-3}$, which is partly due to the large range of coexistence for the two phases.

If one assumes that there is no limit to the thermodynamic stability for one-component liquid (a spinodal line), then the classical theory of homogeneous nucleation gives a bell-shaped T dependence of $\log J_1$; if we neglect the activation energy, then the peak on the curve corresponds to $T_m = T_0/3$ [3]. The fall in the nucleation probability at very low temperatures is due to the low atomic mobility, so the amorphous state may be stabilized when the temperature is very low. If we incorporate the activation energy, the maximum value of $\log J_1$ is reduced, and the peak shifts towards higher temperatures. The solid lines in Fig. 3 show $\log J_1$ as a function of T for germanium. The activation energy was estimated from data on the viscosity [7], in both cases with extrapolation to very high degrees of supercooling. The figure also shows the crystallization points of amorphous germanium films [4-6].

In conclusion we note that the statistical methods used here for determining the nucleation frequency involve defining the distribution for small volumes under isothermal conditions and in continuous cooling mode, and these techniques are applicable to homogeneous and heterogeneous forms of nucleation. The results from the present substances do not conflict with the classical theory of homogeneous nucleation.

## Literature Cited


1. Ya. I. Frenkel', Selected Works, Vol. 3 [in Russian], Izd. AN SSSR, Moscow (1959).
2. V. P. Skripov, V. P. Koverda, and G. T. Butorin, Kristallografiya, 15:1219 (1970).
3. D. N. Hollowman and J. Turnbull, Advances in Metal Physics, Vol. 1 [Russian translation], Metallurgiya, Moscow (1956), p. 304.
4. J. F. Pocza, International Conference on Semiconductor Physics and Chemistry, Heterojunctions and Layer Structures, Budapest, 1970, Vol. 3, Budapest (1971), p. 61.
5. N. T. Gladkich, R. Niedermauer, and K. Spiegel, Phys. Status Solidi, 15:181 (1966).
6. W. Kleber and I. Mietz, Kristall u. Technik., 3:509 (1968).
7. V. M. Glazov, S. N. Chizhevskaya, and N. N. Glagoleva, Liquid Semiconductors [in Russian], Nauka, Moscow (1967).

# THE EFFECTS OF THERMAL STRAINS ON THE ACTIVITY OF NUCLEI IN INDUCED CRYSTALLIZATION IN GLASSES


## I. Gutsov

*Institute of Physical Chemistry, Bulgarian Academy of Sciences, Sofia*


Induced crystallization in a glass produces a variety of glass ceramics (glass sinters, pyroceramics, etc.), and in principle this can be considered as heterogeneous nucleation, in which nuclei are provided by structures of colloidal or supercolloidal size [1-4]. It is usually assumed [1-3] that the initiating capacity of such a nucleus is governed by Dankov's law, which relates to the discrepancy between the lattice parameters of the substrate and the new phase.

Recently, this technologically important process has been examined quantitatively and has also been simulated [4-6] from the crystallization of Graham glass $[NaPO_3]_X$; the nuclei were submicroscopic (0.1-1 $\mu$m) single crystals of platinum-group metals (Ir, Pt, Pd, Rh), and also gold, silver, and copper [6], which were produced within the bulk of the melt at 1000°C. The melt was quenched to room temperature, as in the production of glass ceramics; the resulting induced glass was then isothermally annealed at 300-400°C, so the crystallization temperature lay considerably above the vitrification temperature $T_g$ for Graham glass [275°C).

It was found [6-8] that the nucleation was typically of nonstationary type [9, 10], and this may be interpreted in terms of the theory developed by Frenkel' [11] and Zel'dovich [12] for the general theory of phase formation. It proved possible to use this theory with the measurements to characterize the initiators quantitatively [4, 6]. The experiments also yielded an activity series for the metals (Ir > Pt > Pd > Rh > Ag > Cu > Au), which does not agree with the classical concepts of epitaxy, which involve lattice correspondence [4, 5]. The same conclusion was reached by others [13] from studies on the induced crystallization of technical glasses. Here we show that the activity of a substrate introduced into a glass is substantially dependent on the thermal stresses arising in the nuclei and in the glass on cooling the latter from $T_g$ to room temperature.

The correlation between the activity $\Phi$ of the substrate and the difference $\Delta\alpha$ in the thermal-expansion coefficients of glass and crystal has been reported previously [4, 5]; here we examine the question quantitatively, and we show that the elastic strains arising in the nuclei can result in residual changes that elevate the catalytic activity.

The nonstationary effects cause the nucleation rate I to be a function of time:

$$I(t)=I_0 \exp\left(-\frac{\tau}{t}\right),$$

where $I_0$ is the steady-state nucleation rate, while $\tau$ is the induction period [9, 10].

In the case of heterogeneous nucleation, the corresponding period is $\tau^*$, and the work of nucleation is $A^*_{k_n}$, these two quantities being related to the corresponding quantities $\tau_0$ and $\overset{\circ}{A}_{k_n}$ for the homogeneous case by

$$A^*_{k_n} = A^\circ_{k_n} \cdot \Phi \tag{1a}$$

$$\tau^* = \tau_0 \cdot \xi. \tag{1b}$$

Here $\Phi$ and $\xi$ are functions of the adhesion energy $\beta$ at the substrate surface and of the surface energy $\gamma$ of the interphase boundary, which here is between crystal and melt.

Kaishev [14] has shown as follows for a nucleus of a cubic crystal:

$$\Phi = 1 - \frac{\beta}{2\gamma}. \tag{2}$$

The same model implies [6, 7] for $\xi$ that

$$\xi = \frac{6\Phi}{1+4\Phi}. \tag{3}$$

Therefore, $\tau^*$ and $\tau_0$ can be measured and processed to give $\xi$ from (1b), which gives $\Phi$ by transformation of (3):

$$\Phi = \frac{1}{6}\left(\frac{1}{\xi} - \frac{2}{3}\right)^{-1}. \tag{4}$$

Table 1 gives values for $\Phi$ determined in this way [4, 6] for various substrates with Graham glass; it is clear that the differences in the $\Phi$ for the cubic face-centered metals cannot be explained in terms of the discrepancies d in the lattice parameters between the metals and the crystalline $\alpha$-$Na_3[P_3O_9]$.

Becker [15] found that $\beta$ can be determined to a first approximation from the lattice model if only the first neighbors are incorporated, which gives

$$\beta_0 = \frac{n_2}{N_A} \cdot \frac{Z_2}{Z_0} \cdot \Delta F_0. \tag{5}$$

Here $\Delta F_0$ is the molar energy of solution of the substrate in the new crystalline phase, $n_2$ is the

TABLE 1

| Metal | $\xi$ | $\Phi$ | Discrepancy d in lattice parameters, % | $\alpha \cdot 10^6$, $deg^{-1}$ |
|---|---|---|---|---|
| Ir | 0.032 | 0.007 | +1.17 | 6.5 |
| Pt | 0.056 | 0.009 | —1.15 | 8.9 |
| Pd | 0.122 | 0.022 | —0.13 | 10.6 |
| Rh | 0.282 | 0.058 | +1.97 | 9.8 |
| Ag | 0.282 | 0.058 | —5.02 | 18.7 |
| Cu | 0.282 | 0.058 | — | 16.2 |
| Au | 0.501 | 0.125 | —4.97 | 14.2 |
| $(NaPO_3)_x$ glass | — | — | — | 25.0 |

number of atoms per unit interphase surface, $N_A$ is Avogadro's number, and $Z_2$ and $Z_0$ are the coordination numbers correspondingly at the interface and in the bulk of a hypothetical ideal solution of the substrate. If the latter is elastically or plastically deformed, then the strain energy has to be incorporated into an analogous definition of $\Delta F$, and consequently

$$\beta=\beta_0+\beta_\varepsilon=\frac{n_2}{N_A}\cdot\frac{Z_2}{Z_0}\cdot(\Delta F_0+\Delta F_\varepsilon\cdot V_m). \tag{6}$$

Here, as is usual, the strain energy $\Delta F_\varepsilon$ is referred to unit volume, while $V_m$ is the molar volume of the substrate.

Formula (6) provides a simple relation between the substrate activity and the state of strain; we are now able to determine the change in the potential energy of the nuclei when these are strained by thermal stresses, which arise in glass–metal joints [16, 17], for example, in particular below $T_g$, where the solidified melt can be considered as an isotropically elastic body. The solution to the Lamé problem gives these stresses as proportional to

$$\delta=(\alpha_g-\alpha_m)\cdot(T_g-T_r),$$

where $\alpha_g$ is the coefficient of thermal expansion for the glass, $\alpha_m$ is the coefficient of thermal expansion for the metal core, and $T_r$ is room temperature.

Rigorous solution of this problem would require us to incorporate the fact that the nuclei are elastically anisotropic; here we use the isotropic approximation on the assumption that the elastic modulus E and Poisson's ratio $\mu$ are identical for the two materials: $E_m = E_g$, $\mu_m=\mu_g$, and in that approximation we thus assume that the spherical nuclei and glass matrix differ only in thermal expansion $\alpha$, which is isotropic for a cubic material. Consider now the thermal stresses set up at a distance r from the center of a spherical nucleus of radius a, for which purpose we use Popov's calculation [18] to show that the deformation vector $u_r$ is defined by

$$u_r=-\frac{(1-2\mu)}{M}\cdot\delta\cdot r \tag{7a}$$

for $0 \leq r \leq a$ (within the nucleus) and

$$u_r=\frac{(1+\mu)}{2M}\cdot\delta\cdot\frac{a^3}{r^2} \tag{7b}$$

for $r > a$ (for the glass matrix).

The stress $\sigma$ in the metal core and around it is as follows:

$$\sigma_t=-\frac{E_g\cdot\delta}{M} \quad \text{for} \quad 0\leqslant r\leqslant a, \tag{8a}$$

$$\sigma_t=-\frac{E_g\cdot\delta}{M}\cdot\frac{a^3}{r^3} \quad \text{for} \quad r>a, \tag{8b}$$

these being the principal normal stresses in the tangential direction, and

$$\sigma_r=-\frac{E_g\cdot\delta}{M} \quad \text{for} \quad 0\leqslant r\leqslant a, \tag{9a}$$

$$\sigma_r=-\frac{E_g\cdot\delta}{M}\cdot\frac{a^3}{r^3} \quad \text{for} \quad r>a, \tag{9b}$$

these being the principal normal stresses in the radial direction. If $E_m = E_g$, the factor M takes

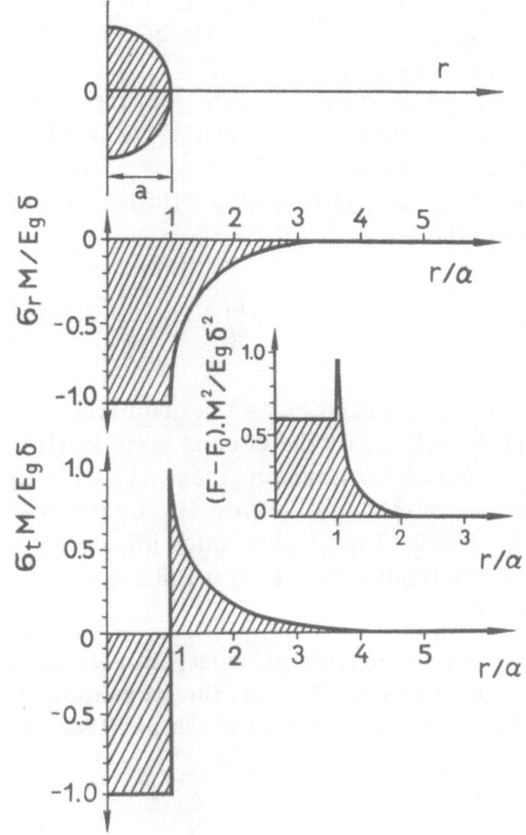


**Fig. 1.** Radial and tangential stresses $\sigma_r$ and $\sigma_t$ in the glass and metal nucleus given by by (8) and (9); the strain energy $\Delta F_\varepsilon = F - F_0$ is given by (11).

the value $^3/_2(1 - \mu)$, and (8) and (9) are shown graphically in Fig. 1 for this case. We clearly have two cases $\delta > 0$ and $\delta < 0$, where the stresses are different in character. As $T_g - T_r$ is always positive, the nuclei are subject to uniform tension for $\alpha_m > \alpha_g$, while for $\alpha_g > \alpha_m$ they are subject to uniform compression.

The strain energy $\Delta F_\varepsilon$ can now be calculated from

$$\Delta F_\varepsilon = \frac{\lambda}{2} \cdot u_{i,i}^2 + \theta \cdot u_{i,k}^2, \tag{10}$$

where $u_{i,k}$ is the strain tensor, $u_{i,i}$ is the sum of the diagonal components, and $\lambda$ and $\theta$ are the Lamé coefficients [19]. We transfer from spherical coordinates to Cartesian ones to express $u_{i,k}$ and $u_{i,i}$ in terms of $u_r$ as in (7a) to get from (10) that

$$\Delta F_\varepsilon = \frac{3}{2} \cdot (1 - 2\mu) \cdot E_g \cdot \frac{\delta^2}{M^2} \tag{11a}$$

for the nucleus ($0 \leq r \leq a$) and

$$\Delta F_\varepsilon = \frac{3}{4} \cdot (1 + \mu) \cdot E_g \cdot \frac{\delta^2}{M^2} \cdot \frac{a^6}{r^6} \tag{11b}$$

for the surrounding unbounded glass matrix (Fig. 1).

The calculations show that $\sigma_r$ and $\sigma_t$ are so large in realistic cases that there are inevitably residual plastic strains in the nuclei; the latter therefore accumulate energy $\varkappa \Delta F_\varepsilon$, where $\varkappa < 1$.

The induced crystallization occurs on repeatedly taking the glass to the crystallization temperature $T_{cryst} > T_g$; it can be shown that even at $T \approx T_\sigma$ the stresses induced in the glass will relax comparatively rapidly, and that $\sigma_r^g \approx 0$ and $\sigma_r^g \approx 0$ when the crystallization temperature is reached. If, however, the melting point $T_m$ of the metal core is comparatively high, and is such as to meet the condition $T_{cryst} < {}^2\!/_3 T_m$, i.e., if $T_{cryst}$ is less than the annealing point of the metal $T_{ann} \approx {}^2\!/_3 T_m$, then the potential energy $\Delta F_\varepsilon$ given by (11b) remains in the nucleus (completely or partially). In that case, (2), (6), and (11a) give the activity for a given melt as

$$\Phi = \left(1 - \frac{\beta_0}{2\gamma}\right) - \varkappa \cdot \frac{1}{2\gamma} \cdot \frac{n_2 \cdot Z_2}{N_A \cdot Z_0} \cdot E_g \cdot \frac{3}{2}(1 - 2\mu)\frac{\delta^2}{M^2} = A - \text{const} \cdot \Delta\alpha^2. \tag{12}$$

Here $\beta_0/2\gamma$ characterizes the chemical and structural factors (for $\Delta\alpha = 0$); the substrate activation is defined by the second term in (12). Figure 2 has been constructed in accordance with (2) with the $\Delta\alpha$ for Graham glass and the various metals mentioned above, which serves to explain the above series in $\Phi$ (Table 1). An interesting point is that the chemically less active metals in Fig. 2 (Au, Rh, Ir) give much higher values for A and $\varkappa$; with $\gamma = 100$ ergs/cm$^2$ for Au, Rh, Ir, we get from (12) that $\beta_0 = 163$ ergs/cm$^2$, while for Ag, Cu, Pd, and Pt we get that $\beta_0$ is 186 ergs/cm$^2$.

This discussion of substrate activation in response to thermal stresses is confirmed by other expressions (Fig. 3); this reveals at least qualitatively the tendency for the inducing effect to increase with $\Delta\alpha$ for the positive case ($\alpha_g > \alpha_m$) and for the negative one ($\alpha_m > \alpha_g$).

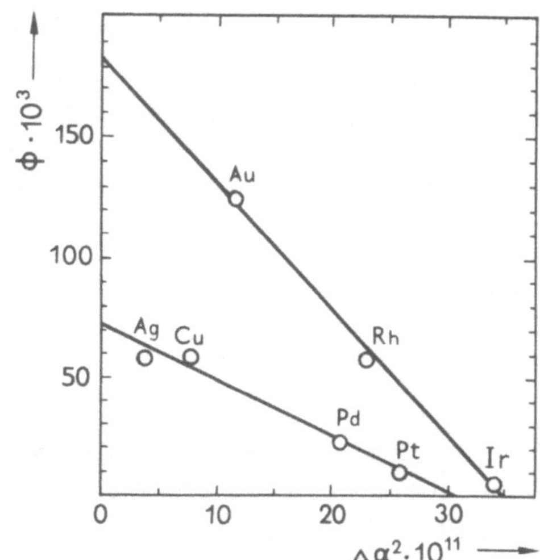


Fig. 2. Experimental results for the activity $\Phi$ for nuclei in $NaPO_3$ glass in relation to $\Delta\alpha^2$ from (12).

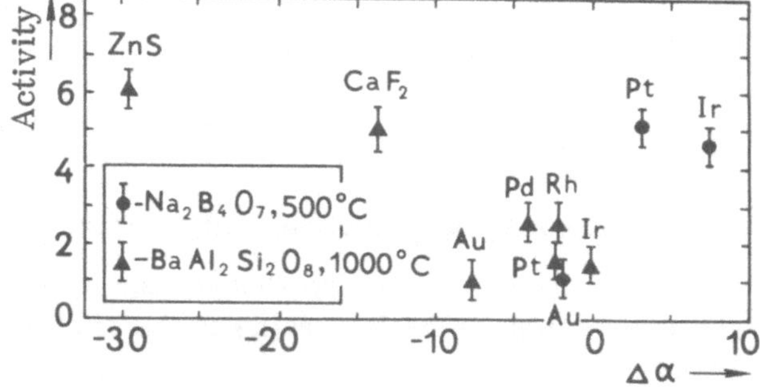


Fig. 3. Activities of crystallization activators in $NaAl_2Si_2O_8$ glass derived from feldspar [13] and in $Na_2B_4O_7$ of borate glass [4] in relation to $\Delta\alpha \cdot 10^6$; the ordinate represents the activity in arbitrary units.

Therefore, this scope for activating nuclei provides a new criterion for selecting crystallization initiators for glass ceramics; Fig. 2 shows the effect. Equations similar to (6) and (12) would be expected also if $\Phi$ is defined by the number of dislocations formed at the interfacial boundary [20]. In that case, $\varkappa \Delta F_{\varepsilon}$ would be a measure of the energy accumulated in the surface layer in the form of dislocations.

The author is indebted to Academician R. Kaishev and Professor B. Popov for discussion of the results, and to Dr. D. Kashchiev for detailed advice.

## Literature Cited


1. S. D. Stookey, Glastech. Ber., 32K:51 (1959).
2. V. N. Filippovich, In: Structural Transformations in Glasses [in Russian], Nauka, Moscow (1965), p. 30.
3. E. M. Rabinovich, In: The Structure of Glass, Vol. 3, Catalyzed Crystallization of Glass, Consultants Bureau, New York (1964), p. 21.
4. I. Gutzov and S. Toshev, In: Advances in Nucleation and Crystallization in Glasses, Proceedings of the Chicago Conference, April 1971, American Ceramic Society, Washington (1972), p. 10.
5. I. Gutsov, E. Popov, S. Toshev, and M. Marinov, In: Growth of Crystals, Vol. 8, Consultants Bureau, New York (1969), p. 78.
6. I. Gutzov, S. Toschev, M. Marinov, and E. Popov, Kristall u. Technik, 3:37, 337 (1968).
7. S. Toshev and I. Gutzov, Phys. Status. Solidi., 24:349 (1967).
8. I. Gutzov and S. Toshev, Kristall u. Technik, 3:485 (1968).
9. S. Toshev and I. Gutzov, Kristall u. Technik, 7:43 (1972).
10. V. Ya. Lyubov and A. L. Roitburd, In: Problems of Metallography and Metal Physics [in Russian], Metallurgizdat, Moscow (1958), p. 91.
11. Ya. I. Frenkel', The Kinetic Theory of Liquids [in Russian], Izd. AN SSSR, Moscow (1959).
12. J. B. Zeldovich, Acta Physicochim. URSS, 18:1 (1943).
13. D. Bahat, J. Mater. Sci., 4:847 (1969).
14. R. Kaishev, Bull. l'Acad. Bulg. Set. (Phys.), 1:100 (1950).
15. R. Becker, Annal. Physique, 32:128 (1938); D. Turnbull, In: Impurities and Defects [Russian translation], Metallurgizdat, Moscow (1960), p. 141.
16. M. L. Lyubimov, Glass–Metal Joints [in Russian], Gosenergoizdat, Moscow (1957).
17. B. Routh, Glass in Electronics [Russian translation], Sov. Radio, Moscow (1969).
18. B. K. Popov, Annuaire d'Inst. Chim.-Techn. (Sofia), 5(2):73 (1958).
19. L. D. Landau and E. M. Lifshits, Theory of Elasticity [in Russian], Nauka, Moscow (1965), pp. 12–27.
20. J. H. van der Merwe, In: Single-Crystal Films [Russian translation], Mir, Moscow (1965), p. 172.

# HIGH-VOLTAGE ELECTRON MICROSCOPY OBSERVATION OF NUCLEATION AND GROWTH OF PRECIPITATES ON DISLOCATIONS IN Al–4% Cu ALLOY

N. Takahashi and T. Taoka

*Faculty of Engineering, Yamanashi University, Kofu, JEOL Ltd., Akishima, Tokyo*

## 1. Introduction

Particles of a second phase grow mainly at dislocations, but it is very difficult to use standard electron microscopes for continuous observation of nucleation and growth at a given point on account of the dimensional effect in thin films [1]. The behavior of the particles and related dislocations in a film is different from that in a bulk crystal. These difficulties can be overcome by high-voltage electron microscopy. A JEM-1000 microscope has been used to image second-phase precipitates similar to those seen in massive Al−4%Cu specimens [2]. For instance, a film of thickness 0.5 μm or more is thick enough to reveal the Widmanstatten structures when the film is heated in the microscope. We have examined nucleation and particle growth at a point in an Al−4%Cu specimen by high-voltage electron microscopy.

## 2. Methods

The specimens were foils of this alloy of thickness 0.1 mm, which were treated in solution at 540°C for 3 hr, followed by quenching in water at room temperature. The dislocations were introduced by scratching the foil with a diamond point of radius of curvature 15 μm under a load of 1 g. The region near the scratch was thinned by jet electropolishing, and then specimens were cut electrolytically for the microscope. These were set up on the heating attachment, which was combined with the goniometer. Contamination and the production of radiation defects were avoided by keeping the specimen out of the beam during the heating time (over 2 hr), with insertion only during the observation. Also, the specimen was inclined during each recording in order to observe the dislocations and the precipitation (it is not always possible to see both at once). It was difficult to observe exactly the same point on each occasion, and therefore we used the junction between three grain boundaries as a reference point. In spite of these precautions, it was extremely difficult to identify the instant of nucleation and thus to obtain recordings (particularly at high magnification), because the image drifted during the heating.

## 3. Results and Discussion

### (a) Nucleation and Growth

We present here a typical series of photomicrographs from one of the numerous experiments. Figure 1 shows the initial state after quenching. There are numerous dislocations

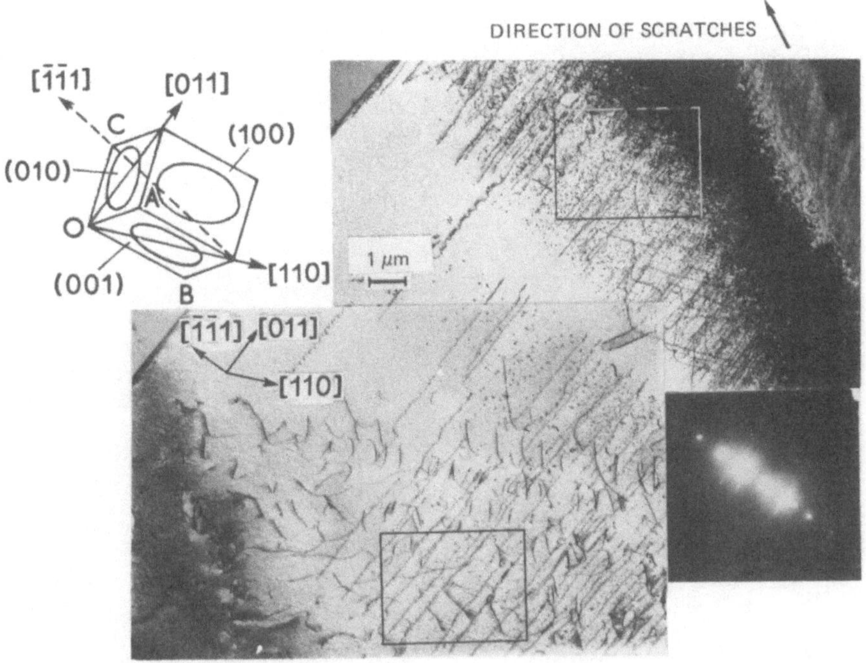


Fig. 1. Quenched Al−4% Cu specimen with scratch showing morphology of θ' phase; here the precipitate was produced on the {100} planes of the matrix. The electron-diffraction pattern corresponds to the upper photomicrograph.

near the scratch, whereas helicoidal dislocations were formed in another area during the quenching. Figure 2a shows that annealing within the microscope at 150°C for 1 hr caused preferential formation of nuclei of the second phase (diameter less than 0.02 μm) at the dislocations, and no nuclei of a type different from those formed at the dislocations were observed. The dislocation density was very high near the scratch, and very few nuclei were observed there. The nucleation probability is not uniform along the length of a dislocation, and therefore the precipitate is unevenly distributed along a dislocation, although nuclei certainly decorate all the initial dislocations, which were not displaced in spite of the heating.

Annealing at 200°C for 5 min caused the size of the nuclei to increase to about 0.05 μm, but no fresh nuclei were produced (Fig. 2b). Further heating at 200°C for 15 min caused the growth to continue, with the nuclei becoming precipitates (Fig. 2c). Three types of platy precipitate are formed on the {100} planes of the cubic lattice of aluminum.* This photograph also shows many particles of the second phase not seen in Fig. 2a. Figure 2d shows the state of the specimen after further annealing at 200°C for 20 min, where it is difficult to identify the area with the part used in the first photograph on account of the extensive growth of the platy precipitates. Although these precipitates decorate the initial dislocations (as could be seen at magnifications between 6000 and 10,000), the dislocations began to move after the growth stage shown in Fig. 2, and the disposition was different from that in the initial stage.

* Films of this alloy of thickness 0.2 μm or less are adequate for use in transmission in an ordinary electron microscope at 100 kV, but the three-dimensional structure of the precipitates is then not seen.

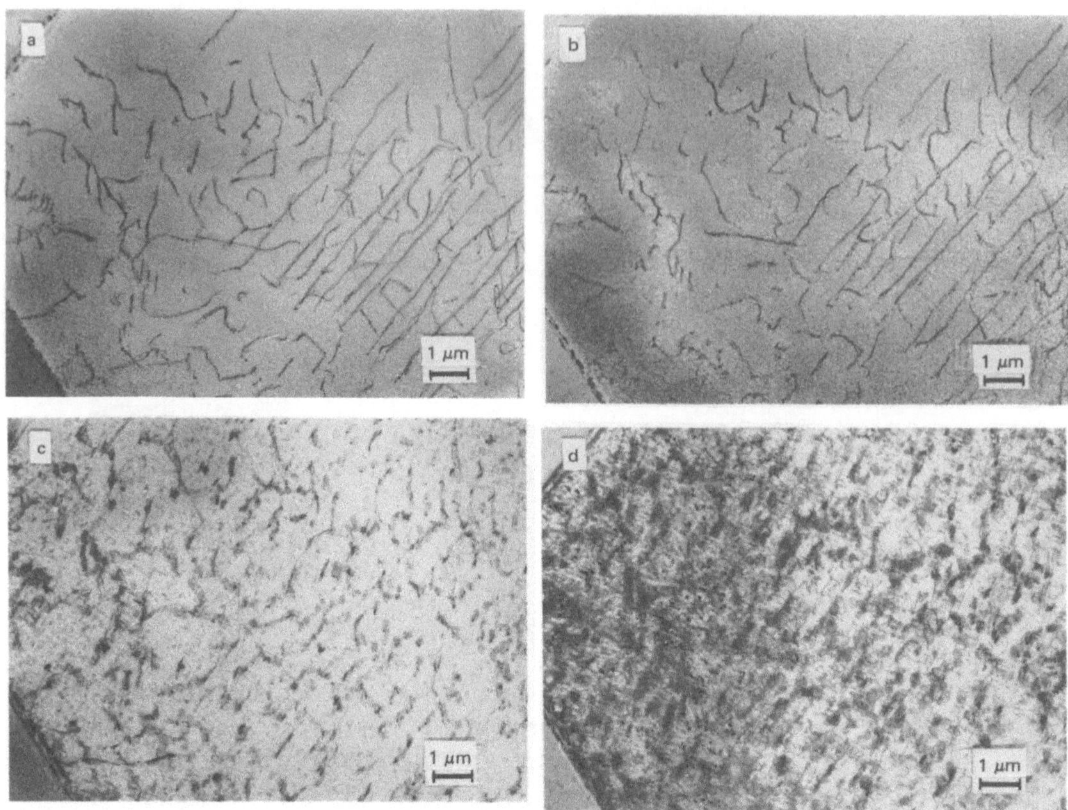


Fig. 2. Electron micrograph showing nucleation and precipitate growth in the same part of an Al–4% Cu specimen on heating within the JEM-1000: (a) nucleation due to aging at 1500°C for 60 min; (b) nucleation and growth on subsequent aging at 200°C for 5 min; (c) growth of nuclei on further aging at 200°C for 15 min after stage a; (d) growth of nuclei by further aging at 200°C for 20 min after stage 1. Magnification × 6000.

## (b) Nucleation

The nucleation and growth processes will be clear from the photomicrographs representing enlarged photographs of the parts enclosed in the frames in Fig. 1. The microdiffraction pattern corresponding to the photograph contained only one system of ⟨111⟩ reflections, so the pattern did not indicate the exact crystallographic orientation of the film; however, it proved possible to establish the orientation almost exactly by means of the two types of helicoidal dislocations seen in Fig. 1, and also on account of the platy morphology of the precipitates in the {100} planes of the matrix. This orientation is shown in Fig. 1. Figure 3 shows nucleation in a part of the first photograph near the scratch. Some of the nuclei have arisen along a large helicoidal dislocation at the top left and on certain other dislocations not far from the scratch, where the dislocation density is very high. It would seem that the nucleation near the scratch differs from that in less deformed regions. Figure 4 shows the nucleation in the region remote from the scratch, where the dislocation density is low. Figure 5 shows that the nucleus indicated by the arrow in Fig. 4 has grown into a plate of size 0.3 μm* in the (100) plane, but at the same time there are other nuclei lying in (010) and (001) planes; in these cases, however, the correspondence between the precipitates in the successive stages is not so clear. There-

* Diameter measured on observation from the side.

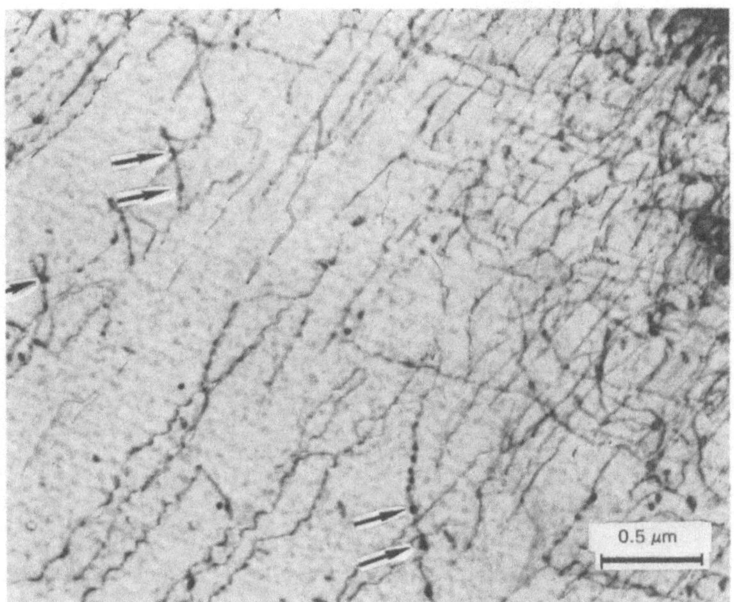


Fig. 3. Nucleation in Al−4% after aging at 150°C for 60 min (the photomicrograph shows an enlarged image of the part enclosed in the frame in the upper part of Fig. 1).

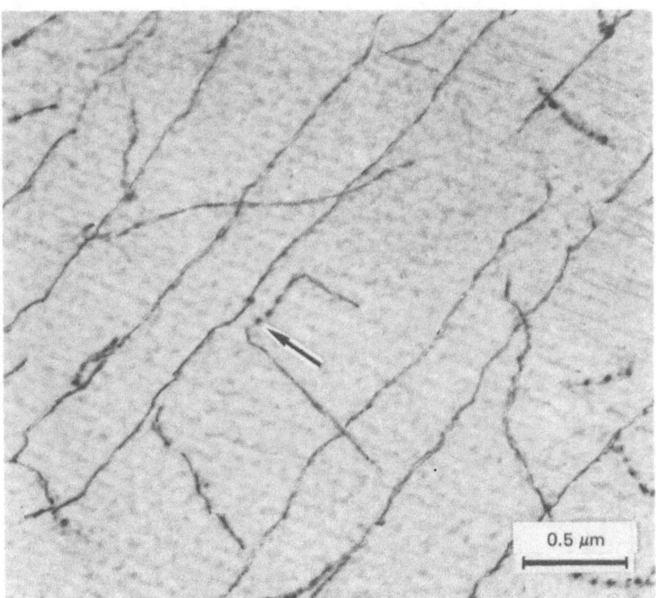


Fig. 4. As Fig. 3, but for the part enclosed in the lower part of Fig. 1.

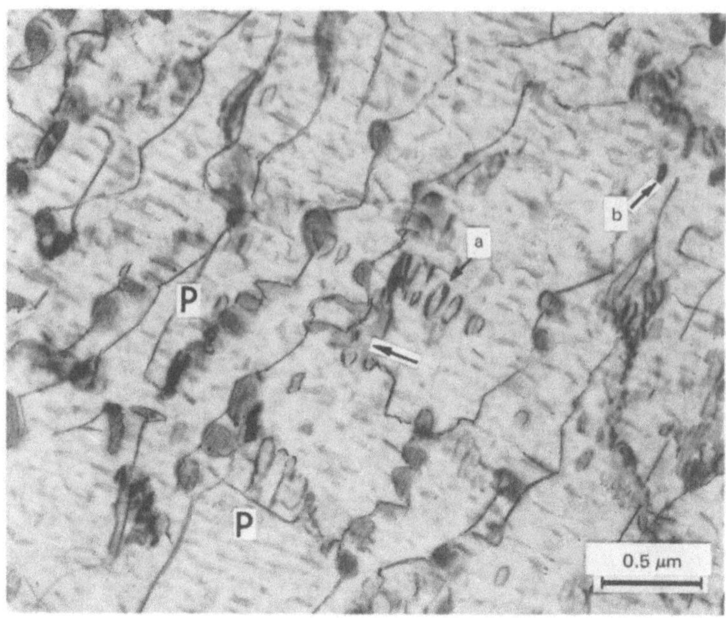


Fig. 5. Growth of precipitates in Al−4% Cu, aging at 150° C, 60 min, followed by 200° C, 50 min. Part shown in Fig. 4.

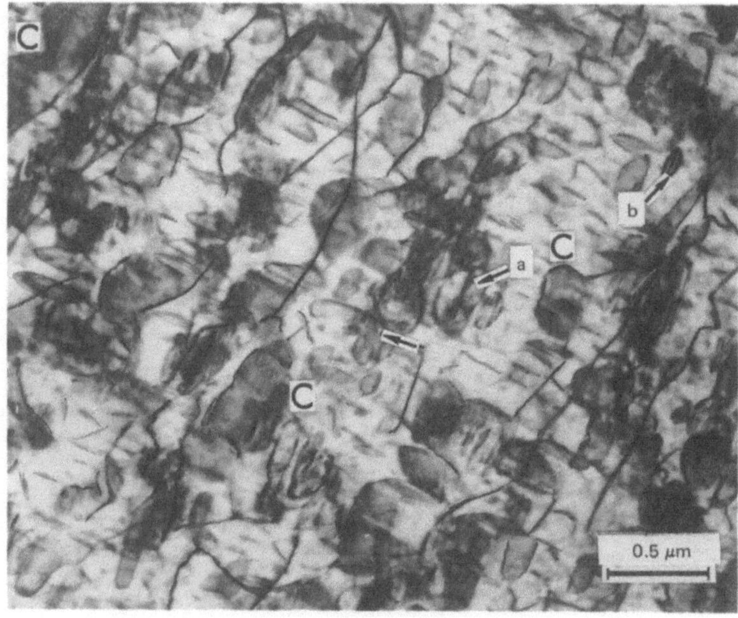


Fig. 6. Further growth of precipitates in Al−4% Cu after aging at 150°C, 60 min, followed by 200°C, 20 min.

fore, one concludes that the nuclei formed on the three cube faces were produced at early stages, although the process develops considerably at much later stages and much higher temperatures. The initial nucleation occurred at large helicoidal dislocations, not at smaller dislocations.

These experimental difficulties made it impossible to establish where the nucleation occurred (on edge or screw dislocations), and also where the precipitate of the second phase developed.

### (c) Precipitate Growth

In the stage shown in Fig. 5, new particles of the thickened phase are formed, as at a and b, which are absent in Fig. 4. Figures 4 and 5 together show that there is also a change in the disposition of the dislocations at this stage. The dislocations now seen may not coincide with initial ones, which have been blocked by the precipitates and become immobile, These precipitates have caused the stresses around the dislocations to relax, but dislocation segments free from nuclei continue to move during precipitate growth, so a new disposition of the dislocations arises. From this viewpoint, the closely spaced nuclei formed at edge dislocations will continue to grow in the previous fashion away from the dislocation, while the initial dislocations without nuclei will be displaced by the growing precipitates, and therefore one gets the disposition of the precipitates as shown in region P in Fig. 5.

The rate of precipitate growth would appear to be the same for all three types of plate appearing on the $\{100\}$ planes, although one sometimes finds large particles arising by fusion of small ones, as at C in Fig. 6.

It is not entirely clear which of the phases $\theta'$ and $\theta''$ is responsible for the particles of the second phase seen here; it is not possible to measure the temperature of a thin film precisely, in contrast to a massive specimen, so the actual temperature may have exceeded the measured value. Also, even our specimens, which were much thicker than those used in ordinary electron microscopy, appear to show much more rapid reaction. Boyd and Nicholson [4] have made a detailed study of the behavior of the $\theta'$ and $\theta''$ phases, and they found no coalescence of $\theta''$ precipitates. It would seem that our particles were most likely of the $\theta'$ phase, although the annealing conditions would lead one to suppose that the $\theta''$ phase was formed.

Figure 6 does not reveal clearly the typical Widmanstätten structure for the $\theta'$ phase; the size of the $\theta'$ precipitate is insufficient to make the structure clearly visible, and each particle is small by comparison with the typical scale of the Widmanstätten structures found in undeformed massive specimens of this alloy. This effect may be related to the influence of the scratch on the precipitation [unpublished results].

### Literature Cited

1. G. Thomas and M. J. Whealan, Phil. Mag., 6:1103 (1961).
2. N. Takahashi and T. Taoka, Compt. Rend., 272B:403 (1971).
3. N. Takahashi, H. Tomita, and Shimizu, Basic Problems in Thin Film Physics, Clausthal, Göttingen (1965), p. 725; N. Takahashi, Compt. Rend., 273C:1328 (1971).
4. J. D. Boyd and R. B. Nicholson, Acta Met., 1379 (1971).

# DISPERSION OF THE NEW PHASE IN THE EARLY STAGES OF MASS CRYSTALLIZATION

D. B. Kashchiev

*Institute of Physical Chemistry, Bulgarian Academy of Sciences, Sofia*

## Introduction

Any detailed kinetic description of mass crystallization requires a knowlege of the volume of the new phase, the total number of particles in the system, and the size distribution, as has been several times demonstrated [1-5]. Kolmogorov [1] and Avraami [2] have used fairly general assumptions to derive the time dependence of the crystalline volume and have obtained expressions for the number of crystallites at a given instant. Appropriate corrections were applied for the reduction in the amount of material available for crystallization and the restriction on crystallite growth due to contacts. Avraami also applied this model in a very general form to find the size distribution.

We have related the number of crystallites to the size for two particular cases of mass crystallization; we have used Avraami's general formula and applied it to the initial stages, where the crystallite growth is not appreciably limited by mutual contact.

## Formulation and Solution

Consider an initial phase of volume $V_0$; at the start, $t = 0$, the phase becomes supersaturated, and nuclei are produced with a frequency $J(t)$ [$cm^{-3}sec^{-1}$]; the nucleation rate J as a function of time t is governed by nonstationary effects [6, 7] and by the change in the conditions for the phase transition. The resulting nuclei begin to grow with a linear speed $\mathbf{v}(\mathbf{n}, \mathbf{r}, t)$, which is dependent on the unit vector **n** for the normal to the surface of the crystal, as well as on the radius vector **r** of a point on that surface at time t. We assume that suitable averaging over all directions will allow us to characterize the crystallite size by means of an effective radius r, and then by $v(r, t)$ [$cm \cdot sec^{-1}$] we denote the linear rate of increase in r, which is also found from **v** by averaging with respect to direction [1].

Very general considerations indicate that v must be a complicated function of r and t [3, 8], but in most cases of practical interest one can consider v as a product of two functions, each of which is dependent on one variable only:

$$\frac{dr}{dt} = v(r, t) = c(t)f(r). \tag{1}$$

The forms of c(t) and f(r) are substantially dependent on the growth mechanism and on the external conditions [3, 8].

Let $r(t, t')$ be the radius of a freely growing crystallite at time t, and at a previous instant t' the crystalline nucleus has a radius $r_c(t')$. If $r_c$ is small enough, we get from (1) that

$$\int_0^r \frac{d\rho}{f(\rho)} = \int_{t'}^t c(\tau)d\tau. \tag{2}$$

We introduce the functions

$$X(r) = \int_0^r \frac{d\rho}{f(\rho)},$$

$$Y(t) = \int_0^t c(\tau)d\tau,$$

and the inverse functions $\Psi[X(r)] = r$ and $\Phi[Y(t)] = t$ to get from (2) that

$$r(t, t') = \Psi[Y(t) - Y(t')], \tag{3}$$

$$t'(r, t) = \Phi[Y(t) - X(r)]. \tag{4}$$

It is clear that a crystallite formed at the start will have the maximum radius $r_m$, i.e.,

$$r_m(t) = r(t, 0) = \Psi[Y(t)]. \tag{5}$$

We further assume that the shape of the crystallite is such that the growth occurs mainly in 1, 2, or 3 dimensions; in that case, the volume $V_n(t, t')$ of a freely growing crystallite formed at a time $t' < t$ will become as follows by time t:

$$V_n(t, t') = k_n r^n(t, t') = k_n \Psi^n[Y(t) - Y(t')], \tag{6}$$

where n = 1, 2, 3 are the dimensions of the growing crystal and $k_n$ is a form factor. For instance, n = 3 and $k_3 = 4\pi/3$ for a sphere of radius r, n = 2, and $k_2 = \pi h$ for a disc of radius r and thickness $h \ll 2r$, and n = 1 and $k_1 = 2\pi R^2$ for a cylindrical needle of length 2r and cross section $\pi R^2$, with $R \ll r$.

We see from (3) and (4) that there is a unique relationship between the radius r at time t and the time of production t'; this means that particles having radii in the range from r to r + dr were formed in the time interval from t' − dt' to t', i.e.,

$$Z(r, t)dr = -V_0 J(t')\theta(t')dt', \tag{7}$$

where $Z(r, t)$ is the distribution of the crystallites by radius and $\theta(t')$ is the proportion of the volume remaining uncrystallized at time t', in which the former is given by the following [1, 2]:

$$\theta(t') = \exp\left[-\int_0^{t'} J(\tau)V_n(t', \tau)d\tau\right]. \tag{8}$$

Since from (2), for t = const,

$$dt' = -\frac{dr}{c(t')f(r)}, \tag{9}$$

equation (7) can be written as

$$Z(r,\,t)=\frac{V_0 J[t'(r,\,t)]\theta[t'(r,\,t)]}{c[t'(r,\,t)]f(r)},\quad 0\leqslant r\leqslant r_m(t), \tag{10}$$

where $t'(r, t)$ is defined by (4); this is Avraami's formula [2] for the crystallite distribution, which applies when the growth is virtually unhindered by contacts. It is clear that $Z(r, t) \equiv 0$ for $r > r_m(t)$.

From the distribution we get mean radius $\bar{r}(t)$ and the most probable radius $r_0(t)$ as functions of time; from (3), (9), and (10) we have the first of these quantities as

$$\bar{r}(t)=\Big(\int_0^{r_m(t)} rZdr\Big)\Big(\int_0^{r_m(t)} Zdr\Big)^{-1}=[V_0/N(t)]\int_0^t \Psi[Y(t)-Y(t')]J(t')\theta(t')dt', \tag{11}$$

while the second is the solution to

$$\left[\frac{\partial Z(r,\,t)}{\partial r}\right]_{r=r_0(t)}=0.$$

In (11), the quantity

$$N(t)=\int_0^{r_m(t)} Zdr=V_0\int_0^t J(t')\theta(t')dt' \tag{12}$$

is the total number of crystallites in the system at time t [1, 2].

We now apply Avraami's formula of (10) to two particular cases of crystallization at constant supersaturation.

1. We assume that $J$ = constant and that the crystals grow at constant rate $v$ = constant, which is independent of $r$ and $t$. This is the case of a crystal growing in a liquid solution for example [3], and also that of a one-component liquid [8, 9] or a vitreous melt [10, 11]. We put $c(t) = v$ and $f(r) = 1$ to get after appropriate steps that

$$Z(r,t)=\frac{V_0 J}{v}\exp\left[-\frac{k_n}{n+1}v^nJ\left(t-\frac{r}{v}\right)^{n+1}\right],\quad 0\leqslant r\leqslant r_m. \tag{13}$$

Figure 1 shows Z as a function of r for two different times $t_1 < t_2$ (curves 1 and 2 respectively). The distribution increases monotonically with r for a given t, and (5) shows that for example we have

$$r_m(t)=vt \tag{14}$$

with the maximum value independent of time and crystal size,

$$Z_m=\frac{V_0 J}{v}.$$

We introduce the dimensionless variable $x_n=[k_n v^n J/(n+1)]^{\frac{1}{n+1}}\,t$, to get from (11) and (13) that

$$\bar{r}(t)=[(n+1)v/k_nJ]^{\frac{1}{n+1}}F_1(x_n), \tag{15}$$

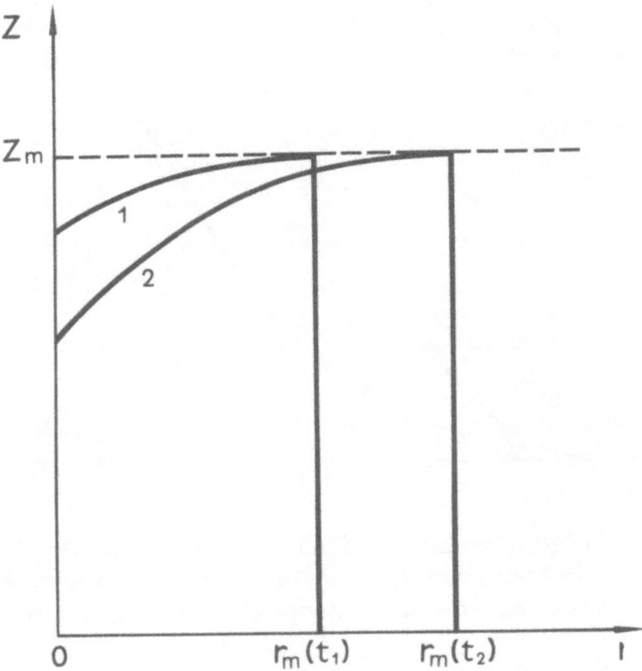


Fig. 1. Crystallite size distributions after times $t_1$ (curve 1) and $t_2$ (curve 2) for J = constant, v = constant, and $t_1 < t_2$.

where $F_1$ takes the form

$$F_1(x_n)=x_n-[\int_0^{x_n}\lambda\exp(-\lambda^{n+1})d\lambda]\,[\int_0^{x_n}\exp(-\lambda^{n+1})d\lambda]^{-1}.$$

It follows from (14) and (15) that $r_m \to \infty$ and $\bar{r}\to\infty$ for $t\to\infty$, because (13) and (10) not be used after some interval t*, since past that point the crystallite growth becomes substantially restricted and ultimately terminates on account of contacts. We can estimate t* by using (13) to calculate V(t) as the total volume crystallized at time t, the result being compared with the exact Kolmogorov–Avraami formula [1, 2] in the form

$$V_{K-A}(t)=V_0[1-\exp(-x_n^{n+1})]. \tag{16}$$

For V(t) we have

$$V(t)=k_n\int_0^{r_m(t)}r^nZdr=(n+1)V_0\int_0^{x_n}(x_n-\lambda)^n\exp(-\lambda^{n+1})d\lambda. \tag{17}$$

From $V(x_n^*)-V_{K-A}(x_n^*)=0.05\,V_0$ with (16) and (17) we get that $x_n^*\approx 0.64;\ 0.75;\ 0.82$ for n = 1, 2, and 3 respectively, and consequently $t^*=x_n^*[(n+1)/k_nv^nJ]^{\frac{1}{n+1}}$: Substitution of $x_n^*$ into (16) shows that the distribution of (13) applies while the crystallized volume is less than $(0.34\text{-}0.36)V_0$.

2. We now consider the case where J = constant but the linear growth rate is dependent on the crystallite radius:

$$v=\frac{a^2}{2r},\quad a=\text{const}. \tag{18}$$

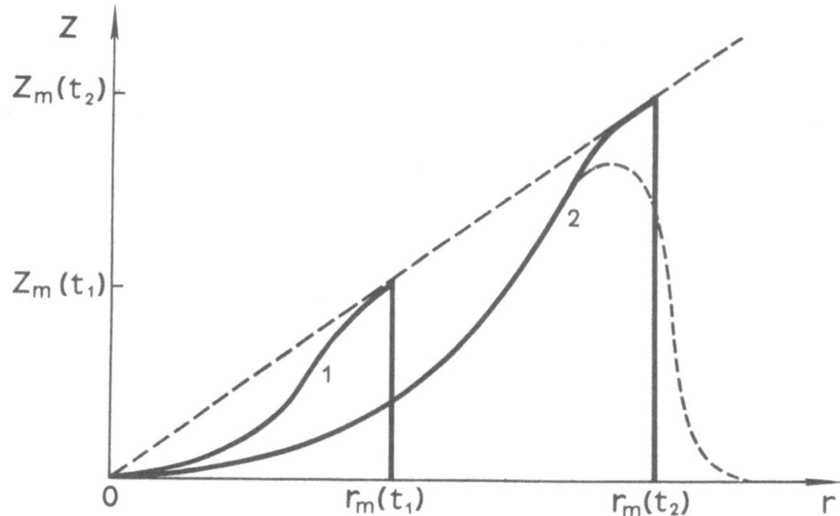


Fig. 2. Crystallite distributions at times $t_1$ (curve 1) and $t_2$ (curve 2) for J = constant, $v = a^2/2r$ and $t_1 < t_2$; the broken line corresponds to nonstationary nucleation, namely J = J(t).

This type of relationship between v and r occurs, for instance, in crystals growing in melts [12] and also ones growing in liquid solutions [3] and solid solutions [13] if diffusion is the rate-limiting step, and also in a one-component liquid when the rate-limiting step is heat transport in the initial phase [8]. We put $c(t) = a^2$ and $f(r) = 1/2r$ to get after appropriate steps that

$$Z(r,t)=\frac{2V_0J}{a^2}\, r\exp\left[-\frac{2k_n}{n+2}a^nJ\left(t-\frac{r^2}{a^2}\right)^{\frac{n+2}{2}}\right],\quad 0\leqslant r\leqslant r_m. \tag{19}$$

Figure 2 shows Z as a function of r for $t_1 < t_2$ (curves 1 and 2 respectively); as previously, the distribution increases monotonically with r for any given t, and from (5) we have for

$$r_m(t)=a\sqrt{t} \tag{20}$$

the maximum in time, but the value is independent of the crystal size:

$$Z_m=\frac{2V_0J}{a}\sqrt{t}. \tag{21}$$

The dimensionless variable $x_n = [2k_n a^n J/(n+2)]^{\frac{2}{n+2}}t$ is then used with (11) and (19) to get

$$\bar{r}(t)=[(n+2)a^2/2k_nJ]^{\frac{1}{n+2}}F_2(x_n), \tag{22}$$

where

$$F_2(x_n)=\left[\int_0^{x_n}(x_n-\lambda)^{1/2}\exp(-\lambda^{\frac{n+2}{2}})d\lambda\right]\left[\int_0^{x_n}\exp(-\lambda^{\frac{n+2}{2}})d\lambda\right]^{-1}.$$

As in the first case (19)-(22) can be used only for the initial stages, i.e., when the crystalline volume is $V < 0.35V_0$; nevertheless, as the process terminates at $X_n \approx 1$, we get approximate expressions from (15) and (22) for the mean radius at the end. As $F_1$ and $F_2$ take values close to 1 for $X_n \approx 1$, we get for cases 1 and 2 above that the mean radii at the end are respectively

$$\bar{r}_1 = \alpha_1[(n+1)v/k_nJ]^{\frac{1}{n+1}},$$

$$\bar{r}_2 = \alpha_2[(n+2)a^2/2k_nJ]^{\frac{1}{n+2}},$$

where $\alpha_1$ and $\alpha_2$ are numbers of the order of unity. The first of these expressions is virtually equivalent to Volmer's formula [14] for n = 3 and to Erofeev's formula [15] for n = 1, 2, and 3.

## Conclusions

The result of crystallite interaction in the Kolmogorov–Avraami theory is growth restricted by the contacts; therefore, (10), (13), and (19) correspond to the size distribution only within the framework of this model. However, as the interactions are unimportant at the initial stages of crystallization, these equations can be applied provided that $V < 0.3V_0$ even although the subsequent growth may follow some altogether different mechanism, e.g., coagulation.

Further, some of the existing crystallites may dissolve and vanish if the supersaturation falls; therefore, the above relationships apply strictly only if the phase transition occurs at constant supersaturation. Nevertheless, they can be used as a good approximation also when the rate of fall in the supersaturation is so slight that the final number of crystallites does not alter substantially as a result of any change in $R_c$ with time.

Finally we note that the nucleation frequency has been assumed to be constant, and for this reason the theoretical distribution is bounded on the right by a vertical straight line in both cases; if, however, one incorporates nonstationary effects, this vertical line becomes a smooth curve (broken line in Fig. 2), and the distribution acquires a typical dome-shaped form. Therefore, if we know that v is independent of r or is a decreasing function of r, then the behavior of the right side of the expression for Z will indicate whether or not there is nonstationary nucleation.

## Literature Cited


1. A. N. Kolmogorov, Izv. AN SSSR, Otd. Mat. Est. Nauk, 3:355 (1937).
2. M. Avraami, J. Chem. Phys., 7:1103 (1939); 8:212 (1940); 9:177 (1941).
3. O. M. Todes, In: Problems in Kinetics and Catalysis, Vol. 7 [in Russian], Izd. AN SSSR, Moscow (1949), p. 91.
4. E. Bauer, A. K. Green, K. M. Kunz, and H. Poppa, In: Basic Problems in Thin Film Physics, R. Niedermayer and H. Mayer, eds., Vandenhoeck and Ruprecht, Göttingen (1966), p. 135.
5. S. Toshev and I. Gutzov, Phys. Status. Solidi, 24:349 (1967).
6. Ya. B. Zel'dovich, Zh. Eksp. Teor. Fiz., 12:525 (1942).
7. D. Kashchiev, Surface Sci., 14:209 (1969).
8. A. A. Chernov and V. Ya. Lyubov, In: Growth of Crystals, Vol. 5A, Consultants Bureau, New York (1968), p. 7.
9. J. B. Hudson, W. B. Hillig, and R. M. Strong, J. Phys. Chem., 63:1012 (1959).
10. I. Gutzov, Silikattechnik, 20:159 (1969).
11. I. Gutzov, A. Razpopov, and R. Kaishev, Phys. Status Solidi (a), 1:159 (1970).
12. N. G. Ainslie, C. R. Morelock, and D. Turnbull, Symposium on Nucleation and Crystallization in Glasses and Melts, M. Reser, G. Smith, and H. Insley, eds., Amer. Ceram. Soc. (1962), p. 97.
13. Ya. I. Frenkel', Introduction to the Theory of Metals [in Russian], GIFML, Moscow (1958), p. 320.
14. M. Volmer, Kinetik der Phasenbildung, Steinkopff, Dresden (1939), p. 206.
15. B. V. Erofeev, Inst. Khimii AN BSSR, Sb. Nauchn. Rabot, 5(1):13 (1956).

# CRYSTALLIZATION AS A MATRIX REPLICATION PROCESS

G. I. Distler

*Institute of Crystallography, Academy of Sciences of the USSR, Moscow.*

The basic concept of this discussion is that a substrate crystal or seed is an electrically active matrix and therefore crystallization is to be considered as a replication in an electrically active real crystal. The active structure is provided by a variety of defects, and point defects are of particular importance in crystallization, as in other heterogeneous processes. The actual structure reflects the features of the ideal structure in a regular fashion, and the basic structure influences many of the structure-sensitive properties via the real (defective) structure rather than directly.

Until recently, crystallization was discussed largely in isolation from other heterogeneous processes such as adsorption, catalysis, and adhesion. We consider it better to consider crystallization as a particular case of a broad class of chemical reactions, particularly surface reactions.

From this viewpoint, the components in the corresponding reactions are active centers (mainly point defects) on the substrate or seed; this approach appears the more justified because point defects are centers active in adsorption, catalysis, luminescence, recombination, and so on, i.e., are components in various heterogeneous and topochemical reactions.

The nature of the nucleation centers is also very important in crystallization; there are three major viewpoints on nucleation. The first is that nuclei are produced at random from a two-dimensional gas, while the second is that nuclei are produced only at active centers, which are mainly point defects, and finally the third view is a compromise, in that nucleation is selective at low supersaturations but random at high ones. Recent new experimental evidence indicates that nucleation occurs mainly at electrically active point defects.

It has frequently been observed in crystal decoration that nuclei form groups (squares, dumbbells, triangles, trapezia, and so on), which reflect the presence of composite active centers [1-3] (Fig. 1). The decorating particles lie at distances of 50-200 Å within such groups, while the groups themselves may or may not be oriented one with respect to another or with respect to the crystallographic directions. Also, the supersaturation determines what type of center is involved (Fig. 1, a-c), which confirms that nucleation is analogous to heterogeneous reaction. In a nucleation, one type of component is provided by an atom, ion, or molecule of the deposited substance, while another component is provided by active centers on the surface (the exact type of center is dependent on the reaction conditions).

Many experiments have shown that a crystal will produce $5 \cdot 10^{10}$-$10^{11}$ cm$^{-2}$ densities for nuclei; this corresponds to the number of electrically active point defects in an ionic crystal, while in a semiconductor it corresponds to the number of fast states at the semiconductor–oxide boundary, which themselves arise from point defects, and this again confirms that point defects produce selective nucleation.

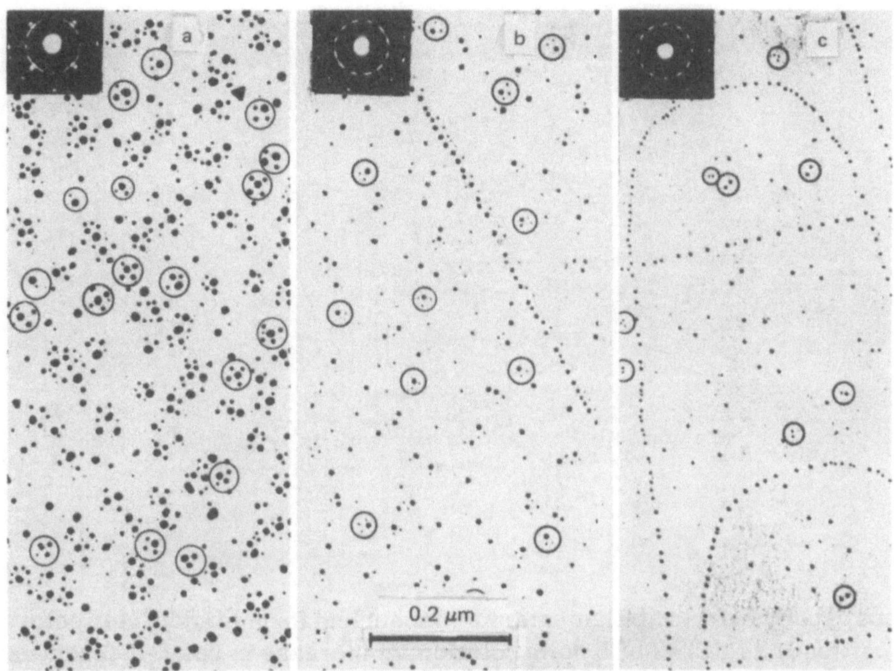


Fig. 1. Selective nucleation of gold at various composite active centers in an NaCl crystal after exposure to gamma rays at supersaturations decreasing from left to right.

Gold decoration of NaCl doped with divalent metals such as lead or cadmium reveals that there are double electrical layers at the interface [4, 5], as Frenkel' predicted [6] and which was subsequently discussed by Lyakhovek [7] and Lifshits and Geguzin [8]. The double electrical layers are detected by the selective nucleation arising at the charged point defects, which in this case occur at the impurity ions. However, double layers are detectable only on cleaved surfaces produced in air, which means that the activity is very much enhanced by the interaction with the atmosphere. It has been shown [9] that only water influences the crystallization of gold in NaCl when the cleavage is performed in air, which means that the polar water molecules are adsorbed and affect the activation energy of the corresponding heterogeneous process. It would seem that the chemically adsorbed water influences mainly the electrically active point defects rather than the ions in the main lattice, and it would seem that the same probably applies for other polar substances.

Composite active centers are extremely important in nucleation via topochemical reactions; measurements have been made on decomposition in $NaCl-PbCl_2$ ($CaCl_2$, $BaCl_2$, $SrCl_2$) solid solutions and in additively colored NaCl crystals, as well as crystals exposed to x-rays or $\gamma$-rays; low dope concentrations and small radiation doses increase the concentrations of composite active centers [1-3, 10]. Clumps of point defects tend to occur as the dope concentration or radiation dose increases, and the complex centers mentioned above often form a basis for such clumps. One consequence of this is that one gets clumps of point defects nearly spherical in form, which may coexist with filamentary structures (Fig. 2a). Such clumps of point defects are the rule rather than the exception, so a real crystal is to be considered as a complex electrically microheterogeneous system. Nuclei of a new phase arise almost always at the center of such a clump under appropriate thermal conditions. The formation of a new phase is frequently preceded by an increaze in the point-defect density at the center of the clump (Fig. 2b), i.e., the new phase arises only when a very high level of point defects occurs

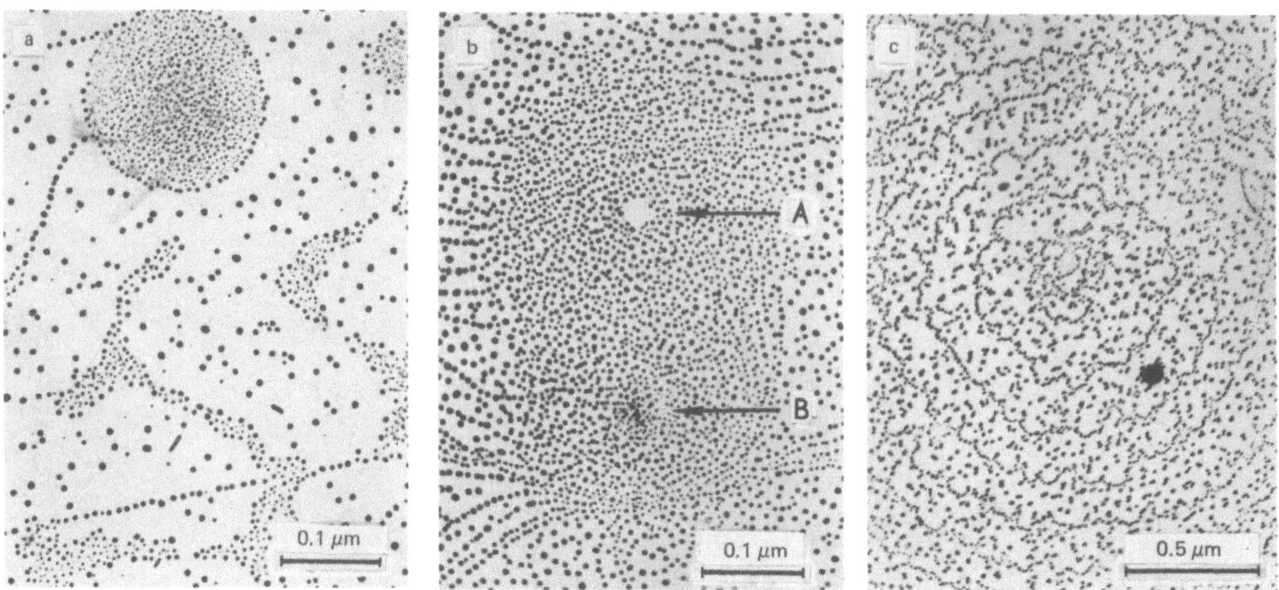


Fig. 2. (a) Clumps of point defects of filamentary and spherical form in additively colored NaCl crystals; (b) decomposition of a $NaCl-PbCl_2$ solid solution; A) increase in point-defect density (formation of prephase) at the middle of a defect clump; B) production of the new phase at the center of the clump of defects; (c) dope clump on a dislocation arising from decomposition in a $NaCl-PbCl_2$ solid solution.

over some comparatively large area of the crystal. In some cases, the dope clumps that precede the formation of a new phase themselves arise at edge dislocations that form Cottrell atmospheres (Fig. 2c), as has been demonstrated [11] for the $NaCl-BaCl_2$ system. The inclusions of the new phase may take various shapes characteristic of the system, and the sizes may be up to several microns, while the particles are frequently oriented along certain crystallographic directions. The clumps of point defects around nuclei of the new phase may be considered as double electrical layers at the interfaces. The new phase is produced by much the same mechanism in all the above cases, namely doped crystals, additively colored ones, and irradiated crystals. One supposes that many topochemical processes, in particular photographic ones, occur in much this way. The clumps of point defects are seen in decoration because the nucleation occurs selectively at the electrically active defects.

Until recently it had been assumed that the primary nuclei were formed with a definite orientation with respect to the substrate, but it has been found that this is not so if a careful distinction is drawn between the nucleation proper and the subsequent coalescence or growth. As an example we consider the nucleation of silver chloride (on thermal evaporation) on cleavage surfaces of lithium fluoride [12] (Fig. 3a, b) and of lead sulfide on triglycine sulfate [13] (Fig. 3c, d). The nuclei become oriented only at the coalescence stage. Masson et al. [14, 15] have observed crystallization directly in the electron microscope and found that the nuclei become oriented by rotation after a certain interval. It seems likely that many who have assumed that nuclei are formed with a definite orientation in fact observed not the nucleation stage proper but instead the stage of coalescence of growth, when the orientation arises.

As the original nuclei are not oriented with respect to the surface, we need to reconsider existing theoretical concepts on nucleation, as has been clearly formulated previously [16]; this applies not only to the phenomenological theory (Bauer [17] and Hirth et al. [18]) but also to atomistic ones (Walton [19]). We also need precise experiments to indicate whether oriented nucleation ever occurs.

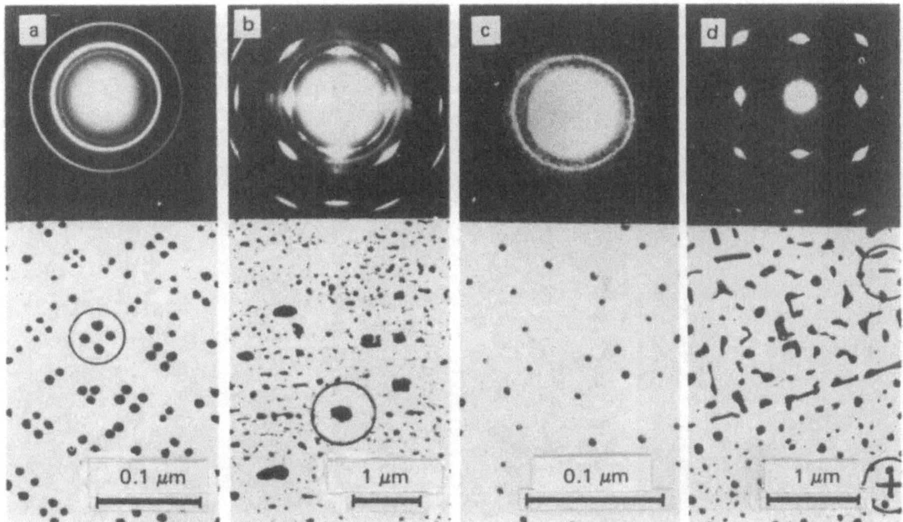


Fig. 3. Unoriented formation of AgCl nuclei of LiF (a) and triglycine sulfate (c); oriented coalescence of AgCl nuclei on LiF (b) and triglycine sulfate (d).

The stage following nucleation is that of coalescence or growth, which results in reasonably large particles; if model substrates are used, such as triglycine sulfate, which is a ferroelectric, and whose domain structure can be determined, then it is found that the rate of oriented coalescence for nuclei or fairly large particles is governed by the electrical microrelief [20, 21]. This oriented coalescence always occurs most rapidly on the negative domains [22], and this at first sight surprising result occurs because the negatively charged areas adsorb thin layers of water, as has been demonstrated directly for NaCl crystals [23, 24]. Sears and Hudson [25] have shown theoretically that thin layers of water act as lubricants that facillitate migration and rotation for fairly large particles.

A real crystal is electrically inhomogeneous, on account of localized clumps of vacancies, interstitial atoms, and impurities, which make themselves felt at the coalescence stage [12, 26]. The surface of a real crystal reflects primarily the electrical heterogeneity, i.e., the crystal acts as an electrostatic mosaic, whose elements are of size 100-1000 Å. A real crystal in addition is macroscopically inhomogeneous, as can be seen clearly at the decoration stage. If triglycine sulphate is decorated with anthraquinone, the positive domains form uniaxial acicular textures, while the negative domains produce biaxial ones [27] (Fig. 4a). Comparison may be made with cleaved surfaces on NaCl and mica, which shows [28, 29] that these materials also have positive and negative regions of size 50-500 μm (Fig. 4b, c). Cleavage steps tend to kink at these boundaries, as has been observed for domains in triglycine sulfate [30] and for p−n junctions in semiconductors [31], i.e., the steps kink as a rule at electrical boundaries. This again shows that regions differing in type of crystallization differ also in electrical structure. The electrostatic mosaics occurring on real crystals have also been demonstrated by probe measurements [32].

The patterns in the early stages of crystallization show that electrical boundaries thus exist on and within crystals; these boundaries can be seen in ferroelectric crystals such as triglycine sulfate, e.g., as domain walls, some of which are electrically less active than the domains themselves (since no selective crystallization occurs there), whereas some part of the walls are electrically more active than any other part of the surface (the largest particles are then produced). If the polarization of a triglycine sulfate crystal is reversed, a charged impurity accumulating at certain walls will remain in place and represent an electrical bound-

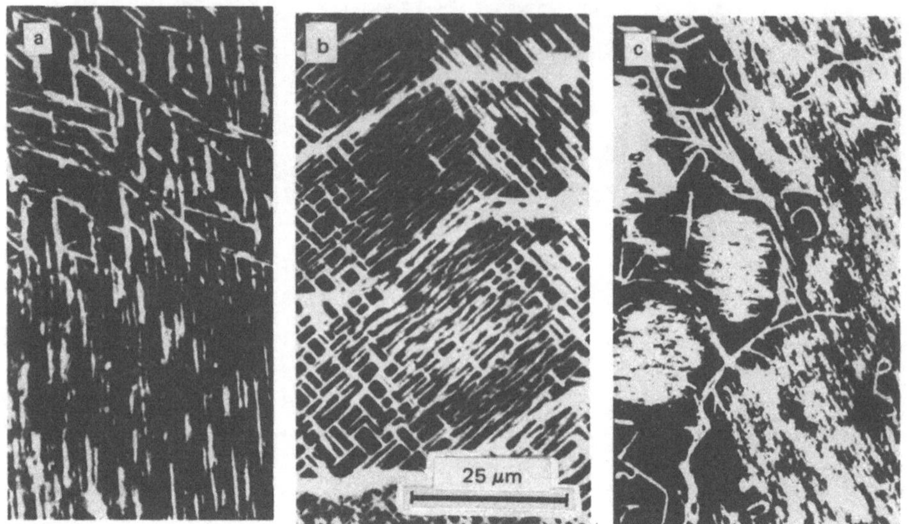


Fig. 4. Uniaxial texture in anthraquinone crystals on positively charged parts of surfaces and biaxial (triaxial) structures on negatively charged parts in (a) triglycine sulfate; (b) NaCl; (c) mica.

ary unrelated to a domain wall (Fig. 5a). A cleavage surface in NaCl shows many cleavage steps, which themselves are often inactive, and there are reasonably broad dead zones where there is no crystallization [33, 34] (Fig. 5b). The low activity of these zones near the steps is due to the electrical neutrality, and the boundaries of these zones represent elements in the electrical relief, which does not necessarily coincide with the geometrical relief. Very often, the steps and the dead zones are electrical boundaries separating areas differing in electrical properties, as is clear from the differences in rates of oriented coalescence in such areas. Filamentary and spherical clumps of point defects frequently also have sharp electrical boundaries, as is clear from the decoration patterns.

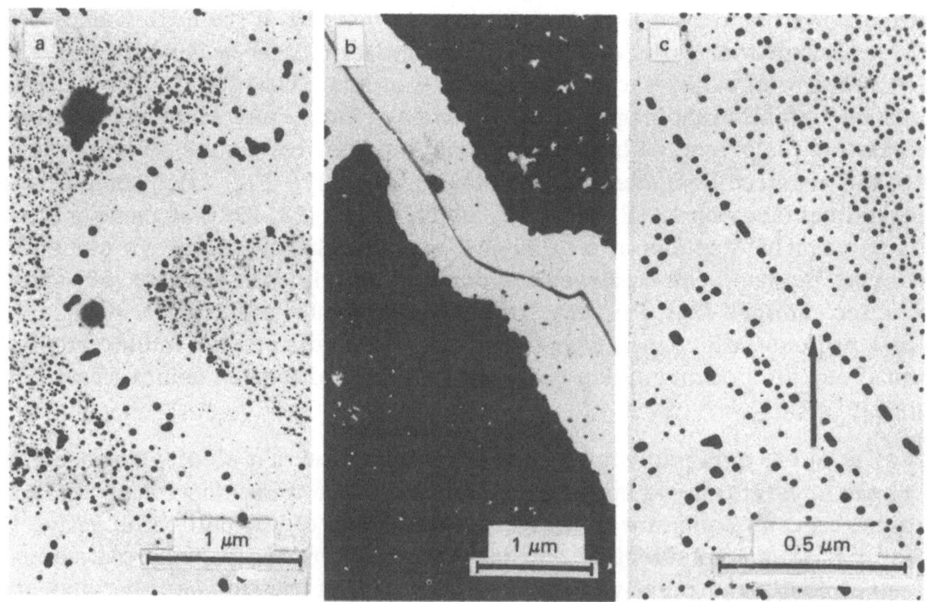


Fig. 5. Various electrical boundaries at surfaces: (a) triglycine sulfate; (b, c) NaCl.

Much importance has been assigned to cleavage, growth, and evaporation steps in studies on decoration and crystallization [35, 36]; however, our experiments show that the geometrical relief has no direct effect on the crystallization [20, 21, 37]. Some cleavage steps show no crystallization at all, while others show uneven crystallization (Fig. 5b, c). This occurs because there is competition between crystallization at the electrical boundaries (in particular, electrically active cleavage, growth, and evaporation steps) and crystallization at electrically active point defects, either single or in clumps, which may occur on the smooth surfaces. Therefore, it is not the steps themselves as elements of the geometrical relief that initiate the crystallization, but instead any electrical parameters, which have to compete with those of adjacent smooth areas.

This evidence shows that the electrical relief is important during crystallization, and this leads us to reconsider the Kossel–Stranski and dislocation theories of crystal growth. These assume that the surfaces have geometrical relief, which is responsible for controlling the crystallization, although this has not been confirmed by experiment. On the other hand, the geometrical relief is active by virtue of any electrical properties it may have (this applies to linear growth steps as well as to points of emergence of screw dislocations). Therefore, only if the electrical parameters of the geometrical relief are favorable can one expect the microgeometry to affect the crystallization.

A very important and interesting case of electrical boundaries in a real crystal is provided by those seen as bridges between nuclei; these have several times been observed [38-41], but no rational explanation has been proposed. We ascribe the bridges to some particular electrical state of the surface (or volume) between differently charged point defects, since the electric field strength (and field bunching) will be greatest at such points. At such sites one gets linear polarization structures, which are electrically more active than are adjacent areas [42, 43]. However, these linear structures are less active than the point defects themselves, since the latter become apparent at an earlier stage in the crystallization. The linear polarization structures are very characteristic of real crystals, and they form major elements governing the early stages of crystallization.

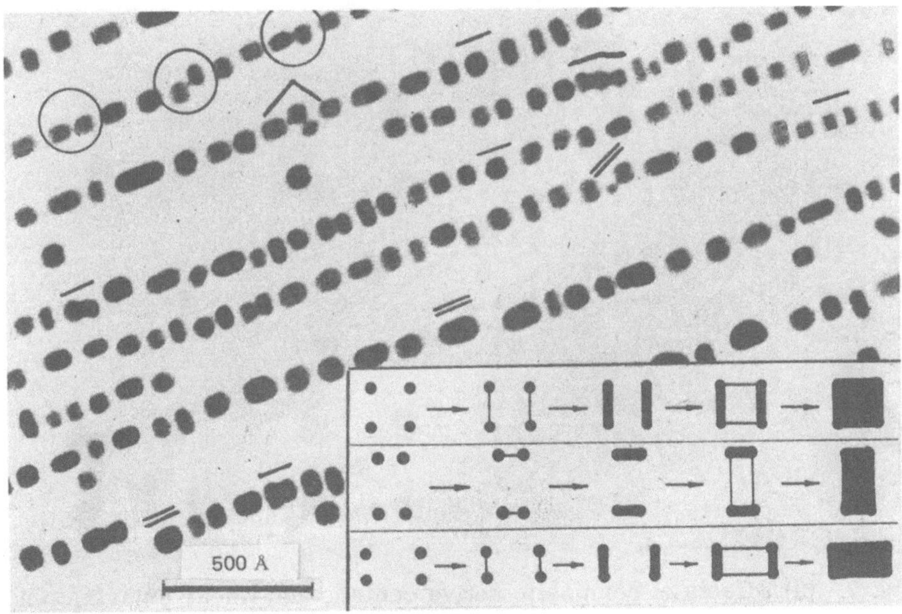


Fig. 6. Decorating particles of gold of various shapes, sizes, and orientations related to the structure and orientation of composite active centers.

These linear bridges are also seen on decorating composite centers; such centers with their bridges provide a new explanation for the deposition of particles of a definite shape, size, or orientation [44]. For instance, dumbell centers most readily produce rectangular particles, whose orientations correspond to those of the dumbbells, while triangular centers produce triangular particles, square ones produce square particles, and so on, i.e., the shape and orientation of the center predetermine the morphology and orientation of the resulting crystals (Fig. 6). This photograph shows clearly that the composite centers, which here lie on cleavage steps, determine the activity of the steps locally rather than the electrical properties of the steps as a whole, with the active centers being point defects at or near the steps. Pentagonal or other unusual shapes result in particles of similar shape and symmetry, which has recently attracted attention [45, 46].

An important effect due to the electrical relief is the long-range effect, in which the bulk character of surface phenomena makes itself felt, in particular in crystallization. An amorphous film, e.g., of carbon or silicon monoxide on NaCl, produces an orienting action on gold, lead sulfide, and other substances deposited on the outside, and it is possible to produce single-crystal films of a high degree of orientation [47-53].

Groups of decorating particles (Fig. 7a) are seen on the outside of such an amorphous film as well as directly on the substrate; these reflect the composite active centers, which also produce bridges between the nuclei in the groups, which reflect the linear polarization structures, and which enable one to determine the sign alternation in the centers (Fig. 7b), since such structures can arise only from point defects differing in charge sign. The sizes of these centers, as seen through amorphous films, are of the same order as those of centers observed directly, which implies that the long-range transmission of the structural information occurs with very little loss of resolution, which means that the amorphous film allows the crystalli-

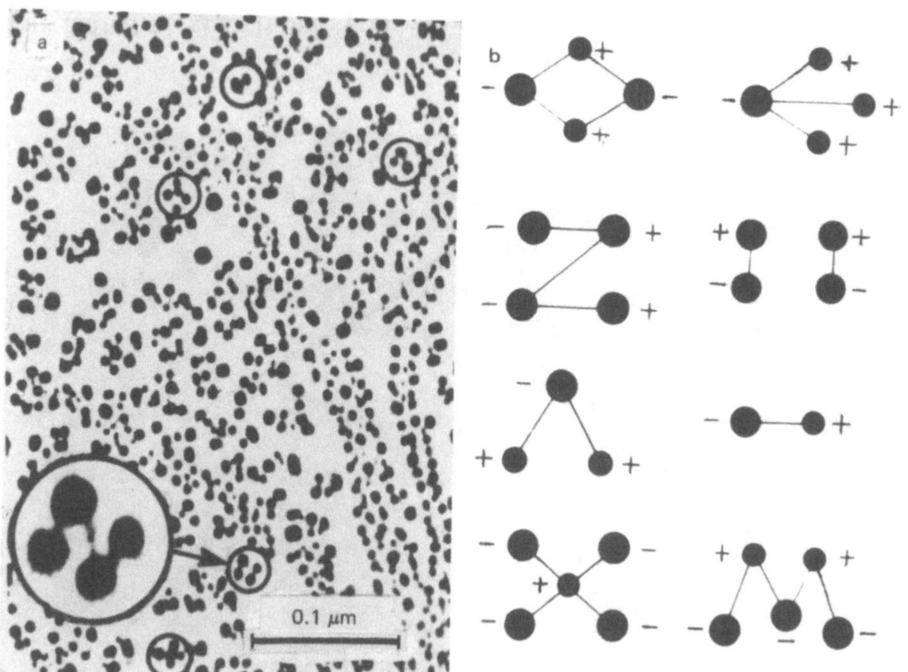


Fig. 7. (a) Effects of composite active centers and linear polarization structures in NaCl acting via a carbon film of thickness 80–1000 Å; (b) models for composite active centers with alternating signs.

zation structure to be transmitted from the electrical relief. These amorphous films can be used as substrates in epitaxy, which shows that the lattice parameter of the underlying crystal and the surface microgeometry cannot be the decisive elements in the crystallization.

The long-range effects have been ascribed to internal polarization of electret type, which transmits the potential relief from the substrate [54]; direct proof of the electret mechanism comes from the use of films of materials of thermoelectret type (PVC) or photoelectret type (selenium or zinc oxide). Anthraquinone crystallizes with a definite orientation, as in the case of direct use of NaCl, when it occurs on the contact side of a PVC film detached from an NaCl crystal only if the film has been set in the thermoelectret state [55, 56]. Therefore, such films of PVC, carbon, or other substances constitute secondary matrices, and their polarization structures retain the crystallization information. Similarly, amorphous films of selenium, which shows photoelectret behavior, store and transmit the structural information in the same way [57, 58]. Orientated crystallization of the anthraquinone occurs also on the outside of a polycrystalline film of zinc oxide, which is also a photoelectret [59, 60]. This shows that the induced polarization structures arise and exist without reference to the crystallographic directions in the boundary layer. The polycrystalline zinc oxide stores the single-crystal information after separation from the substrate. The photoelectret state can be destroyed by illumination, as in selenium films.

These properties of amorphous and polycrystalline films throw considerable light on the nature of such materials generally; such a film, although apparently amorphous or polycrystalline in diffraction, can retain electrical order of single-crystal type (in the case of selenium or zinc oxide, namely photoelectrets, the effect is clearly associated with photoelectrons in deep traps).

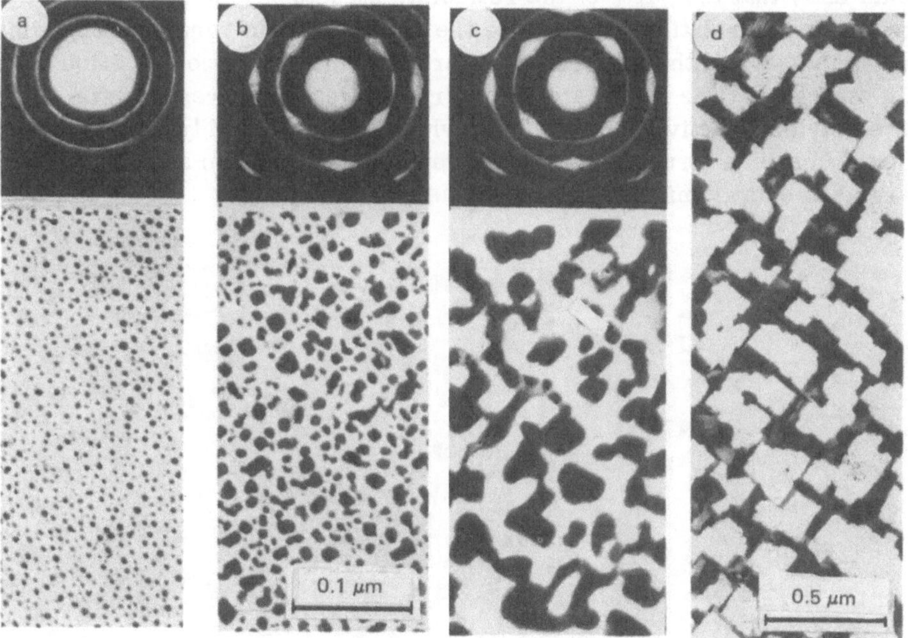


Fig. 8. (a–c) Kinetics of oriented crystallization of AgCl on contact side of PVC film representing a thermoelectret copy of NaCl, with the process terminating in a reticulate structure; (d) reticulate structure of NaCl crystal revealed by sublimation under an electron beam applied to a single-crystal Ag single film made on the contact side of a PVC film representing a thermoelectret copy of the NaCl surface.

This mapping of the electrical structure occurs at various levels, including the point-defect one [42]. Vacuum-evaporated silver chloride will crystallize on a PVC thermoelectret film detailed from NaCl very much as on the NaCl itself. At first, randomly disposed nuclei are formed (Fig. 8a), and then oriented coalescence occurs, with bridges between the particles (Fig. 8b). This means that amorphous films map the electrically active point defects where the nucleation occurs, as well as the linear polarization areas between them, where the bridges occur. A reticulate structure mapping that in the substrate crystal occurs on further deposition (Fig. 8c). This structure is the most stable and is seen also in sublimation (for instance, in response to an electron beam), for single-crystal silver chloride films made directly on NaCl or on the contact side of a PVC film [61] (Fig. 8d). There is a correlation between the surface density of the nuclei and the density of the most persistent films during sublimation, which shows that the parts most resistant to electron bombardment are areas in the silver chloride that arise at the electrically active point defects or at the corresponding electrical copies responsible for the nucleation.

The reticulate structure of a real crystal is readily seen in the early stages of crystallization (Fig. 9a). This structure is influenced by various external factors, e.g., gamma-irradiation of triglycine sulfate (Fig. 9b, c), which reverses the charges in some centers and therefore suppresses certain linear elements [62]. Particular structures similar to those produced by crystallization and sublimation have frequently been observed after chemical or ionic etching [63, 64]. This electrically active structure explains the narrow channels frequently seen in film growth. When crystallization has occurred on the elements of the reticulate structure, the original linear bridges begin to grow laterally and result in channels. The sublimation patterns show that the parts of the film corresponding to the lateral growth lie on the largely inactive parts of the substrate, where the adhesion is less than that on the electrically active parts. The data show that the parts of the reticulate structure (point defects and groups of these) and the linear polarization bridges are the most active in various topochemical and heterogeneous reactions and therefore they differ substantially in composition and properties from the main structure. The reticulate structure implies that a real crystal tends to contain sublattices of electrically active point defects, which often are highly distorted. One assumes that the elements in such a structure serve to transmit information and also act as preferential means of charge and energy migration, as well as diffusion, etc.

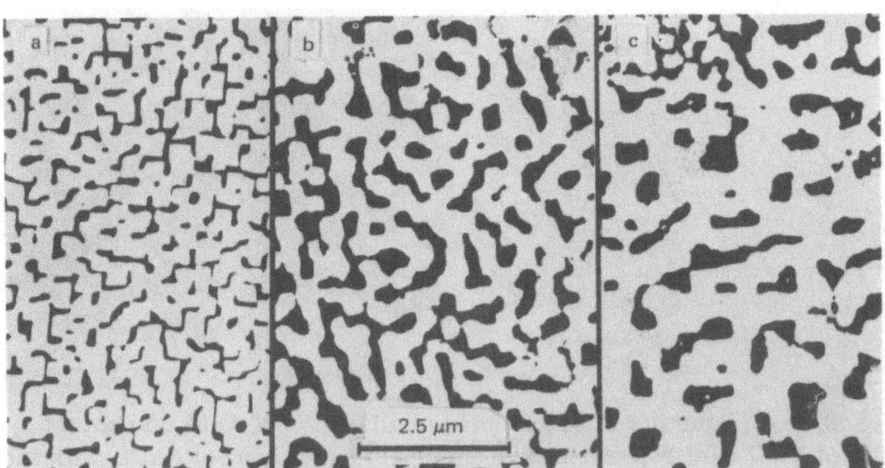


Fig. 9. (a) Reticulate structure of a triglycine sulfate crystal revealed by crystallization of AgCl; (b, c) changes in the reticulate structure of triglycine sulfate due to exposure to gamma rays as seen by crystallization of AgCl.

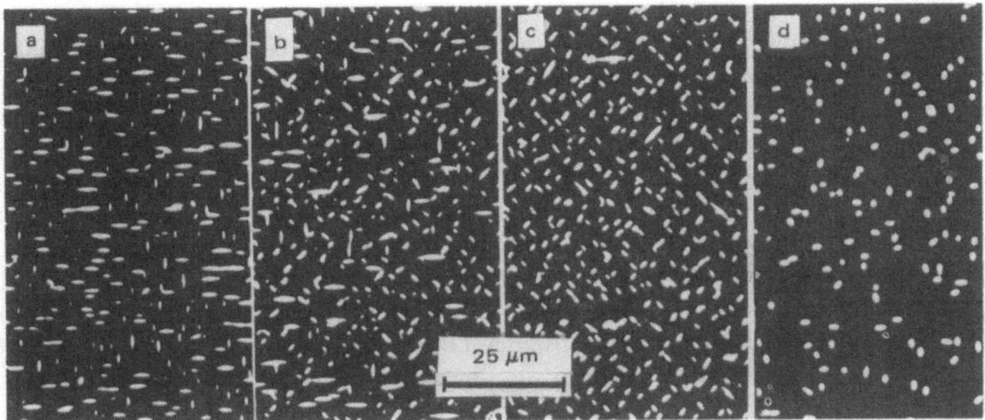


Fig. 10. (a–c) Oriented crystallization of anthraquinone on polycrystalline silver films of various thicknesses on NaCl: (a) texture axes parallel to the ⟨110⟩ and ⟨1$\bar{1}$0⟩ directions in NaCl, Ag film less than 100 Å thick; (b) texture axes parallel to ⟨100⟩, ⟨1$\bar{1}$0⟩, ⟨100⟩, and ⟨010⟩ in NaCl, Ag film thickness 125 ± 25 Å; (c) texture axes parallel to ⟨100⟩ and ⟨010⟩ in NaCl, Ag film thickness greater than 150 Å; (d) oriented crystallization of anthraquinone on the contact side of a polycrystalline silver film detached from NaCl, texture axes parallel to ⟨110⟩ and ⟨1$\bar{1}$0⟩ in NaCl.

This reticulate structure within and on crystals has been observed also with metallic films [65-68]. Until recently, such films had not been made from metals, but it had been assumed (on the basis of the macroscopic properties of metals) that metal films should screen out the potential relief of the crystal surface and therefore could not reflect that relief. However, experiment shows that films of silver, gold, nickel, bismuth, aluminum, and so on made by vacuum evaporation (about $10^{-5}$ mm Hg) can produce oriented crystallization of decorating materials on the outside. Parts a and b of Fig. 10 show biaxial textures due to anthraquinone crystals on the outside of polycrystalline silver films. If the thickness of the silver is greater than 150 Å, the axes of the anthraquinone crystals alter in orientation by 45°, namely, from along ⟨110⟩ to along ⟨110⟩ in the NaCl. The contact sides retain the structural information when the film is removed from the NaCl (Fig. 10d). The texture axes also change in direction when nickel is used in the same way.

These features of metal films can be explained as follows. The nuclei are formed during crystallization selectively at electrically active point defects; the electrical field strength at distances of atomic order may be in the range 100–1000 kV/cm, so the crystallization essentially occurs in an extremely strong microscopic field (and also in the presence of a fair amount of residual gas or vapor, in particular water or oxygen, under these conditions). One therefore assumes that the atoms or ions interact with the residual gas near the electrically active centers to produce local semiconducting and/or insulating compounds of the metals, in which polarization structures of electret type are produced. These information-bearing structures are formed within the metal films in very small quantities and have virtually no effect on the macroscopic properties of the film. The data indicate that a metal single crystal contains an information-bearing semiconducting and/or insulating network, which enables one to discuss from a single viewpoint many processes, in particular crystallization, occurring at the surface and in the volume of nonmetallic and metallic single crystals.

Finally, we must mention the information-bearing properties of water [69], which plays a major part in many crystallization and information-transfer processes. Double electrical layers at interfaces are not detected on gold decoration of cleavage surfaces made under vacuum. When silver is used with such crystals, the particles coalesce in an oriented fashion at the highest rate on areas covered by double layers (if the cleavage is in air), whereas vacuum-cleaved materials do not show double layers. The nuclei migrate and show oriented coalescence most readily on the negatively charged parts of the substrate, where there are physically adsorbed water films, which act as lubricants. Therefore, oriented coalescence of silver and oriented nucleation in gold occur at the same parts of the substrate, namely where there are thin films of water. This means that the gold nucleates at the outside of a thin film of water, which covers local areas and which therefore must have information-bearing properties. There are many references [70-73] to the possible information-bearing structures in water. One assumes that thin films of water are influenced by electrically active elements in the crystal to produce local polarization structures, i.e., thin boundary layers of water act as a medium that will transmit long-range forces. Local polarization structures in the water facilitate elementary acts in various heterogeneous processes, in particular nucleation.

There is a general conclusion to be drawn. Crystallization is always a matrix-replication process controlled by the electrical structure of the substrate or seed; the nucleation is programmed at the point-defect level, particularly point defects most active in this reaction (crystallization). It may be, of course, that nucleation can occur at electrically active elements in the geometrical relief under special circumstances. The growth and coalescence of nuclei or fairly large particles is controlled by the electrical structure, which may include elements of the geometrical relief. The detailed crystallization conditions determine which elements in the electrical relief control the program. Therefore, the formation of a crystal or tin film is a replication process, which reproduces the matrix via the electrical parameters of the latter. There is a far-reaching analogy between processes such as crystallization and biological matrix replication processes.

## Literature Cited


1. G. I. Distler, V. N. Lebedeva, and V. V. Moskvin, Fiz. Tverd. Tela, 10:3489 (1968).
2. G. I. Distler, J. Crystal Growth, 3(4):175 (1968).
3. G. I. Distler, V. N. Lebedeva, and V. V. Moskvin, Kristallografiya, 14:664 (1969).
4. G. I. Distler, V. N. Lebedeva, V. V. Moskvin, and E. I. Kortukova, Fiz. Tverd. Tela, 11:2390 (1969); 12:1149 (1970).
5. G. I. Distler, V. N. Lebedeva, and V. V. Moskvin, 7th International Congress on Electron Microscopy, Grenoble, 1970, P. Favard, ed., Vol. 2, p. 361.
6. Ya. I. Frenkel', Kinetic Theory of Liquids [in Russian], Izd. AN SSSR, Moscow (1945).
7. K. Lehovec, J. Ceem. Phys., 21:1123 (1953).
8. N. M. Lifshits and Ya. E. Geguzin, Fiz. Tverd. Tela, 7:62 (1965).
9. K. Miltama, H. Niyahara, and H. Aoe, J. Phys. Soc. Japan, 23:785 (1967).
10. G. I. Distler, V. N. Lebedeva, V. V. Moskvin, and E. I. Kortukova, Kristallografiya, 15: 1049 (1970).
11. S. Kupsa, M. Hartmanova, and G. Vlasak, Czechosl. J. Phys., 6:789 (1969).
12. G. I. Distler and V. P. Vlasov, Fiz. Tverd. Tela, 11:2226 (1969).
13. V. P. Vlasov and G. I. Distler, Kristallografiya, 16:663 (1971).
14. A. Masson, J. J. Metois, and R. Kera, Surface Sci. 27:463 (1971); 27:483 (1971).
15. A. Masson, J. J. Metois, and R. Kera, In: Advances in Epitaxy and Endotaxy, H. G. Schneider and V. Ruth, eds., Leipzig (1971), p. 103.
16. A. Green, E. Bauer, R. L. Peck, and J. Dancy, Kristall u. Technik, 5:345 (1970).
17. E. Bauer, Z. Kristallogr., 110:395 (1958).
18. J. P. Hirth, S. J. Hruska, and G. M. Pound, In: Single-Crystal Films, Francombe and Sato, eds., Pergamon Press, Oxford (1964).
19. D. Walton, J. Chem. Phys., 37:2182 (1962).

20. G. I. Distler and V. P. Vlasov, Kristallografiya, 14:872 (1960).
21. G. I. Distler and V. P. Vlasov, Thin Solid Films, 3:333 (1969).
22. V. P. Vlasov, Yu. M. Gerasimov, and G. I. Distler, Kristallografiya, 15:346 (1970).
23. W. C. Price, W. F. Sherman, and G. R. Wilkinson, Proc. Roy. Soc., 247:467 (1958).
24. R. A. Ladd, Surface Sci., 12:37 (1968).
25. G. W. Sears and J. B. Hudson, J. Chem. Phys., 30:2380 (1963).
26. G. I. Distler and E. G. Sarovskii, Fiz. Tverd. Tela, 11:547 (1969).
27. S. A. Kobzareva, G. I. Distler, and V. P. Konstantinova, Kristallografiya, 15:510 (1970).
28. S. A. Kobzareva and G. I. Distler, Kristallografiya, 16:601 (1971).
29. S. A. Kobzareva and G. I. Distler, J. Crystal Growth, 10:260 (1971).
30. V. P. Konstantinova, Kristallografiya, 7:748 (1962).
31. V. I. Sokolov, Fiz. Tverd. Tela, 7:295 (1965).
32. V. B. Deryagin and M. S. Metsik, Fiz. Tverd. Tela, 1:1521 (1959).
33. G. I. Distler and E. I. Tokmakova, Kristallografiya, 14:1055 (1969).
34. G. I. Distler, and E. G. Sarovskii, In: Physicochemical Problems of Crystallization, Vol. 2 [in Russian], Alma-Ata (1971), p. 178.
35. H. Bethge, Phys. Status Solidi, 2:1 (1962); 2:775 (1962).
36. H. Bethge, 7th International Congress on Electron Microscopy, Grenoble, 1970, P. Favard, ed., Vol. 2, p. 365.
37. G. I. Distler, V. P. Vlasov, Y. M. Gerasimov, and E. G. Sarovsky, 7th International Congress on Electron Microscopy, Grenoble, 1970, P. Favard, ed., Vol. 2, p. 465.
38. R. F. Adamsky and L. Leblanc, National Vacuum Symposium, American Vacuum Society, New York (1956), p. 453.
39. H. Poppa, Z. Naturforsch. A419:835 (1954).
40. D. W. Pashley, Adv. Phys., 14:327 (1965).
41. D. Lewis and D. J. Stirland, J. Crystal Growth, 3/4:200 (1968).
42. G. I. Distler and E. I. Tokmakova, Thin Solid Films, 6:203 (1970).
43. G. I. Distler, Dokl. Akad. Nauk SSSR, 199:802 (1971).
44. G. I. Distler and Yu. M. Gerasimov, Abstracts of the 8th All-Union Conference on Electron Microscopy [in Russian], Vol. 2, Moscow (1971), p. 7.
45. E. Gillet and M. Gillet, J. Crystal Growth, 13/14:212 (1972).
46. S. Ogawa and S. Ino, J. Crystal Growth, 13/14:48 (1972).
47. G. I. Distler, S. A. Kobzareva and Y. M. Gerasimov, J. Crystal Growth, 2:45 (1968).
48. Yu. M. Gerasimov and G. I. Distler, Kristallografiya, 14:1101 (1969).
49. G. I. Distler and E. I. Tokmakova, Kristallografiya, 16:212 (1971).
50. M. S. Khidr, P. Igmacz, and J. F. Pocza, Proceedings of the 2nd Colloquium on Thin Films, E. Hahn, ed., Akademia Kiado, Budapest (1968), p. 45.
51. A. Barna, P. B. Barna, and J. F. Pocza, Thin Solid Films, 4:R32 (1969).
52. C. A. O. Henning, Nature, 227:1129 (1970).
53. G. I. Distler and L. A. Shenyavskaya, Fiz. Tverd. Tela, 14:1400 (1972).
54. G. I. Distler and L. A. Shenyavskaya, Fiz. Tverd. Tela, 11:488 (1969).
55. G. I. Distler and S. A. Obzareva, Dokl. Akad. Nauk SSSR, 188:811 (1969).
56. G. I. Distler, Kristall u. Technik, 5:73 (1970).
57. G. I. Distler and V. G. Obronov, Dokl. Akad. Nauk SSSR, 191:584 (1970).
58. G. I. Distler and V. G. Obronov, Nature, 224:261 (1969).
59. G. I. Distler and V. G. Obronov, Dokl. Akad. Nauk SSSR, 197:819 (1971).
60. G. I. Distler, J. Crystal Growth, 9:76 (1971).
61. G. I. Distler and E. I. Tokmakova, Kristallografiya, 17:634 (1972).
62. G. I. Distler, V. A. Yurin, Yu. M. Gerasimov, N. V. Belutina, and V. A. Meleshina, Izv. Akad. Nauk SSSR, Ser. Fiz., Vol. 36, no. 6 (1972).
63. Ya. B. Pines and I. G. Ivanov, Kristallografiya, 11:802 (1965).

64. G. S. Gritsaenko, B. B. Zvyagin, R. V. Boyarskaya, A. I. Gorshkov, N. D. Samotin, and K. E. Frolova, Methods in Electron Microscopy of Minerals [in Russian], Nauka, Moscow (1969).
65. G. I. Distler, Y. M. Gerasimov, E. I. Kortukova, and V. G. Obronov, Naturwissenschaften, 58:564 (1971).
66. G. I. Distler, Yu. M. Gerasimov, and V. G. Obronov, Kristallografiya, 17:62 (1972).
67. G. I. Distler, V. G. Obronov, and Yu. M. Gerasimov, Fiz. Tverd. Tela, 14:682 (1972).
68. G. I. Distler, Y. M. Gerasimov, and V. G. Obronov, Thin Solid Films, 10:195 (1972).
69. G. I. Distler and V. V. Moskvin, Dokl. Akad. Nauk SSSR, 201:891 (1971).
70. H. S. Frank, Proc. Roy. Soc., A247:481 (1958).
71. G. Nemethy and H. A. Sheraga, J. Chem. Phys., 36:3382 (1962).
72. J. Klotz, In: Horizons in Biochemistry [Russian translation], Mir, Moscow (1964), p. 399.
73. Structure and Role of Water in Living Organisms [in Russian], Izd. Leningr. Universiteta, No. 1 (1966), No. 2 (1968); No. 3 (1970).

# ELASTIC INTERACTION IN EPITAXIAL EFFECTS


## V. L. Indenbom

*Institute of Crystallography, Academy of Sciences of the USSR, Moscow*


Recent evidence on epitaxial growth laws does not fit within the usual concepts on epitaxy mechanisms and has raised various new problems in the theory of crystal growth. In particular, one has to establish why a nucleus at the surface of a crystal begins to become oriented only above some critical size, and why such nuclei encounter no obstacles from intermediate amorphous or polycrystalline forms of thickness up to $10^2$ or even $10^3$ Å, as well as why nuclei sometimes tend to have regular two-dimensional lattices and so on. Explanations are sometimes sought in reconsideration of concepts on actual crystal structures [1], and here particular interest attaches to experimental checks on such hypothesis, particularly those that explain epitaxy via the existing theories of crystal growth and structure.

The electrical relief of the substrate plays a considerable part in nucleation, as is clear from the undoubted fact that a nucleus is subject to electrical and mechanical interaction with the substrate. Crystals growing in the solid state are dominated by the mechanical interaction with the matrix, whereas in epitaxial growth it is clear that mechanical stresses must be present even in a very simple one-dimensional model [2]. Such models imply that the first monatomic layers grow on the substrate with the lattice parameter of the latter, and the proper lattice parameter will be attained only as the elastic energy increases (this energy is linearly dependent on the film thickness), one result of this being that epitaxial dislocation are formed at the interface.

In such cases, interfacial surface energy is of considerable importance, and the epitaxial nuclei are usually not monatomic in thickness, and may even be isometric. Therefore, the interface with the substrate contains epitaxial dislocations from the start, which weaken the mechanical interaction and may allow a nucleus to migrate over the surface and change in orientation [3]. If the density of the epitaxial dislocations subsequently falls (as is clear from the reduced mobility and the acquisition of a definite orientation), then the mechanical interaction with the substrate must have increased. General considerations [4] lead one to expect that the total elastic energy should be of the same order as the total surface energy of the boundary layer.

The following factors appear to be responsible for orientation of nuclei on homogeneous substrates:

(a) anisotropy in the network of epitaxial dislocations (anisotropy in the boundary energy), and

(b) anisotropy in the stresses due to anisotropy in the shape of the nucleus and elastic anisotropy in the substrate.*

---

* Anisotropy in stress distribution may arise also from anisotropy in the thermal expansion, but this case has not been examined in the experiments discussed below.

The anisotropy in the boundary energy has been discussed in detail [5], and it is very important in the orientation of nuclei directly on the substrate, particularly if the latter is crystalline, but it cannot explain the orientation if there is an intermediate amorphous or crystalline layer (long-range action [6]). The latter effect, however, is readily explained by the second factor. One can calculate the stresses caused by a nucleus in the substrate [4], and it is clear that these vary little over distances R of the order of the diameter D of a nucleus and decrease as $(D/R)^6$ at distances $R \gg D$; correspondingly, the density of the elastic energy is almost constant in the region $D \sim R$ and decreases as $(D/R)^3$ at large distances. This means that any increase in the thickness d of the intermediate amorphous layer causes the energy of the interaction between the nucleus and an anisotropic substrate to fall slowly and linearly for $d < D$, whereas for $d > D$ it vanishes rapidly in accordance with $(D/d)^5$, i.e., the energy of the elastic interaction with the substrate preserves a relationship to the orientation provided that the dimensions are small by comparison with the thickness of the intermediate isotropic layer. It would seem that this consequence of the hypothesis agrees with results on long-range effects, since ranges of action of 100-200 Å have been observed for particles of diameter 10-100 Å, while ranges of $10^3$ Å or so have been observed for particles of 100 Å size and length up to several microns.

The elastic strain caused by an epitaxial nucleus in the substrate may affect the mutual disposition of the nuclei and result in regular configurations of decorating particles on a completely homogeneous surface. This effect is in no way different from the ordered (supercrystalline) disposition of particles of a new phase during the decomposition of solid solutions [7] or from the ordered disposition of pores arising on irradiating various materials [8]. As in both cases the precipitates or pores are produced at random in the crystal, any regular disposition of the decorating particles cannot be treated as due to decoration of some form of center, although a crystal may in fact contain several types of center. In particular, it would be of interest to establish whether these regular patterns of decorating particles arise by motion of the particles over the surface.

This hypothesis does not give an unambiguous explanation of particle orientation on amorphous films lying on crystalline substrates. It would seem fairly clear that such a film cannot be completely amorphous, but the mechanism of structuring or texturing in such a film requires further research. It may be that the film is deformed by the stresses arising during growth, and in this connection we may note that experiments [4] have demonstrated directly that particles can produce plastic strain in the substrate.

I am indebted to A. A. Chernov for valuable discussion on the various mechanisms of epitaxial growth, to S. A. Semiletov for indicating the possible nature of memory in amorphous films, to G. I. Distler for many discussions on the above hypothesis, and to M. A. Yaroslavskii for making some estimates of the anisotropy in the elastic energy for the interaction between particles and substrate.

## Literature Cited

1. G. I. Distler, this volume, p. 44; see also G. I. Distler, Proceedings of the Fourth All-Union Conference on Crystal Growth, Mechanisms and Kinetics of Crystal Growth, Part 1 [in Russian], Erevan (1972), p. 109.
2. F. C. Frank and J. H. Merwe, Proc. Roy. Soc., Ser. A, 198:205, 216 (1949).
3. A. Masson, J. J. Metois, and R. Kern, Advances in Epitaxy and Endotaxy, Leipzig (1971).
4. Ya. E. Geguzin, A. S. Dzyuba, V. L. Indenbom, and N. N. Ovcharenko, Fiz. Tverd. Tela (in press).
5. N. Cabrera, Mem. Sci. Rev. Met., 62:205 (1965).

6. M. L. Frankenheim, Ann. Phys., 37:516 (1936). R. S. Bradley, Z. Kristallogr., 96:499 (1937); see also [1].
7. A. G. Khachaturyan, Zh. Eksp. Teor. Fiz., 58:175 (1970).
8. J. H. Evans, Nature, 299:403 (1971); V. K. Sikka and J. Moteff, Appl. Phys., 43:4912 (1972).

# ELECTRON MICROSCOPIC INVESTIGATIONS OF SURFACE DIFFUSION AND NUCLEATION OF Au ON Ag (111)


## M. Klaua

*Central Institute for Solid State Physics and Materials Science, Institute for Solid State Physics and Electron Microscopy, Halle*


The experiments were carried out on defined Ag (111) surfaces produced by the following special technique. We started with an Ag single crystal sphere 3-4 mm in diameter at which by means of electrolytical growth [1] the low-index which by means of electrolytical growth [1] the low-index growth faces (111), (100), (110), and (211) were produced. For cleaning these faces the crystal was heated in a vacuum of $10^{-6}$ torr at 700-900°C, and after cooling down to 30-160°C a 1/100-1/10 monolayer of Au was deposited at a constant deposition rate. Then a carbon film was deposited and this film, together with the embedded gold nuclei, was dissolved from the Ag face and examined in the transmission electron microscope.

Figure 1 represents a typical picture clearly showing a decoration effect of Au, i.e., atomic steps on the Ag (111) faces are preferentially decorated with gold nuclei. Therefore, one can use the gold decoration method to visualize growth steps, evaporation steps, and glide steps in the same way as in case of alkali halides.

In this paper the surface diffusion and the nucleation of Au are of special interest. Estimating the mean residence time of Au on Ag (111) faces at the applied substrate temperatures of 30-160°C one obtains residence times that are much longer than the times of a few minutes needed for the experiments. From this it follows that the reevaporation of Au is negligible and that all of the deposited gold atoms must coalesce into nuclei, i.e., there is no equilibrium of adsorption between the vapor beam and the atoms adsorbed at the surface. The concentration of adatoms increases up to a critical value at which the nucleation occurs spontaneously. This critical concentration of adatoms was determined by varying the amount of Au deposited at a constant deposition rate at different temperatures, i.e., the critical time for the first nucleation was determined. In accordance with the results obtained by other authors [2] who investigated nucleation on conditions of time-dependent adsorption, the value of $n^* = 2 \cdot 10^{-13}$ atoms $cm^{-2}$ determined in this work proved to be nearly independent of the deposition rate and the substrate temperature. It was found that the observed density of nuclei always corresponds to the saturation density. As the nucleation takes place very quickly, it was not possible to measure the nucleation rate. The observed gold nuclei were always two-dimensional, which could be proved.

The experiments were carried out at substrate temperatures varying between 30 and 160°C, while the deposition rate and the quantity of deposition were held constant. The three examples in Fig. 1 clearly show the temperature dependence of the saturation density in the atomically smooth regions between the steps. At rising substrate temperature the saturation

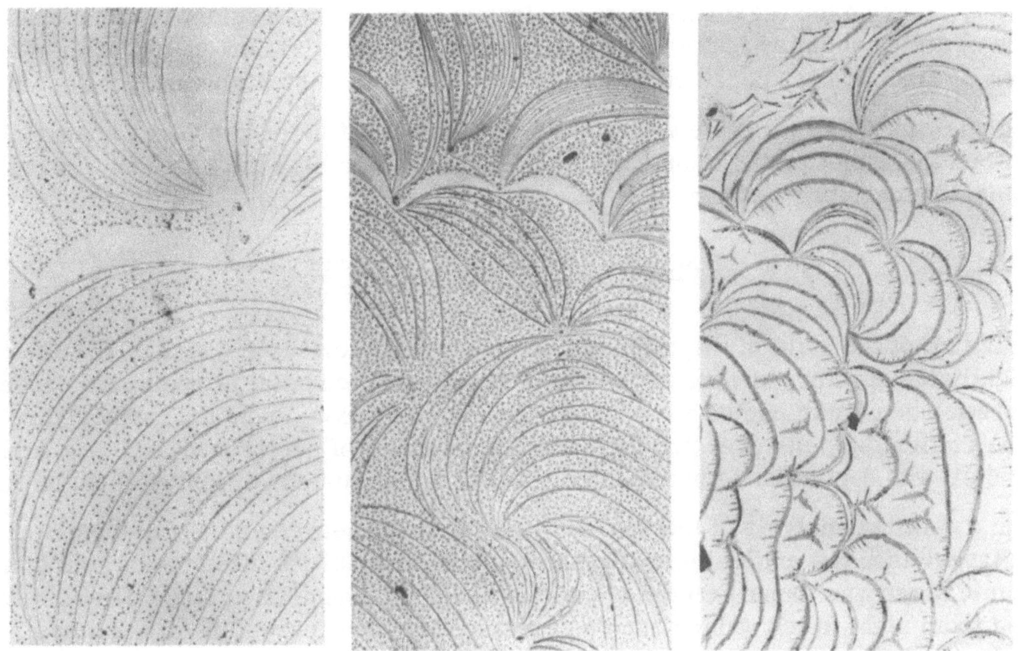

Fig. 1. Nucleation of Au on Ag (111) at 40°C, 70°C, and 159°C substrate temperature (from left to right).

density decreases. Further on, if an atomically smooth surface region is bounded by parallel or nearly parallel atomic steps, one can observe a critical step distance $\lambda^*$, below which no nucleation takes place in the region between the steps. This critical step distance increases with rising substrate temperature. Figure 2 shows the plots of saturation density $N_s$ and critical step distance $\lambda^*$ in dependence on the reciprocal absolute substrate temperature. The heavy lines correspond to a deposition rate of 0.088 Å · $sec^{-1}$, the thinner line to a deposition rate of 0.035 Å · $sec^{-1}$.

In order to explain these results we first examine the simplest case of nucleation between parallel steps some distance below the critical value. From the vapor gold atoms impinge on the surface with a certain deposition rate $J_k$. The gold atoms diffuse to the steps and there they remain fixed. Because of the high density of nuclei observed along the steps (see Fig. 1) we consider the steps to be ideal sinks. Neglecting the movement of the steps we can use the following one-dimensional stationary diffusion equation:

$$\frac{\partial^2 n}{\partial x^2}+\frac{J_k}{D}=\sigma, \tag{1}$$

n being the concentration of the diffusing gold adatoms, x the local coordinate perpendicular to the step, and D the diffusion coefficient. As boundary conditions we set $n = 0$ at the step and $\partial n/\partial x = 0$ for $x = \rho\lambda$. By introducing the asymmetry factor $\rho$ we take into account that the maximum concentration will not necessarily lie in the middle between the steps. If in the general solution of the diffusion equation we set $\lambda = \lambda^*$, then the distribution of adatom concentration between the steps can be determined, where the maximum value is given by the critical concentration of $n^* = 2 \cdot 10^{13}$ atoms $cm^{-2}$ at $x = \rho\lambda^*$. We obtain the following equation for the critical step distance:

$$\lambda^*=\frac{1}{\rho}\left(\frac{2n^*D}{J_k}\right)^{1/2}=\frac{1}{\rho}\left(\frac{2n^*a^2\nu}{J_k}\right)^{1/2}\exp\frac{-\Delta H_{diff}}{2kT}, \tag{2}$$

if the usual expression for the diffusion coefficient is taken. If we neglect the temperature dependence of the asymmetry factor $\rho$ we can determine the activation energy of Au surface diffusion on Ag from the temperature dependence of the critical step distance.

In an analogous way we solve the radially symmetric diffusion problem for surface diffusion between the nuclei on the atomically smooth faces, obtaining the following expression for the saturation density $N_s$ of nuclei:

$$N_s = \left(2\ln\frac{\bar{\lambda}}{2V_0} - 1\right)\frac{J_k}{2\pi a^2 \nu n^*}\exp\left(\frac{\Delta H_{diff}}{kT}\right) \tag{3}$$

As the logarithm of the ratio of the mean distance between nuclei to the size of the nuclei $r_0$ is only slightly temperature-dependent, we also obtain an exponential temperature dependence of the saturation density of nuclei on the substrate temperature. From the temperature dependence of the critical step distance and the saturation density of nuclei (see Fig. 2) a value of $\Delta H_{diff} = 0.5$ eV was determined for the activation energy of surface diffusion of Au on Ag (111).

From a detailed analysis of the distributions of nuclei between steps having distances 3 to 5 times larger than the critical step distance a peculiarity of the diffusion of gold atoms across atomic steps was found. On the right side of Fig. 3 evaporation steps can be seen that have moved from right to left (which can be concluded from the curvature of the steps), i.e., the higher level of the steps is on the left side (which is schematically shown at the bottom of Fig. 3). One can see a region of depletion of nuclei on the lower side of the steps and a region of enrichment on the upper side of the steps. On the left, examples of measured distributions of nuclei for a normalized step distance are shown. At lower temperatures the distribution is asymmetric, at higher ones it is symmetric. As the distributions of nuclei directly reveals the adatom concentration before the beginning of nucleation, the asymmetry factor $\rho$ can be determined from the positions of the maximum in the distribution of nuclei. The reason for the asymmetric concentration at lower temperatures is a partial or complete reflection of diffusing adatoms at the upper side of the step, i.e., there exists an additional activation energy for the jump of gold atoms from the upper side to the lower side of the step. This step effect has already been observed by Ehrlich and Hudda [3], who studied the diffusion of W on W tips at low temperatures in the field-ion microscope.

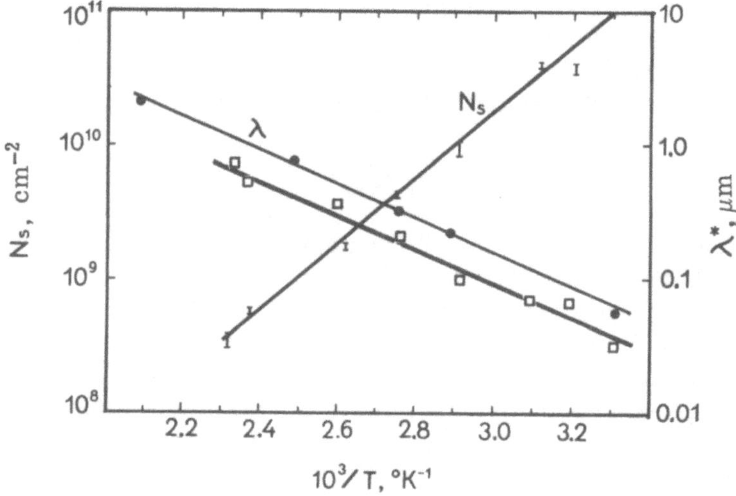


Fig. 2. Measured saturation density $N_s$ and critical step distance in dependence on the substrate temperature.

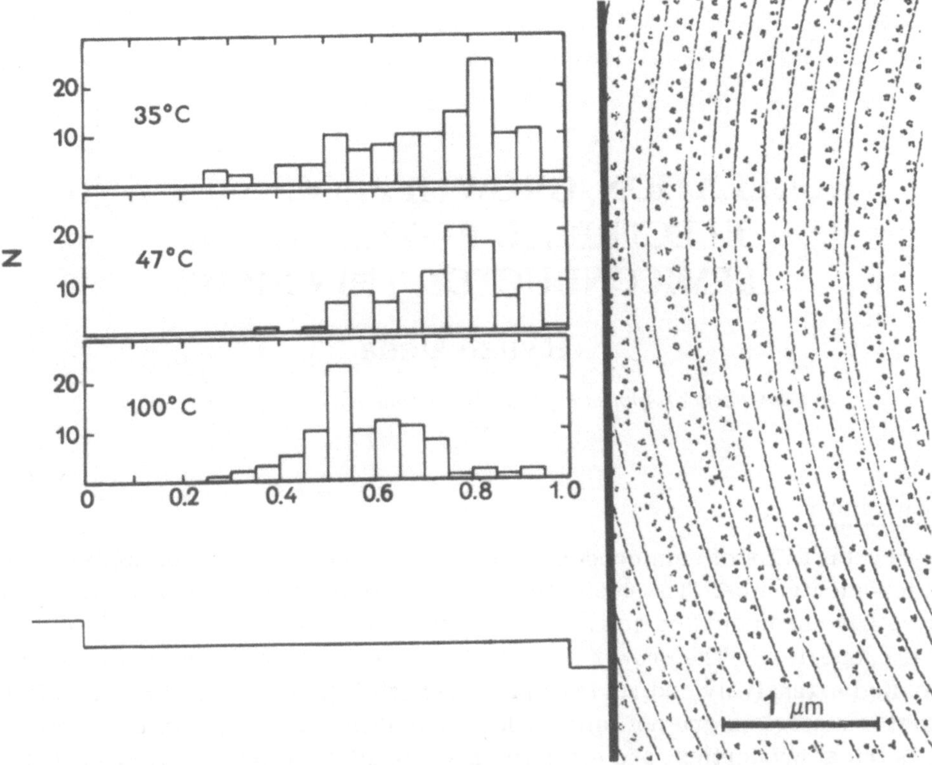


Fig. 3. Asymmetric distribution of nuclei between steps (right), step profile (bottom), and measured densities of nuclei between steps (left).

Considering the diffusion flux from both sides to the step, we can derive the following expression for the temperature dependence of the asymmetry factor:

$$\frac{\rho-\frac{1}{2}}{(1-\rho)\rho}=\left(\frac{J_k}{2n^*\nu}\right)^{1/2}\exp\left(\frac{2\Delta H_s-\Delta H_{\mathrm{diff}}}{2kT}\right). \tag{4}$$

Thus, from the measured temperature dependence of the distribution of nuclei (see Fig. 3) an activation energy of $\Delta H_s = 0.67$ eV was estimated for the jump of atoms from the upper side of the step to the lower side of the step.

## Literature Cited


1. R. Kaishev, E. Budewski, and J. Malinowski, C. R. Acad. Bulg. Sci. 2:29 (1949).
2. R. D. Gretz, J. Phys. Chem. Solids, 27:1849 (1966).
3. G. Ehrlich and F. G. Hudda, J. Chem. Phys., 44:1039 (1966).

# NUCLEATION, GROWTH PROCESSES, AND ELECTRICAL PROPERTIES OF SEMICONDUCTOR THIN FILMS

**Ryuzo Ueda**

*Department of Applied Physics, Waseda University, Shinjuku-ku, Tokyo*

## 1. Introduction

Single-crystal films of semiconductor materials have a variety of applications in active devices of microelectronics. The nucleation and growth processes of these films on crystalline substrates have been studied actively to clarify the relationships between film structures and electrical properties. The growth processes of metal deposits on alkali halide crystals have been studied extensively and reviewed elesewhere [1]. However, the formation processes of semiconductor materials are not fully understood at present owing to their complicated structure, electrical properties, and experimental conditions. This review describes and summarizes the results hitherto obtained, particularly on epitaxial silicon and some of II–VI and III–V compounds.

## 2. Epitaxial Growth of Silicon Films by Vacuum Deposition

Early work on the growth of single crystal films of silicon was performed by chemical vapor deposition using silane and other silicon compounds. The relationships between continuous films and electrical properties have been studied [2-6]. Vacuum deposition techniques were developed a little later, and films with good electrical properties were prepared in ultrahigh vacuum systems [7-10].

Recently, single-crystal silicon films have been obtained by the vacuum deposition method on ($\bar{1}012$) and ($11\bar{2}0$) sapphire and also on (111), (001), and (113) spinel surfaces at higher temperatures in a vacuum of $2 \cdot 10^{-6}$ torr [11, 12]. Reflection electron diffraction revealed that the initial deposits on sapphire contain four preferred orientations, all with (111) parallel to the substrate surfaces. Further deposition leads to the eventual formation of one orientation. A coalescence process between nucleated islands was found to be essential for the growth mechanism of single-crystal films. However, the initial deposits on spinel consist of only one orientation during growth. Electrical properties have been measured and carrier concentrations and Hall mobilities estimated for thicker films.

## 3. Epitaxial Growth of II – VI and III – V Compounds by Vacuum Deposition

Evaporated films of II–VI and III–V compounds have been widely used for thin-film transistors and photoconductive devices. The performance of these devices depends strongly on the film structure. Conditions such as the chemical composition of the vapor phase, the degree of dissociation, the rate of deposition, and substrate structure and temperature have always been important factors in determining film structures and the electrical or optical properties of these compounds. CdS, CdSe, CdTe, ZnS, GaAs, and other compounds have been deposited and studied extensively [13-15].

The epitaxial growth of CdS and ZnS thin films and their structural defects have been studied by transmission electron microscopy and diffraction methods [16]. In the case of CdS layers, both the $\alpha$-hexagonal and $\beta$-cubic forms are produced from the vapor. The coexistence of these two phases is affected by substrate temperature and film thickness. For very thin deposits on (001) rock salt, of less than 10 Å average thickness, the $\beta$-phase is predominant, but the $\alpha$-phase increases in quantity with increasing temperature. The epitaxial relations of CdS layers were found to be: $\alpha$-CdS (0001) ∥ NaCl (001), $\alpha$-CdS $(30\bar{3}4)$ ∥ NaCl (001), and $\beta$-CdS (001) ∥ NaCl (001).

In the case of ZnS thin films, the two phases coexist on substrates such as carbon films, rock salt, and mica. The epitaxial relations are complicated and change with temperature. For example, overgrowth on mica is polycrystalline in texture from room temperature to about 150°C. Biaxial orientation has been observed above 200°C. However, only the $\alpha$-CdS structure was observed on the cleaved surface of $MoS_2$ [17]. In both the CdS and the ZnS cases, lattice defects and stacking faults exist in crystallites contained in the films [18-21].

Quite recently, the epitaxial growth of GaAs by a molecular beam deposition method on a (001) GaAs substrate has been studied in situ in an ultrahigh-vacuum high-energy electron diffraction system [22, 23]. GaAs (001)−C(2 × 8) and (001)−C(8 × 2) surface structures were observed during growth. One is an arsenic-stabilized phase and the other a gallium-stabilized phase. The structure of the overgrowth depended on the molecular beam intensities of Ga and As. The method indicates one of the future possibilities of development in this field.

In conclusion, recent progress in the formation processes of thin films of semiconductor materials are reviewed in conjunction with their structure, epitaxy, and electrical properties. Related problems of other materials are also discussed.

## Literature Cited


1. D. W. Pashley, Adv. Phys., 5:175 (1959); 14:372 (1965).
2. J. B. Filby and S. Nielsen, Brit. J. Appl. Phys., 18:1357 (1969).
3. H. M. Mansevit, A. Miller, F. L. Morritz, and R. L. Nolder, Trans. Met. Soc. AIME, 233:540 (1965).
4. R. W. Bicknell, B. A. Joyce, J. H. Neave, and G. V. Smith, Phil. Mag. 14:31 (1966); J. Appl. Phys., 35:1349 (1964).
5. J. L. Porter and R. G. Wolfson, J. App. Phys., 36:2746 (1965).
6. D. J. Dumin and P. H. Robinson, J. Electrochem. Soc., 113:469 (1966); 39:2759 (1968).
7. B. A. Unvala, Le Vide, 104:109 (1963).
8. F. H. Reynolds and A. B. M. Elliot, Solid State Electronics, 10:1093 (1967).
9. C. T. Naber and J. E. O'Neal, Trans. Met. Soc. AIME, 242:470 (1968); L. R. Weisberg and E. A. Miller, ibid.
10. T. Itoh, S. Hasegawa, and H. Watanabe, J. Appl. Phys., 39:2969 (1968).
11. Y. Yasuda and Y. Ohmura, Japan. J. Appl. Phys., 8:1098 (1969).
12. Y. Yasuda, Japan. J. Appl. Phys, 10:45 (1971).
13. M. Shiojiri and E. Suito, Japan. J. Appl. Phys., 3:314 (1964).
14. Y. Yasuda, Japan. J. Appl. Phys., 7:1171 (1968).
15. D. B. Holt and D. M. Wilcox, J. Crystal Growth, 9:193 (1971).
16. R. Ueda, in preparation.

17. R. Ueda, in preparation.
18. D. B. Holt, J. Mat. Sci., 1:280 (1966).
19. B. A. Unvala, J. M. Woodcock, and D. B. Holt, Brit. J. Appl. Phys. 1:11 (1968).
20. J. M. Woodcock and D. B. Holt, Brit. J. Appl. Phys., 2:775 (1969); J. Mat. Sci., 5:275 (1970).
21. R. Sato, Nature (London) 184:2005 (1959); Acta Cryst., 15:1109 (1962); Japan. J. Appl. Phys., 3:626 (1964).
22. A. Y. Cho, J. Appl. Phys., 42:3074 (1971).
23. A. Y. Cho and I. Hayashi, J; Appl. Phys., 42:4422 (1971).

# DISTRIBUTION AND NATURE OF GOLD CRYSTALLIZATION CENTERS ON SINGLE-CRYSTAL SUBSTRATES

**V. I. Trofimov and E. Kh. Enikeev**

*Institute of Physical Chemistry, Academy of Sciences of the USSR, Moscow*

Although many studies have been performed, the nature and numbers of nuclei on substrates remain debatable; two major reasons are as follows. On the one hand, experiments with defective substrates show clearly that the number of nuclei increases regularly with the defect concentration, which indicates that point defects are preferred nucleation sites [1-4]. On the other hand, there are many observations of a marked temperature dependence in the nucleation density or deposition rate, and these data often agree well with theoretical predictions based on statistical (fluctuation) production of nuclei on ideal substrates free from defects.*

Here we present a new approach, in which a traditional study of the concentration is accompanied by an examination of the spatial distribution, which is compared with that of the point defects on a single-crystal substrate. An advantage of this approach is that it eliminates the uncertainty found in studies of the first type [1-4], namely, that arising from the lack of exact knowledge of the number of defects responsible for the nuclei. It will be clear from what follows that this approach not only allows one to identify the nucleation centers but also to define certain aspects of the condensation mechanism.

1. The spatial distribution was established by statistical analysis (by the method of [6]) of electron micrographs from insular condensates of Au on various single-crystal substrates: KCl, fluorite, and mica. The thicknesses of the Au condensates were 1-10 Å, which corresponded to observation of all the nuclei under these conditions [5]. We calculated $\chi^2$, which characterizes the deviation of the observed distribution from a Poisson distribution, with the latter described [7] by

$$P_k = \frac{\lambda^k}{k!} e^{-\lambda}, \tag{1}$$

where $P_k$ is the probability of finding k nuclei within a given cell, while $\lambda$ is the value of k averaged over many cells, which itself is dependent on the cell size.† The precise deviation from the Poisson distribution may be evaluated via Romanovskii's rule [8]; which indicates that if

$$\alpha \equiv \frac{|\chi^2 - \nu|}{\sqrt{2\nu}} < 3, \tag{2}$$

* See [5] for a detailed literature listing on this topic.

† The field used to count the nuclei contained a reasonably large number of particles (300-3000) and was taken far from any linear imperfections in the substrate (steps and so on), the field being split up into 200-300 cells of equal size.

then the discrepancy is without particular significance, where $\nu = (r - 2)$ is the number of degrees of freedom and r is the sample volume. As the mode of processing may affect the result, we calculated $\chi^2(\alpha)$ for each specimen by varying $\lambda$ and by placing the field in various parts of the photograph with the cell size unchanged.

It was found (Fig. 1) that $\alpha$ showed considerable spread even for a given specimen; however, more careful examination showed, e.g., for Fig. 1a, that all these variations in $\alpha$ cover only a small range, and therefore one can assume that the distribution was of Poisson type for a given specimen made at the fairly high substrate temperature $T_s$ of 440°C. The lower substrate temperature of 330°C (Fig. 1b) caused most of the values of $\alpha$ to exceed 3, which indicated substantial deviation from random. Finally, at the even lower temperature of 150°C all the values of $\alpha$ fell in the significant region, which indicates very substantial deviation from a Poisson distribution (Fig. 1c).

This tendency for $\alpha$ to decrease as $T_s$ increases (Fig. 1) was traced in six experiments, where it was completely confirmed, as Fig. 2 shows, where each point has been derived by averaging not less than six values of $\alpha$, which eliminates much of the spread mentioned above (Fig. 1). Three temperature ranges can be distinguished. At substrate temperatures below about 220°C, all the $\alpha$ are larger than 3, so the spatial distributions deviate substantially from Poisson form. In the transition region (220-380°C), the other parameters (condensate thick-

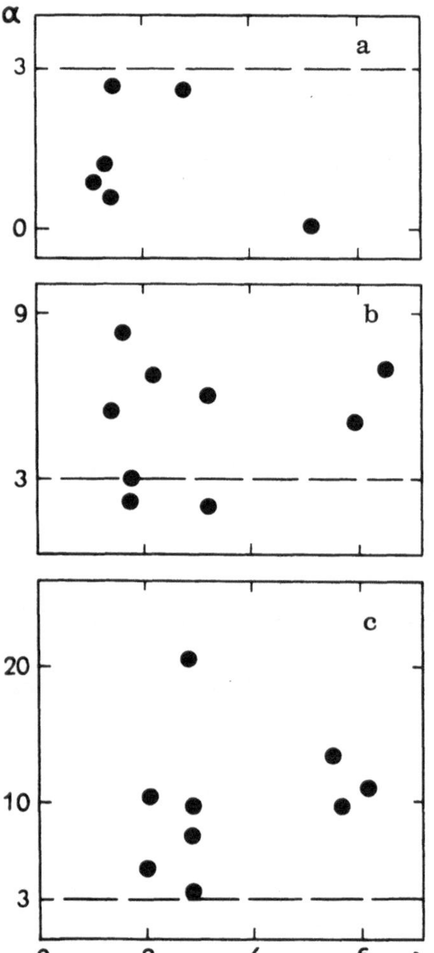


Fig. 1. Results from statistical processing of three Au specimens on KCl at substrate temperatures of (a) 440; (b) 330; (c) 150°C; the thickness h is 3 Å in all cases.

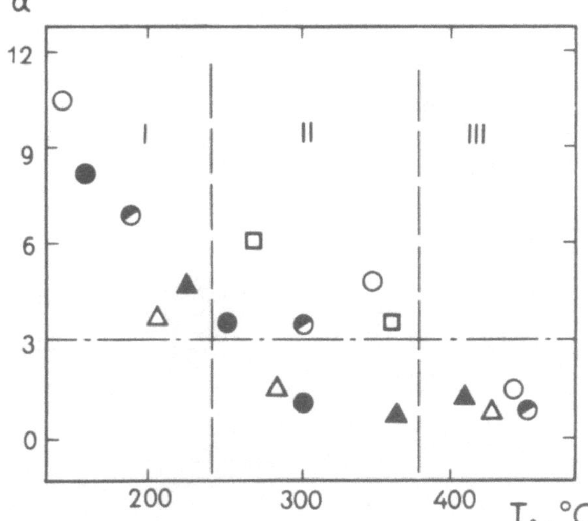


Fig. 2. Temperature dependence of $\alpha$ in the Au–KCl system (○ run 47, h = 3 Å; △ – 64, 4 Å; □ – 72, 8 Å; ● – 76, 1 Å; ◐ – 77, 5 Å; ▲ – 89, 10 Å).

ness, deposition rate, pressure and composition of the residual gas, etc.) cause $\alpha$ to be larger or smaller than 3, with values less than 3 gradually predominating as $T_s$ is raised. Finally, at $T_s \geqslant 380°C$ all the values of $\alpha$ are less than 3, which indicate that a Poisson distribution applies.

At the same time, there is a monotonic fall in the density $N_s$ of the nuclei on the KCl from $N_s = 6 \cdot 10^{11}$ cm$^{-2}$ at 150°C to $N_s = 0.6 \cdot 10^{11}$ cm$^{-2}$ at 440°C [5]. Correspondingly, the mean distance betwen the gold nuclei, $\bar{R} = 1/2(N_s)^{1/2}$ [9], increases from $\bar{R}$ = 70 Å at 150°C to R = 200 Å at 440°C. Therefore, a rise in $T_s$ results in quantitative changes, namely an increase in the mean distance between particles, and also qualitative changes, since the spatial distribution is altered. At low temperatures, when the Au particles lie comparatively close together, ($\bar{R} \leqslant 100$ Å), the positions are correlated, and the extent of the short-range order is indicated by Fig. 1c as extending to 2-3 average separations ($L \simeq \sqrt{\lambda}$, $R \approx 2.5 \cdot \bar{R}$), i.e., out to about 200 Å. High substrate temperatures cause the particles to be much further apart ($\bar{R} \simeq 200$ Å), and there is no correlation in position even in the first coordination sphere.

Results in agreement with those above were obtained also with the Au–$CaF_2$ and Au–mica systems; fluorite under all working conditions ($T_s$ = 300–400°C gave very closely spaced [10] Au particles Au($R \leqslant 50$ Å), and the correlation was strong, i.e., $\alpha$ was large, which applied out to $L \simeq (3\text{-}4)\bar{R} \simeq 200$ Å; in the Au–mica system, this effect was accompanied by a new one (Fig. 3). If the conditions are such that the comparatively low $N_s = 2.6 \cdot 10^{11}$ cm$^{-2}$ applies for $\bar{R}$ = 100 Å, the correlation in particle disposition as $\lambda$ increases up to 10, i.e., $L \simeq \sqrt{\lambda} \cdot R \simeq 350$ Å, becomes substantially less, much as for a solid.

**2.** The concentration of charged point defects within the bulk of an alkali halide is $10^{16}$–$10^{18}$ cm$^{-3}$ [11,12]), which corresponds to a surface density of $10^{10}$–$10^{12}$ cm$^{-3}$; as the Coulomb forces are long-range ones, the defects should interact directly, and therefore one expects that the surface defects will not be randomly disposed but will show an order dependent on the interaction potential. Figure 4 shows results from such a discussion by one of us [13] in accordance with the theory of strong electrolytes [14].

At small distances, the binary correlation function for charged defects differing in sign and having Coulomb interaction takes the following form (curve 1):

$$\Phi(R) = 2\pi R C(R) \exp\left[-2\pi \int_a^R x C(x) dx\right], \quad (3)$$

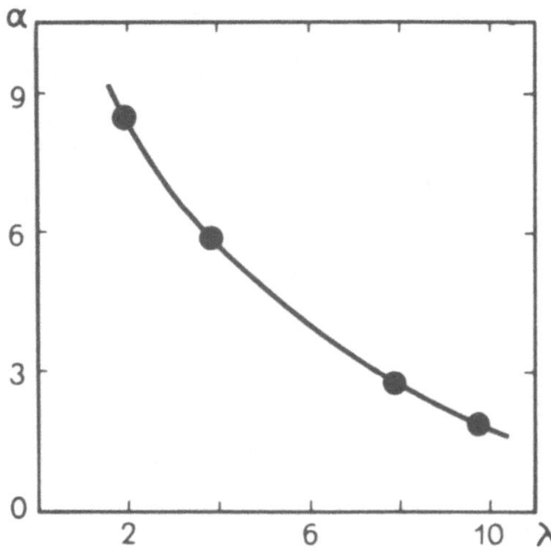


Fig. 3. The λ dependence of α in the Au–mica system.

where

$$C(R)=N\exp(q^2/\varepsilon kTR) \tag{4}$$

describes the concentration as a function of distance, while N is the mean defect density. The interaction can be neglected at large distances, and then (3) takes a different form (curve 2):

$$\Phi(R)=2\pi RN\exp(-\pi R^2N), \tag{5}$$

and this formula describes the distribution of noninteracting defects at any distance. Figure 4 shows the resultant $\Phi(R)$ as a solid line, and the area under this is taken as 1 in accordance with the normalization condition. Then closely spaced defects will have interaction energies greater than the mean kinetic energy, and they are therefore associated into complexes, most often pairs. The separations in such complexes vary from some minimal value of the order of the lattice constant $a$ up to a critical correlation radius $R^*$, which is defined by $kT \simeq q^2/\varepsilon R^*$ and is 15-100 Å. Any defects at distances greater than $R^*$ are randomly distributed in accordance with (5), with the most probable distance to the nearest defect (of either sign) $R_n = 1/(2\pi N)^{1/2}$. If the densities are very high, $N \gtrsim 10^{13}$ cm$^{-2}$, the peak in $\Phi(R)$ as $R = R_n$ degenerates to an inflection.

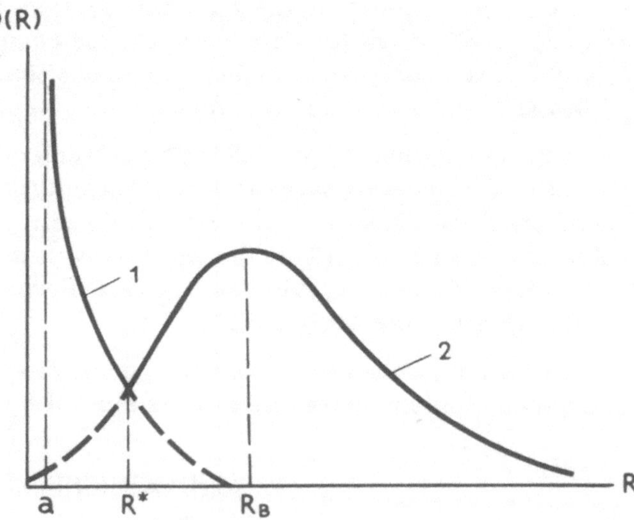


Fig. 4. The binary correlation function for the defect distribution on an insulating single crystal.

3. We assume that the defects in the substrate are either free or associated, and in any case act as nucleation centers [1-4], in which case the trends in the distribution can be compared with the defect distribution; in particular, the Au atoms are highly mobile on KCl at high $T_s$ [5, 15], and therefore they may be transferred [16] to the deepest wells in the potential relief. In that case, the distribution should reflect that of the most active centers. This distribution is random, and therefore under these conditions the nuclei arise at widely separated ($R > R^*$) randomly disposed defects, with the distribution given by (5). The mobility of the Au atoms falls as the temperature is reduced, so the atoms tend to be trapped at less active but more closely spaced defects, so the distribution deviates from random. This agrees with the above numerical estimates. We have already seen that the Au nuclei are randomly distributed for $R \simeq 200$ Å, while for $R < 100$ Å, i.e., distances comparable with the correlation radius, there is a marked deviation from a random distribution. It is also possible that the correlation for high nuclear densities is due to some extent to interaction by diffusion between closely-spaced nuclei.

4. These results give rise to some arguments on the condensation mechanism for Au on ionic crystals. The adsorbed Au atoms falling into the effective field of a charged defect* are polarised and attracted, and these atoms then interact chemically to form a nucleus. This interaction means that the motion of the atoms to the preexisting centers or defects† is not random (contrast an ordinary classical scheme) but directional. This explains why widely spaced large Au particles are produced at high $T_s$, since the widely spaced free defects have the largest polarizing effect on the adsorbed atoms and therefore can trap atoms over larger areas, especially since the mobility of the atoms is high. As defect complexes are favored by energy considerations (although not by entropy), the tendency to associate is accentuated at low temperatures. This tends to increase the concentration of small islands, with the distribution then deviating from random at low $T_s$.

A similar condensation mechanism has been suggested [17] for molecular Fe (Te) beams condensing on ionic crystals, in which the particles in the vapor are directly focused to charged active centers by electrostatic interaction. A difference from [17] is that in our case the focussing occurs in the plane of the substrate with the active participation of surface diffusion. This model not only explains the observed trends in the spatial distribution but also the general temperature dependence of the density. The activation energy for surface diffusion is defined by the slope of the Arrhenius line $\log N_s = f(1/T)$ [5, 15], and this serves to characterize the actual migration process, which incorporates the interaction with the defects. It would seem that this model would explain the extremely rapid and almost unobservable [5] nucleation as well as the feature that values too low by 1-2 orders of magnitude for the density of nuclei at saturation are obtained [5] if one assumes that the adsorbed atoms wander at random without correlation arising from the interaction with the defects. The first feature is due to the rapid flow of the condensed atoms to the defects, while the second is due to the high defect concentration.

The effective field of the original center becomes screened as the nucleus enlarges, so the condensation mechanism may alter. Therefore, although the nucleation is of correlated type, the metal phase in an island may ultimately grow in the usual way by attachment of randomly-wandering adsorbed atoms.

---

* This is the range in which the energy of ion–dipole or dipole–dipole interactions (the latter for associated defects) is of the order of kT.

† This model eliminates the need for concepts such as supersaturation and critical nucleus; an analog of the latter is an isolated defect, which after trapping an absorbed atom acquires a stable configuration (the analog of a stable nucleus).

## Literature Cited


1. G. I. Distler, In: Physicochemical Problems of Crystallization, Vol. 1 [in Russian], Izd. KGU, Alma-Ata (1969), p. 120.
2. V. M. Kosevich, Author's Abstract of Doctoral Dissertation, FTINT AN SSSR, Khar'kov (1959).
3. N. Bethge, J. Vac. Sci. Technol., 6:460 (1967).
4. P. W. Palmberg, C. J. Todd, and T. N. Rhodin, J. Appl. Phys., 39:4650 (1968).
5. V. I. Trofimov and V. M. Luk'yanovich, Fiz. Tverd. Tela, 13:2119 (1971).
6. B. N. Smirnov and T. V. Samoilova, Fiz. Tverd. Tela, 13:2119 (1971).
7. J. Hudson, Statistics for Physicists [Russian translation], Mir, Moscow (1970).
8. A. K. Mitropol'skii, Statistical Calculation Techniques [in Russian], Nauka, Moscow (1971).
9. R. D. Gretz, Surface Sci., 6:468 (1967).
10. V. I. Trofimov, A. E. Chalykh, A. A. Shenk, and V. M. Luk'yanovich, Kristallografiya, 17:690 (1972).
11. M. I. Kornfel'd, Fiz. Tverd. Tela, 10:2422 (1968).
12. A. A. Vorob'ev, Ionic and Electronic Properties of Alkali Halide Crystals [in Russian], Izd. TGU, Tomsk (1968).
13. E. Kh. Enikeev, In: Physicochemical Problems of Crystallizations, Vol. 2 [in Russian], Izd. KGU, Alma-Ata (1971), p. 35.
14. R. M. Fross, Trans. Faraday Soc., 30:967 (1934).
15. K. Hayek and V. Schwabe, Z. Naturforsch. 26a:1879 (1971).
16. A. A. Chernov, In: Physicochemical Problems of Crystallization, Vol. 1 [in Russian], Izd. KGU, Alma-Ata (1969), p. 8.
17. I. D. Nabitovich, Ya. S. Budzhak, V. V. Osipov, and Ya. I. Stetsiv, Fiz. Tverd. Tela, 13:2909 (1971).

# SUBSTRATE-INDUCED STRAIN BY EPITAXIALLY ORIENTED NUCLEI

G. Le Lay, G. Quentel, A. Masson, and R. Kern

*Laboratoire des Mécanismes de la Croissance Cristalline (associé au C.N.R.S.)*
*Université de Provence, Centre de Saint Jérone, Marseille*

## 1. Introduction

When depositing metallic films by evaporation under vacuum it happens that more or less "pathological" growth phenomena, such as whiskers, may be encountered, stressing the instability of a continuous layer subject to interfacial strains and stresses.

Cabrera [1] has underlined that such elastic strains of substrate and deposit may be found as soon as the first nuclei appear. A stress field thus would be linked with surface strains and localized around each nucleus (which partially explains the energetic barrier to be jumped when crystallites coalesce in a thin film).

Under certain conditions [2] (existence of preferential orientations on the substrate, effective possibility of reorientation of the nuclei especially through thermal activation) these stresses may be partially relaxed by epitaxial reorientation of the nuclei. The strain field around each crystallite must be then stable and identical for nuclei of equal size.

For semiconductor surfaces whose structural lability is well known, we may think that such effects must be particularly enhanced and even that the feasibility of long-distance strains may induce structural changes of the free surface as a whole.

As concerns this point we especially know that low-energy electron diffraction (LEED) investigation of the (111) germanium surface reveals different superstructures [3-5] corresponding to atomic positions of neighboring energies [6]. Under thermal effects these structures are labile: the $2 \times 1$ cleavage structure irreversibly reverts to the Ge (111) $2 \times 8$ structure beyond 70°C. A following anneal beyond 300°C leads to a reversible structural transformation Ge (111) $2 \times 8 \rightleftharpoons$ Ge (111) $1 \times 1$. The latter structure has the periodicity of a bulk (111) germanium plane.

Besides these thermal phenomena, we can imagine that other effects, particularly the above-mentioned mechanical effects, can induce new structures. As regards these effects, the study of metal deposits on Ge (111) is interesting. In this domain our investigations differ from the preceding ones by the simultaneous utilization of several techniques.

## 2. Means of Investigation

(a) Obtention of a (111) substrate germanium plane by cleavage in situ under ultrahigh vacuum (UHV). The substrate temperature may be fixed at a value $T_s$ during deposition and then at an anneal temperature $T_R$ for a time $t_R$.

(b) Condensation of Au or Ag thin films in the range $10^{-2} < \Theta < 10^{2}$ under UHV without any pressure rise ($p < 3 \cdot 10^{-10}$ torr) by use of a Knudsen cell as a source. Deposits are controlled by micromass determination and radioactivation measurements (we define the coverage ratio $\Theta$ as $\Theta = 1$ for one deposit atom per germanium surface atom, that is, $7.24 \cdot 10^{14}$ at $\cdot$ $cm^{-2}$).

(c) Observation methods: (i) in situ LEED study between 10 and 100 V; (ii) RHEED (reflection, high-energy electron diffraction) study at grazing incidence; (iii) high-resolution TEM (transmission electron microscopy) and HEED (high-energy electron diffraction) studies of in situ deposited extraction replicas.

## 3. Experimental Results

### (1) LEED Study at Room Temperature of Au Deposits on Ge (111)

In cases A and B (see Table 1), the Ge (111) 2 × 8 pattern becomes blurred and vanishes when $\Theta$ increases from 0 to $10^{-1}$. At $\Theta = 10^{-1}$ the four satellites of 1/8 order around each 1/2 order spot are no longer separated; simultaneously the 1/3 order spots of the $\sqrt{3} \times \sqrt{3}$ structure begin to appear. We note the pattern with diffuse 1/2 order spots Ge (111) 2. Beyond $\Theta = 10^{-1}$ the Ge (111) 2 pattern disappears whereas the Ge (111) $\sqrt{3}$ one becomes more precise.

In cases C and D, the Ge (111) $\sqrt{3}$ pattern does not appear, the fractional order spots gradually vanish, but the integral order spots maintain neighboring intensities: a Ge (111) 1 × 1 pattern remains.

### (2) TEM Investigation

In any A, B, C, or D conditions of Table 1, all the replicas of gold or silver films put into evidence the presence of isolated crystallites. For gold deposits in case A the average size is $d \simeq 150$ Å if $\Theta \simeq 2 \cdot 10^{-2}$ and nearly $d \simeq 3000$ Å when $\Theta = 10^{2}$. In cases A and B, as little as 1% of the total area is covered by nuclei when $\Theta = 2 \cdot 10^{-1}$ and only 20% when $\Theta = 100$. In cases C and D, for a given coverage ratio $\Theta$, the density of nuclei is greater than in cases A and B though the covered substrate area still remains small.

A morphological study up to $\Theta = 10$ shows that if in cases C and D all the crystallites are of roughly hemispherical shape, in contrast to cases A and B two types of individuals are easily noted, the one rod-shaped, the other pyramidal with a triangular or hexagonal basal plane.

TABLE 1

| Exp | $T_S$, °C | $T_R$, °C; $t_R$ | $\Theta$ | | | | |
|---|---|---|---|---|---|---|---|
| | | | 0 | $10^{-2}$ | $10^{-1}$ | 1 | $10^{2}$ |
| A | 20 | 250; 10 min | Ge(111) (2×8) | Ge(111) (2×8) | Ge(111)2+ +Ge(111)$\sqrt{3}$ | Ge(111)$\sqrt{3}$ | Ge(111)$\sqrt{3}$[1] |
| B | 250 | Without any anneal | Ge(111) (2×9) | Ge(111) (2×8) | Ge(111)2+ +Ge(111)$\sqrt{3}$ | Ge(111)$\sqrt{3}$ | Ge(111)$\sqrt{3}$[1] |
| C | 20 | Without any anneal | Ge(111) (2×1) | Ge(111) (2×1) | Ge(111) (1×1) | Ge(111) (1×1) | No pattern |
| D | 20 | Without any anneal | Ge(111) (2×8) | Ge(111) (2×8) | Ge(111) (1×1) | Ge(111) (1×1) | No pattern |

[1]+ Gold pattern (facets).

TABLE 2

| Plane in contact with Ge (111) species | Parallel rows and coincidence path | Misfit | Coincidence mesh in the interface of a single nucleus of the species related to the Ge substrate | Common mesh for the species as a whole related to the Ge substrate |
|---|---|---|---|---|
| (111) Ag | $\langle 1\bar{1}0\rangle$Ag ‖ $\langle 1\bar{1}0\rangle$Ge<br>$4n_{\langle 1\bar{1}0\rangle}$ Ag $\equiv 3n_{\langle 1\bar{1}0\rangle}$ Ge[1]<br>$\langle 10\bar{1}\rangle$Ag ‖ $\langle 10\bar{1}\rangle$Ge<br>$4n_{\langle 10\bar{1}\rangle}$ Ag $\equiv 3n_{\langle 10\bar{1}\rangle}$ Ge<br>and the crystallite[2] symmetrically positioned with $\langle 1\bar{1}0\rangle$ Ge | −3%<br>−3%<br>" | 3×3 | $\sqrt{3}\times\sqrt{3}$ |
| (112) Ag | $\langle 1\bar{1}0\rangle$Ag ‖ $\langle 1\bar{1}0\rangle$Ge<br>$4n_{\langle 1\bar{1}0\rangle}$ Ag $\equiv 3n_{\langle 1\bar{1}0\rangle}$ Ge<br>$\langle 111\rangle$Ag ‖ $\langle 11\bar{2}\rangle$Ge<br>$n_{\langle 111\rangle}$ Ag $\equiv n_{\langle 11\bar{2}\rangle}$ Ge<br>and two equivalent orientations (120° rotated) | −3%<br>+3%<br>" | $\sqrt{3}\times\sqrt{3}$ | $\sqrt{3}\times\sqrt{3}$ |
| (110) Ag | $\langle 1\bar{1}0\rangle$Ag ‖ $\langle 1\bar{1}0\rangle$Ge<br>$4n_{\langle 1\bar{1}0\rangle}$ Ag $\equiv 3n_{\langle 1\bar{1}0\rangle}$ Ge<br>$\langle 001\rangle$Ag ‖ $\langle 11\bar{2}\rangle$Ge<br>$5n_{\langle 001\rangle}$ Ag $\equiv 3n_{\langle 11\bar{2}\rangle}$ Ge<br>and two equivalent orientations (120° rotated) | −3%<br>−1% | $3\sqrt{3}\times 3$ | $\sqrt{3}\times\sqrt{3}$ |

[1]$n_{\langle xyz\rangle}$: nearest neighbor spacing along the $\langle xyz\rangle$ row.
[2]Double-positioning [7].

(3) HEED and RHEED Examination

The HEED examination of gold or silver replicas reveals that in cases A and B the crystallites are epitaxially oriented:

(112) or (110) Au, Ag ‖ (111) Ge for the first type
(111) Au, Ag ‖ (111) Ge for the second type

It has also allowed the determination of the relative orientations of the three species with (112), (110), and (111) basal planes, respectively.

Moreover, the RHEED examination at grazing incidence of Ag layers on Ge (111) has put into evidence the relative orientations of the crystalline rows of the substrate and of the deposit, thus allowing us to establish the epitaxy relationships described in Table 2.

In cases C and D of Table 1, that is, for deposition at room temperature without any anneal, the gold or silver nuclei show no epitaxial orientation. The HEED patterns of replicas still indicate the existence of three species with (111), (112), and (110) planes respectively, in contact with Ge (111). These crystallites are azimuthally misoriented.

## 4. Discussion and Paradox

We give our attention to the appearance of the Ge (111) $\sqrt{3}$ structure.

The LEED examination indicates that the (111) $\sqrt{3}$ pattern, becoming visible as soon as a $10^{-1}$ monolayer of gold has reached the sample surface, is due to the presence of the metal. At this point we could suppose as most authors [8-11] do, that the deposit forms a well-ordered layer of isolated atoms on the germanium substrate in such a way that their network presents a $\sqrt{3} \times \sqrt{3}$ unit mesh compared to the $1 \times 1$ unit mesh of an unreconstructed (111) germanium plane, as long as $\Theta < 1/3$, with crystallites being able to grow on this adsorbed layer as soon as $\Theta > 1/3$.

The TEM observation contradicts such a hypothesis. In fact, it reveals that for very low $\Theta$, and particularly $\Theta = 10^{-1}$, the metal is aggregated, at least for an important part, in discrete nuclei in any of the A, B, C, and D experimental conditions.

It appears, thus, that the driving force tends to agglomerate the metal atoms; it is thus not conceivable that even well-ordered isolated atoms should remain in sufficient quantity on the substrate to lead to a detectable $\sqrt{3}$ pattern.

The HEED examination indicates that at the interface, for those crystallites whose contact plane is (110), there exists a $\sqrt{3} \times \sqrt{3}$ coincidence mesh related to the germanium (111) plane. However it is impossible to think that these interfaces contribute to the LEED pattern since they fill less than 2% of the total area, even when the crystallites are thicker than 100 Å (the LEED gold pattern itself only appears when crystallites bigger than 1000 Å occupy 20% of the total area).

The four investigation techniques thus lead us to a paradox.

## 5. Hypothesis

The solution of this dilemma is suggested if we note that the pattern of the Ge (111) $\sqrt{3}$ structure just appears in LEED if, and only if, the gold crystallites are in epitaxy. Referring to the existence of a strain field around each crystallite according to Cabrera [1], and taking into account the ideas advanced in the introduction of this paper, we are led to formulating the hypothesis that the existence of epitaxially oriented nuclei on (111) Ge induces a structural change of the free surface [Ge (111) $2 \times 8 \rightarrow$ Ge (111) $\sqrt{3}$], the mechanism leading to this structural transformation being the lateral propagation, along the free surface of germanium, of the germanium structure present at the Au – Ge (or Ag – Ge) interface under every crystallite.

Indeed, for the crystallites having a (110) basal plane the lateral propagation of the germanium structure at the interface directly induces the $\sqrt{3}$ structure. Moreover we must note that the ovelapping of the influence areas for two families of crystallites with the same (111) plane in contact with the substrate, but in the double-positioning situation [cf. paragraph 3(3)], as well as for the three equivalent orientations (120° rotated) of the (112) in contact crystallites, leads to the same result.

Stating that the long-range induction falls off at a distance $r_0$ from each crystallite, thus defining its influence area, we can approximately determine the value of $r_0$ from the experimental data. In cases A and B, for a coverage ratio $\Theta > 10^{-1}$, a value for which about $2 \cdot 10^{10}$ crystallites per square centimeter of average size $d \sim 150$ Å are present, we can conclude that $2r_0$ is of the order of 6d, that is, about 1000 Å. This gives a sufficient overlap for the size of the domains presenting the $\sqrt{3}$ structure to exceed the minimum size required to obtain sharp diffraction spots (the coherence width estimated with Scherrer's formula lies somewhere between 500 and 1000 Å).

Admitting that $r_0$ is an increasing function of the size of the crystallites, and thus that the sizes of the domains presenting the $\sqrt{3}$ structure increase with that size, we can interpret, for a given Θ, the temperature dependence of the LEED pattern. A temperature rise sharpens the diffraction spots. The reason for that is that the size of the crystallites increases by coalescence, which widens the domains presenting the $\sqrt{3}$ structure, domains whose size can leap the coherence width.

For example, after a brief annealing at 90°C, for Θ = 4, while d ~ 100 Å, the LEED spots are widened but after a short annealing at 250°C their width is normal though we how have d ~ 400 Å.

Still, we have to interpret the Ge (111) 1 × 1 structure observed at room temperature whereas the crystallites are azimuthally misoriented (conditions C and D of Table 1).

If at the crystallite–substrate interface, for a randomly oriented nucleus, there exists a coincidence mesh, common to substrate and deposit, of order n × m in the Ge (111) 1 × 1 reference system, according to our hypothesis, this mesh must propagate along the free surface. As the (i) crystallites of the collection present every conceivable orientation, resulting in the superposition of (i) $m_i \times n_i$ multiple meshes for the germanium substrate, all these meshes having a common 1 × 1 base in the Ge (111) reference system, an averaging phenomenon occurs in the reciprocal space. It leaves in the diffraction pattern only the diffraction spots common to the (i) coincidence meshes, that is, the integral order spots. Thus no superstructure pattern can be observed.

## 6. Conclusion

To complete this study we intend to present later a model, analogous to that proposed by Haneman [12] to explain the diamond-type semiconductor surface structures, that may describe the propagation along the free surface of germanium of the coincidence mesh existing at the crystallite–substrate interface. We will also show, through energetic considerations, that such a structural change is promoted by putting the crystallites into epitaxy. Finally we think we can support this work by studying other metal–semiconductor systems.

In conclusion we emphasize that we think we have put into evidence the influence of the presence of crystallites in epitaxy, on the free surface structure of the semiconductor substrate.

## Literature Cited


1. N. Cabrera, Mémoires scientifiques. Rev. Metallurgie, 62:205 (15 May 1965).
2. A. Masson, J. J. Metois, and R. Kern, Surface Sci., 27:463 (1971). R. Kern, A. Masson and J. J. Metois, Surface Sci., 27:483 (1971). J. J. Metois, M. Gauch, A. Masson, and R. Kern, Surface Sci., 30:43 (1972).
3. J. J. Lander and J. Morrisson, J. Appl. Phys. 34:1403 (1963).
4. P. W. Palmberg and W. T. Peria, Surface Sci., 6:57 (1967).
5. M. Henzler, J. Appl. Phys. 40(9):3758 (1969).
6. A. Taloni and D. Haneman, Surface Sci., 10:215 (1968).
7. J. W. Matthews, In: Physics of Thin Films, Vol. 4, p. 156, Academic Press, New York, (1967).
8. J. J. Lander and J. Morrison, Surface Sci., 2:553 (1964).
9. H. E. Bishop and J. C. Riviere, Brit. J. Appl. Phys. 2(2):1635 (1969).
10. W. Haidinger and S. C. Barnes, Surface Sci., 20 (1970), 313.
11. K. Spiegel, Surface Sci., 7:125 (1967).
12. D. Haneman, Phys. Rev., 121(4):1093 (1961).

# MIGRATION OF POLYATOMIC GROUPS IN CRYSTAL GROWTH FROM A VAPOR

L. I. Trusov

*Electronic Control Machines Institute, Moscow*

Bassett [1] discovered the displacement and rotation of islets as a whole, which indicates that the effective transport mechanism may be the migration of polyatomic groups in various systems such as metal films on insulators. The activation energy $E_c$ for the migration of such centers is less than the pairing energy.

It is difficult to accept the view that such motion is possible if $E_c$ is equal to the sum of the activation energies $E_0$ for the surface diffusion of individual atoms in an island; in fact, $E_c$ may be much less, and the binding energy for an atom in such an island may be well below the adsorption energy for a single atom, which occurs, in particular, because the valency bonds are taken up by other atoms in the island [2], so the sorption energy is determined by van der Waals forces only. Another reason is [3] the discrepancy between the lattice constants. An atomic chain on a surface with periodic potential relief has some of the atoms raised on barriers, which reduces E substantially. A similar mechanism has been described elsewhere [2, 4, 5].

The angle dependence of the interaction energy for periodic potential relief [2] provides some conclusions on the conditions for rotation and displacements; the dependence is of oscillatory type, so the positions are not equivalent as regards energy. There are several minima, whose depths decrease as the angle of rotation increases, with the result that the energy barrier to rotation becomes constant if the disorientation is sufficiently large, and it is then close to the activation energy for surface diffusion of an isolated atom. This reduction in the activation energy for rotation occurs because a polyatomic group has many of the atoms raised on potential barriers even for a small rotation. On the other hand, in the epitaxial position, when the disorientation angle is zero, $E_c$ is so large that migration probability is negligible.

These islands may move for several reasons. Chopra and Randlett have assumed that forced migration may occur on deposition from a molecular beam on account of collisions with fast beam particles. This has been confirmed by experiment [6].

Estimates have been made [6] of the speed v acquired by an island of diameter 100 Å after collision with an Ag or Au atom moving at a speed of $10^4$–$10^5$ cm/sec which corresponds to a temperature of about 1000°K. If we neglect the friction in the motion of the island over the substrate, the speed from lateral collision should be about 1 cm/sec, so a density of about $10^{16}$ $cm^{-2} \cdot sec^{-1}$ in the beam should result in $10^4$ such collisions per second, and therefore the island may be displaced by about $10^4$ Å between two collisions.

On this basis, Chopra and Randlett explained various qualitative observations; in particular, there is a substantial fall in the coalescence rate on shutting off the beam, and it has also been found that the islands fuse more rapidly when the temperature is raised, and hence with it the kinetic energy of the beam atoms.

The essentially bound state of an island arising from the interaction with the substrate can be incorporated on the basis that the clumps of atoms from the surface are in potential wells [7]. Here one envisages interaction with a beam of atoms arriving at an angle to the substrate and having a Maxwellian velocity distribution corresponding to the temperature of the source.

An expression has been derived [7] for the probability of forced migration. The values of the parameters were those used in [6], and this implies that only about one-tenth of the collisions result in displacement of an island to an adjacent potential well. If one assumes the distance between such wells is of the order of the interatomic distance, then the island will move with a speed of about $10^{-5}$ cm/sec, which is less by several orders of magnitude than the value adopted by Chopra and Randlett.

The above argument is based on quasielastic collision, whereas in fact the blow from a beam atom produces local deformation in the island and subsequent vibrational relaxation, so the deposited energy is dissipated in the substrate via the interface. However, the inelasticity may be substantial only for films obtained by cathode sputtering, since the energy of the incident atoms in that case is large (3-5 eV). A beam produced by thermal evaporation produces largely elastic collision, since Kubo's calculations [8] indicate that it is unlikely that vibrations will be excited.

Finally, these islands are often electrically charged, and this means that the mean energy transfer on collision with such an island may actually exceed the thermal energy of the incident atom substantially [9].

The interactions resulting in migration may also be electrostatic [10].

Particular interest attaches to studies in which the displacement and rotation have been observed by high-resolution cinephotography; in particular, this method has been used [11] to examine the condensation of Ag and Au on $MoS_2$ and MgO. In the case of Ag on MgO, direct observation with a resolution of about 17 Å showed that the disorientation angle fluctuates considerably, the mean deviation being about 1°. At the same time, there is fluctuation in the fringe contrast. These random contrast fluctuations cease whenever the silver beam is cut off. This indicates that the orientations of the isolated islands fluctuate by small amounts in response to the random atomic impact. No such effect was observed for Ag and Au on $MoS_2$.

Finally, it is possible for thermally activated migration to occur, which resembles surface diffusion for molecules. An estimate may be made from the Brownian motion of individual atoms in a N-atom island, and this shows that the probability of such a process is very small [12]. However, $E_c/N$ may be less than $E_0$ (see above), and also the displacements within the island are not independent. In fact, the changes in the positions of the atoms with respect to the potential relief may alter the strength of the interatomic bonds and thus reduce $E_c$ [13].

An analysis has been made [14] of experiments on migration; the recent experiments by Kern et al. [15-17] have defined very closely the range of conditions under which polyatomic groups can migrate.

A characteristic feature of these systems is that the islands interact strongly but wet the surface only weakly. Under certain conditions, contacting islands behave as drops with high surface tension and will fuse or coalesce. A statistical analysis of this is best based on the extended-volume method [18], in which the degree of filling $\eta(t)$ is related not only to the nucleation and growth but also to the coalescence. We introduce a new extended volume

$$V_{ex}(t)=V_e(t)+\int_0^t v_c^3(\tau)(t-\tau)^3\,d\tau, \quad (1)$$

where $v_c$ is the effective coalescence rate, while

$$V_e(t)=\gamma\int_0^t I(\tau)v^3(\tau)(t-\tau)^3d\tau,$$

and $I(\tau)$ is the nucleation rate, with $v(\tau)$ the linear growth rate, and $\gamma$ a geometrical factor that incorporates the shape of a transformed region. Then if each of the processes that govern $V_{ex}$ is random and spatially homogenous, we have

$$\frac{\delta V_c}{\delta V_{ex}}=1-\frac{V_c}{V}, \tag{2}$$

where V is the volume of the system and $V_c$ is the volume filled by crystalline material at time t.

We assume that $V_{ex}$ can be defined as some function of the Avraami extended volume $V_e$ to get

$$dV_c=\left(1-\frac{V_c}{V}\right)\varphi(V_c)dV_e \tag{3}$$

or

$$\eta(t)\equiv\frac{V_c}{V}=1-\exp(-\int\varphi(V_e)dV_e). \tag{4}$$

After expansion of $\varphi$ as powers of $\eta_e\equiv V_e/V$ we get

$$\eta(t)=1-\exp-[a(t)\eta_e(t)+b(t)\eta_e^2(t)+\ .\ .\ .\ ]. \tag{5}$$

The coefficients a(t), b(t) and so on are functions of time and describe the kinetics of the correlation.

If one assumes that the islands are similar to a real gas (a molecule corresponding to an isolated island), then the term proportional to $\eta_e^2$ describes the pair correlation in the spatially homogeneous system. This is the change in the degree of filling due to processes occurring on pair contacts. The function b(t) is related to the fusion rates for pairs of islands. The expansion in powers of $\eta_e$ corresponds to virial expansion for a real gas. Figure 1 shows $\eta(t)$, while detailed exmaples of calculations from (5) are to be found in [14].

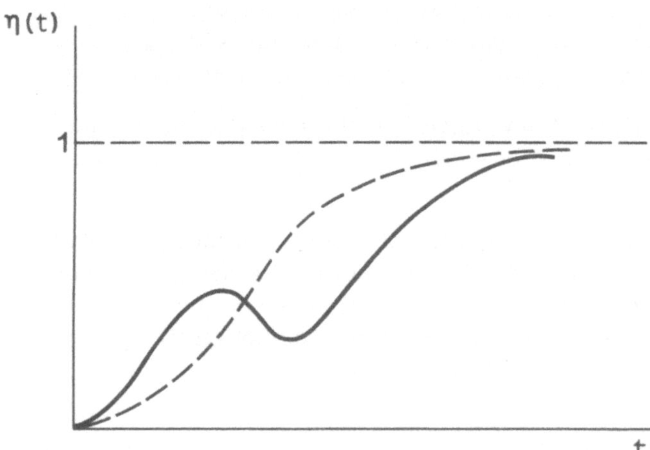


Fig. 1. Degree of filling as a function of time on the basis of coalescence (solid line); the broken line corresponds to the Kolmogorov–Avraami theory.

It is more accurate to use a method based on the size distribution of the islands. On the other hand, we need to alter the form of the kinetic equation substantially in this case; in particular, the Derring–Becker postulates no longer apply (successive attachment of individual atoms). Further, there may be a fall in the number of islands of a given size as a result of coalescence, which should affect the mean of the distribution as well as the variance. The area under the distribution may fall as time passes (this corresponds to a decrease in the proportion of the substrate covered by the islands). These general conclusions agree with the experimental evidence [19, 20].

I am indebted to Professor Ya. E. Geguzin and Professor R. Kern (France) for fruitful discussions.

## Literature Cited


1. G. A. Bassett, Proceedings of the European Regional Conference on Electron Microscopy, Philadelphia (1962), p. 661.
2. H. Reiss, J. Appl. Phys., 39:5045 (1968).
3. G. Bliznakov, In: Growth of Crystals, Vol. 3, Consultants Bureau, New York (1962), p. 23.
4. J. C. du Pieissio and J. H. Van der Merwe, Phil. Mag., 11:43 (1965).
5. V. A. Kholmyanskii, In: Whisker Crystals and Nonferromagnetic Films, Part 2 [in Russian], Voronezh (1970), p. 80.
6. K. L. Chopra and M. R. Randlett, J. Appl. Phys., 39:1874 (1968).
7. V. F. Dorfman and L. I. Trusov, Kristallografiya, 13:562 (1968).
8. R. Kubo, J. Phys. Soc. Japan, 17:975 (1962).
9. V. F. Dorfman and L. I. Trusov, Kristallografiya, 15:788 (1970).
10. A. A. Chernov and L. I. Trusov, Kristallografiya, 14:218 (1969).
11. G. Honjo and J. Jagi, J. Vac. Sci. Techn., 6:575 (1969).
12. M. Gulden, J. Nucl. Mater., 23:1 (1967).
13. L. I. Trusov, Fiz. Met. Metalloved., 28:281 (1969).
14. L. I. Trusov and V. A. Kholmyanskii, Isolated Metallic Films [in Russian], Izd. Metallurgiya, Moscow (1973).
15. A. Masson, J. J. Metois, and R. Kern, Surface Sci., 27:463, 483 (1971).
16. J. J. Metois, M. Gauch, A. Masson, and R. Kern, Surface Sci., 30:43 (1972).
17. A. Masson and R. Kern, J. Crystal Growth, 2:227 (1968).
18. M Avraami, J. Chem. Phys., 7:1107 (1939).
19. K. van Steensel, Philips Res. Repts., 22:246 (1967).
20. I. G. Skofronick and W. B. Philips, J. Appl. Phys., 38:4791 (1967).

# SELECTIVITY AND GROWTH MECHANISMS FOR STAGES OF GROWTH FROM A VAPOR OR GAS

V. F. Dorfman

*Electronic Control Machines Institute, Moscow*

The natural selectivity in the physical mechanisms of crystallization results in a certain amount of self-regulation, which is of particular importance for applications, especially in microelectronics, where nearly all major advances have involved selectivity.* For this reason it is important to examine the relationship between structure, composition, and crystal configuration, since these affect the selectivity in the individual growth processes, and it is therefore important to examine how the mechanisms are affected by the external macroscopic parameters (crystallization temperature, growth rate, etc.).

The abstracts of [2] describe briefly the selectivities of major growth stages from vapors; in particular, there is selectivity in the interaction of an incident particle with a surface, where collision and condensation result in undesirable component segregation, while on the other hand there is improved structural perfection in the film on increasing the particle energy. These conclusions from the earlier work have subsequently been confirmed [3, 4]. Here we discuss particularly the growth stages; the methods of analysis, particularly for competing processes, are those of [5-7]. The results given here are compared with experiment.

## 1. Fluctuation Nucleation and Selective Crystallization at Impurity or Defect Centers

These mechanisms have different effects on the structure and properties of films; there are two possible types of center: (a) those with a radius of capture of interatomic order (as in the fluctuation mechanism), and (b) those with long-range electrostatic interaction. Figure 1a shows that the ordinary (short-range) centers raise the density G of the nuclei at low supersaturations, whereas they have no effect at high supersaturations, although they never reduce G. Conversely, the electrostatic centers reduce the density in a certain range of growth rates, which is dependent on the trapping radius (Fig. 1b), which forms a means of verifying this mechanism. Figure 2 compares the observed grain-size distribution for chromium films [8] with the theoretical result (the probabilities of the elementary acts were determined from the nearest-neighbor model, with the saturation of G at low flux densities).

* In fact, microelectronics owes its origin to advances in selective diffusion, while the latest advance involves selective oxidation; selective epitaxy is one of the major operations in current planar technology, while synchronized selective growth from the vapor state for polycrystalline and single-crystal regions [1] allows one to produce integrated circuits of maximum working power.

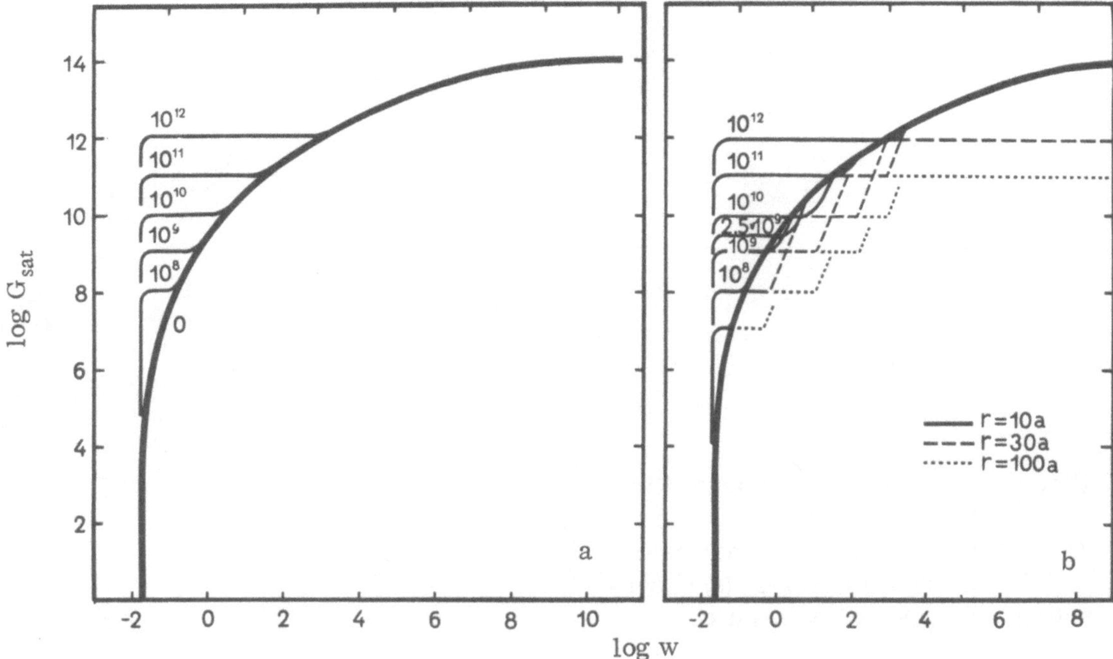


Fig. 1. Saturation nuclear density $G_{sat}$ in $cm^{-2}$ as a function of condensation rate w (atomic layers per second) for various localized-center concentrations; the figures on the curves are the center densities in $cm^{-2}$. The ratio of the bond energy to the thermal energy is $\varphi/kT = 7$: (a) short-range centers, heavy line corresponding to a defect-free surface, thin lines corresponding to surfaces with localized centers; (b) electrostatic centers with various trapping radii r (in terms of the interatomic distance $a$).

The agreement is entirely satisfactory, apart from a small shift along the abscissa, which is readily explained. Calculations on the nucleation kinetics for silver and gold on NaCl are also in good agreement with experiment [9, 10] (throughout the temperature range used with Ag [9], the discrepancies were by not more than a factor 2, while the theoretically predicted behavior of G(T) at the transition from two-dimensional nucleation of Au to three-dimensional nucleation was in agreement with the observations [10]). Whenever a comparison has been made between theory and experiment, there has always been agreement at least as to order of magnitude, and whenever the discrepancies were substantial, the experimental density was higher than the theoretical value. Consequently, one always has either fluctuation nucleation predominant or else nucleation at short-range centers (we here neglect certain cases where some of the molecules are charged in the beam or as a result of electron collision). If electrostatic nucleation effects are to be observed, one clearly requires special conditions or higher measurement accuracy.

## 2. Selectivity in the Formation of Coherent and Defective Grains

The cooperative factor responsible for the selectivity at a given temperature is the density of the two-dimensional gas $R_1$ at the surface; under normal conditions, the growth of a perfect crystal is the basic process determining $R_1$, i.e., the major path. Parts a–c of Fig. 3 show the variation in $R_1$ as a function of dimensionless crystallization time within the framework of the nearest-neighbor model for the minimal growth rate (1 $\mu$m/hr) and the maximum

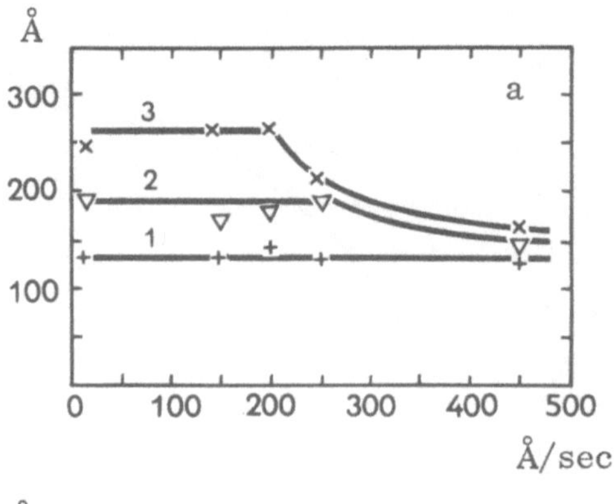


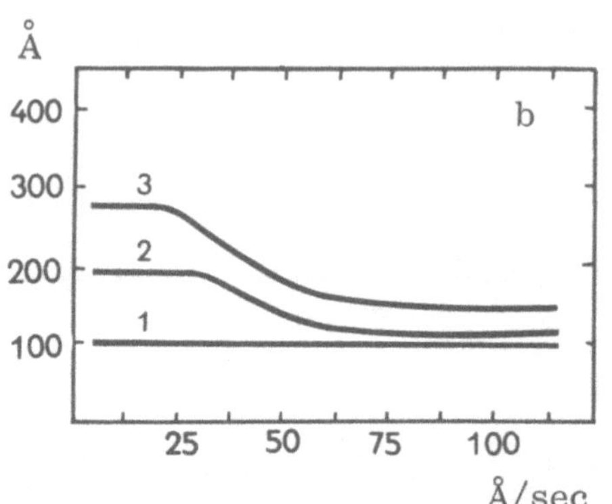


Fig. 2. Grain size in chromium films, Å, in relation to condensation rate Å/sec for substrate temperatures of 1) 25°C; 2) 200°C; 3) 840°C; (a) observations from [8]; (b) calculation.

value (1000 μm/hr), which are the extreme values used in modern technology, and at temperatures such that $\varphi = 1$ eV (germanium and silicon), which correspond to 527°C (minimum working temperature), 927°C (i.e., near the melting point of germanium), and 1227°C (i.e., a typical working temperature for Si). Each grain type corresponds to a certain minimum density (the corresponding values are shown by horizontal lines). The growing crystal retains only the formations whose levels fall below the main path (this corresponds to the right-hand hatched parts of the diagram). The point defects, i.e., equilibrium ones, correspond to the zero level. Therefore, the lattice thermodynamics are reflected in the crystallization, since point defects cannot

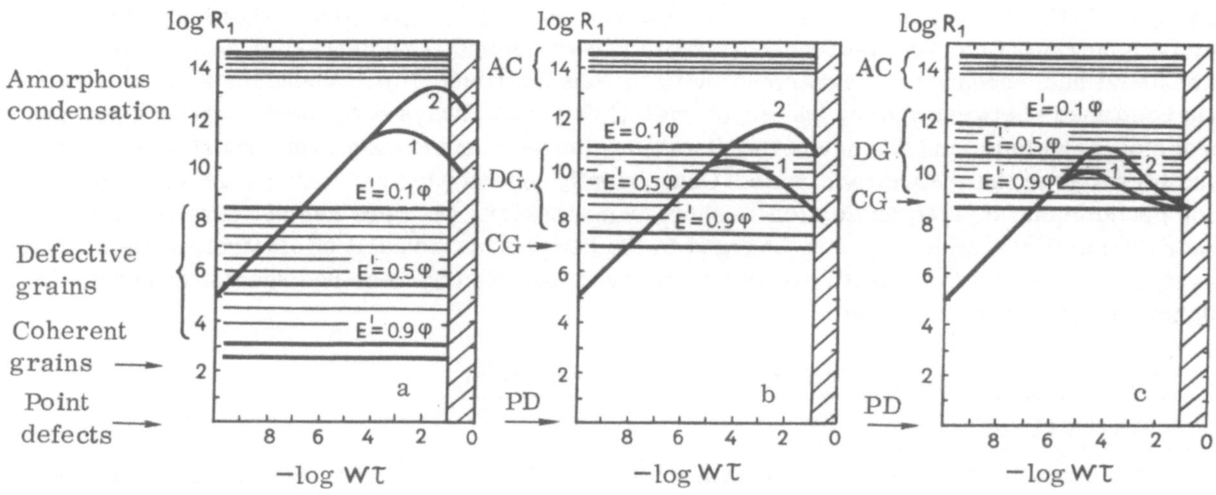


Fig. 3. Characteristic levels of defect formation for various kinetic condensation conditions (The thick lines represent major paths, see text): (a) $\varphi/kT = 15$; (b) $\varphi/kT = 10$; (c) $\varphi/kT = 8$. The curves 1 relate to a condensation rate of 1 μm/hr, while curves 2 relate to 1000 μm/hr.

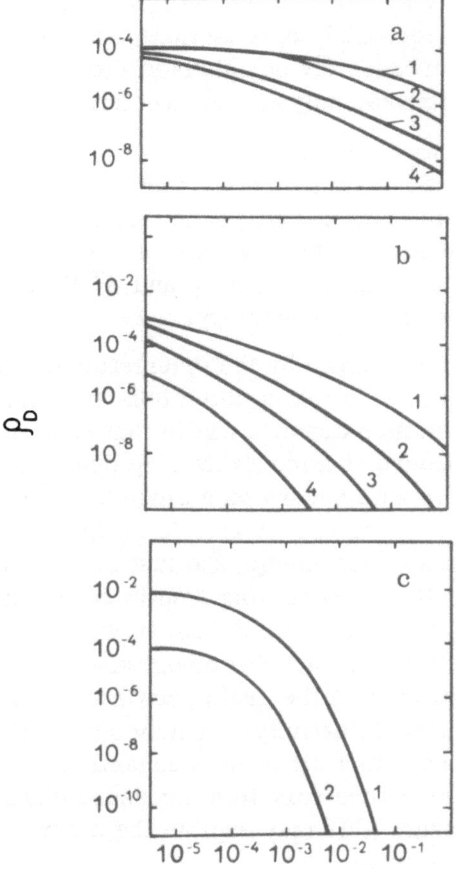


Fig. 4. Relative populations $\rho_D$ of defect positions in relation to tan $\alpha$ (condensation rate w in atomic layers per second): 1) $10^3$; 2) $10^2$; 3) 10; 4) 1.0. (a) $\varphi/kT = 20$; (b) $\varphi/kT = 10$. (c) $\varphi/kT = 8$.

be eliminated under any working conditions. A polyatomic defect corresponds to the middle series of levels (depth of potential well $E' < \varphi$). It is clear that all the defective levels lie below the main path at 527°C, and therefore one cannot produce a defect-free Ge or Si crystal by direct condensation; at 927°C, the defects can be eliminated in the main, while at 1227°C one should obtain a defect-free crystal under suitable working conditions.

If the crystallization is performed on a nonsingular face, the natural elements in the microrelief act as sinks for the two-dimensional gas and lower the main path, which may provide an additional means of improving the structural perfection. This is clear from Fig. 4, which shows the relative populations of the defective positions as a function of disorientation angle on a (111) plane in the diamond lattice. These conclusions agree at least qualitatively with all available experimental evidence.

## 3. Microselectivity in Heterogeneous Chemical Reactions in Gas-Phase Crytallization

The probabilities of the elementary acts in heterogeneous reaction vary as between the free sorption centers and elements in the microrelief, as well as between coherent and defective centers.

Current concepts in the theory of heterogeneous catalysis indicate that the fall in the energy barrier for a reaction becomes more substantial when the binding energy at a sorption center increases.* This increases the probability that free atoms will become attached to the edges of growing grains (which is equivalent to directional condensation), and the main path falls; the flow of atoms to the coherent grains increases (by comparison with the defective ones). Lack of space forces us to omit the quantitative analysis, and we merely give the qualitative conclusions.

The microselectivity in a reaction increases the structural perfection in an epitaxial film; however, the selectivity of the individual acts tends to fall as the temperature is raised, so the argument in section 2 indicates that this effect should be substantial only for comparatively low-temperature methods, which are used for example with Ge and $A^{III}B^{V}$ compounds. However, microselectivity is not so important for silicon technology.

The catalytic microselectivity of a face is determined by the orientation dependence of the growth rate and by the observed growth forms; the following situations are possible: (1) the maximum growth rate coincides with the maximum step density, and in that case the growth figures are vicinal pyramids; (2) the maximum occurs at a moderate step density, and in that case one finds growth hummocks; (3) the maximum rate occurs on a singular face, where vicinal or terrace-type figures are produced; and (4) the growth rate is independent of the orientation and smooth faces are produced. As regards mechanism, the last case is identical with condensation from a molecular beam, where the rate-limiting step is collision with the surface, while the nucleation and grain growth occur by surface diffusion, which is the dominant mechanism at high temperatures and low reagent pressures. The other three cases, on the other hand, are specific to gas-phase crystallization, with the grain growth dependent to some extent on a diffusion-free mechanism (reaction microselectivity). It may be that the most important conclusion from the statistical theory is here that all these mechanisms should be observed within a single system in response to temperature variation, pressure changes, mass-transfer conditions, and hence even equipment design (this may explain the discrepancies in orientation relationships reported by different workers).

The exact sequence of these four major types is determined by the balance between the activation energies and the steric factors for the individual acts (sorption and desorption of initial materials and final products, as well as molecular decomposition), which applies equally to the free centers and the microrelief elements. However, there is a tendency for the mechanism to go over to the one characteristic of molecular beams on raising the temperature and reducing the pressure, as is clear from many experiments [11, 12].

## 4. Phase Selectivity in Multicomponent Crystallization

It has been shown [6] that the phase composition of a film is dependent on the crystallization kinetics; in particular, a two-component eutectic system can produce a second phase even away from the eutectic region if the growth rate exceeds some critical value, with the grains later breaking up or being preserved in the growing crystal, the exact behavior being dependent on the relative position of the main path and the second-phase level. We consider a simple three-phase case, which occurs in the growth of films of $A^{III}B^{V}$ compounds, where the intermetallide may be accompaned by phases representing the pure elements. Here the following point is important: the $A^{III}B^{V}$ crystals contain a single bond type of energy $\varphi_{AB}$ (the analogous bond energy for the pure element A is denoted by $\varphi_{AA}$). However, the lattice of element B is layered, and the interatomic distance in a close-packed layer is much less than

* In complicated cases of multicenter adsorption, the maximum fall in the barrier height may occur at some optimal bonding energy.

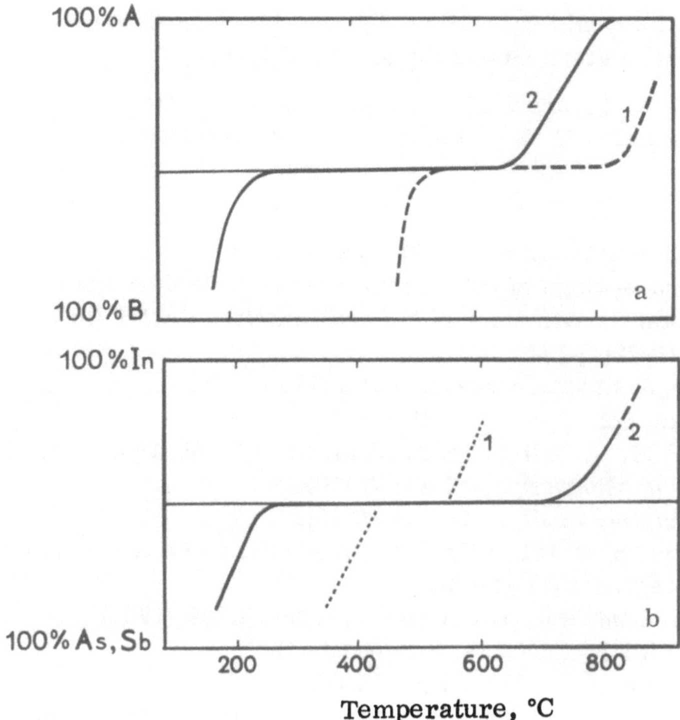


Fig. 5. Composition of a $A^{III}B^{V}$ film as a function of condensation temperature: (1) (InSb): $W_{In} = 20$, $W_{Sb} = 60$ atomic layers per second; (2) (InAs): $W_{In} = 10$, $W_{As} = 500$ atomic layers per second; (a) theory; (b) experiment [13].

the distance between layers, with correspondingly $\varphi'_{BB} > \varphi_{BB}$ (and here $(\varphi'/\varphi'')_{Sb} < (\varphi'/\varphi'')_{As}$). When a $A^{III}B^{V}$ compound condenses, the surface of the substrate acts as a close-packed plane. Then the binding energy of the B atoms to the underlying atoms is $\varphi_{AB}$ if these atoms are in positions of coherent structure for the $A^{III}B^{V}$ lattice, whereas it is $\varphi''_{BB}$ if the positions are not coherent. If a nucleus of the pure phase of element B is formed, then such coherence is inevitable for half of the atoms. This factor determines the instability of such nuclei. On the other hand, nuclei of A are completely stable, and if these are to be suppressed one needs a surplus of B in the beam. Figure 5 compares theoretical results for InSb and InAs films condensing from nonstoichiometric beams with Gunter's measurements on the effects of temperature [13]. The agreement is entirely satisfactory, apart from the high-temperature branch for Sb, which is quite explicable, since this branch runs above the melting point of antimony.

## 5. Selective Growth on Macroscopic Nonuniformity in the Substrate

This form of selectivity is widely used in technology in local epitaxial overgrowth from vapors; the theoretical analysis indicates that stable selective growth can occur even without chemical reaction, e.g., in condensation from a molecular beam, as has been confirmed for various metals condensing on substrates containing insulating, semiconducting, and metal areas. A major point is that simple measures allow one to alter the sign of the selectivity, i.e., to provide condensation on selected parts. On the other hand, selective crystallization on the single-crystal substrate allows one to produce heterostructures free from mechanical stresses (for instance, in epitaxial deposition of Ge on Si), whereas anisotropy in the growth rates allow one to vary the topological pattern during the growth [14].

I am indebted to S. A. Toporovskii and I. D. Khan for participation in the experiments on selective crystallization and to M. S. Belokon' for a useful discussion of the draft.

## Literature Cited


1. J. Kobayashi, IEEE Trans. ED-18:45 (1971).
2. V. F. Dorfman, Proceedings of the Fourth All-Union Conference on Crystal Growth: Abstracts on Crystal Growth Mechanism and Kinetics, Part 1 [in Russian], Izd. AN ArmSSR, Erevan (1972), p. 86.
3. P. Rai-Choudhury, J. Crystal Growth, 7:165 (1970).
4. S. B. Hyder, J. Vac. Sci. Techn., 8:228 (1971).
5. V. F. Dorfman and M. B. Galina, Dokl. Akad. Nauk SSSR, 182:136 (1968).
6. V. F. Dorfman, Kristallografiya, 15:435 (1970).
7. V. F. Dorfman, Kristallografiya, 18:154 (1973).
8. R. E. Tun, In: Physics of Thin Films, Volume 1 [Russian translation], Mir (1967), p. 224.
9. D. Walton, Phil. Mag., 7:1671 (1962).
10. V. Lewis and D. S. Campbell, J. Vac. Sci. Techn., 4:200 (1967).
11. E. I. Givargizov, Fiz. Tverd. Tela, 6:1804 (1964).
12. D. W. Shaw, J. Electrochem. Soc., 115:405 (1968).
13. K. G. Gunter, In: The Use of Thin Films in Physical Investigations, J. C. Anderson, ed., Academic Press, London (1966), p. 213.
14. V. F. Dorfman, Gas-Phase Micrometallurgy of Semiconductors [in Russian], Metallurgiya, Moscow (1974).

# MOLECULAR-BEAM CONDENSATION AND ACCOMMODATION: SODIUM CHLORIDE ON TANTALUM

A. M. Zatselyapin, V. I. Mikhailov, Yu. A. Gel'man, Yu. N. Lyubitov, and A. A. Chernov

*Institute of Crystallography, Academy of Sciences of the USSR, Moscow*

Alkali halides and many other substances contain [1] not only simple molecules but appreciable numbers of polymers [2], which have been observed in the vapor state and in molecular beams resulting from sublimation [3], The latter indicates that polymerization occurs at the surface. The monomer or polymer composition of the vapor may be unimportant to crystallization [4], so we have examined the role of dimers in film condensation by using sodium chloride, which has relatively high volatility below the melting point (1073°K), which is convenient in examining molecular-beam crystallization. Further, the sublimation flux at about 1000°K contains about 30% dimers (the total vapor pressure is about $10^{-2}$ mm Hg). We have examined the condensation of sodium chloride films from such beams, and also the energy transfer between the substrate and the incident particles under conditions that enabled us to vary the intensity and composition of the incident beam as well as the substrate temperature.

We measured the beam intensity, condensation temperature, and speed distribution in the reflected beam for various speed distributions in the incident beam. The substrates were polycrystalline tantalum strip or else crystalline films of sodium chloride formed on the latter.

## 1. Methods

Figure 1 shows the system; the substrate 9 is maintained at a set temperature, and the evaporator 13 provides the molecular beam; the specularly reflected flux (at about 45° to the substrate) is recorded by a mass spectrometer modified from a standard MV2302 that allows one to measure the velocity distribution as well as the mass distribution [5]. The vacuum in the working chamber was 1-2 · $10^{-3}$ mm Hg, which was provided by mercury-vapor and ion-getter pumps, together with zeolite pumps in the forevacuum line. The zeolite pumps reduce the hydrocarbon background by more than an order of magnitude.

Between the evaporator and the substrate there was sometimes the rotor 12 or mechanical velocity selector, which was taken from a hysteresis motor [6]. The speed of the rotor was up to 72,000 rpm, which enabled us to use a selector with a rotor diameter of 64 mm and a length of only 10 mm [7]. The edge had 288 slots of width 0.3 mm and depth 2 mm, whose axes lay at 6° to the axis of rotation.

The mean speed of a sodium chloride molecule in the selected beam at the substrate could be adjusted between 2.6 · $10^4$ and 7.3 · $10^4$ cm · $sec^{-1}$, which is equivalent to a range in evaporator temperature $T_e$ of −40°C to 1500°C. The mass spectrometer recorded the $NaCl^+$

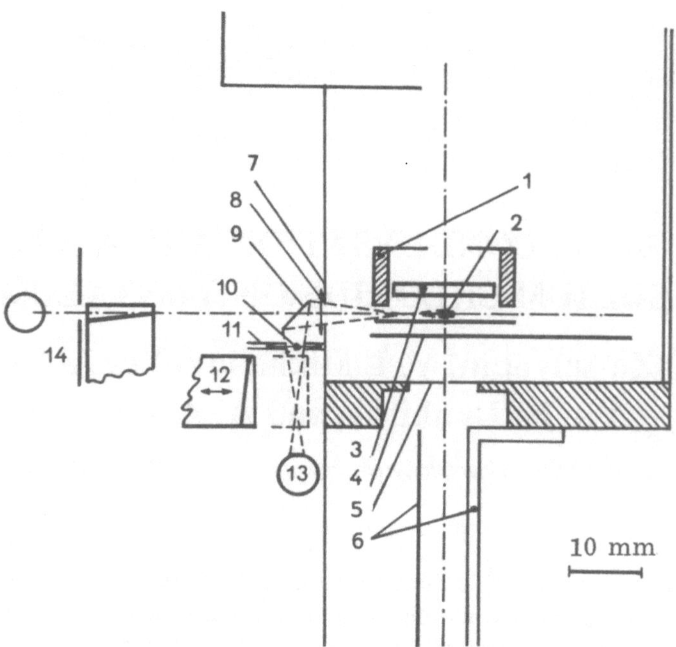


Fig. 1. The apparatus: (1) ionization space in mass spectrometer; (2) ion-production region; (3) extraction electrode; (4, 5) extraction and acceleration slits; (6) reflection plates in velocity analyzer; (7, 11) stops; (8, 10) slides; (9) substrate; (12) selector rotor (the broken line shows the rotor in the position out of the beam); (13) tube evaporator; (14) selector with evaporator in calibration position.

and $Na_2Cl^+$ currents, which are denoted here as $I^{58}$ and $I^{81}$, which correspond to NaCl monomers ($NaCl^+$, m = 58) and $Na_2Cl_2$ dimers ($Na_2Cl^+$, m = 81). The methods of identifying the monomeric and dissociated components have previously been described [8].

The absolute fluxes were determined as follows. An NaCl film was deposited on the substrate under specified conditions; then the incident flux was cut off, the film was evaporated, and the resulting $I^{58}$ was recorded as a function of time t. The operation was repeated three times, while the fourth time the film was not evaporated but instead was removed from the instrument and weighed. The mass was determined spectrophotometrically. The calibration was performed by comparing the area under the $I^{58}(t)$ curve (which was reproduced within a few per cent) with the mass of the film. During the calibration, the sodium chloride was evaporated isothermally at 1140°K, where the proportion of dimers in the vapor does not exceed 8%. See [7] for details of the procedure.

## 2. Results

### (a) Condensation Temperature

Figure 2 shows typical curves for $I^{58}$ and $I^{81}$ in relation to substrate temperatures $T_s$ for incident fluxes as follows: $2 \cdot 10^{14}$ monomers $\cdot$ cm$^{-2}$ $\cdot$ sec$^{-1}$ for the lower pair of curves and $10^{13}$ monomers $\cdot$ cm$^{-2}$ $\cdot$ sec$^{-1}$ for the upper pair. The curves were recorded with an XY recorder, with the abscissa receiving the signal from a thermocouple (VR-5 or VR-20) attached to

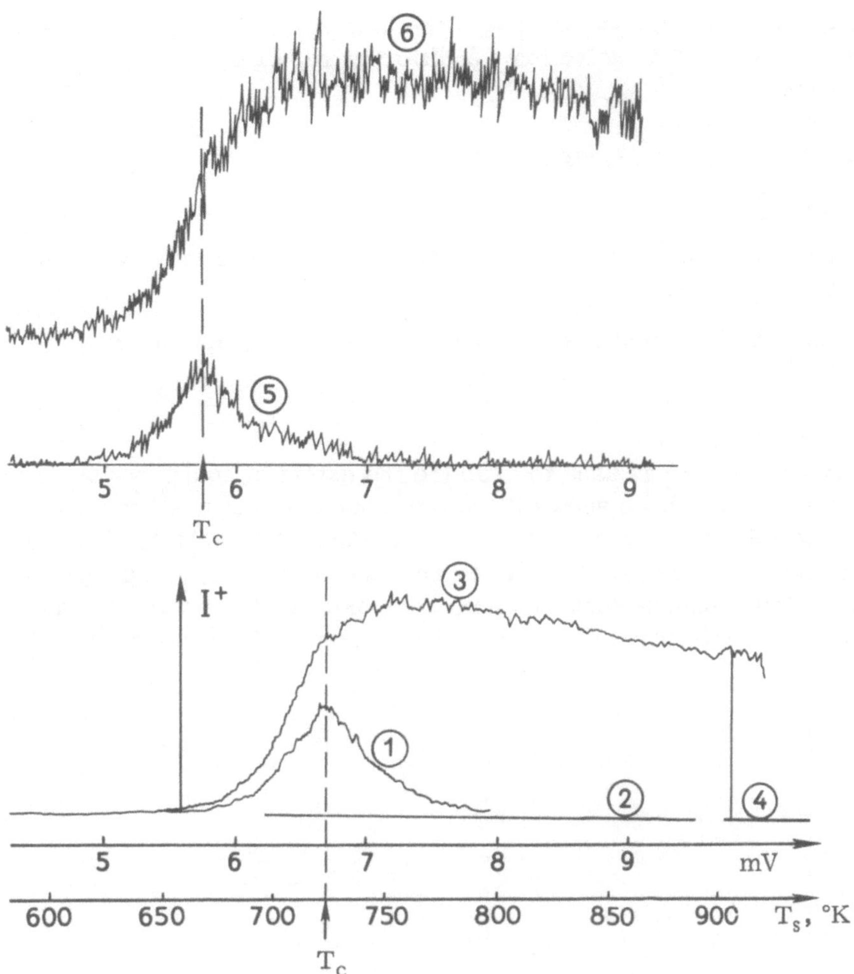


Fig. 2. Typical curves for ion current as a function of substrate temperature; (1, 5) $I^{81}$ for $Na_2Cl^+$, m = 81; (3, 6) $I^{58}$, $NaCl^+$, m = 58; (2, 4) background at the $I^{81}$ and $I^{58}$ lines with slide 8 closed. The two abscissa calibrations indicate the thermocouple readings.

the substrate, while the ordinate was the ion current recorded by the mass spectrometer. The recordings were made on reducing the substrate temperature under the incident beam.

Curves 1 and 5 correspond to dimer evaporation and peak at $T_c$, which corresponds to a peak previously found in the surface ionization of sodium chloride on tungsten [9]. We found that $T_c$ was the limiting condensation temperature corresponding to a given beam intensity. In fact, if the temperature was maintained such that $T_s < T_c$, then the deposit was a crystalline sodium chloride film. No film was produced if the substrate temperature was higher than $T_c$. The condensation was detected by direct observation after removing the substrate, and also by heating the substrate rapidly and recording the corresponding ion-current bursts [7]. We also found that for $T_s > T_c$ a previously deposited film evaporated even in the presence of an incident beam. The error in measuring $T_c$ was not more than 1.5°, and therefore the upper bound to the critical supersaturation in the adsorption layer does not exceed about 8%. Previous discussions [10] have indicated a critical supersaturation of the order of $10^6$% or more, while recent advances in experimental technique have given values in the range 10-100% [10]. Our result comes even closer to indicating that there is no critical supersaturation, i.e., no potential barrier to nucleation.

This result is clearly a consequence of the high defect density at the substrate surface, which substantially facilitates nucleation. A different result is to be expected for a perfect single-crystal substrate.

## (b) Monomer Association and Dimer Decomposition in an Adsorbed Layer

NaCl molecules adsorbed on a surface and diffusing there may combine into groups, which can also split up again.

In particular, the following reaction occurs in an adsorption layer:

$$2NaCl \rightleftarrows Na_2Cl_2. \quad (1)$$

The velocity selector was used to examine the extent of equilibration during the time spent on the surface; it has been shown [7, 8] that such a selector allows one to adjust the proportion of dimers in the beam over the range 15-50%. The selected beam from the oven is directed on to the substrate, and the dimer reevaporation flux is recorded, e.g., as in curves 1 and 3 of Fig. 3. If we assume that the beam composition at the oven was as in measurements of the incident-beam composition [7, 8], then the agreement between the coordinates of the

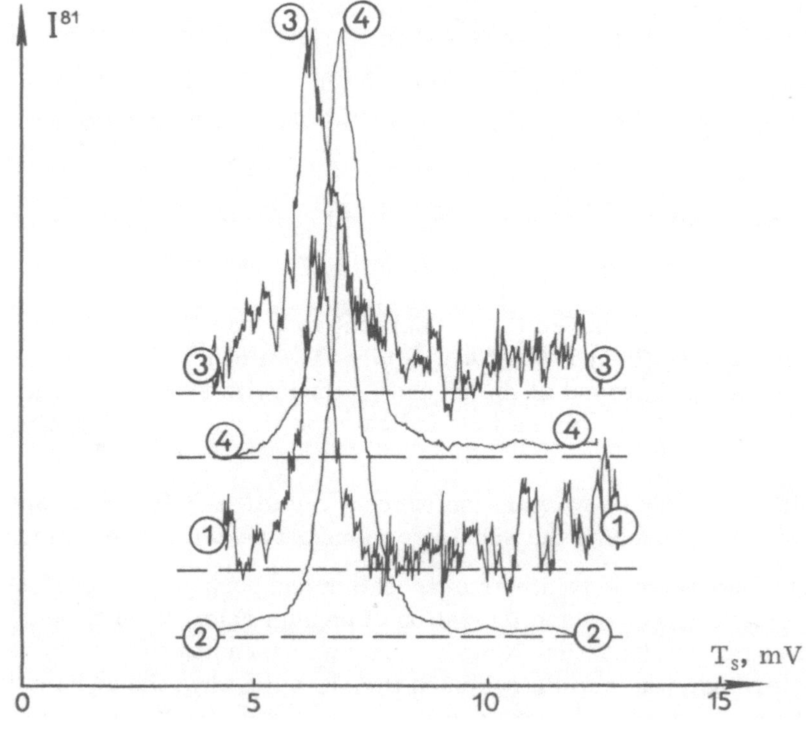


Fig. 3. Dimer–monomer equilibrium at a surface at $T_c$ and dimer desorption flux as a function of substrate temperature: (1) selected low-speed incident beam (8000 rpm), maximum proportion of dimers (about 50%); (2, 4) unselected incident beam; (3) selected high-speed beam (27,000 rpm), minimal proportion of dimers (about 15%). The intensities of the selected beams were about $5 \cdot 10^{13}$ particles $\cdot$ $cm^{-2}$ $\cdot$ $sec^{-1}$. See Fig. 2 for the conversion from mV to °K.

peaks on curves 1 and 3 in Fig. 3 indicates that the steady-state reevaporated flux composition is independent of the composition of the incident beam if $T_s = T_c$. The latter in turn means that equilibrium is established between the monomer and dimer forms during the time spent on the surface.

As $I^{58}$ and $I^{81}$ were measured and the instrument was calibrated from the weight of the evaporated film, we could determine the absolute sublimation fluxes and the corresponding Langmuir vapor pressure of the monomer $P_{NaCl}$; the corresponding curves for $P_{Na_2Cl_2}$ for the dimer were obtained after using the data of [11] on the vapor composition for calibration purposes.

Figure 2 shows the fluxes of reevaporated monomers (circles) and dimers (squares) calculated after such calibration from Fig. 4. Arrows 1 and 2 indicate two condensation temperatures corresponding to incident fluxes of $10^{13}$ cm$^{-2}\cdot$sec$^{-1}$ and $10^{14}$ in the same units. The lines M and D denote the partial pressures of the monomer and dimer components of the saturated vapor as indicated by the literature [3, 12, 13]. Figure 4 shows that the slopes of the parts of the curves corresponding to $T_s < T_c$ and thus to condensate on the substrate agree satisfactorily with the heats of evaporation for the monomer (about 51 kcal/mole) and the dimer (about 60 kcal/mole).

The curves of Fig. 2 were also used to construct $K_e \approx (I^{58})^2/I^{81}$ as a function of $T_s$ in arbitrary units; the result is given in Fig. 5a for an incident flux of $10^{13}$ cm$^{-2}\cdot$sec$^{-1}$ (from curves 5 and 6 of Fig. 2) and in Fig. 5b for the higher flux of $2 \cdot 10^{14}$ in the same units (from curves 1 and 3 of Fig. 2). The arrows, as in Fig. 2, indicate the $T_c$ corresponding to these fluxes. For comparison, Figs. 5a and b show straight lines of slope corresponding to about 45 kcal/mole, which corresponds to published values [14-17] for the vapor-state dissociation of the dimer. If the rate of reaction (1) at the surface is higher than all the other rates, in particular the dimer desorption rate, then the activation energy for $K_e$ is equal to the energy of reaction (1) in the volume. If on the other hand it is impossible to neglect other conceivable reactions at the surface, then there may be a deviation from logarithmic temperature dependence in $K_e$ [18].

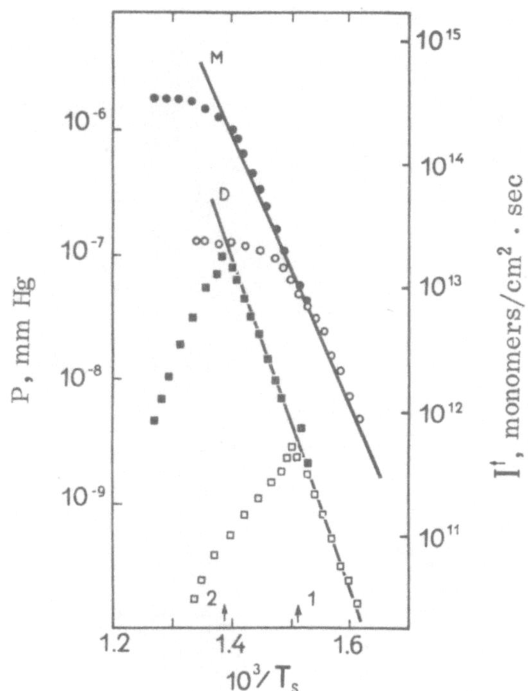


Fig. 4. Temperature dependence of the monomer desorption flux (circles) and dimer flux (squares); arrows 1 and 2 denote $T_c$ for the low flux, about $10^{13}$ particles $\cdot$ cm$^{-2}\cdot$ sec$^{-1}$, and for the high flux, about $2 \cdot 10^{40}$ in the same units (filled symbols).

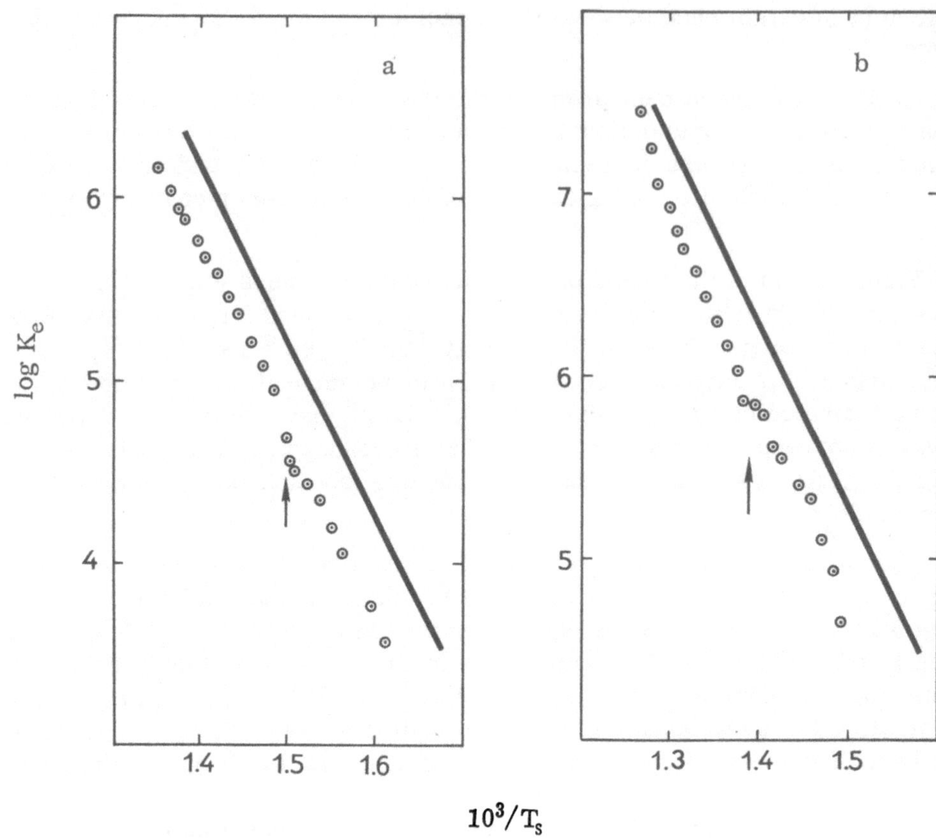


**Fig. 5.** Equilibrium constant $K_e \approx (I^{58})^2/I^{81}$ for the reaction $NaCl_2 \rightleftharpoons 2NaCl$ in arbitrary units in terms of substrate temperature.

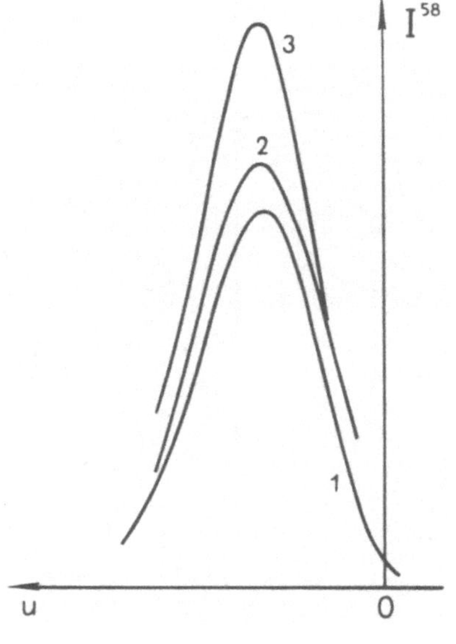


Fig. 6. Observed curves for the distribution of $NaCl^+$ as regards velocity projection on the axis of the molecular beam for various incident-beam temperatures and a constant substrate temperature: (1) Maxwellian incident beam, source temperature about 1100°K; (2) selected beam (8000 rpm), effective temperature about 230°K; (3) selected beam (27,000 rpm), temperature about 1800°K, u is the voltage on plates 6 (Fig. 1) in the velocity analyzer [5].

It may be that Fig. 5 shows such effects.

An important point is that there are narrow gently sloping ranges when the condensation temperature is approached from the low-temperature side, i.e., from the side where there is already a crystalline film of NaCl; it may be that these are due to a fall in the supersaturation and hence to a fall in the density of the steps on the surface of the NaCl near equilibrium [19, 20]. The step density is high far from the equilibrium temperature and the surface is essentially atomically rough. The accelerated exchange between the adsorption layer and the kinks when the density of the latter is high may affect the conditions for surface dimerization.

Figure 5 shows log $K_e$ as a function of $1/T_s$ for temperatures far from $T_c$, which differs substantially from previous results [12, 13]; the reasons for the discrepancy are currently being examined.

It is also of interest to examine the density of the adsorption layer for $T_s = T_c \approx 700°K$, the incident beam was rapidly shut off, and the decaying monomer current was recorded in the absence of the condensate. The fall was exponential with a relaxation time $\tau$ of about 3 sec [7]. As the incident flux was about $10^{14}$ $cm^{-2} \cdot sec^{-1}$, the surface density was of the order of $3 \cdot 10^{14}$ $cm^{-2}$, i.e., of monolayer order. This result agrees with those of [21, 22].

## (c) Thermal Accommodation

Thermal accommodation was examined by measuring the velocity distribution in the reevaporated beam with a selected incident beam. The distribution in the reevaporated beam was Maxwellian, and the mean speed did not vary when the mean speed of the incident beam was adjusted with the selector over roughly a range representing a factor 3. The temperature distribution in the emitted beam corresponded to the substrate temperature. The measurements were made for $T_s > T_c$ in the absence of a condensate on the substrate. Figure 6 shows typical curves. The results for dimers and monomers were analogous. Therefore, there was complete accommodation of the translational energy at the substrate.

The two major results are therefore complete thermal accommodation in this system and definition of the equilibrium conditions for dimerization in the adsorption layer, both of which confirm current views on condensation and adsorption. Support comes also from recent evidence on the cosine law for the evaporation and reevaporation fluxes for alkali halides [23] and zinc [24].

## Literature Cited


1. N. I. Ionov, Dokl. Akad. Nauk SSSR, 59:467 (1948).
2. M. Ya. Gen and Yu. I. Petrov, Usp. Khim., 38:2249 (1969).
3. G. Rothberg, M. Eisenstadt, and P. Kush, J. Chem. Phys., 30:604 (1959).
4. V. M. Bulakh, J. Crystal Growth, 5:243 (1969).
5. Yu. N. Lybitov, Yu. A. Gel'man, A. M. Zatselyapin, V. I. Mikhailov, and L. V. Suman, Prib. Tekh. Eksp., No. 3, 218 (1969), VINITI, No. 525-69 Dep.
6. A. M. Zatselyapin, Prib. Tekh. Eksp. No. 6, 163 (1969).
7. A. M. Zatselyapin, Candidate's Dissertation, Moscow, IKAN (1972).
8. V. I. Mikhailov, Yu. A. Gel'man, A. M. Zatselyapin, and Yu. N. Lyubitov, Zh. Fiz. Khim., 45:1260 (1971).
9. N. I. Ionov and M. A. Mittsev, Zh. Teor. Fiz., 35:1863 (1965).
10. J. B. Hudson and J. S. Sandejas, Surface Sci., 15:27 (1969).
11. L. N. Gorokhov, Yu. S. Khodeev, and P. A. Akishin, Zh. Neorg. Khim., 3:2597 (1958).
12. Yu. A. Gel'man, A. M. Zatselyapin, Yu. N. Lyubitov, and V. I. Mikhailov, Dokl. Akad. Nauk SSSR, 195:1313 (1970).

13. Yu. N. Lyubitov, V. I. Mikhailov, A. M. Zatselyapin, and Yu. A. Gel'man, Zh. Fiz. Khim., VINITI, No. 2811-71 Dep., Moscow (1971).
14. B. H. Zimm and J. E. Mayer, J. Chem. Phys., 12:362 (1944).
15. R. C. Miller and P. Kush, J. Chem. Phys., 25:869 (1956).
16. S. H. Bayer, R. M. Diner, and R. F. Porter, J. Chem. Phys., 29:991 (1958).
17. T. A. Milner and H. M. Klein, J. Chem. Phys., 33:1628 (1960).
18. Yu. N. Lyubitov and V. I. Mikhailov, Zh. Fiz. Khim., 46:2968 (1972).
19. R. Jaecket and W. Peperle, Z. Phys. Chem. (Leipzig), 217:321 (1951).
20. H. Dabringhaus and H. Mayer, J. Crystal Growth, 16:17,30 (1972).
21. N. D. Konovalov and E. A. Tishin, Ukr. Fiz. Zh., 16:474 (1971).
22. B. Ya. Konesnikov, Candidate's Dissertation, Moscow, MGU (1971).
23. W. Marx, J. Sell, and J. E. Lester, J. Chem. Phys., 55:5835 (1971).
24. Yu. A. Gel'man, V. F. Vinogradov, and Yu. N. Lyubitov, Kristallografiya, 18:421 (1973).

# AUTOEPITAXIAL NUCLEATION IN IONIC CRYSTALS

A. Smakula

*Crystal Physics Laboratory, Massachusetts Institute of Technology*

Two factors are essential in all methods of crystal growth: nucleation and growth rate. The nucleation comprehends the initial stage until a nucleus reaches the critical size, while the growth rate is related to the formation of regular crystal form. Both factors depend on many variables which are not yet well established and seldom under proper control; therefore most synthetic crystals are of relatively poor quality.

The simplest conditions seem to be in crystal growth from the vapor phase by the epitaxial method when the substrate is a perfect crystal of the same material. This method I would like to call "autoepitaxy."

The nucleation does not start randomly, but only on "specific points" of the surface. These points are probably connected with some defects such as vacancies, dislocations, impurities, or others. For detection of defects, thermal etching in high vacuum but not too high temperature has been used. In Fig. 1a the etch pits on the cleaved (100) face of KBr are shown. Several features are evident. The etch pits have a square shape and are almost of square size. The edges of the pits coincide with the cubic edges of the crystal. The steps along the edges are sharper at the corners than in the central part of the edges. The surface between the pits does not show any distortion. Occasionally, irregular pits appear, as shown in Fig. 1b. The number of pits per square unit depends on the temperature and the size of the pits depends on the time.

More interesting results have been obtained on single crystals of TlBr. These crystals are very soft and cannot he cleaved. In Fig. 2 the etch pits on a (110) face are shown. The irregular shape of the pits is caused by surface distortion during cutting and grinding. After recrystallization by proper annealing and careful repolishing, the etch pits have regular form, as shown in Fig. 3. On the (100) face we see squares, on the (110) face rhombs, and on the (111) face triangles. In TlBr crystals the habit faces are dodecahedron faces (110). The different shapes of the etch pits on (100), (110), and (111) faces are caused by the negative projection of dodecahedra in three different directions.

The reverse process of evaporation is condensation. In Fig. 4 are shown crystallites of TlBr grown on (100), (110), and (111) faces of annealed single TlBr crystals. On the (100) face we see squares, on the (110) face rhombs superimposed on hexagons, and on the (111) face hexagons with three noticeable spikes. In all three cases we have crystallites of the same dodecahedron shape, observed from three different directions. Most of the crystallites have the orientation of the substrate crystals; but sometimes there are misfits, as can be seen in Figs. 4a and 4b. We have to assume that the misfits are caused by some defects on the surface of the substrate crystals. Which defects we do not know. One is sure they are not mecha-

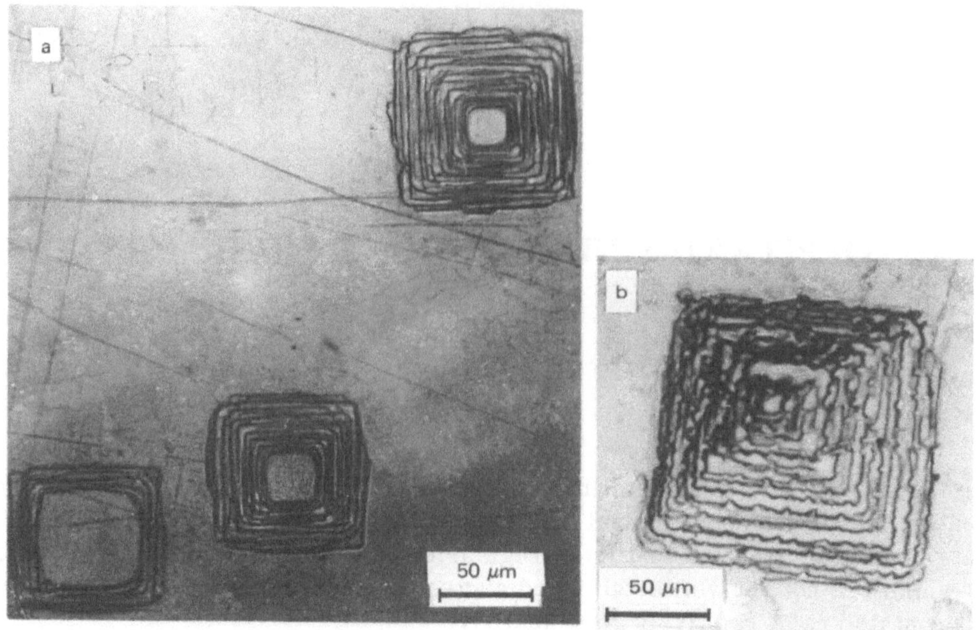


Fig. 1. Thermal etch pits (300° C, 20 h) on a (100) cleavage surface of potassium bromide: (a) regular pits; (b) irregular pits.

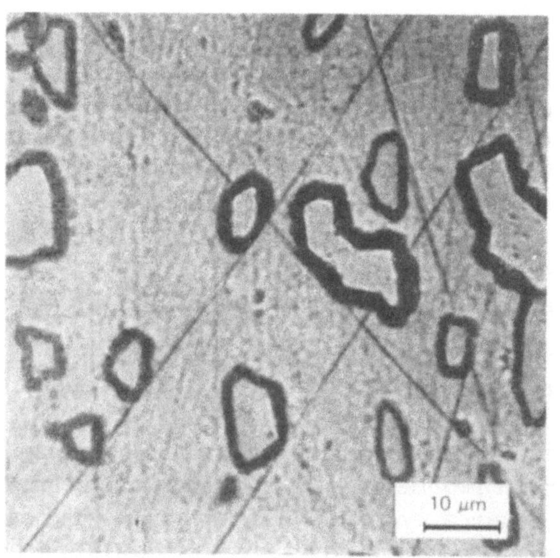


Fig. 2. Thermal etch pits (300°C, 30 h) of irregular shape on a (110) face of thallium bromide, surface deformed during grinding.

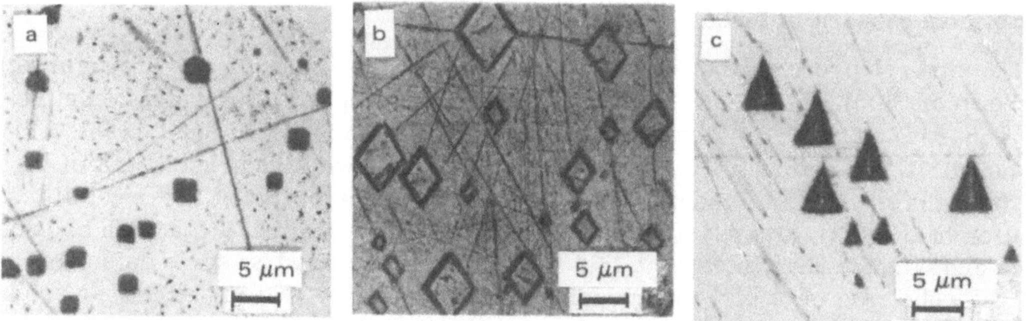


Fig. 3. Thermal etch pits (300° C, 16 h) on faces of a carefully annealed thallium bromide crystal: (a) (100); (b) (110); (c) (111).

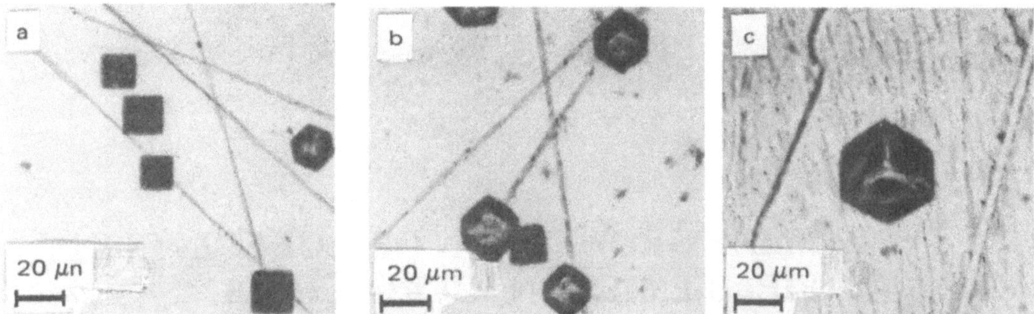


Fig. 4. Thallium bromide crystals grown from the vapor state on a thallium bromide substrate (300°C, 24 h), faces: (a) (100); (b) (110); (c) (111).

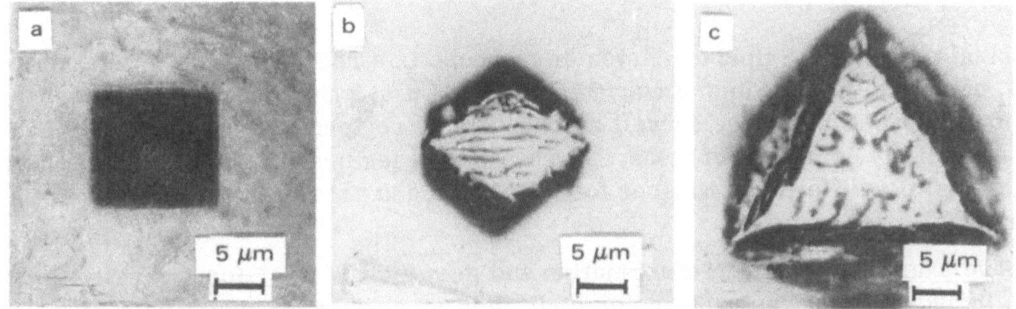


Fig. 5. Thallium bromide crystals grown on a thallium bromide substrate (300°C, 24 h), faces: (a) (100); (b) (110); (c) (111).

nical defects since the visible scratches are not preferred places for formation of crystallites. In Fig. 5 crystallites grown on (100), (110), and (111) faces are shown under high magnification. One can see that the surfaces are quite rough. This indicates that the growth process does not proceed in regular steps as theoretically is generally assumed.

## Conclusions

Nucleation of evaporation and condensation has been studied by thermal etching on NaCl and TlBr at moderate temperature and by autoepitaxial deposition on TlBr single crystals. In both crystals the evaporation starts at specific points and proceeds in width and depth, forming regular etch pits, but leaving the remaining surface area unchanged. On a distorted surface of TiBr the etch pits are irregular. The apparent shape of the regular etch pits depends on the crystal orientation, and the size depends on the time of the heat exposure. The reverse process, condensation, starts also on specific points of the substrate crystal, and the apparent shape of the crystallites depends on the substrate orientation. Some orientational deviations of crystallites and roughness of the surfaces have also been observed, which may cause defects in grown crystals.

# CRYSTALLINE GROWTHS ON OBJECT POINTS IN FIELD-EMISSION MICROSCOPES

**O. L. Golubev, B. M. Shanklin, and V. N. Shrednik**

*Institute of Technical Physics, Academy of Sciences of the USSR, Leningrad*

A typical object in the field-emission microscope is a rounded single crystal of a refractory metal, namely a tip of radius of curvature of 0.1-1 μm, which usually bears several types of almost ideal close-packed face as well as transitional regions having high Miller indices, which are essentially composed of steps. The structural features of such a surface mean that there are essentially different conditions for nucleation and growth on various atomic layers in a single material.

We have examined the mode of formation and growth of W on W for a variety of external conditions, including temperature T and electric field strength E. The experiments were performed under field-emission conditions at a pressure of $10^{-9}$-$10^{-11}$ mm Hg (controlled by the adsorbed gases), as well as under ion-microscopy conditions, with helium in the chamber and the point cooled with liquid or solid nitrogen. Overgrowths of all types were prepared at such pressures and were carefully examined in electron emission, and then were examined in more detail with ion emission.

Figures 1 and 2 show a series of emission pictures for crystalline films of W formed by condensation of 8 or 24 monatomic layers on a W point at various T; above about 600°K, bright rims appear around the {110}, {112}, and {100} faces, which were first observed by Muller [1]. The shape has been explained [2, 3] as due to preferential directions of linear diffusion on such faces, in conjunction with anisotropy in the reflecting barriers at the edges of the planar nets, as well as migration barriers along such edges. For instance, a {112} face has a channel structure, and the channels lying between the close-packed rows of atoms along the ⟨111⟩ direction are paths for one-dimensional displacement where the activation energy is much less than that for the perpendicular direction. Therefore, a {112} face is surrounded by two half-moon arcs, which are broader in the direction of the adjacent {111} face, since reflecting barrier at a {112} boundary on the {111} side is less than that on the {100} side.

The differences between these barriers are most pronounced at low T, but in any case the barriers are high enough for the migration to be substantially influenced in all cases. Correspondingly, Figs. 1 and 2 show the most highly variable structures in the range in T from 600 to 800°K. At higher T, the structures become smoother and are partially eliminated. In the case of the 8-layer film at 1680°K (Fig. 1f), the picture is very similar to that of the initial rounded point. If the amount of deposited material is doubled (Fig. 2), a temperature of 1610°K is inadequate to smooth the surface completely, and the highly overgrown faces require a higher temperature in order to acquire a more nearly equilibrium form.

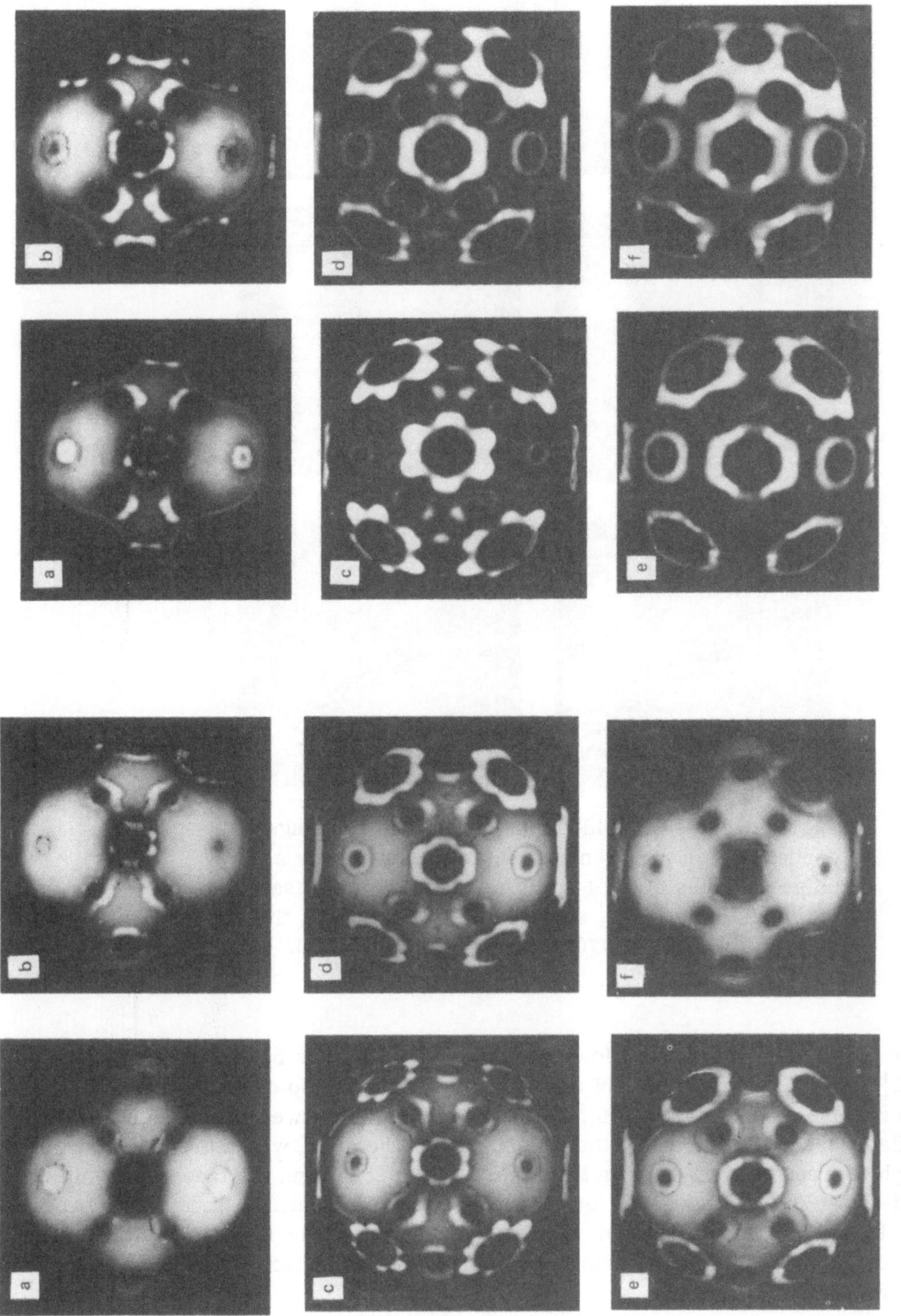

Fig. 1. Field-emission patterns of overgrowths arising by condensation on a heated single-crystal tungsten point (radius of curvature about 0.7 $\mu$m), thickness 8 atomic layers of tungsten, substrate temperatures in °K: (a) 580; (b) 640; (c) 750; (d) 910; (e) 1280; (f) 1680.

Fig. 2. Series of patterns recorded from the point of Fig. 1 with 24 monatomic layers of tungsten, substrate temperatures in °K: (a) 600; (b) 630; (c) 750; (d) 970; (e) 1280; (f) 1610.

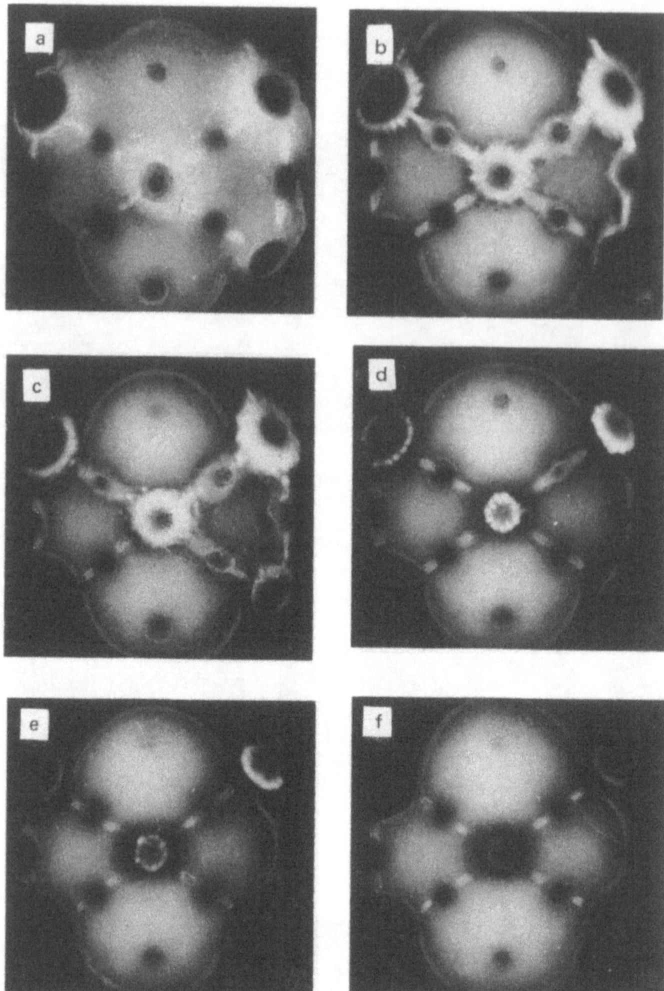


Fig. 3. Series of field-emission pictures of tungsten condensing on tungsten at low temperatures for a point of radius about 1 μm, thickness of condensate 0.3 monatomic layer, substrate temperature in °K: (a) 100; (b) 240; (c) 370; (d) 400; (e) 450; (f) 550.

Substrate T below 500°K does not allow adsorbed atoms to pass beyond face edges even on {110} and {112} faces, where the heat of migration is least, and so one finds randomly distributed clumps of atoms on such faces (Fig. 3). The ion studies showed that such structures are completely disordered, in contrast to those resulting in the arcs, which are highly ordered and represent high degrees of perfection in the edges of the net around the faces [4]. Any state of the type of that shown in Fig. 3 does not produce these arcs on heating, since highly ordered growth requires the adsorbed atoms to be adsorbed essentially one by one into the lattice. Large clumps produced by condensation below 500°K block the migration of single atoms and essentially alter the surface structure.

It is possible also for growths of W on W to be formed without external deposition; if a point is heated in a field of $E = (2\text{-}20) \cdot 10^7$ V/cm, the rounded form (the annealed form) becomes unstable on account of the surface tension being more than balanced by the field forces.

The stepped parts of the initial crystal, in particular the regions of the {111} and {100} faces, then give rise to overgrowths; this can be represented as follows. The atoms in the adsorbed phase on the heated rounded point are polarized by the external field and drawn into the strongest field regions at the tip [5, 6]. The fluxes of such atoms become substantial at high T and E, and nuclei arise, which are followed by three-dimensional overgrowths on the stepped parts. These formations themselves tend to raise the field at the vertex and thus increase the growth rate. The only limitation on the process is the evaporation of the atoms from the vertex in response to the field [6], or else competition from the Laplace pressure if the external field as a function of the reciprocal of the radius of the point increases less rapidly than would be implied by a square-root law.

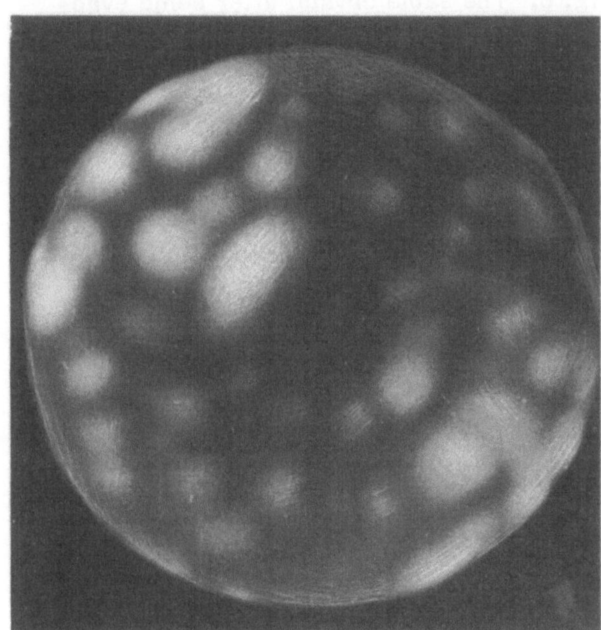

Fig. 4. Field-emission picture of tungsten on tungsten produced at 2200°K in a field of 48 MV/cm in a few minutes, point radius 0.6 $\mu$m, potential V = 2.4 kV.

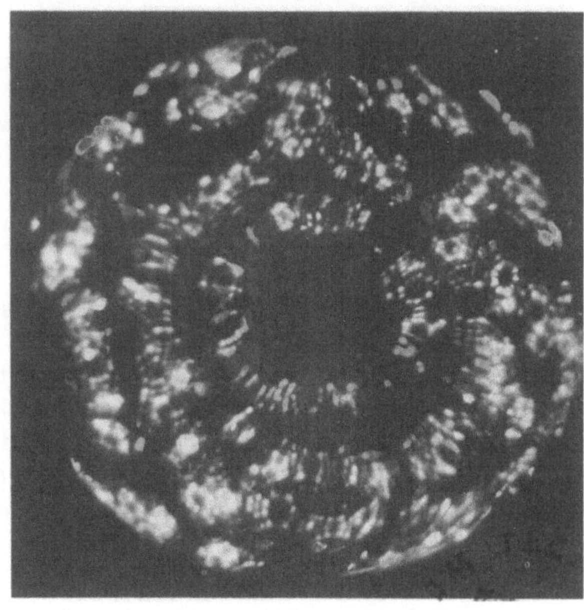

Fig. 5. Helium-ion image of the same structure (Fig. 4) at V = 27 kV, point temperature 65°K.

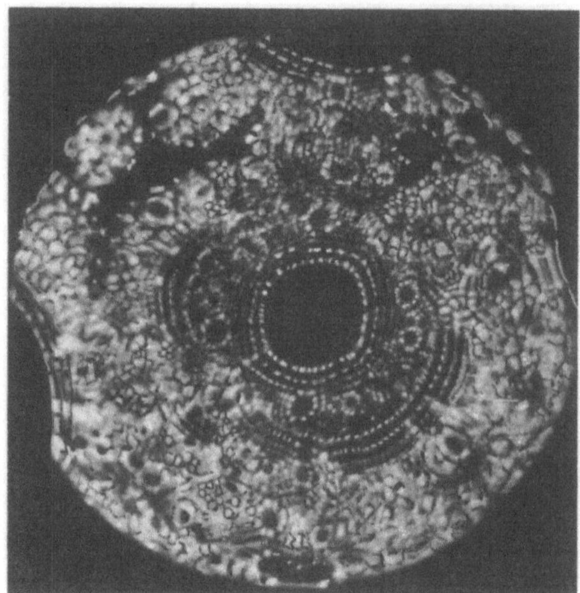

Fig. 6. The same crystal after some evaporation at V = 36 kV.

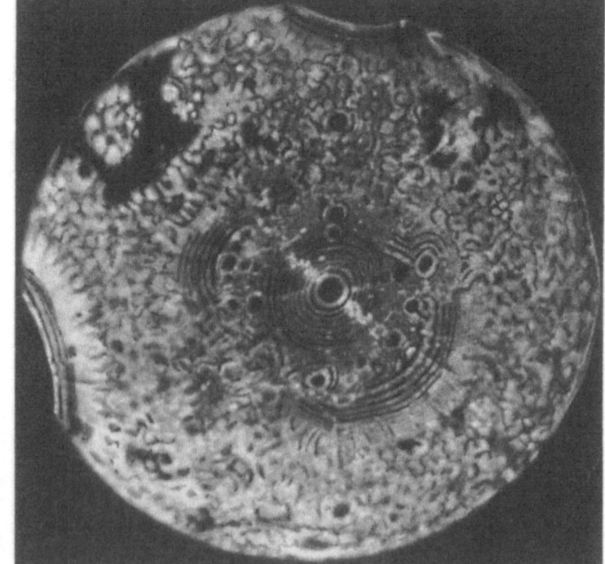

Fig. 7. The same crystal after further evaporation at V = 46 kV.

Figure 4 shows the initial electron-emission pattern from such a structure for a point at 2200°K and E = 48 MV/cm over a period of 2 min. There are clearly many spots over the surface. Figure 5 shows the helium pattern of this structure, and it is clear that the atoms have an ordered disposition. Figures 6 and 7 show successive stages in the field evaporation for the structure as of Fig. 5. Overgrowths joined at the base produced numerous defects and dislocations, irregular rows of atoms, and so on. Figure 7 shows the overgrowths completely eliminated in the central part of a point near a (110) face, so one sees here the final field-evaporation form of W.

At relatively low T and E, we get not a large number of small overgrowths but a comparatively small number of large ones; the electron emission (Fig. 8) at 2000°K and E = 25 MV/cm

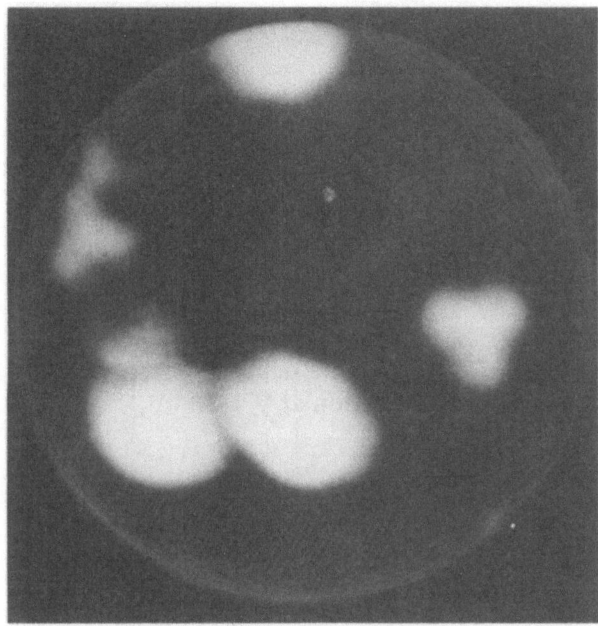

Fig. 8. Films of tungsten on tungsten produced at 2000°K at 25 MV/cm in 30 sec as revealed in electron emission at V = 2.8 kV, radius of initial point 0.5 $\mu$m.

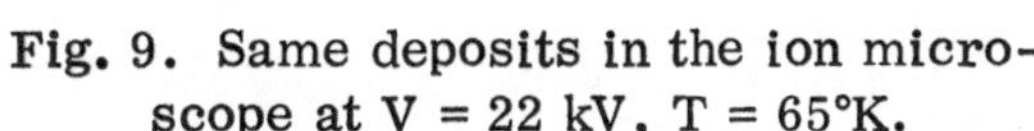

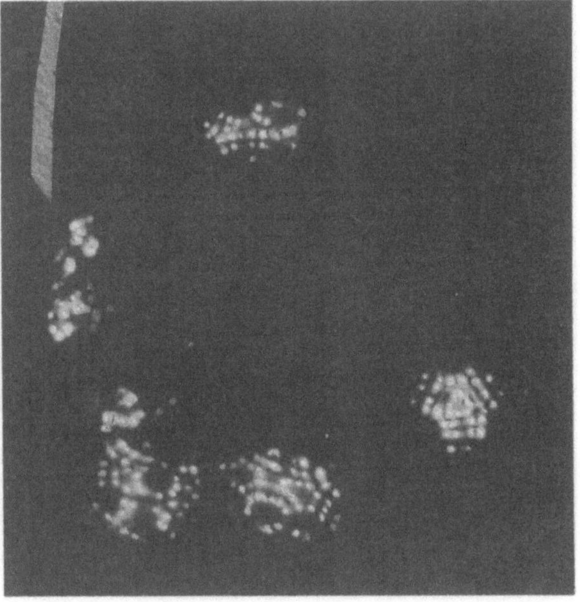

Fig. 9. Same deposits in the ion microscope at V = 22 kV, T = 65°K.

confirms this. Figure 9 shows the ion pattern, while Fig. 10 shows the most detailed picture of the structure. The overgrowth on the extreme right extends into the region of a (111) face and has a characteristic threefold symmetry, with the individual atoms along the face clearly resolved. An overgrowth with threefold symmetry has grown only along the ⟨111⟩ axis of the initial growth, so we suppose that these overgrowths continue to the structure of the substrate.

Overgrowths in the neighborhood of {100}, which are joined at the base, also result in many defects and irregularities; Fig. 11 shows this best for the state after extensive evaporation under the field, where the bases themselves are visible. Overgrowths around {111}, which fuse at the base with those around {100}, produce a clear pattern, with the atomic rows in the two overgrowths displaced one relative to the other. The numerous defective sites between the overgrowths are visible also on other parts of the points. These defective structures arise be-

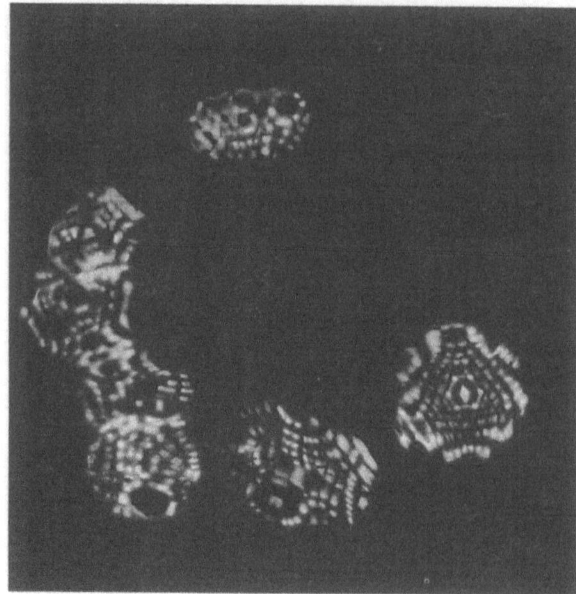

Fig. 10. The same deposits after field evaporation at V = 25 kV.

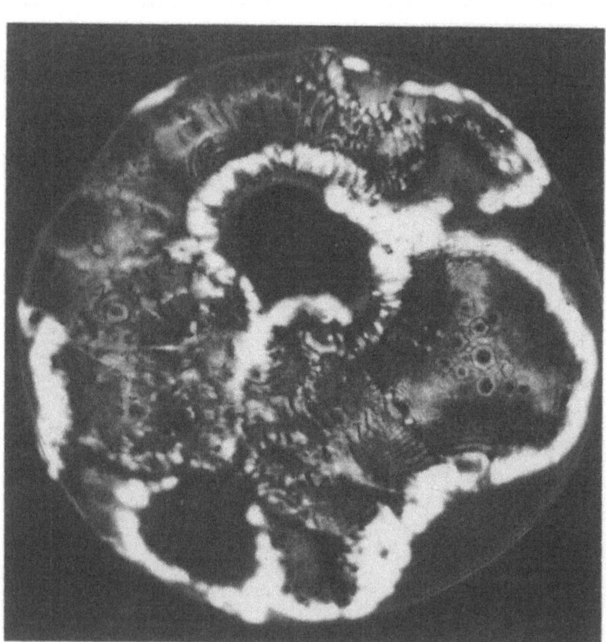

Fig. 11. After extensive field evaporation of deposits to reveal base of growths at V = 36 kV.

cause these projections were deformed during the growth, being stressed by the field forces, which lie along the axis of each projection. These axes do not coincide in direction, so we get discrepancies in the positions of the atoms relative to the ideal unperturbed crystal, which results in structure defects.

## Literature Cited

1. E. W. Muller, Z. Physik, 126:642 (1949).
2. M. Drechsler, Z. Elektrochem., 58:334 (1954).
3. M. Drechsler and R. Vanselow, Z. Kristallog., 107:162 (1956).
4. V. N. Shrednik, G. A. Odisharia, and O. L. Golubev, J. Crystal Growth, 11:249 (1971).
5. I. L. Sokol'skaya, Zh. Tekh. Fiz., 26:1177 (1956).
6. M. Drechsler, Z. Elektrochem., 61:48 (1957).

# STRUCTURE AND DESORPTION ENERGY FOR MONOMOLECULAR BARIUM OXIDE FILMS

T. A. Tumareva and T. S. Kirsanova

*Kalinin Polytechnical Institute, Leningrad*

We have examined the formation of films ranging in thickness from a single monolayer up to some dozens of layers on points in field-emission microscopes, particularly as regards the effects of deposition conditions (deposition rate and amount deposited), as well as the effects of point orientation and temperature. The film structure is related to the desorption characteristics. We chose the W + BaO system because this provides high-contrast field-emission pictures and also conditions favorable to examining nucleation in BaO in the microscope [1]. Barium oxide is an efficient emitter, so Becker's method can be used to examine desorption [2]. The sensitivity is high, and the emission current alters by an order of magnitude when the thickness of the coating changes by only 0.1 monolayer.

Figure 1 shows a series of patterns for various substrate temperatures T°K and film thicknesses (or degrees of coating $\Theta$ as measured in monolayers). The temperature determines the thickness at which an ordered structure appears, so far as this can be judged from such patterns.

Figure 2 shows the next series of patterns recorded at various substrate temperatures and deposition rates; the structure was deduced from these pictures.

At comparatively low temperatures (curve 1, 300°K), the film grows by formation of subcritical nuclei; the structure is finely divided, no matter what the deposition rate. The linear dimensions of the subcritical nuclei on (110) face are estimated as 200-300 Å and are virtually unaltered up to $\Theta = 10$.

At higher temperatures (650-850°K, curve 2), the picture is different, and we conclude that a film grows from critical nuclei, with the size of the latter ranging from 300-400 for $\Theta$ around 3 up to 1000 Å for $\Theta$ around 5.

Finally, at even higher temperatures and deposition rates that are not too high one obtains patterns representing single-crystal films of BaO on W (curve 3). Such a film is produced only after fairly prolonged deposition. A distinctive feature of the picture is that there are eight dark spots of the same shape around the [100] pole (no such pattern has ever been observed for a clean tungsten point). This pattern is not accompanied by an increase in the emission, so a smooth single-crystal film of the adsorbate must be produced.

The structure of the film is dependent on the amount deposited and on the substrate temperature; the thickness or degree of covering $\Theta$ is defined as $t_d/t_T$, where $t_d$ is the deposition time and $t_T$ is the time needed at substrate temperature T to produce a monolayer.

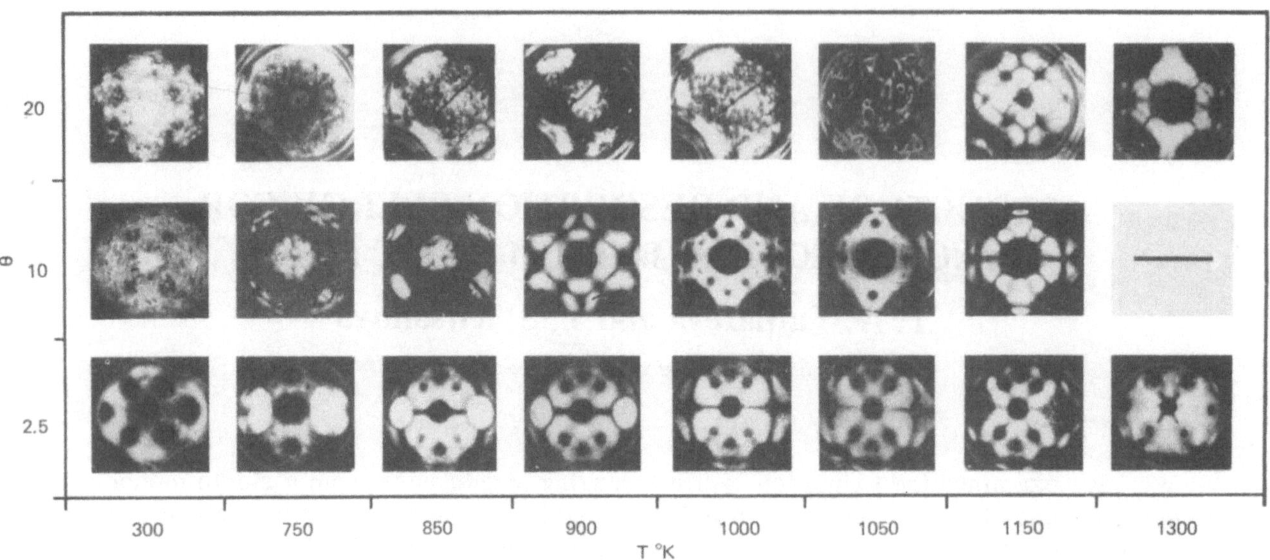


**Fig. 1.** Field-emission images of barium oxide films on tungsten at various thicknesses (degrees of covering) and various temperatures.

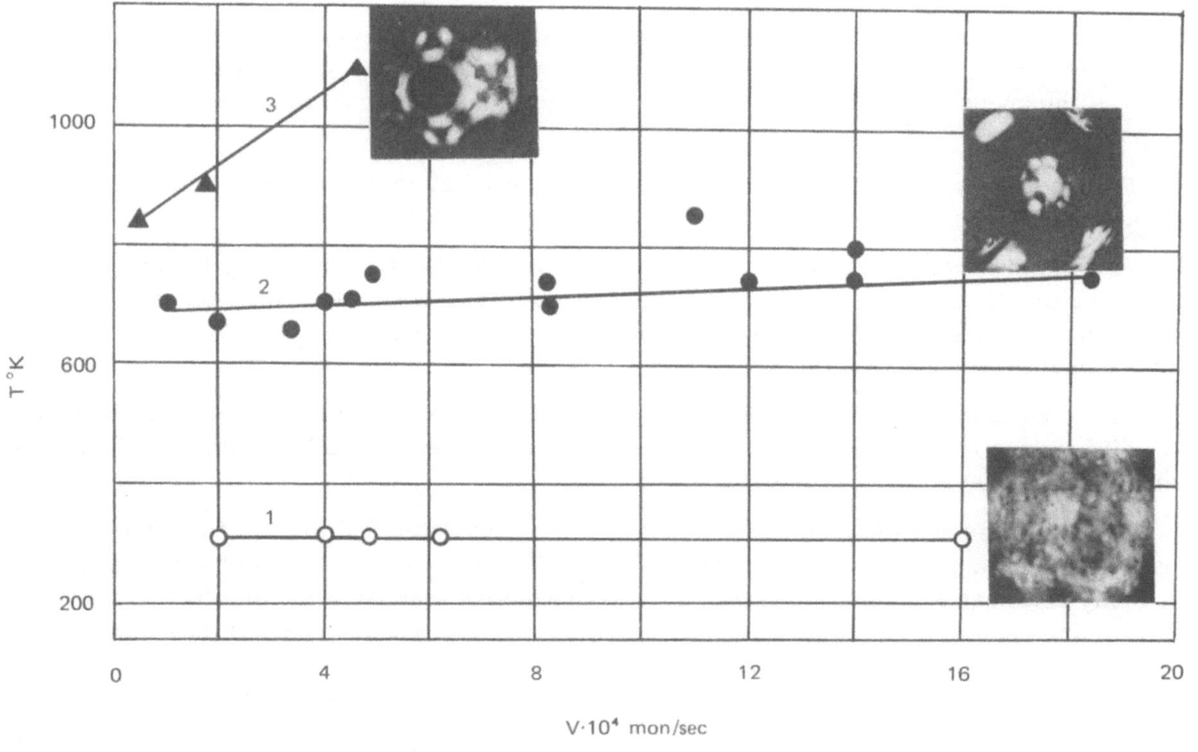


**Fig. 2.** Structure of barium oxide films on tungsten at various temperatures for various deposition rates.

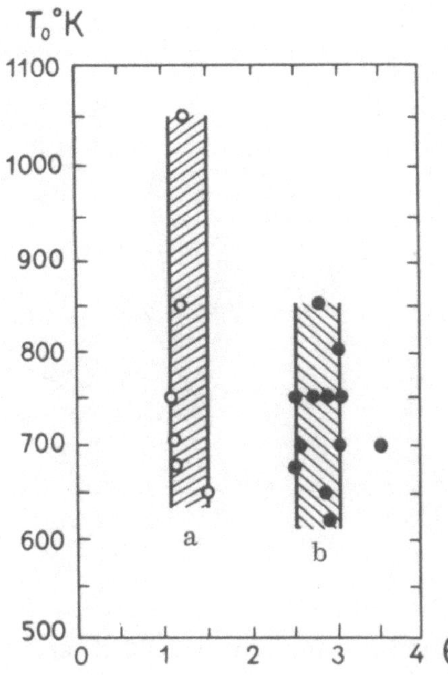


Fig. 3. Production of barium oxide crystallite in relation to thickness and substrate temperature during evaporation: (a) (112) face; (b) (110) face.

Figure 3 shows the occurrence of growing nuclei (crystallites) for various $T_0$ and Θ; the points correspond to various deposition rates. Crystallites are seen on (110) faces only for Θ > 2.5, whereas for (112) faces they are seen for Θ > 1; the temperature range that resulted in films (300-1100°K) was examined in some detail (by steps of 50°), which serve to define closely the ranges in which the particular states occurred: BaO crystallites on (110) are at 650°850°K, as against 650-1000°K on (112).

The critical thickness as a function of deposition rate (i.e., incident flux) was determined by measuring the time t* required to produce stable nuclei; the linear result for $1/t^* = f(v_d)$ (where $v_d$ is the deposition rate of BaO) at 650, 750, and 800°K indicates that the critical covering for onset of nucleation is independent of the deposition rate [3]. This critical covering is given by the slope as about three monolayers. A similar result has been reported [4] for silicon monoxide.

Our results on the growth of BaO appear to indicate that the film grows by fusion of subcritical nuclei; the process appears to occur as follows: the subnuclei are always produced on depositing the material on the point, and under certain conditions (T > 650°K, Θ > 2.5) it becomes favorable for the nuclei to fuse into larger structures, which are clearly seen in the microscope. This occurs when a certain critical concentration is attained, and this is seen in the image as a sharp rise in the emissivity. The model also does not conflict with another observation, namely that barium oxide crystallizes preferentially on closepacked faces and surrounding steps. On such parts of the surface, the bond from the adsorbate to the substrates is weaker than on rougher areas, so one naturally expects that particles will join up into larger clumps on (110) and (112) faces.

Nothing has been published on the heats of desorption for these various films; however, we have been able to correlate the desorption characteristics of the barium oxide on (110) faces of tungsten with the film structure for coatings thicker than a monolayer. Figure 4 shows the amount of evaporated material in terms of monolayers as a function of temperature for the BaO–W (110) system. The films were heated successively to higher temperatures in the range 800-1600°K, and at each temperature only part of the material evaporated although the heating was reasonably prolonged. For instance, in one case the initial thickness of the film was Θ =

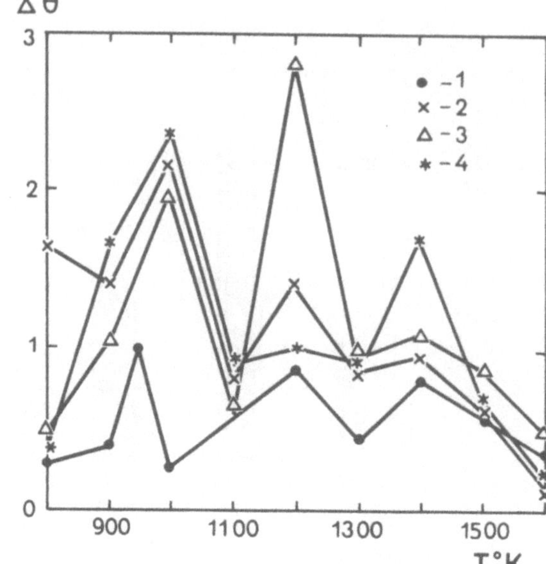


Fig. 4. Amount of evaporated material in terms of monolayers as a function of heating temperature, for initial Θ of: (1) 6; (2) 10; (3) (13); (4) 15.

10, and 1.5 monolayers were evaporated at 900°K, and two further at 1000°K, and so on. Figure 3 shows that the evaporation curve consists of a series of straight lines, with the 1000, 1200, and 1400°K zones corresponding to different desorption phases.

A study has also been made [2] of the amount of evaporated material as per cent of the film thickness for a W(110) single crystal; at 1300-1400°K the film is most resistant to evaporation when it is about three monolayers thick, while a film of only two monolayers evaporates more rapidly. We determined whether these effects are due to chemical interaction between barium oxide and the tunsten by examining the BaO–Pt system, where no reaction is likely. Here again the heat of evaporation and the amount of evaporated material were substantially dependent on the film thickness [5].

These results were processed in terms of the reaction rate $\Delta\Theta/\Delta t = AT\Theta e^{-\lambda/kT}$, where $\lambda$ is the heat of evaporation, Θ is film thickness, and A is a constant. The heat of desorption was estimated from the equation and from the Arrhenius straight line over a narrow temperature range. The heats of desorption given by the equation were very much dependent on the temperature range used, and the result was regularly reproducible, since the structure of the film becomes more perfect at higher temperatures, so the resistance to evaporation improves. It proved difficult to estimate the heat of desorption from the slope of the Arrhenius line, since the results are represented essentially by broken straight lines. A negative slope for $\log[(\Delta\Theta/\Delta t)/T] = f(1/T)$ was found for the region of minimal evaporation, which means that temperature elevation does not facilitate desorption but in fact hinders it.

The following results for desorption in the BaO–W(110) system were obtained: for the low-temperature phase (800-1000°K) $\lambda \approx 2.7$-$3.3$ eV (from the equation), or $\lambda \approx 0.9$ eV (from the slope of the Arrhenius line), for the medium-temperature phase (1100-1200°K) $\lambda \simeq 3.8$-$4.2$ eV (from equation) or $\lambda \simeq 1.5$ (from the Arrhenius line), and for the high-temperature phase (1300-1500°K) $\lambda \simeq 4.4$-$5$ eV.

The Arrhenius lines gave substantially lower heats of desorption than did the equation, which is probably because it is incorrect to use Arrhenius's method here because the film structure is very much dependent on the temperature.

These field-emission photographs showed that the transition from a fine-grained or amorphous structure to an insular (crystalline) one occurs at 600-800°K, so the first (low-temperature) phase corresponds to a crystalline structure, and the evaporation occurs from a film consisting of crystallites. After the film has been heated to 1000-1100°K, the pattern (Fig. 1, $\Theta$ = 10) indicates that a single-crystal film is formed, which develops around the (100) faces of W, and this structure persists on heating the film in the range 1000-1200°K, since in that range there is a sharp fall in the amount of barium oxide evaporated from a (110) surface of W. A single-crystal epitaxial film of BaO is formed at $T_0 \sim 1000$°K, since Gorodetskii found a similar structure in low-energy electron diffration [6]. Therefore, the second phase of desorption (~1200°K) is due to evaporation of barium oxide from the single-crystal film. The marked fall in the amount of material evaporating at even higher temperatures (1300°K) is due to the reaction between barium oxide and tungsten [7]. The increase in evaporation around 1400°K and the fall in the work function are probably due to the production fo free barium, which activates the film.

In conclusion we may say that the change in film structure, and in particular the transition from the crystalline (insular) structure to a single-crystal one, is reflected in considerable change in the heat of evaporation.

## Literature Cited

1. T. S. Kirsanova and T. A. Tumareva, Proceedings of the Conference on Electronic Technique, Series Technology of Electrical Vacuum Devices [in Russian], Vol. 9 (15), p. 85, Elektronika, Moscow (1969), Fiz. Tverd. Tela, 11:1441 (1969).
2. A. R. Shullman, G. S. Kirsanova, and A. A. Osinin, Trudy LPI im. M. I. Kalinina, No. 311, 63 (1970).
3. R. Gretz and G. M. Pound, In: Condensation and Evaporation of Solids, New York (1964), p. 575.
4. I. L. Sokol'skaya and S. A. Shakirova, Fiz. Tverd. Tela, 13:319 (1971).
5. A. R. Shul'man, T. S. Kirsanova, L. N. Karpova, and V. M. Belova, Zh. Fiz. Khim., 45: 2288 (1971).
6. D. A. Gorodetskii and Yu. P. Mel'nik, Izv. Akad. Nauk SSSR, Ser. Fiz., 33:461 (1969).
7. J. A. Cope and E. A. Coombes; In: Efficient Thermionic Cathodes, Vol. 4 [Russian translation], Izd. Energiya, Moscow (1964), p. 166.

Part II

# GROWTH KINETICS AND SURFACE MORPHOLOGY

# COMPUTER MODELING OF CRYSTAL GROWTH PROCESSES

**Kenneth A. Jackson**

*Bell Laboratories, Murray Hill, New Jersey*

Computer modeling is capable of providing powerful insights into crystal growth processes. In this paper, some of the recent work in this area will be reviewed. This is an active and rapidly expanding field of activity. The number of papers on this topic in print is likely to double within the next year or so, and this review may be premature. But I hope to indicate below where this subject is going as well as where it has been.

There are two basic types of computer calculation which are relevant to crystal growth processes. One of these is the analysis of the statistical processes involved in crystal growth. The other involves the calculation of the positions and energies of atoms or molecules in various clusters or configurations. In this paper I will discuss the former type of analysis.

The statistical analyses treat the crystal in what is known as a quasi-chemical approximation. The atoms or molecules are assumed to occupy lattice sites, and to have energies which depend on the occupancy of neighboring sites. In the real world, of course, the atoms are in thermal vibration, and do not possess fixed energies; but rather, atoms in a certain type of configuration have some energy on average.

The models which are treated in the computer are extensions of the simple model of the surface of a crystal which contains adatoms, clusters of atoms, holes, ledges, kinked ledges. The interactions between these various configurations can be examined in a much more realistic manner in a computer, simply because more configurations can be treated. The growth rates and growth anisotropies of crystals have been discussed qualitatively for many years in terms of these models. The basic assumption underlying these models has been that the anisotropies observed in nature reflect the structure and geometry of the surfaces of the crystals. And the basic goal of the computer modeling is to relate the geometry and structure of the surface to the observed anisotropies. There is considerable justification for this point of view. The large anisotropies observed during crystal growth cannot be accounted for in terms of the directional properties of the atoms or molecules which make up the crystal (as many of the directional properties can) but rather must depend on the cooperative nature of the crystal growth process. For example, the growth rate of naphthalene from the melt differs by as much as a factor of 1000 in two different directions, and the growth rate of ice, by a factor of 100 or more. But none of the other properties of naphthalene or ice is so anisotropic. This means that these differences in growth rate are associated with cooperative processes such as surface nucleation, which take place at the crystal surface. These depend on the structure and geometry of the surface, and to a lesser extent, on the details of the bonding. The computer modeling of crystal growth processes using statistical techniques has the major goal of elucidating the link between surface structure and growth, and indeed, making quantitative predictions of growth characteristics based on the models.

The basic problem is the one to which Burton, Cabrerra, and Frank [1] addressed themselves in the third part of their famous paper on screw dislocation growth. The first part of that paper contains the classic description of the screw dislocation mechanism. The second part deals with cooperative processes in a monolayer, and applies the exact Onsager solution to the two dimensional Ising model. This treatment suffers from the fact that it treats only a monolayer. The third part of that paper attempts to treat the multilevel interface. But that was before the time of computers, and the problem was too complex.

In this paper I will outline the advances which have been made in treating this problem. Most of these have involved extensive use of the computer. I will concentrate on the treatment of single component crystals and will not discuss the outstanding work which has been done by Chernov [2] and others on computer modeling of binary alloys and chains.

There are two basic approaches which have been taken to this problem. One is to use a combination of analytical methods and the computer. The analytic methods are used to describe the model in the form of equations which can be solved in the computer. There have been various approaches, involving greater or lesser use of the computer, and it is encouraging to note that the various methods give similar results. The second class of methods, generally known as Monte Carlo methods, involve direct computer simulation of the growth process. The actual structure of a small segment of a crystal surface is modeled in the computer for some small period of time, and rules are provided by which atoms can jump. A particular atom will jump if a random number exceeds a calculated jump probability, hence the name Monte Carlo. The analytic methods raise doubts about how well the equations take the multiple-atom interactions into account, since these can be treated only approximately. The Monte Carlo methods contain uncertainties arising from the small sizes and short times which can be treated in the computer.

The computer modeling is attempting to answer the question: if we know the chemical bonding and the chemical potentials of the phases at the interface, can the rate of motion of the interface be calculated? Until this question can be answered affirmatively, the roles of adsorption and of defects will not be understood properly. The answer to this question involves a detailed knowledge of the structure of the surface. At this time we have reliable models for the structure of the surface, good qualitative agreement between the structure and the growth characteristics, and in the simplest cases, quantitative agreement between calculated and experimental rates.

## Interface Structure

The first analyses of interface structure were based on one-layer models. The Langmuir [3] adsorption isotherm is one of these. Their treatment included the entropy of the adsorbed atoms but it ignored the interactions between adsorbed molecules. Fowler and Guggenheim [4] added an approximation to the interaction between adsorbed atoms. Burton, Cabrerra, and Frank [1] presented the exact solution to the one layer model. Jackson [5] used a one layer model similar to that of Fowler and Guggenheim to discuss the structure of the crystal–melt interface. For simplicity, let us briefly discuss this model. The free energy change produced by adding atoms to an initially plane face of a crystal is given by

$$\frac{\Delta F_s}{NkT_E} = -\frac{L\Delta T}{kT_E} x + \alpha x(l-x) + \frac{T}{T_E}[x \ln x + (l-x)\ln(l-x)], \tag{1}$$

where $\alpha = (L/kT_E)\xi$, N is the number of sites on the surface, kT has the usual meaning, $T_E$ is the temperature of equilibrium between the two phases, $\Delta T = T_E - T$, x is the fraction of sites which are filled, L is the change in binding energy per atom associated with crystallization, $\xi$ is the fraction of the total binding energy associated with the layer parallel to the interface. $\alpha$ is related to the adatom energy divided by the thermal energy.

The first term in (1) represents the free energy decrease due to adding Nx atoms to the crystal; the second term represents the free energy increase due to the incomplete bonding in the partially filled layer; the last term represents the entropy contribution due to random mixing of the atoms in the surface layer. At equilibrium, the first term is zero. For large $\alpha$, the free energy has minima given by

$$x = \exp(-\alpha) \quad \text{and} \quad x = 1 - \exp(-\alpha). \tag{1a}$$

This corresponds to a few adatoms and a few holes in the layer below. The number of adatoms and holes increases for smaller $\alpha$, and, for $\alpha$ smaller than 2 or so, it becomes apparent that the one-layer model is inappropriate. These features are also present in exact solution [1].

It is well established that $\alpha$ can be used to classify and predict crystal growth morphologies [6], but I will not elaborate on this aspect here.

The problem was then how to treat the multilevel case, and in particular, what to do about the entropy. Temkin [7] proposed a model for the multilevel surface. He assumed that the atoms in each layer were arranged in a random fashion on the atoms in the layer below. This permitted him to obtain expressions for the energy and entropy of each layer of atoms. The resultant change in free energy from an initially flat interface was given by a sum over all the layers (see Fig. 1),

$$\frac{\Delta F_S}{NkT_E} = \frac{L\Delta T}{kT_E^2}\left[\sum_{i=-\infty}^{0}(1-x_i) - \sum_{i=1}^{\infty} x_i\right] + \alpha \sum_{i=-\infty}^{\infty} x_i(1-x_i) + \frac{T}{T_E}\sum_{i=-\infty}^{\infty}(x_{i-1}-x_i)\ln(x_{i-1}-x_i), \tag{2}$$

where the index i labels a layer parallel to the interface. Each of the terms has the same significance as in (1); the first term gives the free energy contribution due to the net number of atoms added; the second term is the sum of the interaction energies in each layer, and the third term is the sum of the entropy contributions, where there has been some cancellation of terms in the sum. Far from the interface in the solid, $x_i = 1$ and in the liquid or vapor, $x_i = 0$ so that there is no contribution to the free energy due to layers far from the interface.

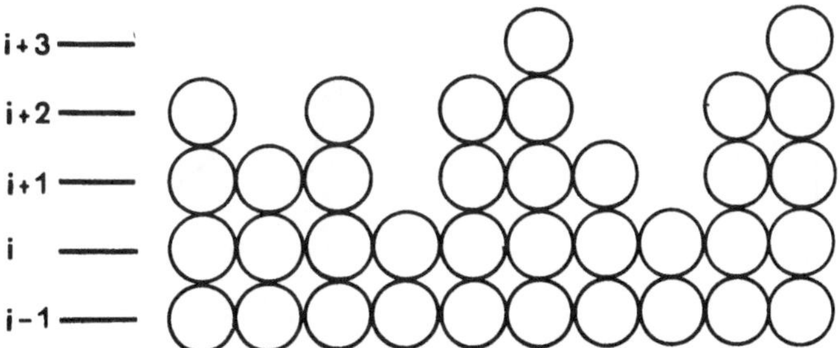


Fig. 1. Schematic view of the structure of an interface. The index i labels a plane of atoms parallel to the interface.

The minimum free energy is given by the recursion relationship

$$\frac{x_i - x_{i+1}}{x_{i-1} - x_i} = \exp\left[\alpha \frac{T_E}{T}(2x_i - 1) + \frac{L\Delta T}{kT_E T}\right]. \tag{3}$$

This recursion relationship can be used to calculate $x_{i+1}$ if $x_i$ and $x_{i-1}$ are known. Note for example that $x_{-1} = 1$, $x_0 = 1 - e^{-\alpha}$, $x_1 = e^{-\alpha}$, $x_2 = 0$ is the solution for $T = T_E$, $e^{-\alpha} \ll 1$, in agreement with the one-level model for large $\alpha$.

For the general case, using the computer, a set of $x_i$ can be determined subject to $x_i = 1$ at $i = -\infty$ and $x = 0$ at $i = +\infty$. Interface profiles calculated in this way are shown in Fig. 2.

This model does not take clustering on the surface into account. It assumes that the atoms in each layer are randomly distributed. This is the zeroth-order model, and is based on the Bragg–Williams [8] treatment of order–disorder. Clustering can be treated using various cluster expansions. The simplest of these is the 1st order, which is also known as the quasi-chemical model, and is equivalent to the Bethe–Peierls [8] treatment of order–disorder. The 1st order model of the interface was treated by Leamy and Jackson [9]. The basic equation is similar to (2), the only difference being that the second term is more complicated, and involves an order parameter $\beta_i$ for each layer. There are also several other auxiliary equations relating the $\beta_i$ to the pair probabilities and so on, but these need not concern us here. The entropy term is the same. This entropy term is equivalent to assuming that the only atoms in layer i which can evaporate are those which are not covered by atoms in layer i + 1. It is thus independent of the degree of clustering.

It is also possible to derive the interface profiles by writing down an expression for the net evaporation rate for each layer, based on the number of atoms with one, two, three, etc. nearest neighbors (these numbers can be expressed in terms of $x_i$ and $\beta_i$). In the computer some arbitrary profile is used to start, then atoms are redistributed according to the calculated net rates until all the net rates are zero. This process converges rapidly to the same interface profile calculated from the recursion relationship such as (3).

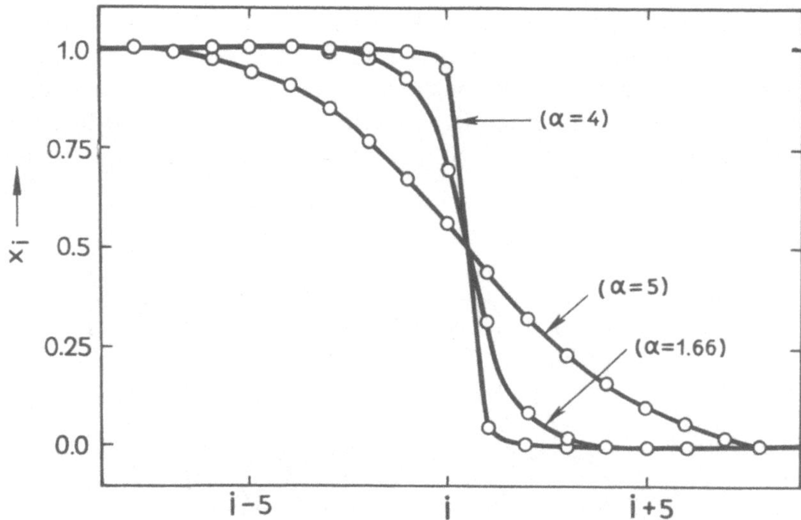


Fig. 2. Interface profiles calculated for the zeroth-order model, showing the fraction of sites filled in the various layers parallel to the interface.

Direct Monte Carlo simulation has also been used by Leamy and Jackson [9]. Here a replica of the crystal surface is stored in the computer. Atoms evaporate if a random number exceeds the evaporation probability for the atom (which depends on the number of neighboring atoms). The evaporated atoms are recondensed at random. Pair exchange schemes have also been used where atoms are interchanged according to certain rules, again using random numbers.

The interface profiles derived from zeroth-order, from first-order, and from Monte Carlo calculations are shown in Fig. 3. They are strikingly similar. The zeroth order does not include clustering, the first order includes pair probabilities, and the Monte Carlo treats clustering correctly, although there is some uncertainty in the result due to the finite sample size and the statistical nature of the calculation.

The surface roughness R is plotted in Fig. 4. R is defined as the number of unsatisfied bonds per site parallel to the surface. A single adatom would contribute four and so would a hole. The simple one-layer model, for which the number of holes and adatoms are each $e^{-\alpha}$, would predict $R = 8e^{-\alpha}$ as shown in Fig. 4. This is extrapolated far beyond the range of $\alpha \ll 1$, on which the calculation is based. It is surprising how little difference all the clustering and configurational complexity makes to the roughness of the surface. The one-level model (exact solution) has a roughness limited to one layer.

Figure 5 compares the roughness of $\{100\}$, $\{110\}$, and $\{111\}$ faces of a simple cube based on the zeroth-order model. The roughness is different for each face, but still rather similar.

The models analyzed above are all "column" models: atoms can evaporate or condense only from the top of each column. There can be no vacancies in the crystal. An atom cannot evaporate from or condense on the side of a spike: it can only sit on an atom in the layer be-

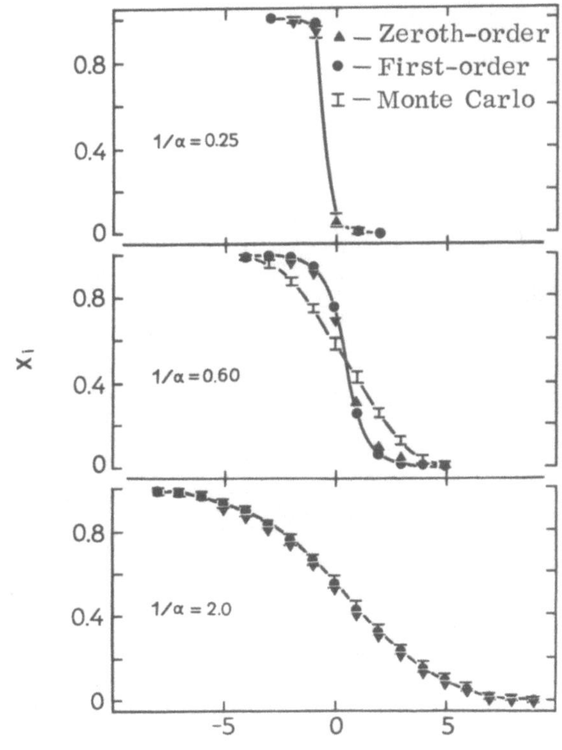


Fig. 3. Comparison of interface profiles for zeroth-order, first-order, and Monte Carlo calculations.

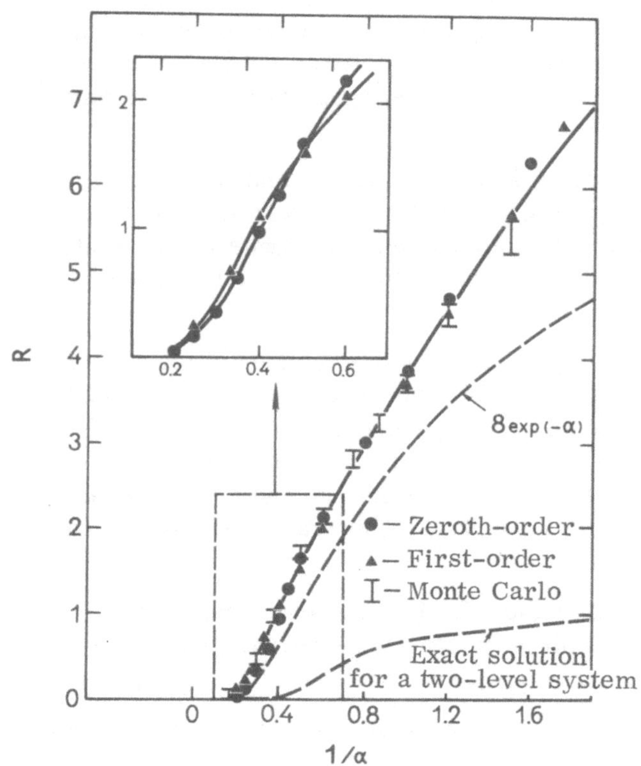


Fig. 4. Comparison of interface roughness R for various models of the interface.

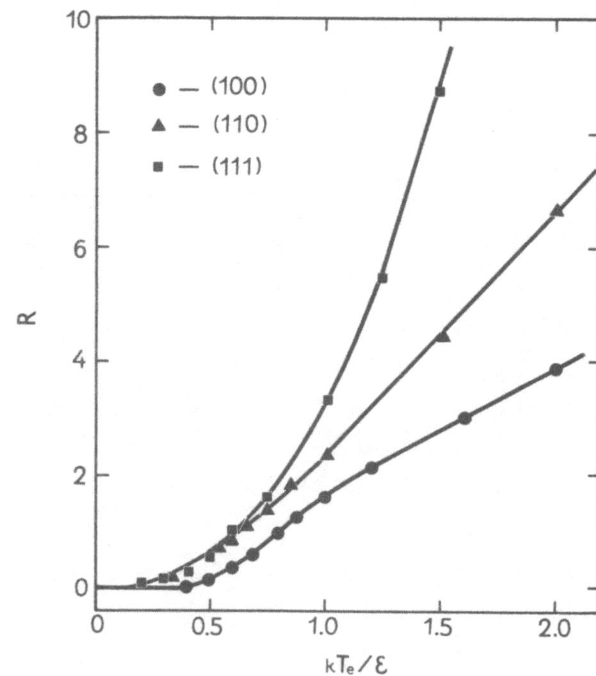


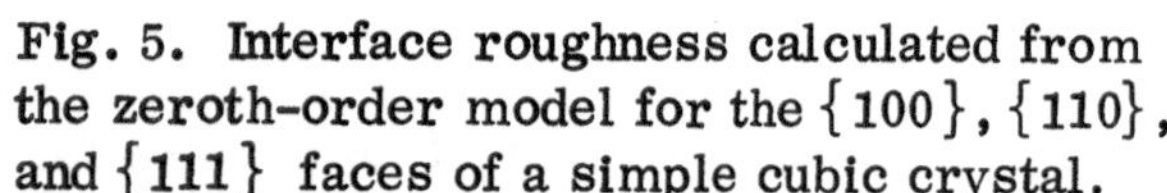

Fig. 5. Interface roughness calculated from the zeroth-order model for the {100}, {110}, and {111} faces of a simple cubic crystal.

low. Noncolumn models have not been treated analytically, but they have been examined using the Monte Carlo method. There are differences: the noncolumn high-index faces tend to break up into small segments of low-index faces. There are detailed differences depending on $\alpha$ and orientation. But the overall roughness of the surface is still quite similar.

Calculations have also been done using Monte Carlo methods for fcc crystals using both the column and noncolumn models, to obtain polar plots of surface roughness. The similarities in the surface roughness for all these calculations indicate strongly that the calculated surface configurations are realistic. There are significant differences between zeroth-order, first-order and Monte Carlo calculations for homogeneous order–disorder calculations, and as will be shown below, there are indeed important differences between these models for surfaces. But they all predict similar surface roughness.

It has been pointed out [10] that the decrease in surface free energy due to roughening results in a negative surface free energy if it is subtracted from the surface free energy calculated by counting broken bonds across a plane interface, for simple cubic crystals. In spite of the questionable nature of this calculation, it is worth noting that the same calculation for a face-centered cubic crystal does not give a negative surface free energy for any orientation, for values of $\alpha$ associated with real crystals.

## Calculation of Crystal Growth Rates

Crystal growth rates have been calculated by two different methods which are in fact equivalent. The method due to Jackson [11] requires calculations of the free energy of the interface for various positions of the interface relative to the lattice planes. This can be determined from equations such as (2). Once the $x_i$ have been determined, they can be inserted into (2) to obtain $\Delta F_s$. Such calculations have been performed (using Lagrange multipliers to fix the position of the interface with respect to the lattice planes). The free energies obtained for

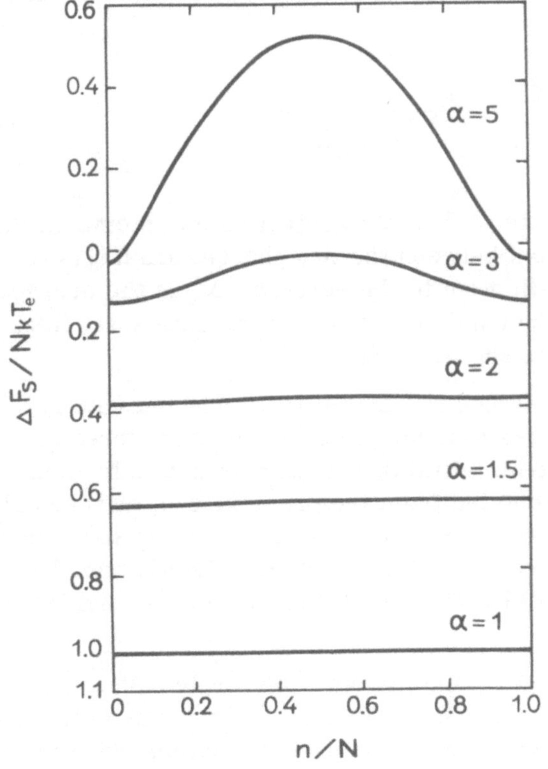


Fig. 6. Variation of interfacial free energy with position of the interface with respect to the lattice planes, for various values of $\alpha$, for the zeroth-order model.

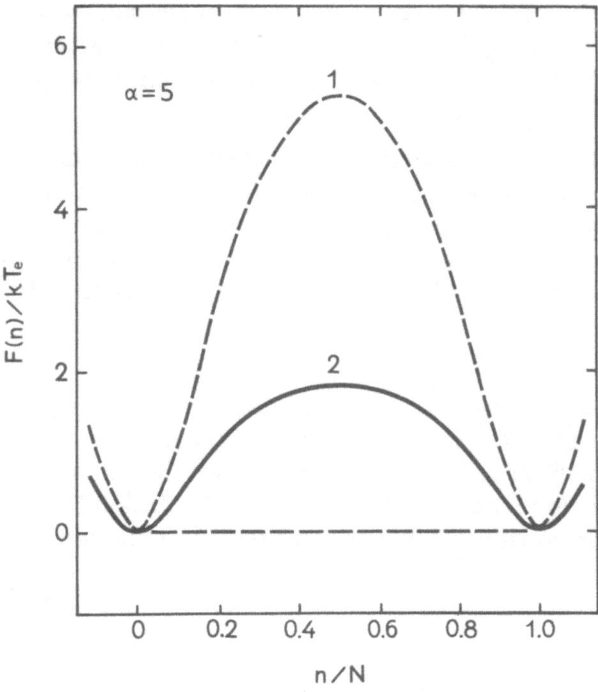


Fig. 7. Comparison of the variation of interfacial free energy with position for the zeroth- and first-order models. A similar ratio of about a factor of 3 in peak height was found between the two models for various values of $\alpha$.

the zeroth order are shown in Fig. 6, and are compared with the first-order model in Fig. 7. Although the minimum free energy in the two models (corresponding to the equilibrium position of the interface) is the same, the maximum free energy is different by about a factor of three. Clustering makes a big difference when a smooth (high $\alpha$) surface is half way between atom planes.

The growth rate can be calculated by taking the derivative of $\Delta F_s$ with respect to interface position and inserting it into

$$V = \frac{P^{+}N[1-\exp(-N\Delta f_0/kT)]}{N+\sum_{n=0}^{N-1} p_n + \sum_{n=0}^{N-1} p_n p_{n+1} + \ldots + \sum_{n=0}^{N-1} p_n p_n+1 \ldots p_{n-2}} \tag{4}$$

where $p_n = \exp(-\Delta f_n/kT)$, to obtain V, the growth rate. $P^+$ is the arrival rate of atoms on the surface, $\Delta f_0$ is the difference in free energy per atom between the two phases, and $\Delta f_n$ is the change in free energy associated with adding the n-th atom to the surface: $\Delta f_0$ is the average value of $\Delta f_n$. This method of calculation permits the number of sites on the substrate, N, to be varied, which has important consequences as will be shown below.

The other method of calculating growth rates is due to Temkin [12]. The net rate at which atoms are added to each layer can be expressed in terms of the number of atoms in each layer, and the order parameters for each layer. Some initial distribution of atoms in each layer is assumed and then a new distribution is calculated from the previous one, and the calculation is then repeated. The distribution of surface atoms soon settles down. In some cases, no growth occurs, in others, there is steady-state motion of the interface. Calculations have been done by this method for the zeroth-order model by Temkin [12] and by Leamy and Jackson [13] for the first-order model.

These two methods, one based on thermodynamics and the other on a rate equation approach, seem quite different, but in fact are based on the same models and give the same results. Indeed, for equilibrium the rate equations reduce to the recursion relations such as (3).

In the direct rate equation method, the interface fluctuates in speed with the period of the lattice as it moves from plane to plane. The actual growth rate is determined from the time interval required to move a repeat distance. Using (4), these fluctuations are buried in the mathematics: the p factors are reciprocally related to the instantaneous growth rate. The rate calculated from (4) gives directly the average time to move a repeat distance.

The results of such calculations are shown in Fig. 8. These growth rates were calculated from (4), but similar results are obtained by the Temkin method. The broken lines are the zeroth-order model, the solid lines are the first-order model. The calculations were made for large N. The free energy curves used in the calculations for the two models are different (Fig. 7) and so it is not surprising that the growth rates are different. The uppermost curve represents the limiting growth rate, where there is no barrier to the growth process. In this limit, which corresponds to a rough surface, the net growth rate for the crystal is just the sum of growth rates on each site. The growth rate is given by the classical theory [14]. There are no nucleation barriers to continued growth and defects are unnecessary. At small undercooling, the growth rate is linear with undercooling. This corresponds to the growth of metals or other low-entropy-of-fusion crystals growing from the melt.

The other growth rate curves show one feature in common: a region at small undercooling where no growth occurs. The reason for this can be seen most readily from (2). The second and third terms in (2) represent a periodic free energy barrier to growth, as shown in Figs. 6 and 7. The first term in (2) adds a negative term linear in number of atoms added.

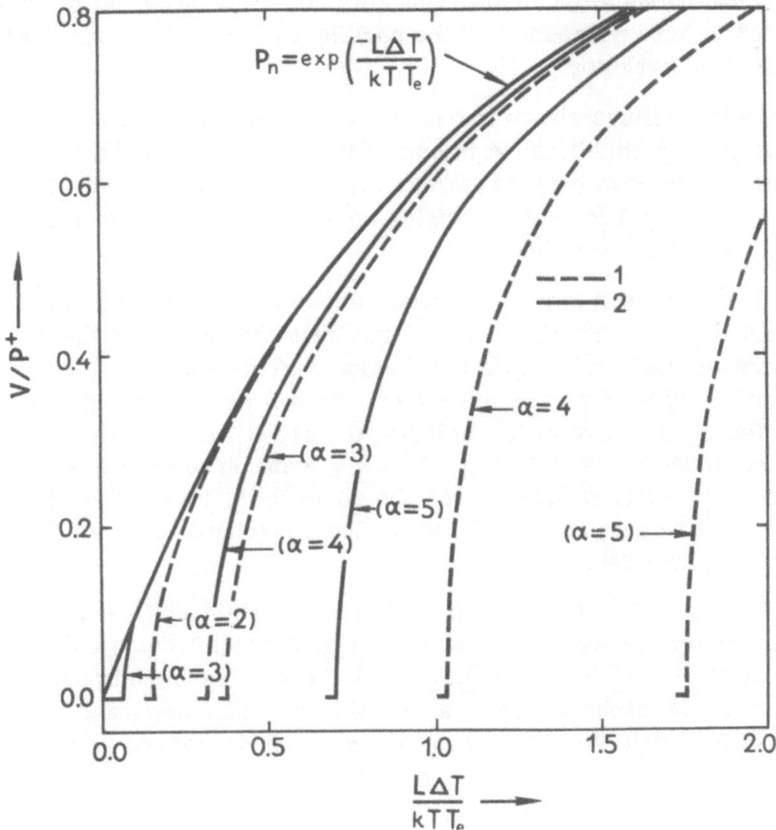


Fig. 8. Growth rates calculated from (4) for the zeroth- and first-order models. Various values of $\alpha$ for the two models are shown.

At small undercooling, there is a decrease in free energy on adding a layer, but the barrier, represented by the bump in Fig. 6, is still present, so no growth occurs. At large enough undercooling, the free energy decreases continuously as x goes from 0 to 1, so that growth can occur. When this occurs, the growth rate approaches the upper curve, or classical limit, in Fig. 8.

The analysis assumes that the interface remains planar during growth, and that the interface position can be described by a single parameter x. Because the interface is assumed to be very large in area, its energy is well defined, and there is no possibility that the interface can surmount even a small energy barrier by a fluctuation. This is, of course, not true in nature but is true in the analysis of the models because of implicit assumptions.

Thus, in Fig. 8, there is no growth until the undercooling is sufficient to overcome the barrier, so that the interface can move as a plane through the lattice. For both the zeroth-order models, the critical undercooling increases with $\alpha$. But the critical undercooling depends on the model as well. The free energy barrier is higher for the zeroth-order model than for the first-order model (Fig. 7), and so the critical undercooling is correspondingly larger. The reason for this discrepancy is that the first-order model treats clustering better than the zeroth-order model. So it represents a first step in the direction of including nucleation effects. An nth-order model would treat nucleation much better, but the mathematical complexity increases rapidly with order. N. Kikuchi [15] has treated models up to sixth order for homogeneous alloys, and finds that the barrier continues to decrease with increasing n, and nucleation behavior is present for n = 6. The interface is much more complex to analyze than a homogeneous alloy, and so this is not a promising way to proceed. But we can anticipate by analogy with the homogeneous alloy results that for models treating larger clusters the barrier height would continue to decrease with increasing order of the model, to be replaced, at high order, by nucleation behavior.

The critical undercoolings shown in Fig. 8 have no significance in predicting crystal growth rates. They do have significance in one sense. The zeroth-order model indicates the maximum effect the barrier can have on growth rates. It therefore indicates the limiting behavior. The interface can go around the barrier by various mechanisms, but the barrier to plane front growth is there in principle.

The theory which has been reviewed briefly above is similar in principle to that outlined by Cahn and co-workers [16]. Temkin [17] has recently extended this theory. Cahn predicted the presence of a critical undercooling in the theory of growth, and divided the growth curve into two regions, with a transition region between. At low undercoolings, growth would be by nucleation or on defects. At large undercoolings, the growth rate would approach the classical growth rate curve (as indeed it does in Fig. 8). Cahn went on to derive, on the basis of a thermodynamic model, a relationship between the roughness, or width of the interface and the critical temperature. However, as we have seen, the roughness does not depend on the model and the critical temperature does.

So we are left with the basic problem: which is the right critical temperature to use? Or indeed, does the critical temperature have any significance? For small $\alpha$, a critical temperature is not predicted by any of the models. For large $\alpha$, growth by nucleation or on defects dominates over most of the experimentally accessible undercooling range. In between, a theory which modifies classical nucleation theory to take account of surface roughness is needed. The nucleation process is obviously an inherent and intrinsic part of the growth process, and should not be added on after the fact as is necessary in the present treatments.

The role of nucleation, as well as some of the difficulties associated with direct Monte Carlo simulation of the growth process, show up in the dependence of the growth rate on the size of the surface in (4). Figure 9 shows growth rates calculated from (4), using the zeroth-order model, for $\alpha = 2$. For $N \geq 10^6$, the break in the curve is evident, as shown also in Fig. 8.

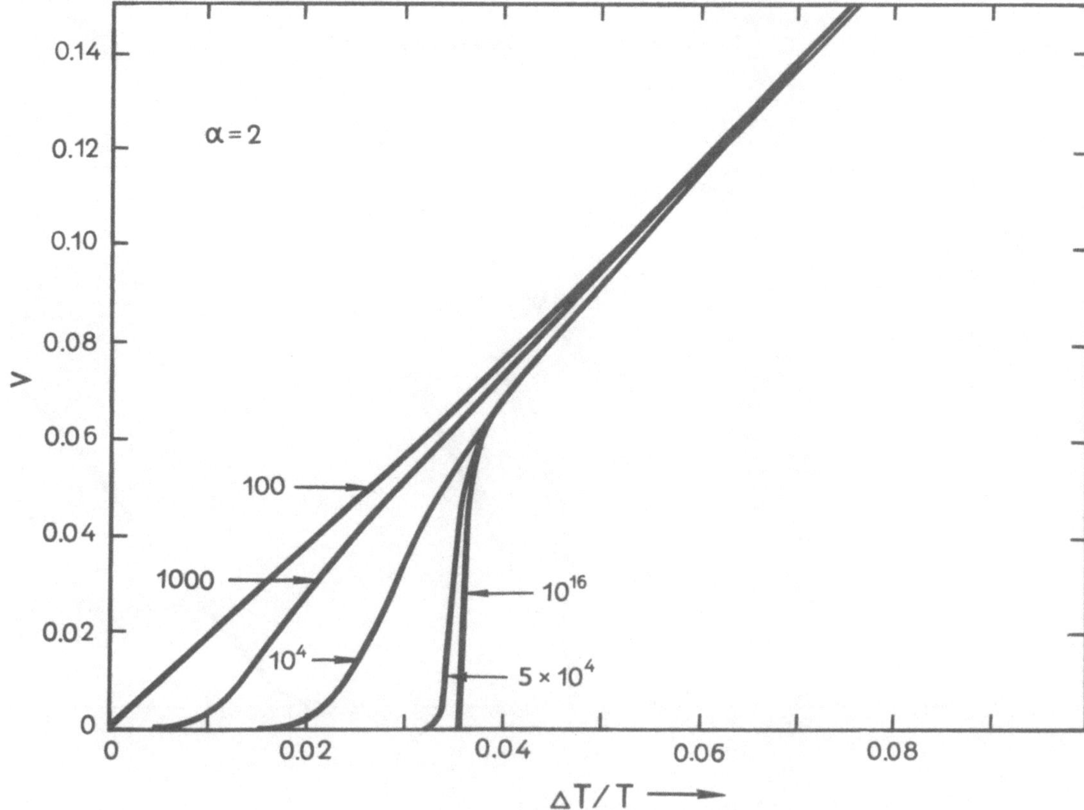


Fig. 9. Effects of surface size N on the growth rate given by (4) for the zeroth-order model.

For $N \leq 100$, the break is not present. The reason for this result is evident from (4). The sharpness of the melting point is defined by the N in the exponent in the numerator of (4). If the surface is large, the melting point is well defined. If N is small, the average energy of the surface atoms has some statistically uncertainty, as does the melting point. When the statistical uncertainty of the energy of the surface atoms is greater than the energy barrier to the motion of the interface, then the barrier will not impede interface motion.

The data of Fig. 9 are replotted in Fig. 10, assuming that the N atoms are on the tip of a whisker, and that the equilibrium temperature of the tip is depressed below that of an infinite interface. A nominal value of the surface free energy based on the bond energy was used to calculate the depression in the equilibrium temperature. It is evident that with this correction, the large interface grows most rapidly at larger undercooling. However, there are regions where a smaller surface can grow more rapidly than a large one, shown for example by the crossing of the curves labelled $10^4$ and $5 \cdot 10^4$. To the left of the break, the $10^{16}$ curve predicts negligible growth; the $10^4$ curves shows small, but finite growth in the region to the right of its melting point at $\Delta T/T = 0.02$. A large surface can be subdivided into smaller areas. Each of these could grow faster at small undercooling, but at the expense of creating more surface area. And this sounds a lot like surface nucleation. As was stated above, the curves in Fig. 10 were shifted an amount based on an approximate bulk surface energy, which is a maximum value. The edge energy of a disc on a surface is likely to be considerably less than this, so the curves have probably been shifted too far. Calculation of the proper edge free energy on a partially rough surface is needed to determine the proper amount to shift the curves. The growth rate would then be the upper envelope of such curves. But this same edge free energy can be used in con-

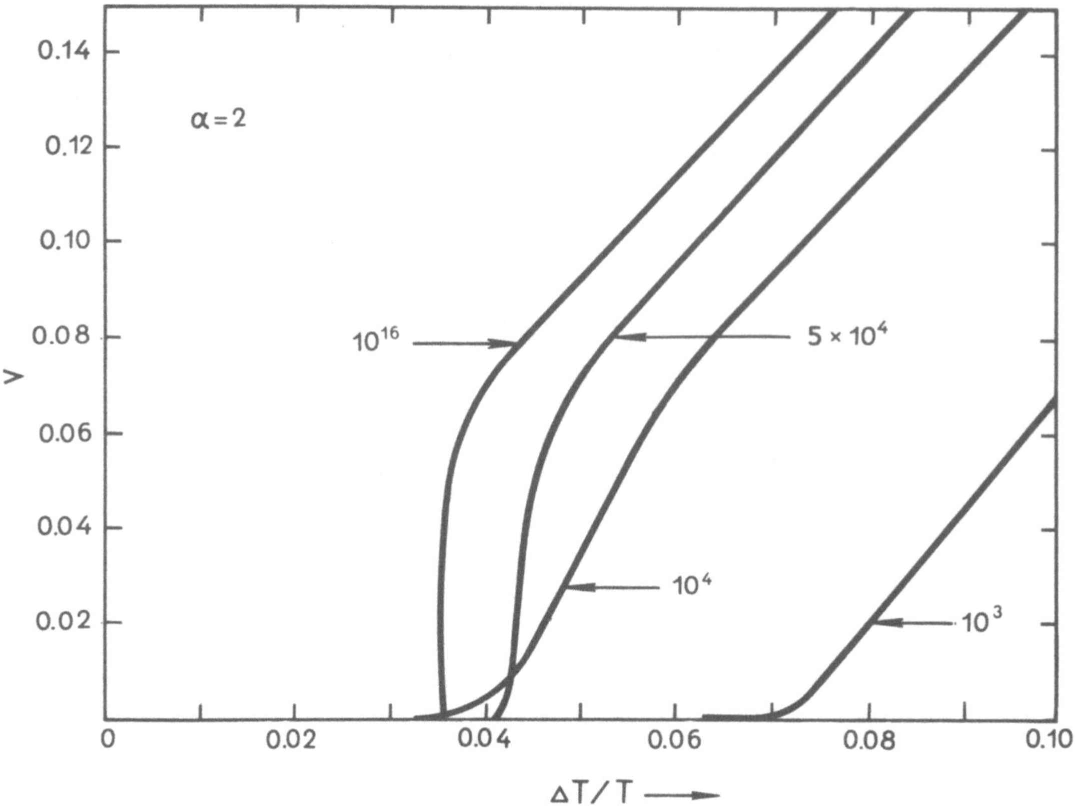


Fig. 10. The curves of Fig. 9 replotted, taking into account the effect of finite surface.

ventional surface nucleation theory. It is possible that these two approaches will prove to be different but equivalent ways of treating the same problem.

The other consequence of Fig. 9 is its implications for Monte Carlo calculations. These are typically done using 20 × 20, 30 × 30, or 40 × 40 arrays. Computer costs place a stringent upper limit here: 100 × 100 arrays are 10 times more expensive than 30 × 30 arrays. Figure 9 shows that none of these correspond to the large surface limit, and so the sharp break would not be observed even if it were present. The first-order model predicts a break at a smaller undercooling, and so still larger surfaces would be required to show the break in the growth curve. Since we don't know where the break should be, Monte Carlo calculations on small surface areas must be viewed with suspicion.

This has been confirmed by Monte Carlo calculations for small $\alpha$ [18]. The Monte Carlo calculations show statistical uncertainly in the position of the interface, which fluctuate back and forth as the calculation proceeds. Its roughness does not change significantly, but the net motion is masked by the fluctuations.

For large $\alpha$ this problem should not arise. Provided that the substrate is large enough so that a statistical sampling of clusters of critical size is present, the Monte Carlo calculations should treat the nucleation process properly. For simulation of nucleation from the vapor phase, Adams and Jackson [19] were able to obtain agreement with the time dependence of various cluster sizes as predicted by Zinsmeister [20]. Differences were noted between the analytic and Monte Carlo results, which were attributed to approximations in the analytic treatment. The computer can be used to generate motion pictures to illustrate the surface

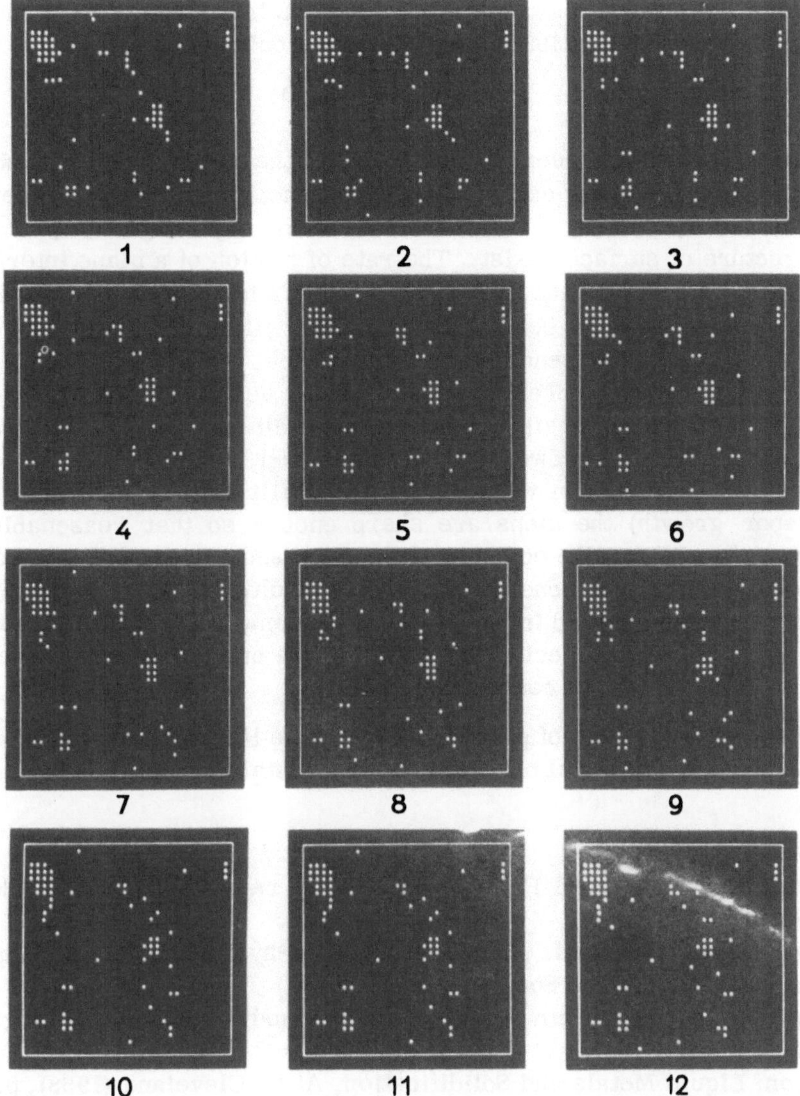

Fig. 11. Frames from a computer-generated movie, showing clustering effects and the motion of atoms on a substrate. The motion of individual atoms can be followed from frame to frame.

nucleation process, as shown in Fig. 11, which contains several frames from such a movie. The motion of individual atoms and their clustering can be followed from frame to frame.

In the intermediate range of $\alpha$ between, say, 2 and 5, there is some uncertainty about the size of surface needed to obtain significant results. Some calculations have shown too much statistical scatter, others have not. It is not clear at this time whether Monte Carlo calculations will converge to the intermediate curves of Fig. 9, or whether such curves are necessarily masked by fluctuations. Gilmer and Bennema [21] have made extensive Monte Carlo calculations for $3.5 \le \alpha \le 5$, simulating solution growth. The fact that their results do not show large statistical scatter argues that their surface area was large enough. Their results can be fitted well to nucleation theory. The calculated growth rate increases when steps are introduced on the surface. They also observed that surface diffusion increases the growth rate in a manner consistent with theory. This means that their surfaces are not in equilibrium configurations, even though the roughness that they observed was similar to other calculations.

More work of this kind is needed, particularly for the smaller values of $\alpha$ to tie these results in to the size dependence of the statistical mechanics calculations.

## Conclusions

Computer calculations have been of great value in the recent development of crystal growth theory. Computer modeling can be used in conjunction with analytic methods to facilitate the analysis of various models of interfaces and of the crystal growth process. Reliable models for the structure of surfaces exist. The rate of motion of a plane interface has been calculated for several models, but the motion of nonplanar interfaces involves the structure of steps on the surface which have not been treated in detail. Both the nucleation of new steps and the motion of existing steps depends on their structure, and a detailed understanding of crystal growth depends on these processes. Calculations dealing with the structure of steps on surfaces are proceeding. For small $\alpha$ (e.g., metal−melt interface), interface is rough and so steps are unimportant. The growth rate is rapid and isotropic. Here, the computer modeling has done little more than verify the applicability of classical theory. For large $\alpha$ (e.g., most vapor growth) the steps are sharp enough so that reasonable modeling can be done using small areas. The computer has been used successfully, but only to a limited extent. Much more remains to be done. For example, multilevel islands, adsorption, and the role of defects have not been treated in detail, although some initial attempts have been made. Computer simulation will prove especially valuable where analytic methods become intractable, as is frequently the case with these complex problems.

The computer seems capable of providing a valuable link between our models of the growth process and the experimental observations of crystal growth in nature.

## Literature Cited

1. W. K. Burton, N. Cabrera, and F. C. Frank, Phil. Trans. Roy. Soc. (London), 243A:299 (1951).
2. A. A. Chernov, Proc. Int. Conf. Crystal Growth, J. Phys. Chem. Solids, Suppl. 1, 25 (1967).
3. I. Langmuir, J. Amer. Chem. Soc., 40:1361 (1918).
4. R. Fowler and E. A. Guggenheim, Statistical Thermodynamics, Cambridge University Press (1965), p. 430.
5. K. A. Jackson, Liquid Metals and Solidification, ASM, Cleveland (1958), p. 174; Growth and Perfection of Crystals, Doremus et al., eds., Wiley, New York (1958), p. 319.
6. K. A. Jackson, Progress in Solid State Chemistry, H. Reiss, ed., Vol. 4 (1967), p. 53.
7. D. E. Temkin, In: Crystallization Processes, N. N. Sirota et al., eds., Consultants Bureau, New York (1966), p. 15.
8. K. Huang, Statistical Mechanics, Wiley, New York (1967), p. 329.
9. H. J. Leamy and K. A. Jackson, J. Appl. Phys., 42:2121 (1971).
10. D. Nason and W. A. Tiller, J. Crystal Growth, 10:79 (1971).
11. K. A. Jackson, J. Crystal. Growth, 5:13 (1969).
12. D. E. Temkin, Kristallografiya, 14:417 (1969); Soviet Phys. – Crystallography, 14:344 (1969).
13. H. J. Leamy and K. A. Jackson, in press.
14. H. A. Wilson, Phil. Mag., 50:238 (1900); J. Frenkel, Phys. Z. Sovjet Union, 1:498 (1952).
15. R. Kikuchi, J. Chem. Phys., 47:1664 (1967).
16. J. W. Cahn, Acta Met., 8:554 (1960); J. W. Cahn, W. B. Hillig, and G. W. Sears, Acta Met., 12:1421 (1964).
17. D. E. Temkin, Kristallografiya, 15:877 (1970); Sov. Phys. – Crystallography, 15:767 (1971).
18. D. deFontaine and K. A. Jackson, unpublished.

19. A. C. Adams and K. A. Jackson, J. Crystal Growth, 13/14:144 (1972).
20. G. Zinsmeister, Vacuum, 16:529 (1966); Thin Solid Films, 2:497 (1968); ibid. 4:363 (1969).
21. G. Gilmer and P. Bennema, J. Crystal Growth, 13/14:148 (1972).

# MONATOMIC LAYER PROPAGATION RATE AND THE ELECTROLYTIC DEPOSITION MECHANISM FOR SILVER

V. Bostanov, R. Rusinova, and E. Budevski

*Central Electrochemical Current Sources Laboratory, Bulgarian Academy of Sciences, Sofia*

The faces representing the equilibrium shape of a crystal grow by layers, namely by growth-front propagation. Crystal-growth theory envisages two mechanisms for formation of such fronts: (a) on the dislocation-free faces of a step by two-dimensional nucleation, and (b) via a persisting spiral growth front if there is a screw dislocation, which gives rise to a spiral step.

When a front has arisen, it propagates, and the speed of propagation largely determines the crystal growth rate. There are two major step propagation mechanisms: (a) the atoms enter growth positions after surface diffusion, and (b) the atoms enter the growth position directly.

Electrochemical deposition and dissolution are topics closely related to crystal growth theory, and the electrolytic growth of a crystal occurs with the electrode potential deviating from the equilibrium value, the deviation being termed an overvoltage, which is similar to the supersaturation. The current density determines the growth rate.

If the growth kinetics may be controlled by surface diffusion, then the theory gives the following relationship between the current I and the overvoltage $\eta$ [1]:

$$I = i_{0,ad} 2L\lambda_0 [e^{\alpha zF\eta/RT} - e^{-(1-\alpha)zF\eta/RT}] \tanh(x_0/\lambda_0), \tag{1}$$

where $i_{0,ad}$ is the density of the adsorbed-atom exchange flux with the electrolyte, $\lambda_0$ is the surface-diffusion penetration depth, $x_0$ is half the distance between adjacent steps, $\alpha$ is a conversion factor, z is valency, and F is the Faraday number. The total step length L on a face is $(S/2x_0)$, where S is the geometrical area of the face.

The relation between the current and the overvoltage for direct entry to a growth site [2] is

$$I = i_{0,st} 2r_0 L [e^{\alpha zF\eta/RT} - e^{-(1-\alpha)zF\eta/RT}], \tag{2}$$

where $i_{0,st}$ is the current density representing the atomic flux to steps from the ions in solution and $r_0$ is the radius of an atom in the lattice.

A major aspect of the kinetics and mechanisms of electrolytic deposition is the mode of lattice construction.

Here we examine the speed of a monatomic step in relation to face growth rate in the presence of dislocations and the mode of entry of atoms into growth sites for electrolytically deposited silver.

## 1. Methods

We used 6 N $AgNO_3$ solution and 45°C with electrodes whose surfaces represented isolated single-crystal silver faces. These electrodes were produced by a capillary method [3]. A silver single crystal was grown electrolytically, with the crystal gradually entering a glass capillary and filling the latter. The single-crystal filament growing in the capillary was bounded by faces of the equilibrium form. The crystallographic orientation was chosen such that a (100), (110), or (111) face was perpendicular to the axis of the capillary. If the electrolysis conditions are chosen appropriately, such a face gradually increases in size and fills the entire cross section. This natural face was subsequently used in examining the electrolytic deposition.

However, the most important point is that the cubic and octahedral faces in the capillary may grow with a steadily diminishing level of surface defects if the electrolysis conditions are appropriate, and in particular with a reduction in the number of screw dislocations [3, 4]. This enables one to produce atomically smooth cubic and octahedral faces on silver single crystals, which may be entirely free from dislocations.

## 2. Monatomic Step Propagation Rates

The propagation speed of a monatomic step is best examined on a dislocation-free face [4]. Such a face does not grow at a low overvoltage, i.e., no current passes, and the nuclei for a new layer are produced by a brief overvoltage pulse, after which the layer propagates over the entire face in response to a low overvoltage. The current-time curves observed after the production of a new layer are shown in Fig. 1 for a circular capillary; the shape of the curve indicates the site of occurrence of the nucleus and (to some extent) the shape. The area under the curve gives the amount of charge equivalent to a monolayer, and the value indicates that the film actually was a monolayer.

One can obtain more detailed information on the shape and speed of a monatomic layer by means of a rectangular capillary; the method is as described in [3, 4] for a circular capillary.

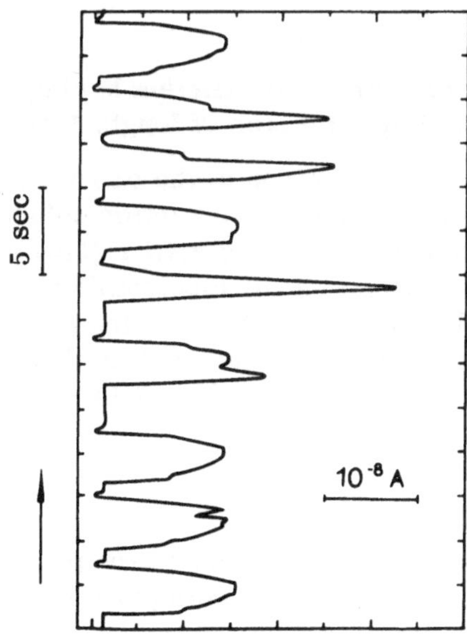


Fig. 1. Current–time curves for constant overvoltage during propagation of two-dimensional nuclei on a (100) face bounded by a circular capillary.

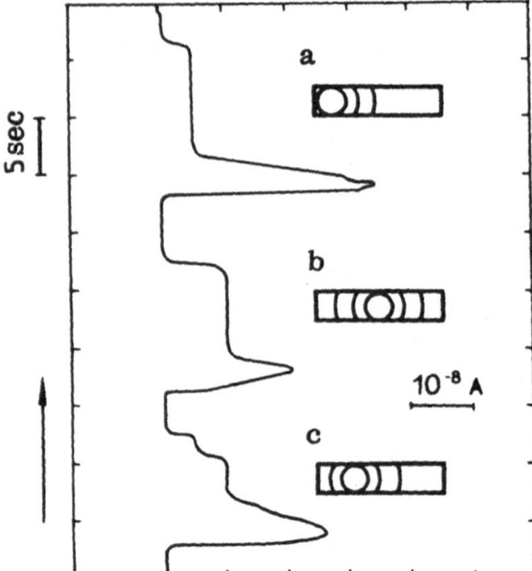


Fig. 2. Current–time curves at constant overvoltage for propagation of a two-dimensional nucleus on a (100) face bounded by a capillary of rectangular cross section (0.1 × 0.4 mm).

Figure 2 shows current-time curves following the application of a pulse sufficient to initiate nucleation on a cubic face bounded by a rectangular capillary; as in the case of Fig. 1, the current is linearly dependent on the length of the growing step. The two-dimensional nucleus extends until the steady-state current is reached, when the propagating monolayer attains the capillary wall. From this point onwards, the growth step maintains a constant length, so the propagation speed is constant, and therefore there is a plateau on the current-time curve. The current falls to zero when the monolayer is completed. The current thus has two steady-state values for a given overvoltage, with a ratio of 1:2 between them (curves a and b of Fig. 2). In the case of curve 1, a nucleus is formed at one end of the face and only one step propagates. If a nucleus is formed at the center of the face, one gets two steps, and therefore the current is twice as large, as curve b shows. Curve c in Fig. 2 illustrates the formation of a nucleus between the middle of the face and the edge.

Figure 3 shows the steady-state current as a function of overvoltage for single steps; the linear relationship can be represented as

$$I = æL\eta, \tag{3}$$

where L is the length of the growing step, while æ is a kinetic coefficient having the dimensions $\Omega^{-1} \cdot cm^{-1}$ (this can be considered as the conductivity of a step of length 1 cm).

Curves a and b of Fig. 3 were obtained with two orientations of the cross section with respect to the crystal; in one case, the longitudinal axis of the cross section coincided with a ⟨100⟩ crystallographic axis (curve a), while in the other it coincided with a ⟨110⟩ axis (curve b). The ratio of the slopes of the two lines is $\sqrt{2}$, which means that the monatomic growth fronts on a (100) face are polygonal (square). Consequently, the length of a growing step is known in both instances. In the case of curve b, this is precisely equal to the width of the rectangular cross section. Then the slope of the curve gives us the kinetic coefficient in (3), which is found as æ $= 2 \cdot 10^{-4} \Omega^{-1} \cdot cm^{-1}$.

## 3. Growth of Faces with Screw Dislocations

Frank has shown [5] that a face grows by means of spiral layers in the presence of screw dislocations; the fronts may be polygonal or circular, and one then gets pyramids or cones. Figure 4 shows such pyramids for a cube face of a silver single crystal.

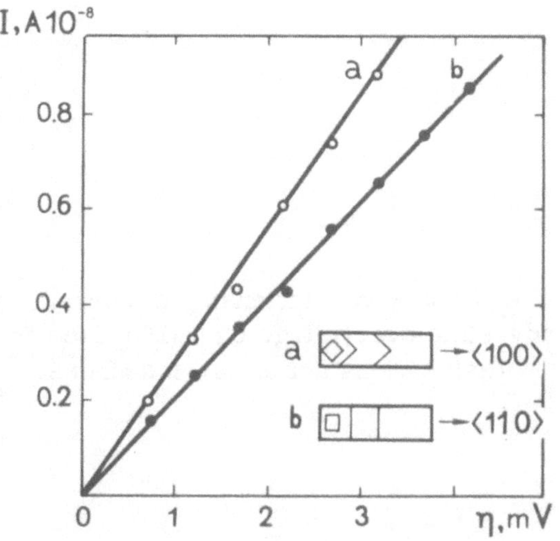


Fig. 3. Relation of steady-state current to overvoltage for propagation of a monatomic layer of a given length on a (100) face.

Barton et al. [6] found that the distance d between two adjacent spiral steps is proportional to the critical size of a two-dimensional nucleus for a given supersaturation; in electrolytic deposition, with the spiral formed by square steps, the following expression applies [7]:

$$d = 4l_0 = 8V_m\varepsilon / zFh\eta, \tag{4}$$

where $l_0$ is the length of a side of a critical square two-dimensional nucleus for a given overvoltage, $V_m$ is the molar volume of the electrolytically deposited metal, $\varepsilon$ is the specific edge energy of a step, and the height h is determined by the component of the Burgers vector of the screw dislocation normal to the surface.

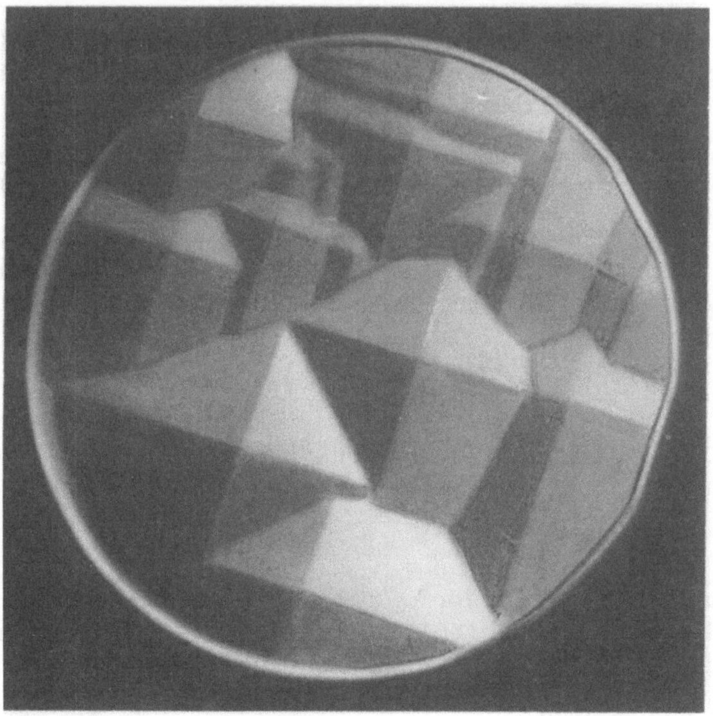

Fig. 4. Pyramids formed by spiral-layer growth on a (100) face.

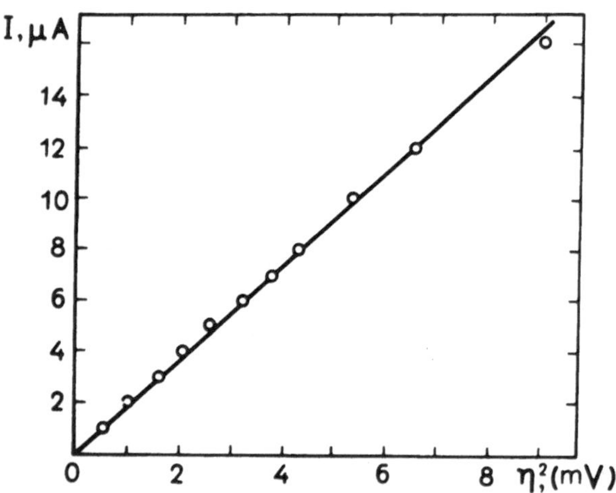


Fig. 5. Observed relationship of current to steady-state overvoltage on a (100) face when screw dislocations act as layer sources.

Under steady-state conditions, the distance between spiral steps will be the same throughout the face; the total step length L is given by S/d, where S is the face area. Then (3) and (4) imply that the current as a function of the steady-state overvoltage is as follows for spiral growth:

$$I = æ S \frac{zFh}{8V_m \varepsilon} \eta^2. \tag{5}$$

We examined this for silver deposited on faces of silver single crystals [8], and Fig. 5 shows the results for a cube face, which confirm the theory given by Barton et al. and indicate that the empirical relationship of current to overvoltage and step length applies.

## 4. Electrolytic Deposition Mechanism for Silver

High-frequency impedance measurements on (100) faces with atomically smooth surfaces or relief due to spiral growth [9] show that silver atoms enter the growth sites directly; the result for exchange between steps and the solution was $i_{0,st}$ = 170A.

The following conclusion is drawn on the electrolytic deposition mechanism from the relationship between propagation speed and overvoltage for monatomic layers of precisely defined length.

If the growth mechanism is governed by surface diffusion, then at low overvoltages ($\eta < RT/zF$) the theoretical relationship of current to overvoltage becomes

$$I = i_{0,ad}\, 2L\lambda_0 \frac{zF}{RT} \tanh\left(\frac{x_0}{\lambda_0}\right) \eta. \tag{6}$$

If the atoms enter the crystal directly, the following applies for low overvoltages:

$$I = i_{0,st}\, 2r_0 L \frac{zF}{RT} \eta. \tag{7}$$

Then (7) resembles (6) in agreeing with (3) for $\lambda_0/x_0 < 1$; it is clear that the kinetic coefficient in this linear relation will contain a variety of kinetic parameters, the precise ones being dependent on which of the two mechanisms applies for the electrolytic deposition.

If surface diffusion applies, then

$$æ = i_{0,ad}\, 2\lambda_0 \frac{zF}{RT}.$$

We follow the arguments of [9] to calculate the value and compare it with the observed one; the exchange current was measured by an impedance method on a (100) face of a silver single crystal free from growth steps [10]. The result was $i_{0.ad} = 0.06\ A \cdot cm^{-2}$, while the range in surface diffusion $\lambda_0$ may be deduced from the fit between the $I \sim \eta^2$ relation and the observation throughout the range from 0 to 3 mV in overvoltage. As the quadratic relationship implies $\lambda_0/x_0 < 1$, one can derive $\lambda_0$ from half the minimum distance between adjacent spiral steps. The minimum distance for 3 mV as calculated from (4) is 270 Å.* Therefore, $\lambda_0$ must be less than or the order of 130 Å.

In the case of the surface-diffusion mechanism, the calculated result was $æ = 4.7 \cdot 10^{-6}\ \Omega^{-1} \cdot cm^{-1}$, whereas the observed one was $æ = 2 \cdot 10^{-4}\ \Omega^{-1} \cdot cm^{-1}$; consequently, the contribution from surface diffusion is less than 2.5%. The current during step growth is governed primarily by the direct entry to the growth site, in accordance with (7); the observed value for æ gives the exchange current for a step as $i_{0,st} = 200\ A \cdot cm^{-2}$, which is of the same order as in the experiments of Vitanov et al., so this current is larger by four orders of magnitude than the current for adsorbed atoms on the surface generally.

## Literature Cited


1. W. Lorenz, Z. Naturforsch., 9A:716 (1954).
2. M. Volmer, Kinetik der Phasenbildung, Steinkopff-Verlag, Dresden (1939); Zh. Fiz. Khim., 5:319 (1934).
3. E. Budevski and V. Bostanov, Electrochim. Acta, 9:377 (1964); V. Bostanov, A. Kodeva, and E. Bulevski, BAN, Izv. Inst. Fizikokhim., 6:35 (1964).
4. E. Budevski, V. Bostanov, T. Vitanov, Z. Stoinov, A. Kotzeva, and R. Kaishev, Phys. Status Solidi, 13:577 (1965).
5. F. C. Frank, Disc. Faraday Soc., 15:48 (1949).
6. W. K. Burton, N. Cabrera, and F. C. Frank, Phil. Trans. Roy. Soc., 243A:299 (1951).
7. R. Kaischev, E. Budevski, and J. Malinovski, Z. Phys. Chem., 204:348 (1955).
8. V. Bostanov, R. Rusinova, and E. Budevski, BAN, Izv. Otd. Khim. Nauki, 2(4):885 (1969).
9. T. Bitanov, A. Popov, and E. Budevski, J. Electrochem. Soc., in press.
10. T. Vitanov, E. Sevast'yanov, V. Bostanov, and E. Budevski, Elektrokhimiya, 5:451 (1969); T. Vitanov, E. Sevast'yanov, Z. Stoinov, and E. Budevski, Elektrokhimiya, 5:238 (1969).


* In this expression we take the height of the spiral steps h as monatomic, since screw dislocations with 1/2 ⟨110⟩ Burgers vectors are the most favorable for the face-centered lattice of silver. Also, ε was determined for monatomic steps on a (100) face via the kinetics of two-dimensional nucleation.

# DETERMINATION OF KINETIC CRYSTALLIZATION COEFFICIENTS IN EXPERIMENTS WITH WHISKERS

E. I. Givargizov

*Institute of Crystallography, Academy of Sciences of the USSR, Moscow*

## 1. Introduction

The growth rate as a function of the driving force (supersaturation or supercooling) is important in any research on crystallization mechanisms and kinetics; the kinetic crystallization coefficient provides a quantitative parameter [1]. Although this quantity is very important in such research, the data are rather scanty, while virtually nothing has been published on the temperature dependence. A review [1] gives a list of values, which relate in the main to aqueous solutions and range from about $10^{-5}$ to $10^{-3}$ cm/sec for various substances.

The difficulty in determining this quantity arises in the main from measurement of the supersaturation or supercooling directly at the crystallization front; a method has been described [2] for determining the supersaturation and the kinetic coefficient together provided that the surface energy of the crystal face is known. The essence of the method is as follows. The growth rate of a whisker crystal growing from a vapor is measured as a function of the diameter d in the size range (below 1 $\mu$m) where the Gibbs–Thomson effect begins to make itself felt, i.e., the equilibrium vapor pressure is dependent on the radius of curvature. Then the supersaturation at the crystallization front is $\sigma = \sigma_0 - (4\Omega\alpha/kT) \cdot (1/d)$, where $\sigma_0$ is the effective supersaturation in the gas, $\Omega$ is the specific volume of an atom, and $\alpha$ is the surface energy, while k and T have their usual meanings. An important point is that there is a wide range of supersaturations at the crystallization front for whiskers of different diameters even for the given supersaturation in the main gas phase, which makes the technique extremely effective. We have used this approach [2] for silicon growing by the vapor–liquid–crystal mechanism [3], which is a form of crystallization from solution in a melt (in the present case, from solution in gold). The results for various supersaturations at around 1000°C for silicon were all around $10^{-4}$ cm/sec.

This technique develops a method described previously [1] in the following respects:

(a) the temperature dependence of the coefficient has been measured for Si at 930-1090°C, which has given the activation energy for crystallization;
(b) kinetic coefficients have been measured for Ge, GaAs, and SiC; and
(c) the kinetic coefficient has been determined for polar faces of GaAs, which has confirmed the previous conclusion on the type of rate-limiting step in crystallization.

## 2. Methods

The crystallization was performed in a vertical quartz system by a method previously described [4]. As a rule, we used a chloride–hydrogen process in a flow system. The Si whiskers were grown on single-crystal Si substrates of (111) orientation, which were kept at temperatures between 850 and 1100°C; the $SiCl_4$ concentrations in the $H_2$ ranged from 1 to 5%. The Ge whiskers were grown on Ge substrates of (111) orientation at temperatures between 650 and 900°C with $GeCl_4$ concentrations in the $H_2$ between 0.2 and 2%. The growth of the Ge whiskers was initiated by adding small amounts of arsenic, which was added to the gas mixture as $AsCl_3$ [5]. The GaAs whiskers were grown on A(111) substrates, i.e., gallium faces, and B($\bar{1}\bar{1}\bar{1}$) faces, i.e., arsenic ones, using the $H_2$ + $AsCl_3$ gas-transport system at 600-800°C.

The SiC whiskers were grown on $\{0001\}$ single-crystal plates of $\alpha$-SiC in a gas-transport system in which the C source was provided by graphite and the Si by a Si + $SiO_2$ mixture (it would seem that volatile SiO is the material transported here, this being reduced at the substrate by the hydrogen or carbon). The crystallization temperature was 1300-1350°C.

We also grew GaP whiskers by a method similar to that previously described [6] for GaAs and CdSe whiskers.

The metallic solvent in all cases was gold; in addition, SiC was crystallized with an Au + Al alloy.

The crystals were examined in a JSM-2 scanning electron microscope; we measured the diameter d and height h. The growth rate v was calculated from the duration of the crystallization (tests showed that all the crystals on a given substrate grew simultaneously). The results were plotted as $v^{1/2}$ against 1/d, on the assumption of a quadratic relationship, namely $v = \beta\sigma^2$, where $\beta$ is the kinetic crystallization coefficient. It has been pointed out [2] that the resulting straight lines indicate that the basic assumption is sound. Figure 1 shows a typical pattern for the crystals, while Fig. 2 shows typical results for $v^{1/2}$ as a function of 1/d. The intercepts on the 1/d axis determine the crystal diameter $d_{cr}$ such that the Gibbs–Thomson effect on the chemical potential is sufficient to halt the growth. If the surface energy $\alpha$ is known, one gets the effective supersaturation in the gas phase as

$$\sigma_0 = \frac{4\Omega\alpha}{kT} \cdot \frac{1}{d_{cr}}$$

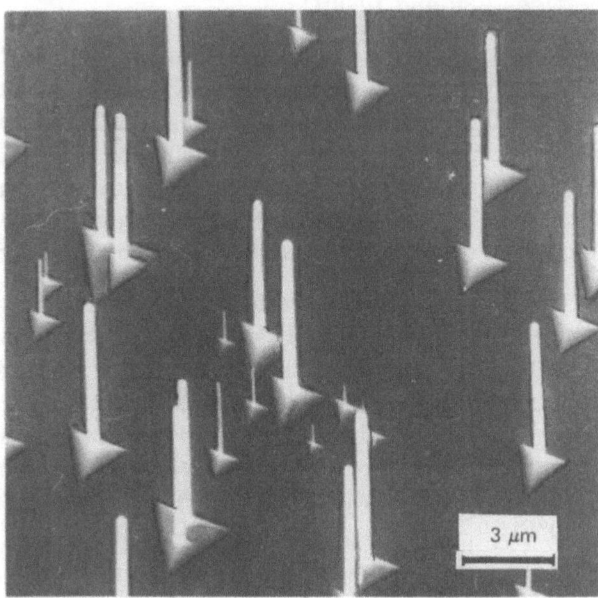


Fig. 1. Typical pattern of oriented silicon whiskers on a (111) Si substrate.

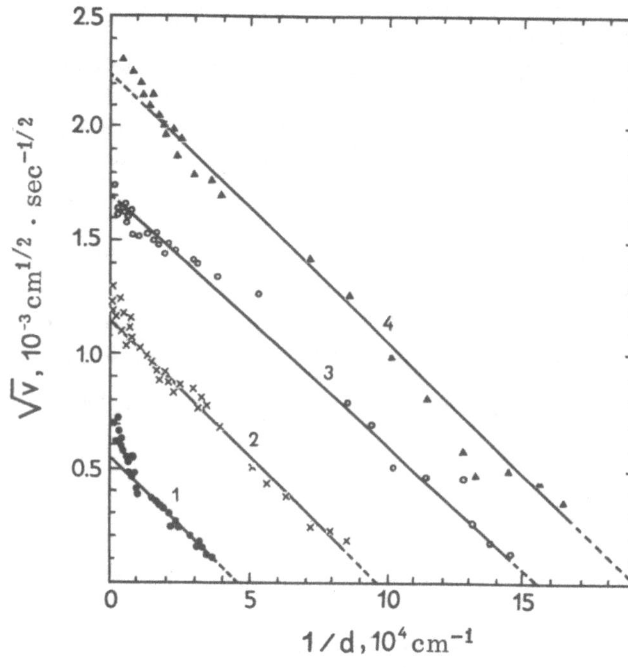


Fig. 2. Relationship of $\sqrt{v}$ to 1/d for Si whiskers grown at 1020°C; curves 1-4 correspond to the following $\sigma_0$: (1) 3%; (2) 6%; (3) 10%; (4) 12%.

Then the slope of the line defines the kinetic coefficient:

$$\beta = \left[\frac{kT}{4\Omega\alpha} \cdot \tan\varphi\right]^2,$$

where $\varphi$ is the slope angle.

## 3. Results and Discussion

### 1. Temperature Dependence of the Kinetic Coefficient for Silicon

The above method was applied to the kinetic coefficient for silicon for temperatures between 930 and 1090°C.

At relatively low temperatures (900-1000°C), the partial oxidation of the substrate means that there is a danger that the crystallization front will be blocked, and therefore that the results will not be reproducible. This difficulty was avoided by brief (2-3 min) growth at elevated temperatures (about 1100°C), and then the supply of chloride was cut off and the system was cooled to the working temperature, whereupon the growth was continued. The result was that the whiskers were formed on pedestals (Fig. 3), which indicated clearly the start of the second stage, and the results were highly reproducible.

The $SiCl_4$ concentrations in the gas (and hence $\sigma_0$) were chosen arbitrarily for a given temperature, since $\beta$ is independent of the supersaturation, as has been confirmed elsewhere [2]. The crystals were measured on four or five areas of a given substrate. For each part we constructed a graph of the type of Fig. 2, and the averaged graphs were referred to 1/d = 1 (Fig. 4). The results for $\beta$ are given in the second column of Table 1. These agree as to order of magnitude with published values [1]. In the present case we had fairly concentrated solutions (the Si : Au ratio for a saturated solution at 930°C is about 48 : 52, as against 59 : 41 at 1090°C).

The driving force (supersaturation) can be defined in different ways for a given process, e.g., as the absolute supersaturation $\Delta c = c - c_0$ (where $c_0$ is the equilibrium concentration and

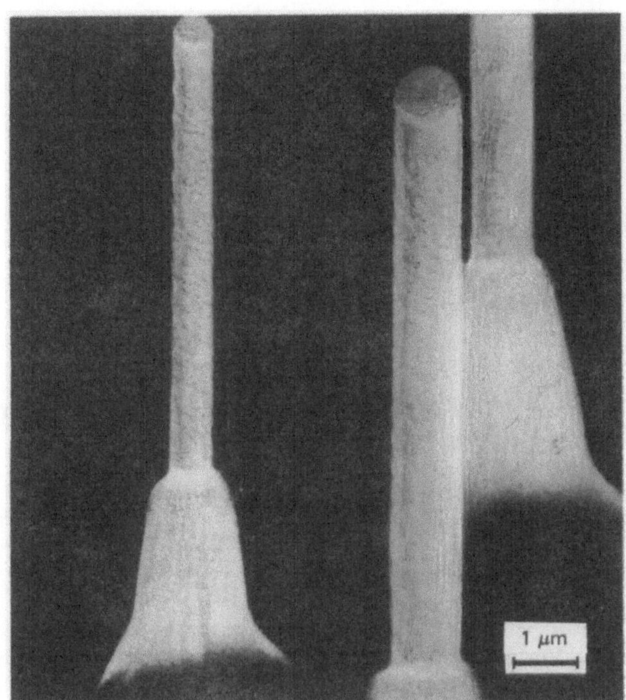


Fig. 3. Si whiskers on pedestals.

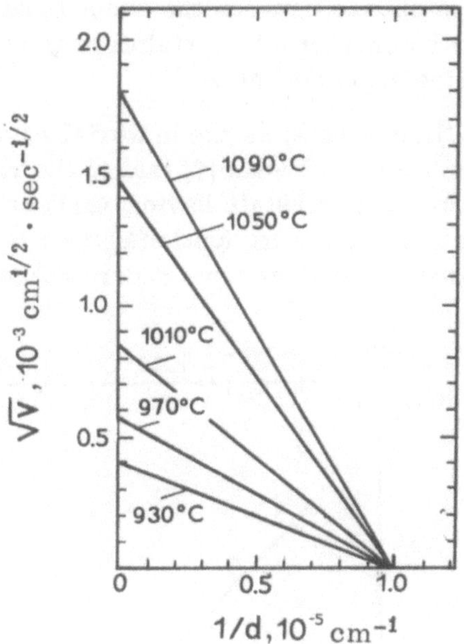


Fig. 4. Plot of $\sqrt{v}$ against 1/d for various temperatures as referred to 1/d = 1.

TABLE 1

| T°C | β, $10^{-5}$ cm/sec | β′, $10^{-5}$ cm/sec |
|---|---|---|
| 930 | 3.7 | 4.0 |
| 970 | 7.9 | 7.2 |
| 1010 | 16.2 | 12.0 |
| 1050 | 49.2 | 28.8 |
| 1090 | 80.0 | 40.0 |

c is the actual concentration) or as the relative supersaturation $\sigma = \Delta c/c_0$; correspondingly, the kinetic coefficient takes different values:

$$v = \frac{\beta}{c_0^2}(\Delta c)^2 = \beta'(\Delta c)^2.$$

This must be borne in mind in determining the temperature dependence, since $c_0$ is itself dependent on temperature, i.e.,

$$\beta'(T) = \frac{\beta(T)}{c_0(T)}.$$

The Si–Au phase diagram then gives the true temperature dependence of the kinetic coefficient; the third column of Table 1 gives $\beta'$, with $c_0$ taken as the Si : Au atomic ratio.

The temperature dependence of the kinetic coefficient was plotted as $\log\beta'$ against 1/T; Fig. 5 shows that the points fit fairly closely to a straight line.

The result for the activation energy is $\Delta E \approx 48$ kcal/mole, which clearly characterizes the rate-limiting step.

We now discuss this aspect in more detail. We have previously [2] examined the vapor–liquid–crystal mechanism as regards the rate-limiting step, and found that there were four stages: (1) transport in the gas phase, (2) deposition at the vapor–liquid boundary, (3) diffusion in the liquid, and (4) deposition at the liquid–crystal boundary, of which the first and third occurred most readily. It therefore then remained to choose between the second and fourth. However, it is now possible to make such a choice with considerable certainty, namely the fourth possibility. The following experimental evidence supports this.

Figure 6 shows GaS whiskers grown on an A(111) substrate, as the material crystallizes with the sphalerite structure and tends to twin, and it has been shown [7] that in the vapor–liquid–crystal process it produces twinned acicular or strip crystals having ⟨211⟩ as growth direction. A twin boundary is an active growth source, since the nucleation energy is lower there, so a twinned whisker will grow more rapidly than an ordinary one under otherwise equal

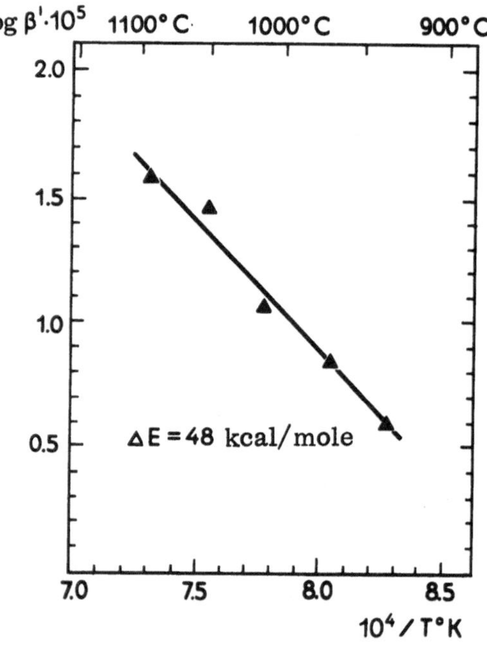


Fig. 5. Temperature dependence of the kinetic coefficients and determination of the activation energy for crystallization of Si.

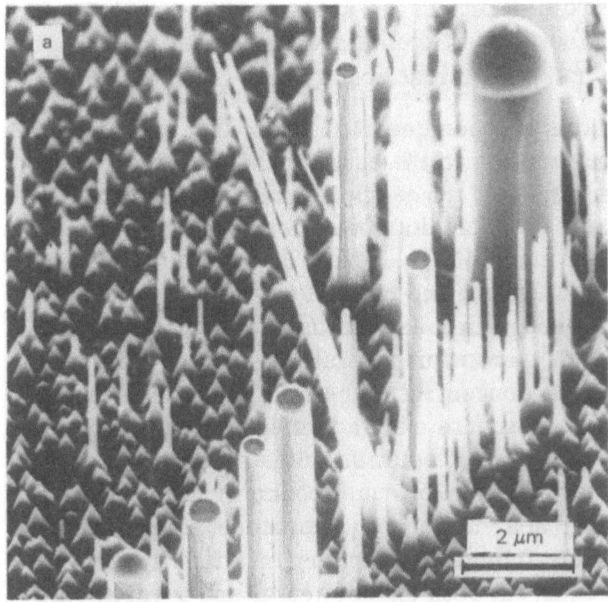


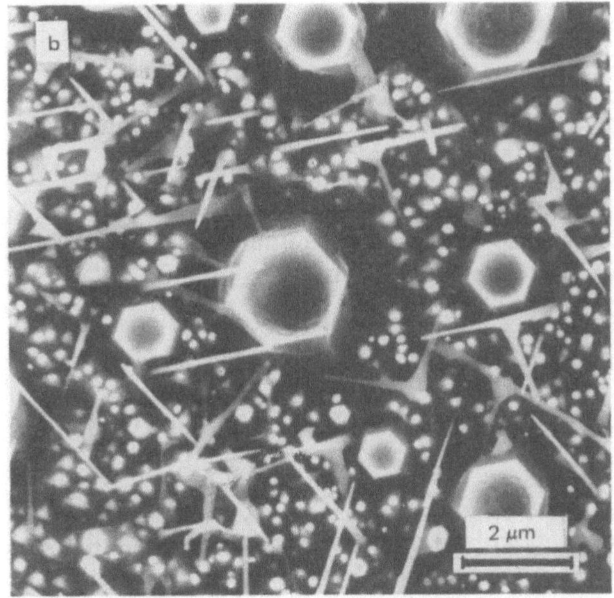


Fig. 6. Relative growth rates of GaAs whiskers perpendicular to the surface or inclined to it: (a) angle of incidence $\theta$ of scanning beam in microscope 45°; (b) $\theta = 0$.

conditions. Figure 6a shows not only crystals perpendicular to the substrate but also inclined ones growing along ⟨211⟩ (indicated by the arrow, and see also Fig. 6b, which was recorded from an analogous part at zero angle of indidence in the scanning electron microscope). While crystals perpendicular to the substrate follow the general laws (i.e., the thicker ones grow more rapidly than the thin ones), the inclined crystals grow appreciably faster, although their diameters are very much smaller than those of the perpendicular ones. This is particularly clearly seen in Fig. 6b, where an inclined crystal rises above the perpendicular one as indicated by the arrow. One therefore concludes that the structure of the crystallization front has a decisive effect on the growth rate of a whisker, i.e., that the fourth stage is the rate-limiting one.*

* In this argument, it is not essential for the structure of the boundary to be twinned; it is sufficient for the structure of the boundary in an inclined crystal to be different from that for a perpendicular one.

A further piece of evidence for the decisive role of processes at the liquid–crystal boundary is given in section 3 of this paper, where the kinetic coefficients are compared for the A(111) and B($\bar{1}\bar{1}\bar{1}$) polar faces of gallium arsenide under identical conditions.

To conclude this section we note that it remains an open question as to what are the processes at the liquid-crystal boundary, and the same applies to the cause of the relatively high $\Delta E$ (the two topics are clearly related). A relation of $V \sim \sigma^2$ type applies to growth via screw dislocations, but Wagner has shown [8] that whiskers of small diameter (less than 1 $\mu$m) grow free from dislocations even on substrates containing many dislocations. It therefore remains to choose between layer growth (two-dimensional nucleation) and normal growth. It is considered [9, 10] that a crystal–melt phase boundary is rough in many instances, i.e., the normal mechanism applies, while Voronkov and Chernov have shown for crystals in contact with solutions [1] that a smooth-rough transition can occur, the concentration being the decisive feature. In our case, the solution is fairly concentrated, so one expects such a transition to be dependent on the temperature, i.e., on the concentration. As the drop of solution takes a roughly hemispherical form near the crystallization front, it is not an area of equal concentration, since the supersaturation (the concentration) will be higher near the edge. Therefore, this transition to kinetic roughness [12] may set in gradually, namely from the periphery of a face toward the center. This means that further research is required, both experimental and theoretical.

## 2. Measurement of Kinetic Coefficients for Germanium and Silicon Carbide

We have used the above method to determine kinetic coefficients for these materials at certain temperatures.

(a) Kinetic Coefficient for Ge. Germanium whiskers resemble Si ones in growing along [111] and being bounded by {211} side faces, so the surface energy was taken as $\alpha_{\{211\}} = 1495$ ergs/cm². The crystallization temperature was 750°C, and the plot of $v^{1/2}$ against 1/d gave

$$\tan\varphi = 2.4 \cdot 10^{-9} \frac{\text{cm}^{3/2}}{\text{sec}^{1/2}}.$$

Calculation then gave

$$\beta = 7.5 \cdot 10^{-6} \text{ cm/sec}.$$

(b) Kinetic Coefficient for Silicon Carbide. Silicon carbide crystallizes in two modifications: cubic (sphalerite type, or $\beta$-SiC) and hexagonal (wurtzite type, $\alpha$-SiC). If the temperature is relatively low ($\lesssim$1500°C), $\beta$-SiC is the usual product, but it has not yet proved possible to make large crystals of $\beta$-SiC, so we used plates of $\alpha$-SiC as substrates for oriented growth of whiskers, although under our conditions (temperature ~1300–1350°C) the product should be $\beta$-SiC.

We grew oriented whiskers, but these were extremely small ($d \sim 0.1$–$1\ \mu$m), so it proved impossible to identify the modification; very general considerations on the growth of SiC whiskers [13] indicated that the crystals should be of the $\beta$ modification. It also proved impossible to identify the side faces, but analogy with Si whiskers indicated that the {211} and {110} face systems should be present, with six faces from each system. Then we have the following estimate for the face energy:

$$\bar{\alpha} = \frac{\alpha_{\{211\}} + \alpha_{\{110\}}}{2} = \frac{3990 + 3450}{2} = 3720 \text{ ergs/cm}^2.$$

The data for $\alpha_{\{211\}}$ and $\alpha_{\{110\}}$ were taken from [13].

With a crystallization temperature of 1350°C, the graph of $v^{1/2}$ against 1/d gave log $\varphi$ = $2.4 \cdot 10^{-9}$ $cm^{3/2} \cdot sec^{-1/2}$, and therefore

$$\beta_{SiC} = 3 \cdot 10^{-6} \text{ cm/sec.}$$

We may note here that SiC crystals grow in one of the ⟨111⟩ and ⟨0001⟩ polar directions.

## 3. Determination of Kinetic Coefficients for Polar Faces of GaAs

Epitaxial films grow on A(111) or B($\bar{1}\bar{1}\bar{1}$) (gallium or arsenic) faces in essentially different ways; on A(111) the growth is difficult, although it is possible to produce single-crystal films under optimal conditions, but the defect level is always high. It was therefore of some interest to examine the growth of whiskers in these directions, the more so since the vapor–liquid–crystal mechanism usually occurs on smooth faces and therefore the choice of possible growth directions is extremely limited.

We found that the modes of growth in the polar directions differ morphologically; along B[$\bar{1}\bar{1}\bar{1}$], the crystals grow in a far more ordered state, and most of them are perpendicular to the substrate over a fairly wide range of conditions (Fig. 7). Along A[111], one usually gets several systems of inclined crystals, whereas the rest of the crystals are perpendicular to the substrate under optimal conditions, i.e., those on a smooth A(111) face, as in Fig. 8.

A detailed description of the morphology of GaAs whiskers falls outside the scope of the present study, and here we merely report the kinetic results.

In both cases, the side facets consisted of six of the {211} planes, so we took $\alpha_{\{110\}}$ = 1270 erg/cm².

The substrates were two parts from a single plate, which had been processed in the same etching agent and coated with gold in the same operation. The whiskers were grown in the same run, so the A(111) and B($\bar{1}\bar{1}\bar{1}$) substrates were under identical conditions. The crystallization temperature was 670°C.

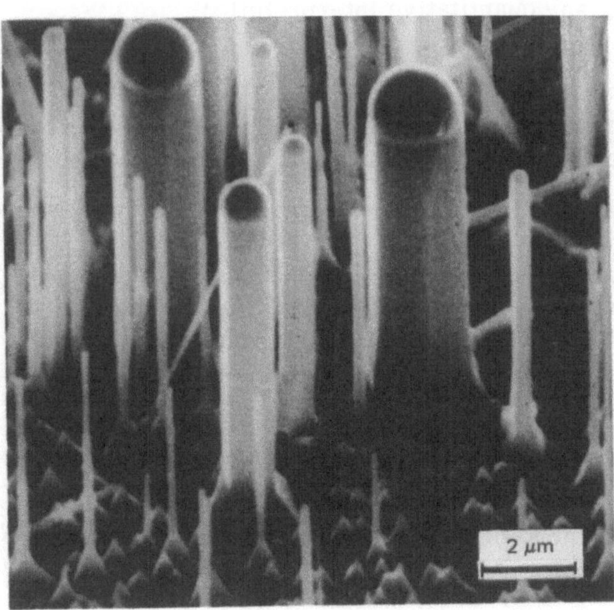


Fig. 7. GaAs whiskers on a B(111) face.

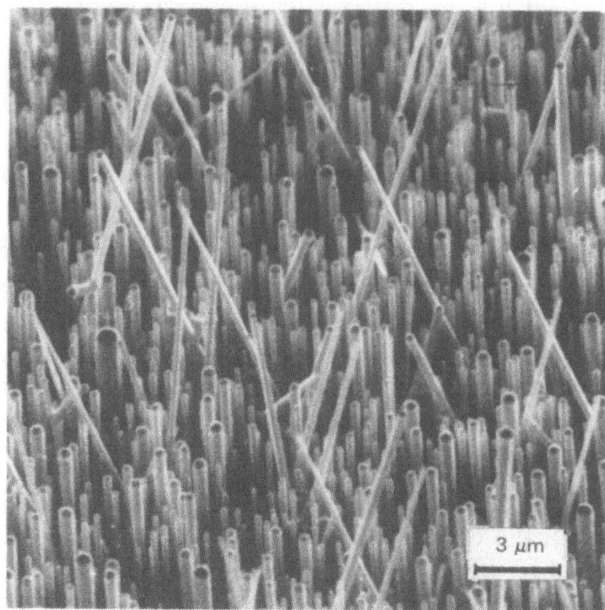


Fig. 8. GaAs whiskers on a A(111) face.

The plots of $v^{1/2}$ against 1/d were straight lines, which gave

$$\beta_{A(111)} = 1.0 \cdot 10^{-2} \text{ cm/sec},$$
$$\beta_{B(\bar{1}\bar{1}\bar{1})} = 5.4 \cdot 10^{-2} \text{ cm/sec},$$

i.e., the kinetic coefficients for the polar faces differ by more than a factor 5, which again indicates that the rate-limiting step is the incorporation at the liquid–crystal boundary, not the vapor–liquid one.

## 4. Other Substances

Generally similar trends were found also for GaP and CdSe, but lack of information on the surface energies led us to avoid attempting any quantitative interpretation.

## 4. Conclusions

Quantitative studies on whiskers provide an efficient means of determining kinetic characterostocs amd thus of elucidating growth mechanisms; two features are important here:

(a) it is possible to examine the steady-state growth in a size range (below 1 μm) where the Gibbs–Thomson effect is important;
(b) the growth conditions for whiskers are highly reproducible, since the liquid cap of alloy used in feeding the crystal is virtually unchanged in size (for comparison, we may note that ordinary microcrystals show conditions changing continuously at the crystallization front), and consequently it is possible to judge the microscopic phenomena from the macroscopic effects along the length.

These features of whiskers have enabled us to devise a method of measuring the kinetic coefficients, and the following results have been obtained:

(a) we have determined kinetic crystallization coefficients for various important materials (Si, Ge, GaAs, SiC);

(b) the temperature dependence of these coefficients has been determined for silicon over the range 930-1090°C, and the activation energy for the process has been found to be 48 kcal/mole;

(c) we have determined kinetic coefficients for smooth polar faces in the sphalerite structure, namely A(111) and B($\bar{1}\bar{1}\bar{1}$) in GaAs, and this serves to confirm the previous conclusion on the rate-limiting step in crystallization.

## 5. Conclusions

We are indebted to N. N. Sheftal' and A. A. Chernov for discussion of various aspects of the present work, and also to L. N. Obolenskaya, P. D. Tkachev, and V. I. Muratova for assistance in the experiments.

## Literature Cited

1. A. A. Chernov, Kristallografiya, 16:842 (1971).
2. E. I. Givargizov and A. A. Chernov, Kristallografiya, 18:147 (1973).
3. R. S. Wagner and W. C. Ellis, Appl. Phys. Letters, 4:89 (1964).
4. E. I. Givargizov and Yu. G. Kostyuk, In: Growth of Crystals, Vol. 9, Consultants Bureau, New York (1975), p. 276.
5. E. I. Givargizov and N. N. Sheftal', Kristall u. Technik, 7:37 (1972).
6. E. I. Givargizov, Dokl. Akad. Nauk SSSR, 211:322 (1973).
7. R. L. Barns and W. C. Ellis, J. Appl. Phys., 33:2296 (1965).
8. R. S. Wagner, J. Appl. Phys., 38:1554 (1967).
9. J. W. Cahn, Acta Met., 8:554 (1960); J. W. Cahn, W. B. Hillig, and G. W. Sears, Acta Met., 12:1421 (1964).
10. K. A. Jackson, D. R. Uhlmann, and J. D. Junt, J. Crystal Growth, 1(1) (1967).
11. V. V. Voronkov and A. A. Chernov, Kristallografiya, 11:662 (1966).
12. A. A. Chernov, In: Physicochemical Problems in Crystallization [in Russian], Alma-Ata (1969), p. 21.
13. G. A. Bootsman, W. F. Knippenberg, and G. Verspui, J. Crystal Growth, 11:297 (1971).

# BULK AND INTERFACE EFFECTS IN CRYSTAL GROWTH BY THE MOVING-SOLVENT METHOD

V. N. Lozovskii, G. S. Konstantinova, V. Yu. Gershanov, E. I. Kireev, and V. S. Zurnadzhyan

*Novocherkassk Polytechnical Institute*

The gradient in the thermodynamic potential arising in the moving-solvent method produces various bulk and surface processes at the interface between crystal and liquid, which influence the growth rate to various extents. Here we examine the effects of these various processes for zone recrystallization under a temperature gradient.* This form of the moving-solvent method has been applied to a wide range of systems and has been extensively researched [2].

In the linear approximation for a two-component conservative system in the absence of convection in the liquid, we have that the growth rate v for a planar zone is given by [3]:

$$v=\frac{\alpha_1\left(1-AD''\frac{dC}{dT}\right)G-\alpha_2\left(D^s\frac{dC_c^s}{dx}+D'^sC_c^sG^s\right)}{1+\frac{\alpha}{lv^i}\Gamma_{ij}+\alpha_1\frac{\Delta H}{\lambda}}, \tag{1}$$

where

$$\alpha_2=\frac{N}{N^s}\frac{1}{C_c-C_c^s},\qquad \alpha=\alpha_2 D\frac{dC}{dT},\qquad \alpha_1=\alpha\left(1+\frac{D'}{D}C_c\frac{dT}{dC}\right),$$

$\Gamma_{ij} = \Gamma_i + v^{j-i}\Gamma_j$, N and $N^s$ are the numbers of atoms per unit volume in the liquid and solid phases, dC/dT is the slope of the liquidus line, $l$ is the thickness of the liquid zone, $D^s$, $D'^s$, D, and D' are the diffusion and thermal-diffusion coefficients for the solid and liquid phases, D" is the Dufour coefficient, $\Delta H$ is the latent heat of dissolution, $\lambda$ is the thermal conductivity of the liquid, and $\Gamma_i$ and $\Gamma_j$ are expressed in terms of the atomic-kinetic dissolution and crystallization coefficients.

The superscripts or subscripts i and j take the values 0, 1/2, and 1 for the following growth mechanisms: normal growth, growth on screw dislocations, and growth by two-dimensional nucleation, and then $\Gamma_0 = 1/\mu_1$, $\Gamma_{1/2} = 1/(\mu_2)^{1/2}$, $\Gamma_1 = \mu_1/\ln(\mu_3/V)$, where $\mu_i$ are the atomic kinetic coefficients.

* The method is also called the temperature-gradient zone-melting method [1].

Equation (1) is the most important for this method, since it provides basic information on the processes during zone displacement; the effects of the various detailed processes are dependent on the working conditions, but certain effects can be neglected. The $AD''(dC/dT)$ and $\alpha_1\Delta H/\lambda$ terms represent the reduction in the temperature gradient G in the zone arising from the Dufour effect and heat release at the interphase boundaries; any fall in G retards the diffusion in the zone and reduces v. In the case of melts, $A < 3 \cdot 10^7$ sec · deg$^2$ · cm$^2$, $D'' < 10^{-6}$ cm$^2$ · sec$^{-1}$ · deg$^{-1}$, $dC/dT < 10^{-3}$ at.% · deg$^{-1}$, $\Delta H < 10^4$ cal · mole$^{-1}$, $\lambda \sim 0.1$ cal · cm$^{-1}$ · sec$^{-1}$ · deg$^{-1}$, $D < 10^{-4}$ cm$^2$ · sec$^{-1}$, and $C_c > 0.1$. So the $AD''(dC/dT)$ and $\alpha_1\Delta H/\lambda$ terms are small by comparison with 1, so the effects of the corresponding processes can be neglected. The term containing $\alpha_2[D^S(dC_c^s/dx) + D'^S C_c^s G^S]$ gives the flux of atoms from the crystallizing boundary into the solid due to diffusion and thermal diffusion. However, $D^S$, $D'^S$, $C_c^s$ are usually less by several orders of magnitude than D, D', $C_c$, so these processes can also be neglected. The $(D'/D)C_c(dT/dC)$ term in the expression for $\alpha_1$ incorporates the atomic transport to the crystallization boundary due to thermal diffusion in the liquid. The values of D'/D lie in the range $10^{-5}$–$10^{-3}$ deg$^{-1}$, $C_c \sim 0.5$; $dT/dC \sim 10^3$ deg at.%, so $(D'/D)C_c(dT/dC)$ in some cases is comparable with unity, which means that thermal diffusion has an appreciable effect. Then (1) can be put as

$$v = \frac{\alpha_1 G}{1 + \alpha\Gamma_{ij}/l v^i}. \tag{2}$$

The $\alpha\Gamma_{ij}/lv^i$ term expresses the relative importance of diffusion in relation to the interphase kinetics; the importance of the latter increases as $l$ is reduced and D increases. If $\alpha\Gamma_{ij}/lv^i \gg 1$, the zone speed is determined by the interfacial processes (the kinetic state). In that case, $v^{1-i}\Gamma_{ij} = Gl\alpha_1/\alpha$, whereas the zone speed is completely determined by mass transport in the liquid if the converse applies and is $v_m = \alpha_1 G$ (diffusion state). Consequently, the growth rate is constant for large $l$ and should increase with $l$ when the latter is small.

This theoretical conclusion does not conflict with the observed $v(l)$, but one obtains quantitative agreement with (2) only if the activation energy U for diffusion in the zone is ascribed values quite excessive for liquids (U > 10 kcal/mole). The $D_0$ factor in $D = D_0 \exp(-U/RT)$ also appears excessive ($D_0 > 10^{-2}$ cm$^2$/sec), and a new point of view has recently been expressed [4] in that the fall in $v(l)$ for small thicknesses is due not to kinetic processes at the interface but to contaminants such as oxides. The thinner the zone, the greater the oxide contamination at the interface, and hence the greater the hindrance to mass transport through the liquid.

We have examined the importance of the interface effects on $v(l)$ for the Si–Al systems with various orientations for the specimen and seed; curve 1 of Fig. 1 shows results for specimens of (110) orientation, while curve 2 corresponds to (111). The seeds in both cases were of (110) orientation, while the specimen preparation and the working conditions were otherwise identical. It is clear that $v(l)$ is due to the kinetics of the dissolution, and any possible effects from differences in oxidation of the Si as between the (111) and (110) orientations were established by measuring the speeds in sandwiches with various seed and specimen combinations. It was found that a specimen oriented on {110} and a seed on {111} always gave a growth rate higher than that when the specimen and seed were correspondingly oriented on {111} and {110} ($l$ = 15 μm), so the fall in growth rate at low thicknesses is due to interface (kinetic) processes, not to oxide contamination. Accidental contaminants and oxides only alter the kinetic coefficients and shift the region where $v(l)$ falls rapidly.

The {111} planes are the closest-packed in silicon and should have lower kinetic coefficients than {110} planes; the region of the sharp fall in curve 1 in Fig. 1 is displaced to smaller $l$, which agrees with (2).

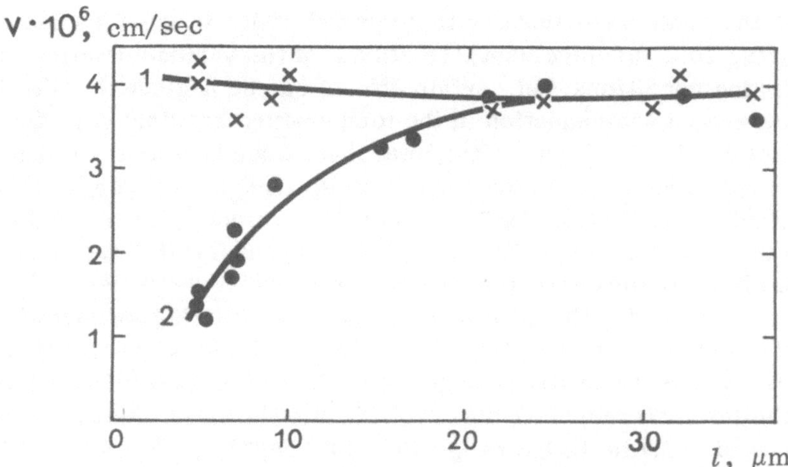


Fig. 1. Speed of an aluminum zone in silicon as a function of thickness: (1) (110) orientation of dissolving crystal; (2) (111). Orientation of seed in both cases (110). Mean zone temperature 1090°C.

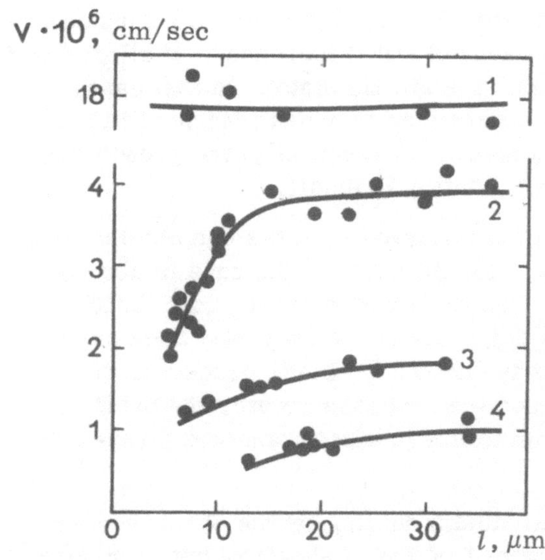


Fig. 2. Speed of an aluminum zone in silicon as a function of thickness for various temperature gradients G: 1) $4G_0$; 2) $G_0$; 3) $G_0/2$; 4) $G_0/4$; mean zone temperature 1090°C, orientation of dissolving crystal (111).

We also tested a theory by examining the effects of G on $v(l)$ for the Si–Al system; Fig. 2 shows curves for four values of G. The curves shift toward larger $l$ as G increases, in agreement with the theory.

Finally, the theory implies that the observed $v(l)$ as constructed in terms of $v/v_m$ against $l/l_{0.5}$ should be described by the universal function $L = 2^{m-1}\eta W^m/(1-W)$ for any system and any working conditions, where $W = v/v_m$, $L = l/l_{0.5}$ ($v_m$ is the asymptotic value of v, while $l_{0.5}$ is the zone thickness such that $v = 0.5v_m$), while m = 1, 1/2, or 0 for normal, dislocation, and nucleation interfacial processes respectively and $\eta$ is a constant that can be taken as 1. Figure 3 shows that this conclusion from the theory is closely confirmed.* The values for the kinetic coefficients found on constructing the curves of Fig. 3 are reasonable.

* Figure 3 also shows that the uniform growth mechanism does not occur in any case.

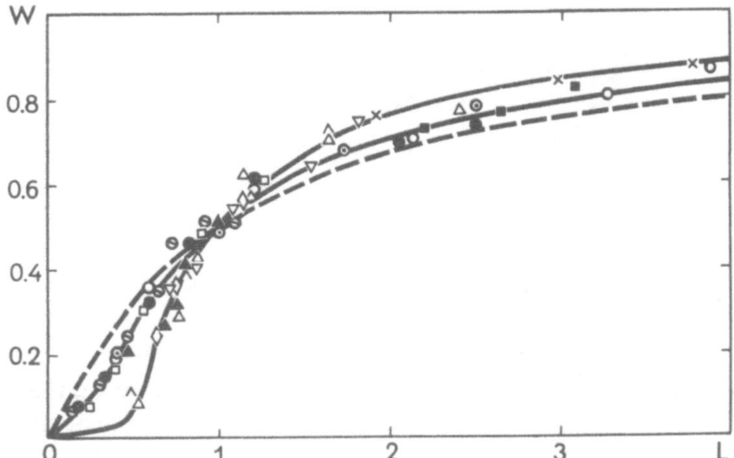


Fig. 3. Values of $v(l)$ for various systems in W and L as coordinates: ● SiC–Cr, 1750°C [5]; ○ Si–Al, 1000°C, [6]; ◇ Si–Ag, 1150°C [7]; Δ Si–Ag, 1300°C [7]; ^ InSb–In, 435°C [8]; ▲ InSb–Pb, 435°C [8]; □ Si–Sn, 1200°C [9]; ■ Si–Al + Sn, 800°C [10].

Therefore, the experimental evidence indicates uniquely that the characteristic relationship between growth rate and zone thickness is due to kinetic processes, not to contaminants such as oxides. It is then necessary to establish why U and $D_0$ derived from these data are usually larger than those found by other methods. We suppose that the thin layer of solution in the melt under normal conditions is not a homogeneous liquid; the heterogeneous inclusions may be thermodynamically equilibrium complexes of cluster type, which occur in virtually any liquid near the point of crystallization or else they may be foreign inclusions or metastable quasicrystalline ones.

In this connection we examined the effects of brief high-temperature treatment on the kinetics; we used Si–Au melts, for which U = 26 kcal/mole, $D_0 = 4.4 \cdot 10^{-1}$ cm$^2$/sec, and the zone does not move under ordinary conditions up to 1000°C (the eutectic temperature is 370°C). The liquid phase was treated in a uniform temperature distribution directly in the zone equipment. Figure 4 shows the results. The zone is accelerated by increasing the treatment temperature, i.e., the diffusion in the bulk is accelerated, while the onset of appreciable motion occurs at lower temperatures. The temperature dependence of the diffusion coefficient derived from this study at 1260°C agrees well with data from the capillary method (U = 8 kcal/mole, $D_0 = 2.5 \cdot 10^{-3}$ cm$^2$/sec).

The result of such heat treatment is almost independent of the treatment time provided that the latter is more than 2 min (it was impossible to use shorter times on account of the lag in the heating system). An interesting point is that the result of the heat treatment is remembered by the zone even if the latter is allowed to cool to room temperature after the treatment. Moreover, if the zone is a heat-treated Si–Au alloy of eutectic composition made by melting under a flux and subsequent rolling (curves in Fig. 5), then the zone speed is close to the value given in Fig. 4.

The accelerated zone motion occurs mainly in the diffusion state, while the effect is reduced in the transition state and is virtually absent in the kinetic mode. The speed is also independent of the heat treatment if heat-treated Si–Au eutectic alloy is used, so we consider

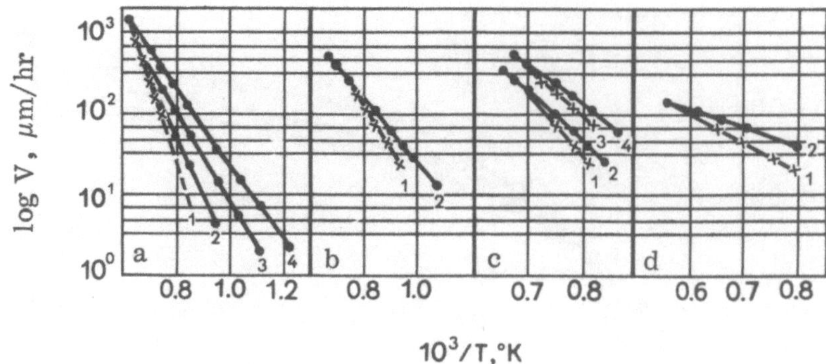


Fig. 4. Effects of heat treatment of the liquid phase on the zone speed in silicon-based systems: (a) Si–Au: (1) no heat treatment; (2-4) treated at 1100, 1200, and 1260°C respectively; (b) Si–Al: (1) without heat treatment; (2) treated at 1260°C; (c) Si–Cu: (1) no heat treatment; (2) treated at 1260°C; (3) without heat treatment in convective mode; (4) heat treatment at 1260°C in convective mode; (d) Si–Pt: (1) without heat treatment; (2) treated at 1260°C.

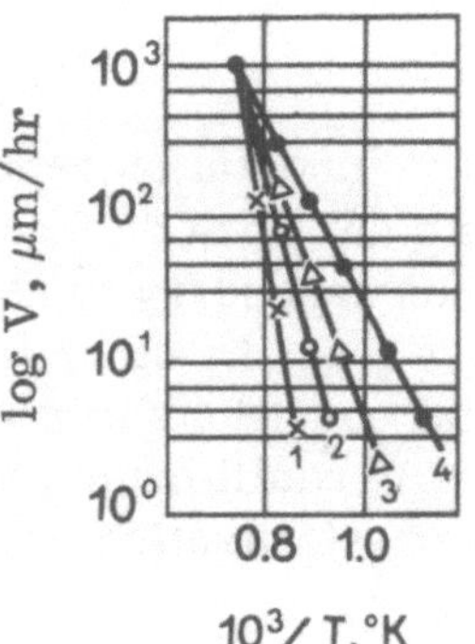


Fig. 5. Temperature dependence of the zone speed with zones consisting of heat-treated Si–Au alloys containing the following wt.% Si: (1) 1; (2) 2; (3) 3; (4) 4.

that oxides and other contaminants are not the cause of the hysteresis in the temperature dependence of the kinetics.* When Si–Au is used, the effects due to oxides and contaminants should be at least comparable with those for pure gold.

We are inclined to ascribe the hysteresis to the contact-melting stage, which is one that virtually always precedes formation of the liquid. Studies on contact melting [12, 13] have shown that the liquid phase is then substantially inhomogeneous. The production and destruction of the nonuniformity serve to explain the temperature dependence of the zone speed. From this viewpoint, results on the kinetics of zone melting in the Si–Au system are of interest, particularly when heat-treated alloys with various silicon contents are used as the zones (Fig. 5). The melting point of the zone is reduced as the silicon content increases, and therefore the importance of contact melting in producing a liquid of liquidus composition is reduced. Correspondingly, the liquid becomes more homogeneous and the rates of diffusion processes in the bulk increase.

The Si–Au system is not the only one for which hysteresis has been observed in the temperature dependence of the zone speed; we have recorded similar results for the Si–Al, Si–Cu, and Si–Pt systems (Fig. 4). The effects of heat treatment on the kinetics vanish in

* A similar effect from oxides has been reported previously for the Si–Al system [11].

these systems at temperatures of 1000, 1150, and 1250°C respectively, whereas structural nonuniformities persist in the melt up to much higher temperatures in the Si–Au system.

The heat treatment affects the zone speed also when the main mass transport mechanism in the bulk is convection (Fig. 4); this also agrees with the above model, since large structural groups in the bulk of the liquid should increase the viscosity and consequently retard the convective mixing.

The lifetime of the nonuniformities in the liquid is related to the metastability, since the composition and temperature of the zone are very close to the values on the liquidus line (the superheating in the zone cannot exceed a fraction of a degree). The heat of mixing is negative in each of the above systems, so the structural groups arising by contact melting that hinder the diffusion are irreversibly eliminated by zone annealing.

## Literature Cited

1. W. J. Pfann, Zone Melting [Russian translation], Mir (1970).
2. V. N. Lozovskii, Temperature-Gradient Zone Melting [in Russian], Metallurgiya (1972), p. 240.
3. V. I. Lozovskii and G. S. Konstantinova, Sb. Trudov Novocherkasskogo Politekhn. Inst., Vol. 259 (1972).
4. M. Kumagawa, M. Ozeki, and S. Yamada, Japan. J. Appl. Phys., 9:1422 (1970).
5. A. A. Kal'nin and Yu. M. Tairov, Izv. Leningrad. Elektrotekhn. Inst., 61:26 (1966).
6. A. I. Udyanskaya, Sb. Trudov Novocherkasskogo Politekhn. Inst., 170:31 (1967).
7. E. A. Nikolaeva, Sb. Trudov Novocherkasskogo Politekhn. Inst., 170:42 (1967).
8. R. W. Hamaker and W. B. White, J. Appl. Phys., 38:1858 (1968).
9. V. N. Lozovskii and A. I. Kalinyuk, In: Physics of Condensed Media [in Russian], Rostov on Don (1970), p. 87.
10. V. N. Lozovskii, A. I. Kalinyuk, and N. F. Politova, Sb. Trudov Novocherkasskogo Politekhn. Inst., 208:34 (1970).
11. V. N. Lozovskii, V. P. Popov, and A. S. Sushchik, Elektronnaya Tekhnika, Ser. 2, No. 3, 41 (1969).
12. L. K. Savitskaya and P. A. Savintsev, In: Surface Phenomena in Melts and Powder Metallurgy Processes [in Russian], AN Ukr. SSR, Kiev (1963), p. 273.
13. I. G. Berezina, L. K. Savitskaya, and P. A. Savintsev, In: Surface Phenomena in Melts and Powder Metallurgy Processes [in Russian], AN Ukr. SSR, Kiev (1963), p. 288.

# FACES WITH HIGH INDICES ON IONIC CRYSTALS AS A CONSEQUENCE OF LAYER GROWTH

**P. Hartman**

*Geologisch en Mineralogisch Instituut der Rijksuniversiteit, Leiden*

Natural crystals sometimes show faces with high indices (h, k, or $l$ larger than 2 or 3, not vicinal faces), that have unexpectedly high persistencies. Examples are {017} and {2.3.19} of anatase [1], {5161} of quartz [2], and {135} of magnetite [2,3]. It has been shown [4, 5] that such faces are often parallel to a direction in which the interaction energy of parallel ionic chains running parallel to the face is near to zero. An explanation was suggested under the assumption that the crystals show the equilibrium form and that the occurrence of the faces is due to a lowering of the specific free energy, because of adsorption of foreign ions. In this contribution an explanation in terms of growth forms will be investigated.

Consider a cubic model with an S face (016) between the F faces (001) and (010). Figure 1 shows the profile of the face. The elementary step height is $d_{001} = a$, the unit cell edge length. Thus, A, B, C, etc. represent parallel identical ionic chains [100], each seen end on.

Suppose now that growth occurs under near-equilibrium conditions. In that case the volume and surface diffusion processes are fast compared with the step velocity. The number of kinks filled in per unit of time is only slightly larger than the number of kinks created per unit of time. Hence the step velocity is determined by the rate of kink formation, and therefore by the energy necessary for this process. Let $E_A$ be the energy necessary for the creation of a kink in the step at A, when the other steps at K, P, etc. are not present. In the case where both the steps at A and P are present, the corresponding energy $E_P$ is

$$E_P = E_A + E(n, 1), \tag{1}$$

where n and 1 are the lattice periods that separate the parallel chains in the b and c directions, respectively. In Fig. 1 n = 6. $E_A$ has a large negative value (attraction), so that, if E(n, 1) is also negative, the formation of a kink at P is more difficult than in the case where step A is not present. Therefore, the step velocity at P is decreased and the profile tends to be maintained when E(n, 1) is negative. On the other hand, if E(n, 1) is positive, the step velocity at P is increased and the profile tends to be coarsened in the sense of Stranski. Maintenance of the profile implies that the face grows layer after layer, first ...K, A, P,..., then ...L, B, Q,..., then ...M, C, R,..., etc.

The growth rate of the face is determined by the value of $E'_A$, the energy necessary to create a kink in step A when other steps in the same layer, like K and P, are also present:

$$E'_A = E_{cr} + E_r + \sum_{k=1}^{\infty} E(kn, k), \tag{2}$$

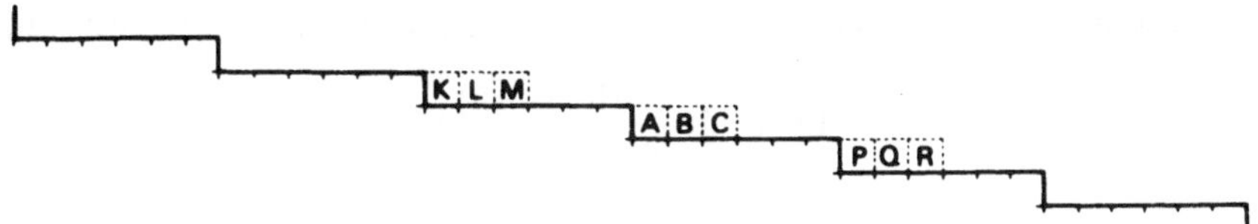


Fig. 1. Profile of the S face (016).

where $E_{cr}$ is the energy necessary to bring a molecule from a kink position into the disordered phase and $E_r$ the energy to detach a molecule from a half row. Taking only the first term of the series,

$$E'_A = \text{const} + E(n, 1). \tag{3}$$

Because the constant always has a negative value, the growth rate is smallest when E(n, 1) has its minimum value. In order to evaluate E(n, 1) we suppose that each ionic chain is electrostatically neutral, without a dipole moment perpendicular to its direction. Then the interaction energy can be written approximately as [4]:

$$E(n, 1) = K \sin 4(\Theta - \Theta_0)/R^4, \tag{4}$$

where R is the distance between the parallel chains and K is a constant dependent on the atomic positions:

$$K = 3e^2 Q/(p \cos 4\Theta_0), \tag{5}$$

where e is the electronic charge, p the period of the chain, and Q is

$$Q = 4 \sum z_i X_i Y_i \left(\sum z_i X_i^2 - \sum z_i Y_i^2\right), \tag{6}$$

with $z_i$ the valence of the i-th ion, and X and Y the coordinates of an ion along the b and c axes of Fig. 1, respectively. The summation extends over all ions in the chain per period p. For

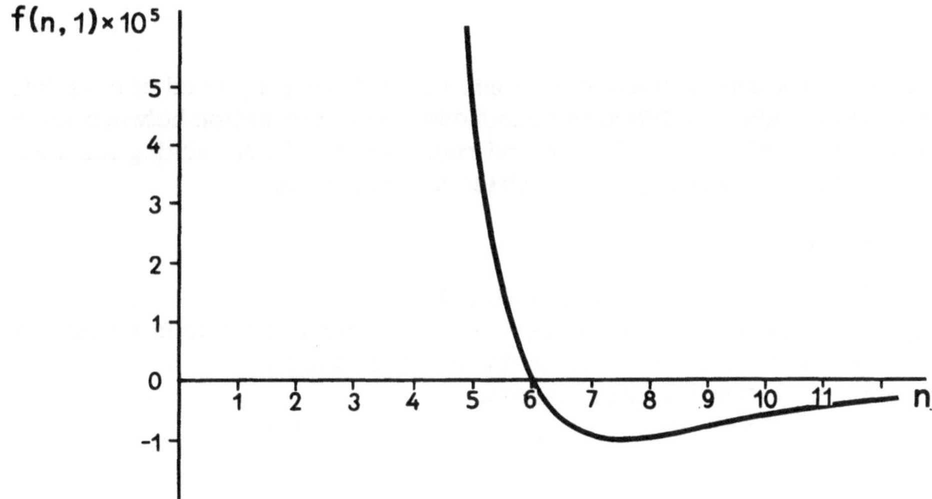


Fig. 2. The interaction energy E(n, 1) of two parallel ionic chains expressed as $f(n, 1) = E(n, 1)a^4/4K$.

small values of the angle $4(\Theta-\Theta_0)$ we can write in the case of the cubic model

$$E(n,\ 1)=\frac{4K(n_0-n)}{a^4 n n_0(n^2+1)^2}, \tag{7}$$

where $1/n_0 = \tan\Theta_0$. The function $f(n, 1) = E(n, 1)a^4/4K$ is shown in Fig. 2 for $n_0 = 6$. The minimum in the curve occurs approximately at $n = 5n_0/4$. A typical value of $4K/a^4$ is about $1/e^2 \cdot Å^{-1}$ (= 331.83 kcal/mole). For $n_0 = 3$, the minimum value occurs near n = 4 and $E(4, 1) = -0.00025\ e^2 \cdot Å^{-1} = -83$ cal/mole.

When $E_A$ is considered as an activation energy, the growth rate of the S face can be written in the form

$$v=C\exp\{E(n,\ 1)/RT\}\approx C\{1+E(n,\ 1)/RT\}, \tag{8}$$

where R is the gas constant, T the absolute temperature, and C a preexponential coefficient independent of n.

The question whether the face with the minimum value of E(n, 1) actually appears on the crystal depends on the angles it encloses with its neighboring faces and the growth rates of these faces. Preliminary calculations showed that for $n_0 = 3$ the face (014) will occur on the growth form.

The growth of the F face itself may be considered as the growth of an S face with very high indices, because at very low supersaturations the distance between steps is very large. As a consequence the S face with lowest growth rate may even become the largest face on the crystal as indeed is sometimes observed on mineral crystals.

During the normal growth process of the F face the S faces can not appear, because the distance between the steps does not decrease. It is therefore necessary that a period of slight dissolution occurs, during which the edge between two F faces is rounded off and many S faces are generated. When growth starts again, the S face with the smallest growth rate will be developed.

The results obtained so far are not restricted to S faces; they also apply to K faces making small angles with an F face.

## Conclusion

High-index faces may develop on an ionic crystal after a period of dissolution and subsequent growth under near-equilibrium conditions. The interaction between ionic chains must be attractive for faces close to an F face and repulsive for faces making larger angles with the F face. No special influences of impurities are required.

## Literature Cited

1. R. L. Perker, Z. Kristallog., 58:522 (1923).
2. P. Niggli, Lehrbuch der Mineralogie, Vol. 2, Borntraeger, Berlin (1926), pp. 138, 414.
3. R. M. Aliev, Dokl. Akad. Nauk SSSR, 172(5):1161 (1967).
4. P. Hartman, Acta Cryst., 12:429 (1959).
5. P. Hartman, Bull. Soc. Franç. Minér. Crist., 82:335 (1959).

# MACROSCOPIC STEPS ON VICINAL GROWTH SURFACES IN GERMANIUM

A. A. Tikhonova

*Institute of Crystallography, Academy of Sciences of the USSR, Moscow*

The surface of an autoepitaxial germanium film grown from a molecular beam under vacuum between 600 and 750°C produces a regular system of macroscopic steps, which appear when the surface is oriented with respect to (111) at more than 0.5°; these steps are parallel and cover the entire surface (Fig. 1). There is no obious source of the steps.

The height of these steps increases with the thickness of the film, the maximum height being 600-900 Å for a 70-μm film, whereas a 160-μm film gives steps of height up to 1000-1200 Å. The mean distance between steps increases as the disorientation angle decreases. The straight parts of the steps extend along $[1\bar{1}0]$ directions, and the normal to the end face lies along $[\bar{1}\bar{1}2]$. The steps tend to split up if the deviation from this orientation is large. Silicon films deposited from the vapor also [1] produce such steps if the orientation is appropriate, so it would seem that these steps provide stable film growth.

The surface state is independent of the residual gas pressure, at least in the range from $5 \cdot 10^{-6}$ to $5 \cdot 10^{-9}$ mm Hg. The interference microscope also shows that the surface shape reproduces that of the substrate, which means that the normal growth rate of the film is independent of the orientation in a macroscopic sense.

Here we discuss a possible growth mechanism for these steps on germanium, which involves collective interaction between the steps when the motion is controlled by atomic diffusion on the adjacent terraces. A computer has been used to simulate the motion of the elementary steps.

The kinematic theory [2, 3] gives the macroscopic change in surface profile during growth, whereas the description in terms of step displacements [4-7] allows one to determine the step interactions and the distribution over the moving front as a function of time. A detailed study has been made [4] of the latter feature in response to special conditions at the leading edge, and it has been found that perturbations propgate along the chain of equidistant steps from the front under certain conditions, and this can cause the steps to pair. However, the analytical approach used there allowed the authors to trace the step motion only up to the latter point. A simulation free from these restrictions allows one to consider the surface changes over longer time intervals.

Before we write the equations for the displacements of the fronts, we must make some comments on the condensation of germanium films. The vapor pressure of germanium over the solid is small, so films can grow under vacuum at appreciable rates only at very high supersaturations ($10^3$-$10^{10}$). In that case, the surface can develop by the attachment of ad-

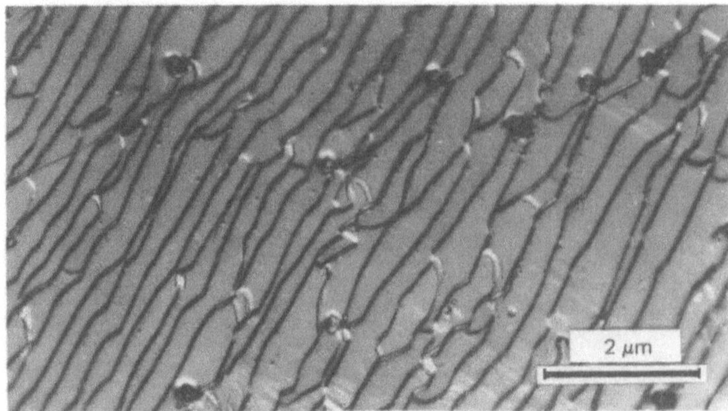


Fig. 1. Photomicrograph of part of a film deposited at about 750°C on a substrate oriented at about 1° with respect to (111).

sorbed atoms to steps only if the surface supersaturation is kept below a critical value on the steps by the action of sinks. We consider the simple case of a quasistationary symmetrical distribution of the adsorbed atoms along the terraces, which allows us to determine the maximum distance $\bar{l}$ between the steps, this maximum being the one immediately preceding nucleation. If the characteristic growth rate is about 0.1 μm/sec and the crystallization temperature is about 700°C, then $\bar{l} = 3 \cdot 10^{-4} \lambda_s$, where $\lambda_s$ is the effective migration length for an adsorbed atom before reevaporation, the result being about 100 times the interatomic distance. Therefore, the growth on a vicinal surface for shorter distances between steps, i.e., when the deviation from (111) is more than 0.5°, can be considered as due to displacement of the component steps. This estimate is in agreement with the observed fluxes of macroscopic steps for mean disorientations more than 0.5°.

Consider part of a vicinal surface composed of N parallel steps (Fig. 2); here the quasistationary and symmetrical distribution will be determined by the following conditions: (1) reevaporation can be neglected because $l_1 \ll \lambda_S$; (2) the step speed is proportional to the halfwidth of the adjoining terrace and is determined by the flux of material condensing there; (3) if a macroscopic step is formed from elementary steps, then the same flux of adsorbed atoms from adjacent terraces will provide displacement of an h-fold step with a speed inversely proportional to the height; and (4) a macroscopic step can remain stable during subsequent growth if the ends are perpendicular to $[\bar{1}\bar{1}2]$ directions. This last condition can be met by a variety of factors, e.g., by rearrangement of the bonds at the ends of multiple steps to give more stable configurations [1], or, if the sink capacity at a step is restricted, by uniform distribution of the incorporated particles between all constituents of a macroscopic step.

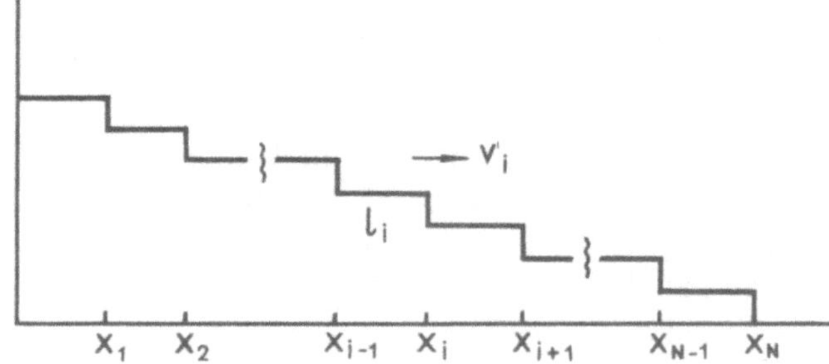


Fig. 2. Schematic representation of part of a vicinal surface, with symbols.

The deviation from uniform conditions for step motion is specified by selecting an initial shape or else by choosing appropriate boundary conditions at the edge of the front. In this example we envisage the case where the edge steps receive a constant flux from some part of length $\Delta/2$, with the lower edge of the front moving over an unbounded plane. Then the boundary conditions give us that the front moves in a way described by the following differential equations:

$$\frac{dx_1}{dt}=\frac{1}{2}\frac{1}{h_1}\frac{R}{N}[\Delta+(x_2-x_1)]$$

$$\frac{dx_i}{dt}=\frac{1}{2}\frac{1}{h_1}\frac{R}{N}[x_{i+1}-x_{i-1}] \qquad 2\ll i\ll N-1 \tag{1}$$

$$\frac{dx_N}{dt}=\frac{1}{2}\frac{1}{h_N}\frac{R}{N}[(x_N-x_{N-1})+\Delta],$$

where R is the intensity of the incident beam and N is the atomic density in a close-packed face. At the start, $t = 0$, we have $h_i = 1$, and $x_i(0)$ is governed by the initial surface shape.

This system was solved numerically by integration for various initial step distributions. Figure 3 shows a characteristic initial shape and the changes occurring during growth up to a mean thickness of 76.5 monolayers. The initial front consists of two parts having different slopes ($h_{i0}/l_{i0}$ = 1/3 and 1/$l$) and contains 100 elementary steps; we put $\Delta/2 = 5$ in the boundary conditions. The solid line joins the midpoints of the single steps and has been drawn on the scale of the multiple steps.

The form of the solution to (1) provides various conclusions that can be compared with experiment:

1. If the initial step density lies near the point where the changes are substantial during the motion, e.g., near the point of inflection on the initial shape or at the upper part of the front, then the individual steps show substantial mutual displacement, which can result in fluctuations in the step density or the formation of macroscopic steps, and the latter can also arise on reasonably smooth parts of the front, e.g., at a surface with a sinusoidal profile, if the magnitude of the perturbation propagating along the chain from the singular points exceeds the distance between some two steps. The macroscopic steps arise thus by a self-maintaining process, since such a step provides additional nonuniformity for adjacent elementary steps and increases the perturbation in the disposition.

2. The distance between macroscopic steps increases as the inclination of the surface to a singular face decreases; also, the mean height of the macroscopic steps increases with

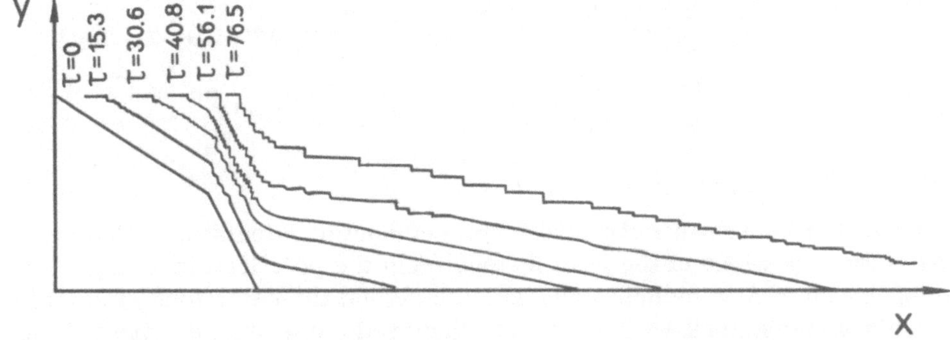


Fig. 3. Results from computer simulation of shape change during growth.

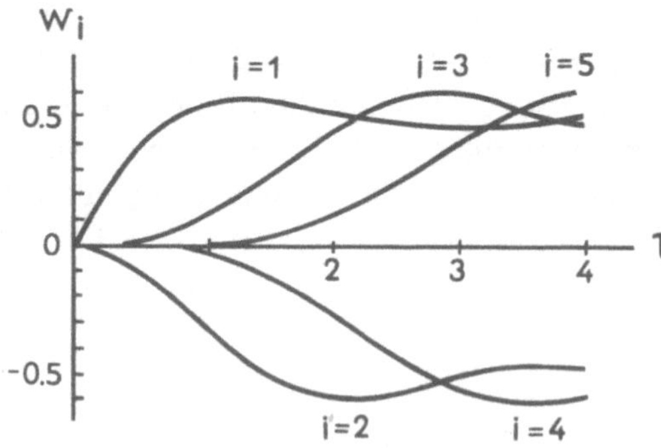


Fig. 4. Change in mutual disposition of steps on propagation of a perturbation along the front.

the film thickness. It follows from (1) that $x_i(t)$ is determined only by the layer thickness and is independent of the growth rate.

3. The mean macroscopic inclination of the surface persists during layer growth, i.e., the mean growth speed along the normal to a close-packed face is independent of the surface orientation. Figure 3 shows that the initial inclination of each of the two linear parts near the point of inflection remains the same until the effects of the boundary conditions propagate to the corner.

The conservation of the macroscopic shape is in accordance with the conclusions from the kinetic theory, since the flux of steps through any section of the surface should be constant if the speed of the steps is linearly related to the separation. However, macroscopic steps are formed although the mean inclination of the surface remains constant as a consequence of the interaction between steps, which is also incorporated in (1).

In order to trace more clearly the interaction between steps, we consider, as in [4], how a perturbation propagating over a semiinfinite chain of equidistant steps will behave when it arises from a stationary nonuniformity at the edge of a front. This nonuniformity corresponds to conditions where the front is bounded at the top by a sufficiently large area of a singular plane, as in the case of (1) above. If the supersaturation exceeds the critical value, condensation occurs on this flat part, so the conditions at the edge of the front should vary with time. However, if the influx to the top step does not change too greatly during the time required to produce two or three monolayers, then the change in the boundary conditions will not affect the propagation of the perturbation along the chain appreciably. In the case of dissolution or low supersaturation, the conditions formulated as above for the edge of the front enable one to judge how perturbations will propagate and also what will be the stationary distribution of steps along the front. Here the width of the sink on the top step on the singular-plane side will be $\lambda_s$.

In this case, (1) can be rewritten as

$$\frac{dx_1}{dt}=\frac{1}{2}\frac{R}{N}\left[\Delta+(x_2-x_1)\right]=\frac{1}{2}\frac{R}{N}(\Delta+l_1),$$
$$\frac{dx_i}{dt}=\frac{1}{2}\frac{R}{N}\left[x_{i+1}-x_{i-1}\right]=\frac{1}{2}\frac{R}{N}(l_i+l_{i+1}), \tag{2}$$

subject to the initial conditions $x_i(0) = il_0$.

The following results were obtained from a computer calculation. Figure 4 shows the time course of the deviations of the first 6 steps from the positions in an equidistant system during the deposition of four monolayers. It is clear that the deviations of adjacent steps are in opposite senses, being positive for the odd-numbered steps and negative for the even ones. At some instant, which is dependent on the distance of the step from the start of the front, a

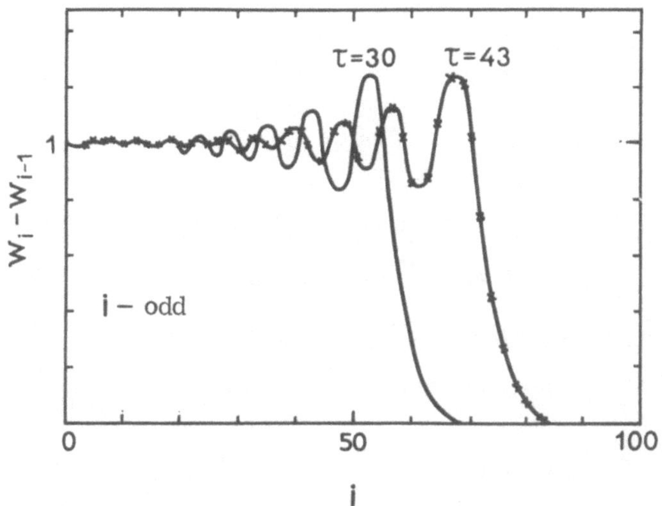


Fig. 5. Length change in the odd terraces in a chain of equidistant steps at two instants.

step is either accelerated or retarded by comparison with other steps, and then it continues to move at about the speed it had at the start. A perturbation propagates over the chain of steps with a constant speed. These conclusions agree with the results from analytical solution [4].

Figure 5 shows the length change in the uneven terraces as referred to the difference $(\Delta - l_0)$: $w_i - w_{i-1} = (l_i - l_0)/(\Delta - l_0)$ for two instants. The lengths of the even terraces vary in the same way, but the values are negative. It is clear that the displacement of adjacent steps one with respect to the other has a peak. If the steps are not to fuse, we must have for the maximum deviation $w_i - w_{i-1} \cong 1.3$ that $l_0 > u_i - u_{i-1}$, which can occur only if $l_0 > 0.56\Delta$; if the distances between terraces is less than this limiting value, then the steps will fuse at the start. If on the other hand $l_0 > 0.56\Delta$, the steps move in pairs, with the distance between the steps fluctuating around a constant value of $\Delta$ for the odd steps and $2l - \Delta$ for the even ones, tending ultimately to a constant value.

## Literature Cited


1. H. C. Abbink, R. M. Broudy, and G. P. McCarthy, J. Appl. Phys., 39:4673 (1968).
2. F. C. Frank, In: Growth and Perfection of Crystals, Wiley, New York (1958), p. 411.
3. A. A. Chernov, Uzp. Fiz. Nauk, 73:277 (1961).
4. W. W. Mullins and J. P. Hirth, J. Phys. Chem. Solids, 24:1391 (1963).
5. B. J. Mason, G. W. Bryant, and A. P. Van den Heuvel, Phil. Mag., 8:505 (1963).
6. T. Surek, G. M. Pound, and J. P. Hirth, J. Chem. Phys., 55:5157 (1971).
7. A. A. Chernov, Dokl. Akad. Nauk SSSR, 117:983 (1957).

# SIMULATION OF MODES OF VIBRATIONS IN TRAINS OF STEPS

**P. Bennema and R. van Rosmalen**

*Laboratory of Physical Chemistry, Delft University of Technology*

## 1. Introduction

Soon after the development of the spiral growth theory of Burton, Cabrera, and Frank [1], in which these authors assumed that steps emitted from step sources are equidistant trains, it became clear that this supposition does not hold in general. As a result of perturbations, all kind of bunching phenomena occur, which among others result in macrosteps. Therefore, Frank [2], and Cabrera and Vermilya [3] applied the traffic flow theory of Lighthill and Whitham [4] to the motion of steps across the surface. This theory was extended by Chernov [5], modified by Mullins and Hirth [6], and applied by several authors to explain observed bunching phenomena. Hulett and Young [7] were the first to carry out computer experiments to simulate their (electrochemical) etching experiments. Very recently Surek [8, 9] carried out extensive computer simulations of step trains. Schwoebel [10, 11] studied the effect of differences in flows entering the left- and the right-hand side of steps and showed that because of this "Schwoebel effect" step trains become unstable under certain well-defined conditions.

The traffic flow theories, based on the concept of kinematic wave, are general and vague in character. The theory of Schwoebel is more specific because it describes the relation between the relaxation times for entering the step and the stability of step trains. This theory however, is ad hoc in character, because it is not integrated in the BCF theory.

In this paper, we will first discuss the theory of Gilmer [12] in which a formalism is developed where small perturbations in the spacing between steps are considered as small deviations from the equidistant step train. Secondly, we will present more results of computer simulation experiments on step trains, which were carried out in order to check the validity of the theory of small deviation for large deviations. Finally, the effects of a varying supersaturation will be described briefly. A more detailed description of theory and experiments will be published elsewhere [13].

## 2. Gilmer's Formalism

Upon applying a Taylor expansion to the first derivative only, Gilmer found that

$$u_n(t)=u(0)e^{iqnl}\,e^{bt+iwt}, \qquad (1)$$

where $u_n(t)$ is a small deviation of the n-th step in reference to the equidistant step train and $u(0)$ is the amplitude of the initial deviation, $l$ the distance between steps, and q a constant. Further

$$b=[f'_+(l)-f'_-(l)]\,[\cos(ql)-1] \qquad (2)$$

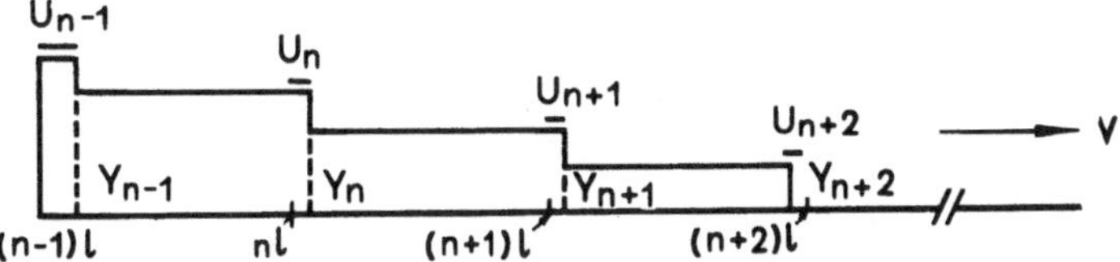


Fig. 1. $y_n$ = position of the n-th step; $u_n$ = deviation of the n-th step from the equidistant step train; v = direction of advance of the step train.

and

$$w=[f'_+(l)+f'_-(l)]\sin(ql). \tag{3}$$

Here $f(l)$ is the functional dependence of the step velocity on the distance between steps, $f'(l)$ is the first derivative (the subscripts + and − mean respectively in front of and behind the step, in reference to the direction of advance).

Three cases can be distinguished:

a) $f'_+(l)=f'_-(l)$; $b=0$; the perturbation continues unaltered;
b) $f'_+(l)>f'_-(l)$; $b<0$; the perturbation decays;
c) $f'_+(l)<f'_-(l)$; $b>0$; the perturbation increases.

The time $\tau$ required for the amplitude of a perturbation to grow or decay by a factor of e is given by

$$\tau=\frac{1}{b}=\frac{1}{(f'_+(l)-f'_-(l))\,(\cos ql-1)} \tag{4}$$

Allowing for periodic boundary conditions, the unspecified constant q can attain different values $q = 2\pi/\lambda$ corresponding to the wave numbers of modes, where $\lambda$ is the corresponding wavelength.

## 3. Computer Simulation

In the computer program, which is similar to the programs of Hulett and Young [7] and Surek [8, 9], we used first as trial functions the following series of coupled differential equations:

$$\frac{dy_n}{dt}=f_+(y_n-y_{n-1})+f_-(y_{n+1}-y_n)=K_1\tanh\frac{y_n-y_{n-1}}{2\lambda_s}+K_2\tanh\frac{y_{n+1}-y_n}{2\lambda_s}, \tag{5}$$

where $y_n$ is the position of the n-th step, $\lambda_s$ the mean displacement, and $K_1$ and $K_2$ constants. (In the case $K_1 \neq K_2$, we have the Schwoebel effect [10, 11].) In the experiments to be described below we used periodic boundary conditions (i.e., "steps on a circle"). By using small step trains on a circle we saved computer time and moreover obtained all the information needed for specific modes of vibration out of an "infinite" step train.

In order to study the modes of vibration in a systematic way we note that as in the theory of lattice waves [14, 15] a system of n coupled particles or n steps has n modes of vibration. For the case of four steps on a circle we have:

$$q=0,\ \pm\frac{2\pi}{4l},\ \frac{4\pi}{4l}\quad\text{and}\quad\lambda=\infty,\quad\lambda=2l,\quad\lambda=4l. \tag{6}$$

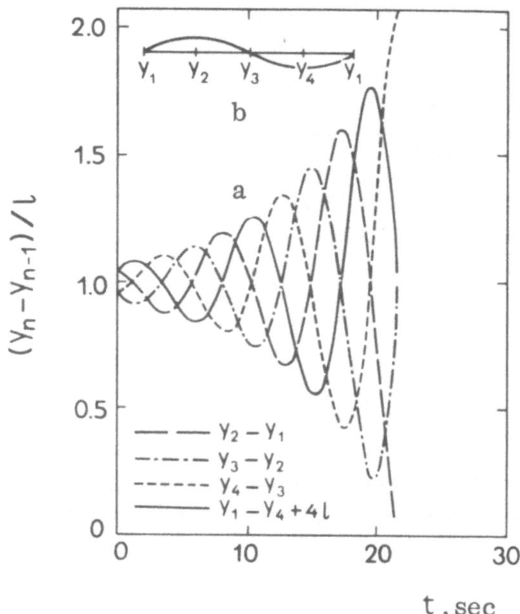


Fig. 2. (a) Distances between steps in units of $l$ as a function of time t. For the imposed perturbation see Fig. 2b. (b) Imposed $u_n$ (5% of $l$) presented as a transverse wave with $\lambda = 4l$.

In the experiments presented in Figs. 2 and 3, we have chosen $K_2 = 0.7K_1$ and $\lambda_s/l = 0.9$ for a system of four steps on a circle. It follows from (5) that $f_+^1(l) < f_-^1(l)$ so that the perturbation increases [case (c)]. We only need to consider the two modes corresponding to $\lambda = 2l$ and $\lambda = 4l$.

In Fig. 2a, b it is shown that we have imposed a very small deviation from the equidistant train as a sinusoidal perturbation with $\lambda = 4l$. It can be seen that the step train disintegrates by oscillations which increase in amplitude. In Fig. 3a, b results of another experiment are presented. Here we imposed a perturbation corresponding to $\lambda = 2l$. We see a

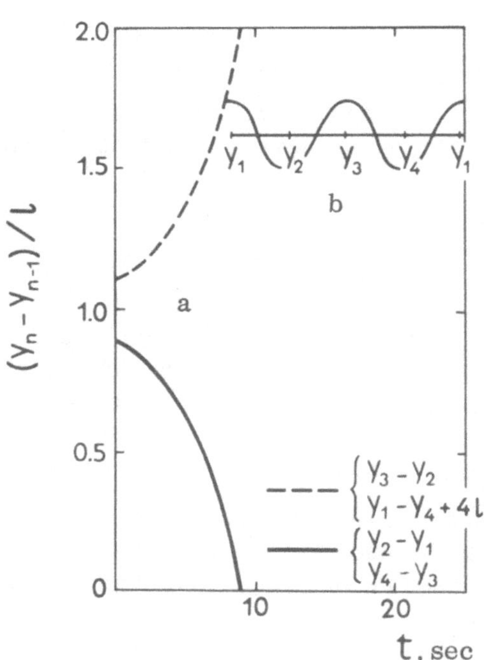


Fig. 3. (a) Distances between steps in units of $l$ as a function of time t. For the imposed perturbation see Fig. 3b. (b) Imposed $u_n$ (5% of $l$) presented as a transverse wave with $\lambda = 2l$.

disintegration in two times two double steps. However, no amplified oscillations occur. Obviously, this is because for $q = \pi/l$, $w \Rightarrow 0$ [see (3)].

In odd step trains, nonoscillating catastrophes do not occur since there is no mode making $w = 0$.

In Fig. 4 it is shown that in the case $K_1 = K_2$ an originally arbitrary perturbation of large amplitude propagates unperturbed as a superposition of two sinusoidal waves, with the two "eigenvibrations" corresponding to $\lambda = 2l$ and $\lambda = 4l$. Obviously, the theory for small perturbations also holds for large perturbations.

In Fig. 5 results of an experiment for the case $K_2 > K_1$ are presented. In contrast to the previous presentations the $u_n$ values are plotted against time t. It can be seen that the imposed mode with $\lambda = 2l$, having a large amplitude, eventually decays since all the $U_n$'s approach the same negative value. We thus observe that during the perturbation the flux of steps is less than the flux of steps for the unperturbed case.

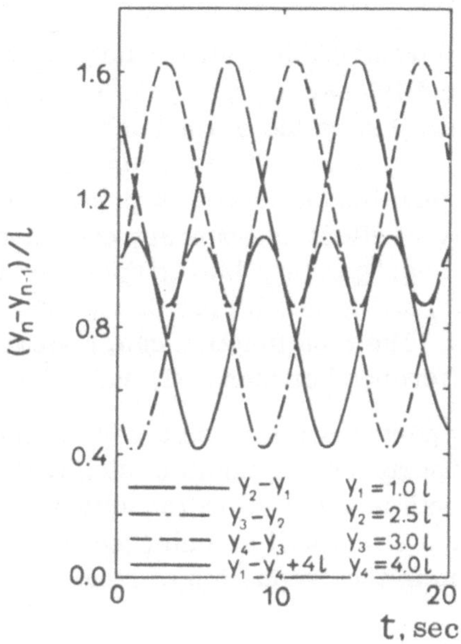


Fig. 4. $K_1 = K_2$. Distances between steps in units of $l$ as a function of time t. Originally at time t = 0 there is only a perturbation of 50% in the position of the second step.

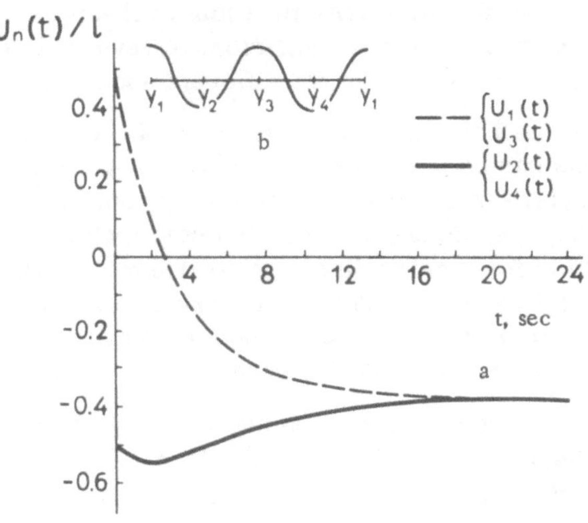


Fig. 5. (a) $K_1 = 0.7K_2$. In contrast to Figs. 2a, 3a, and 4, the deviation of the n-th step from the equidistant train $u_n(t)$ in units of $l$ is plotted against the time t. (b) Imposed $u_n$ (50% of $l$) presented as a transverse wave with $\lambda = 2l$.

## 4. Summary of the Results

The following conclusions may be drawn from Gilmer's formalism and the computer simulation of modes of vibration.

1. Every step train with periodic boundary conditions has its own modes of vibration. The conclusion for small step trains also holds for "infinite" step trains. These have an "infinite" number of modes of vibration. From this "infinite" number of modes, any shape of perturbation (peaks, etch pits, shock waves, etc.) can in principle be represented by a Fourier sum of modes. In principle the history of a particular perturbation can then be predicted on the basis of the knowledge of separate modes. This is the importance of our approach.

2. Small imposed arbritrary perturbations on a large step train of 100 steps decay, and only very small deviations from the equidistant step train remain in case (a) mentioned above. Large arbitrary perturbations in large step trains, however, give rise to collisions for the cases (a), (b), and (c). This is not due to the periodic boundary conditions, because the collisions did occur before the perturbations spreading over the circle in two opposite directions met each other.

3. Gilmer's theory for small deviations holds reasonable well for large deviations. In case (c) the normal modes eventually give rise to collisions. By using the hyperbolic tangent functions for the advance velocity of steps, the theoretical value of the time $\tau$ of (4) proved to give a good approximation of the average of the observed $\tau$. If, however, $f_+^!(l)$ is quite different from $f_-^!(l)$, we must use for $f_\pm(l)$ the more complicated "exact" expressions (4.20) and (4.21) of ref. 12, obtained by introducing the Schwoebel effect as boundary conditions into the BCF theory. In this case the observed $\tau$ is on the average much larger than the theoretical $\tau$ of (4). Thus, step trains are more stable in case (c) and give a larger decay time in case (b) than predicted by the theory of small perturbations. The tanh function can, however, be used as an approximation in the case of a very small difference between $f_+(l)$ and $f_-(l)$.

4. For the three cases mentioned above, any perturbation gives a reduction of the flux of steps as compared with the flux of steps of an unperturbed equidistant step train. Obviously the "equidistant" BCF theory, in which only the zeroth mode of vibration with $\lambda = \infty$ is taken into account, gives the maximal rate of growth. This effect will be enlarged if the perturbation arriving at the source of the steps gives a reduction in the production of steps, which seems to be highly probable.

5. When a train of steps passes a temporal zone of higher or lower supersaturation, a mixture of modes is excited and behaves as described above. However, if there is a permanent zone of different supersaturations, collisions will occur eventually for the three cases mentioned above. Further collisions always will occur at those places, where a sharp discontinuity appears in the zone of the different supersaturations.

6. The physical significance of the modes of vibration for the cases of even and odd step trains can be explained as follows. The nonoscillatory mode (see Fig. 3b) means for case (a) that pairs of steps form a stable system, because all the steps have the same catchment areas. Such modes do not occur in odd step trains, since if, for example, a fifth step is added to a step train of four, pairs disintegrate due to an interaction with the fifth step. This gives then an oscillatory mode. The oscillatory mode of Fig. 2b for four steps can be considered as a single step and a triplet of steps which continuously exchange two steps, so that the single step becomes a triplet, and vice versa.

We note that Mullins and Hirth [6] found similar results from their formalism. We want to thank Drs. A. A. Chernov, V. V. Voronkov, and T. Surek for their criticism and stimulating discussions during the Conference in Tsachkadsore. We also want to thank Dr. F. H. M. Mischgofsky for his interest and discussions.

## Literature Cited


1. W. K. Burton, N. Cabrera, and F. C. Frank, Phil. Trans. Roy. Soc. A243:299 (1951).
2. F. C. Frank, Disc. Faraday Soc., 5:48 (1949).
3. N. Cabrera and D. A. Vermilyea, Disc. Faraday Soc., S:393 (1949).
4. M. J. Lighthill and G. B. Whitham, Proc. Roy, 229A:281 (1955).
5. A. A. Chernov, Sov. Phys. – Uspekhi, 4:129 (1961).
6. W. W. Mullins and J. P. Hirth, J. Phys. Chem. Solids, 24:1391 (1963).
7. L. D. Hulett, Jr. and F. W. Young, Jr., J. Electrochem. Soc., 113:410 (1966).
8. T. Surek, G. M. Pound, and J. P. Hirth, J. Chem. Phys., 55:5157 (1971).
9. T. Surek, J. Crystal Growth, 13/14:19 (1972).
10. R. L. Schwoebel and E. J. Shipsey, J. Appl. Phys., 37:3682 (1966).
11. R. L. Schwoebel, J. Appl. Phys., 40:614 (1969).
12. P. Bennema and G. H. Gilmer, Theory of Crystal Growth, Chapter 4 in the Proceedings of the Summer School in Noordwijkerhout, North Holland, Amsterdam (1973).
13. R. van Rosmalen and P. Bennema, Amsterdam (1973).
14. C. Kittel, Introduction to Solid State Physics, 3nd ed, Wiley, New York (1968), p. 170.
15. J. M. Ziman, Principles of the Theory of Solids, Cambridge University Press (1965), Chapter 1, 2.

# THE DISTORTION MECHANISM FOR PHOTOLITHOGRAPHIC PROJECTIONS IN THE EPITAXY OF SILICON

A. F. Volkov and N. S. Papkov

The photolithographic relief becomes distorted during epitaxial growth on silicon, e.g., in integrated-circuit manufacture, and the initial relief, which usually takes the form of ridges or depressions of rectangular shape and of height 0.05-0.2 μm, tends to be displaced and deformed during the growth [1-3]. It has been found that the extent of this distortion is dependent on the angle and direction of the surface with respect to the close-packed planes, as well as on the working temperature. It has not proved possible to explain the effect in terms of growth-rate anisotropy for macroscopic planes, since (a) the face of a step is formed not by a (111) plane, which has the minimal growth rate, but by a vicinal plane at 15'-35' to (111) [3], and (b) the observed displacements of the faces at small angles exceed considerably those expected from the relationship between growth rate and angle [5]. The temperature dependence implies that this effect should be considered from the viewpoint of the microscopic processes at the surface [2]. Here we explain the effect in terms of the theory of layered growth.

We examined the distortion of the initial relief on spherical specimens of large radius of curvature (500 mm), this relief taking the form of ridges of 0.4 × 0.4 mm square cross sections. Figure 1a shows the general pattern, while Fig. 1b is the corresponding scheme. The disorientation angle on any direction increased from the central (111) zone to the periphery. Vicinal pyramids grew in the central zone, and the photolithographic pattern was completely eliminated. The disorientation with respect to (111) increased away from the central zone, while the distortion of the relief became less. The various forms of distortion indicate that there are two particular disorientation directions: (1) with respect to the nearest (110) plane (the $[11\bar{2}]$ direction) and (2) with respect to the (100) plane (the $[\bar{1}\bar{1}2]$ direction). The features of the relief can be explained if we assume that the epitaxial film grows by a layer mechanism as in the model due to Barton et al. [4]. In that model, the crystal can grow if there is a constant source of steps, and the speeds of the atomic steps are determined by the overlap between the diffusion fields.

It would seem that the sources of the steps in the central zone were screw dislocations emerging on the surface, or other lattice defects, which result in vicinal growth pyramids (Fig. 1); the shape of these pyramids indicates that there is anisotropy in the attachment of adsorbed atoms to the growth steps, and the speed of the steps along $[11\bar{2}]$ is minimal, while it is maximal in the opposite direction and in a direction perpendicular to $[11\bar{2}]$.

This agrees well with the behavior of the free bonds on the atomic steps, which move along the (111) planes in various directions.

At a certain angle, which is comparable with the angle of inclination of the faces on the vicinal pyramids, the step density becomes determined not by the dislocations or other de-

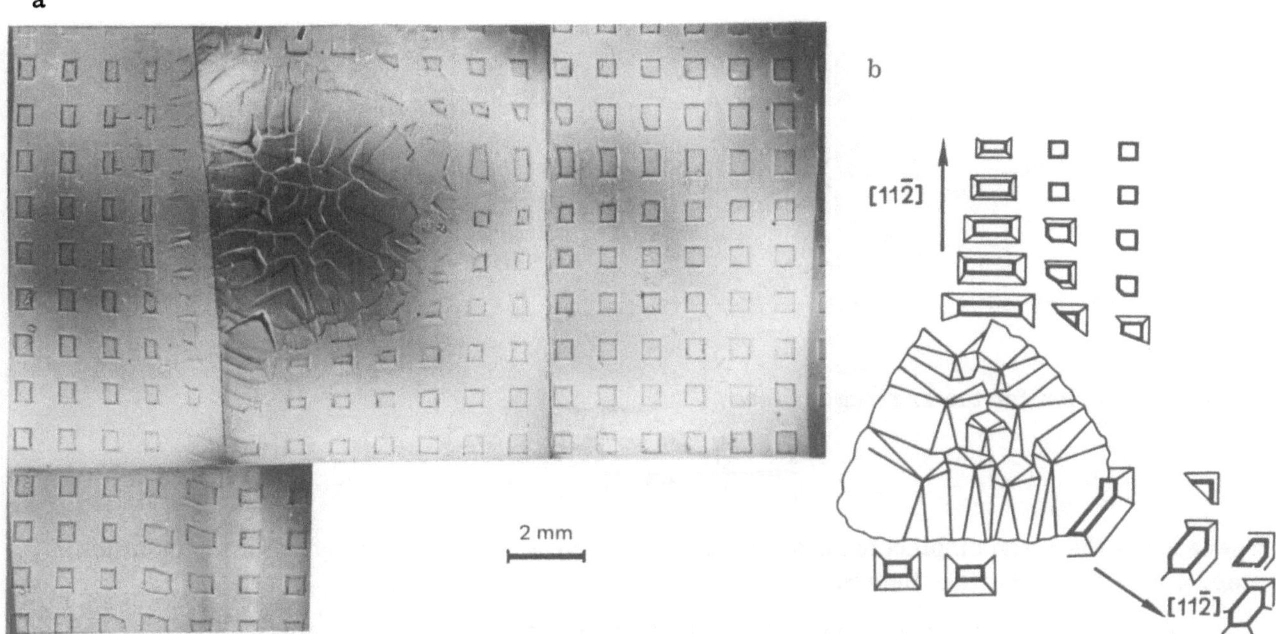


Fig. 1. (a) General pattern of ridge distortion on a spherical specimen; (b) scheme for distortion on a spherical specimen.

fects but by the general inclination of the surface to the close-packed plane. If such a surface has photolithographic ridges, a (111) plane is initially formed on the incident-step side, and then some vicinal plane, which lies at 20'-30' to the close-packed plane.

The resulting ridge has leading and trailing edges formed by systems of steps, whose distances apart on the front face are less than those on the rear face. During the growth, the atomic steps on the front face move more slowly than those on the rear one, so the square ridge becomes smaller along the step direction.

Also, the sizes of the square steps are reduced in the perpendicular direction. This size reduction is more pronounced when the disorientation is in the direction of minimum growth rate, namely [11$\bar{2}$]; this is due to interaction between the diffusion fields of the steps in the two mutually perpendicular directions, and the step speeds perpendicular to the disorientation direction will be the larger the less the competitive effect from the step motion in the disorientation direction. From this viewpoint, one can readily explain also the formation of hexagonal ridges with respect to the [121] direction, which is the direction of minimum step speed.

We have seen above that the steps are reduced in size along this direction and enlarged in the perpendicular direction, which results in hexagons; therefore, the layer-growth model provides a qualitative explanation of the effect and would appear to reflect the actual growth mechanism.

In a quantitative examination, we restrict ourselves to the case where the step source is the inclination of the surface with respect to the close-packed plane. For simplicity, we consider a two-dimensional system in which the surface of the crystal is described by the coordinates x and y. The y direction is taken perpendicular to the close-packed plane, while the x direction is perpendicular to the fronts of the monatomic steps (Fig. 2).

We assume that the curvature of the surface does not change substantially over the interval between two steps, in which case the expression for the step speed [4] can be put as a dif-

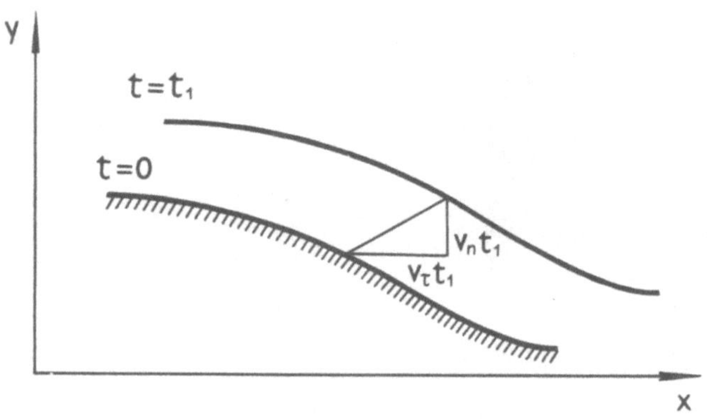


Fig. 2. Scheme for displacement of growing surfaces in x, y coordinates.

ferential equation for the growing surface:

$$\frac{\partial x}{\partial t}=\frac{2\lambda_s H}{h}\tanh\left(-\frac{h}{2\lambda_s}\frac{\partial x}{\partial y}\right), \tag{1}$$

where $\lambda_s$ is the mean displacement of an adsorbed atom and h is the height of a monatomic step, where H is a certain factor.

We envisage a particular case where the part of the growing surface has a range of variation in $\partial x/\partial y$ such that the step speed as a function of the derivative is described approximately by a linear relationship:

$$\frac{\partial x}{\partial t}=\frac{2\lambda_s H}{h}\left(-\frac{ch}{2\lambda_s}\frac{\partial x}{\partial y}+d\right), \tag{2}$$

where c and d are constants whose meanings will be clear from Fig. 3. The method of characteristics enables us to solve (2) in parametric form:

$$y=Hct+y_0,\quad x=\frac{2\lambda_s Hd}{h}t+x_0, \tag{3}$$

where $y_0$ and $x_0$ describe the initial relief; (3) shows that the initial surface is displaced parallel to itself, and $V_n = Hc$ and $v_\tau = 2\lambda_s Hd/h$ define the rates of the normal and tangential displacements relative to the close-packed plane (Fig. 2).

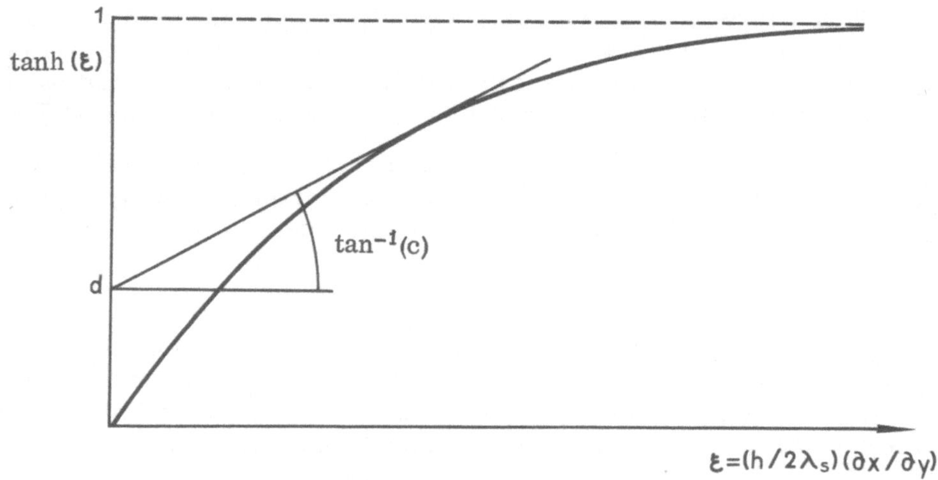


Fig. 3. Graphical interpretation of constants c and d.

It is clear that $v_\tau$, the tangential growth speed, may be considered as a distortion characteristic. Figure 4 shows $v_\tau$ as a function of $\lambda_S$ and $\partial x/\partial y$, and it is clear that $v_\tau$ increases with $\partial x/\partial y$ for any $\lambda_S$, i.e., as the disorientation angle is reduced, or, in other words, the overlap between the diffusion fields of adjacent steps falls the curve representing $v_\tau$ as a function of $\lambda_S$ at a constant disorientation angle has rising and falling branches; as $\lambda_S$ is also dependent on the temperature, we can say that the temperature dependence of $v_\tau$ will also have rising and falling branches.

We consider now the case where the growing surface has a range of inclinations such that one cannot neglect the nonlinearity in the inclination dependence of the growth rate.

Then the different parts of the surface have different speeds $v_n$ and $v_\tau$, and therefore the initial relief will be distorted during the displacement; under these conditions, (1) describes the motion, and the parametric solution takes the form

$$x=\left\{\frac{2\lambda_s H}{h}\tanh\left(-\frac{h}{2\lambda_s}\frac{1}{\partial y_0/\partial \alpha}\right)+\frac{H}{\partial y_0/\partial \alpha}\mathrm{ch}^{-2}\left(-\frac{h}{2\lambda_s}\frac{1}{\partial y_0/\partial \alpha}\right)\right\}t+x_0(\alpha)$$
$$y=H\mathrm{ch}^{-2}\left(-\frac{h}{2\lambda_s}\frac{1}{\partial y_0/\partial \alpha}\right)t+y_0(\alpha), \tag{4}$$

where t is time and $\alpha$ is a parameter corresponding to x for t = 0; (4) describes the motion of parts of the surface having $\partial y/\partial x$ constant.

In the x, y plane, the motion occurs along straight lines whose inclinations are determined by $v_n/v_\tau$; it is clear that (1) describes correctly the motion only for the time preceding the formation of the shock waves, i.e., of parts of the surface where there are discontinuities in $\partial x/\partial y$.

The motion described by (4) may be compared with the observed picture (Fig. 1) by taking the initial surface as a projection on a plane disoriented with respect to the close-packed plane (Fig. 5). The formation of shock waves in the initial stage is eliminated by taking the initial surface such that the sign of $\partial x/\partial y$ does not change at any point.

The rear edge of a step then has a face close to a close-packed plane; the surface that satisfies these conditions may be specified by a system of parametric equations:

$$x=\alpha;\quad y=k\alpha+l\left[1+\left(1+e^{\frac{\alpha-a_1}{b_1}}\right)^{-1}-\left(1+e^{\frac{\alpha-a_2}{b_2}}\right)^{-1}\right], \tag{5}$$

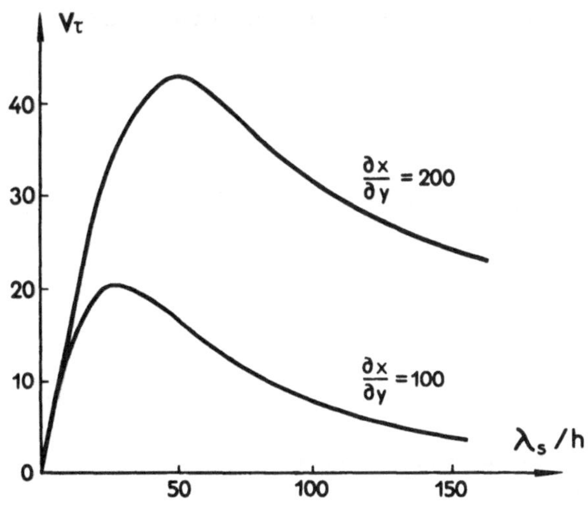


Fig. 4. Lateral-displacement rate as a function of $\lambda_s$ and disorientation angle.

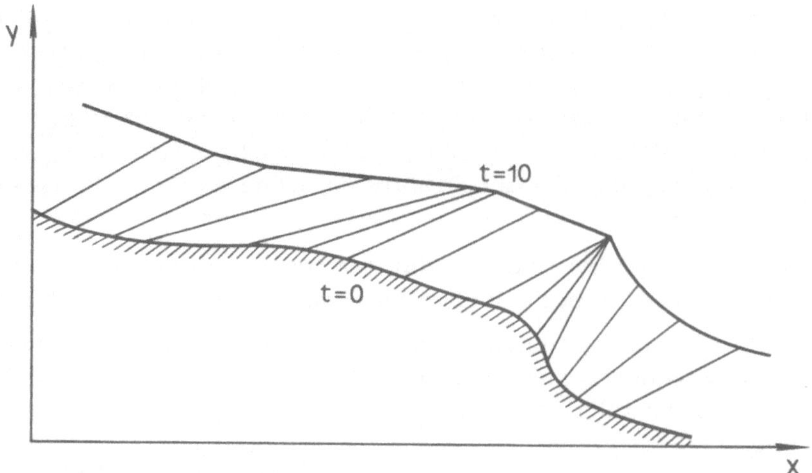


Fig. 5. Displacement and deformation of a step during growth.

where k is the tangent of the angle of inclination, $l$ is the height of a step, and the constants $a_1$, $b_1$, $a_2$, and $b_2$ define the positions of the leading and trailing edges of a ridge and the inclinations of these. The expressions resulting from substituting (5) into (4) are complicated, so computer calculations were employed. Figure 5 shows the shape of the initial surface and the shape at a time preceding the formation of shock waves at the leading edge. It is clear that the initial ridge is displaced and takes up a shape such that the upper flat area becomes narrower, while the leading edge becomes elongated. The calculated form is in general agreement with the observations.

A complete study of the motion involves simulating the initial stage of growth, which precedes the formation of the surface taken as the initial one in Fig. 5; here we are interested in the origin of the region where $\partial x/\partial y$ changes sign. This was also simulated by computer, the initial surface formed by a system of steps (Fig. 6). The motion of each step was described by equations, which in difference form take the following style:

$$x_i(t+\Delta t) = x_i(t) + \frac{\lambda_s H}{h}\left[\tanh\left(\frac{x_i(t)-x_{i-1}(t)}{2\lambda_s}\right) + \tanh\left(\frac{x_{i+1}(t)-x_i(t)}{2\lambda_s}\right)\right]\Delta t,$$

where i is step number and $\Delta t$ was chosen to give stability in the solution.

During the motion, there is overgrowth on the regions adjoining the rear edge and development of close-packed planes; the subsequent motion results in a vicinal surface disoriented with respect to the close-packed plane.

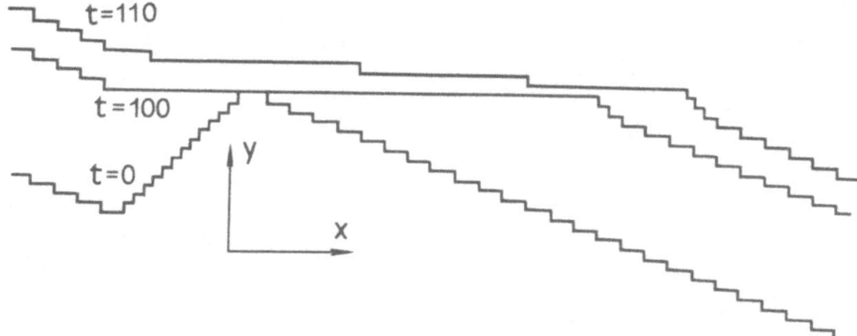


Fig. 6. Successive stages in the motion of a surface derived by simulation.

A type of projection arises at the point where the vicinal surface intersects the upper flat surface, which is due to the difference in speeds of the steps on the intersecting planes. A similar ridge has been observed on epitaxial silicon films where close-packed planes and disoriented planes intersect. Measurements on the deflection of interference fringes indicate that the height of such a ridge is of the order of 0.1 $\mu$m.

An interesting point is that the surface deviates from flat during the motion of these atomic steps, and pairs of steps fuse into double steps, with a subsequent tendency for macroscopic steps to arise. A more detailed analysis of such step coalescence is being prepared for publication.

Therefore, the distortion of the relief during epitaxy of silicon is described satisfactorily within the theory of layer growth. The resulting mathematical model serves to correlate the trends in the displacement of the individual parts with the parameters characterizing the motion of atomic steps.

## Literature Cited

1. C. M. Drum and C. A. Clark, J. Electrochem. Soc., 115:664 (1968).
2. C. M. Drum and C. A. Clark, J. Electrochem. Soc., 117:1405 (1970).
3. K. A. Valiev et al, Proceedings of the First All-Union School on Epitaxy [in Russian], MIET (1971).
4. W. Barton, N. Cabrera, and F. Frank, In: Elementary Processes in Crystal Growth [Russian translation], IL (1959), p. 11.
5. S. Mendelson, In: Single-Crystal Films [Russian translation], Mir (1966), p. 282.

# EFFECTS OF DOPES ON THE ANISOTROPY IN THE RATE OF GROWTH OF GERMANIUM FROM THE VAPOR STATE

A. N. Stepanova

*Institute of Crystallography, Academy of Sciences of the USSR, Moscow*

We have examined the effects of traces of $AsCl_3$ and $PCl_3$ on the growth rates of faces of germanium, which was deposited on single-crystal substrates of various orientations by reduction of $GCl_4$ in flowing hydrogen. The equipment was vertical and the heating was of induction type [1]. The substrate was at 850°C, apart from any cases specially mentioned, while the gas flow speed in the cold part of the reaction tube was 1.4 cm/sec, and the molar concentration of $GeCl_4$ in relation to hydrogen was $M_{GeCl_4/H_2} = 2.6 \cdot 10^{-3}$. The traces of $AsCl_3$ and $PCl_3$ were introduced into the gas in controlled amounts.

## 1. Growth Rate as a Function of Dope Concentration

Figure 1 shows the growth rate as a function of the molar concentration of $PCl_3$ with respect to $GCl_4$ and Fig. 2 does the same for $AsCl_3$.

It is clear that the two dopes have similar effects, but $PCl_3$ is effective at lower temperatures; it has no effect at 850°C. The fall in growth rate at high dope concentration ultimately gives way to etching, which is due to thermodynamic causes. However, there is another feature: above a certain dope concentration, a (110) face grows appreciably less rapidly than a (111) or (100) face. This anisotropy is growth rate is not essentially related to any general fall in growth rate, since it can occur at lower concentrations (Fig. 1). These data show that the growth rates of germanium faces are the same in the absence of dope or at low dope concentrations (under the conditions used here), and growth-rate anisotropy appears only in response to doping.

All the above tests were performed with plates whose orientations deviated from a low-index given face by 1–2°; we used spherical substrates in order to make measurements on surfaces with more closely defined orientations.

## 2. Measurement of the Minimum Growth Rate for a Smooth Face

We measured the ratio of the growth rate V for a surface deviating from a given face to the growth rate $V_0$ for that face; the measurements were made as follows. The films were grown on spherical substrates containing the orientations (111), (110), (100), and (311); in the first few experiments, the growth on each substrate was performed in the absence of dope, and this followed several runs in the presence of $AsCl_3$ ($M_{AsCl_3/GeCl_4} = 3.5 \cdot 10^{-2}$). After each run the face was photographed, and Fig. 3 shows an example of a (111) face in a germanium film grown without dope on a spherical substrate.

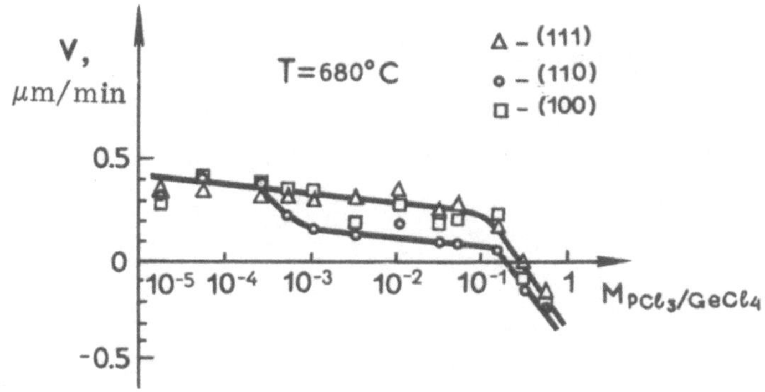


Fig. 1. Growth rate of germanium as a function of $PCl_3$ concentration in the gas phase.

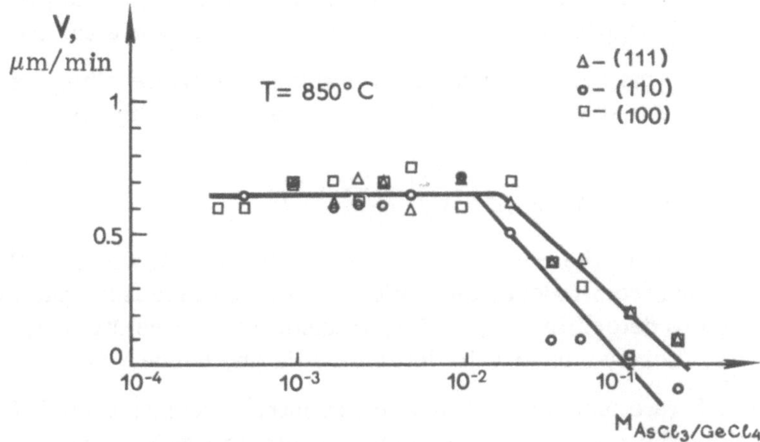


Fig. 2. Growth rate of germanium as a function of $AsCl_3$ concentration in the gas phase.

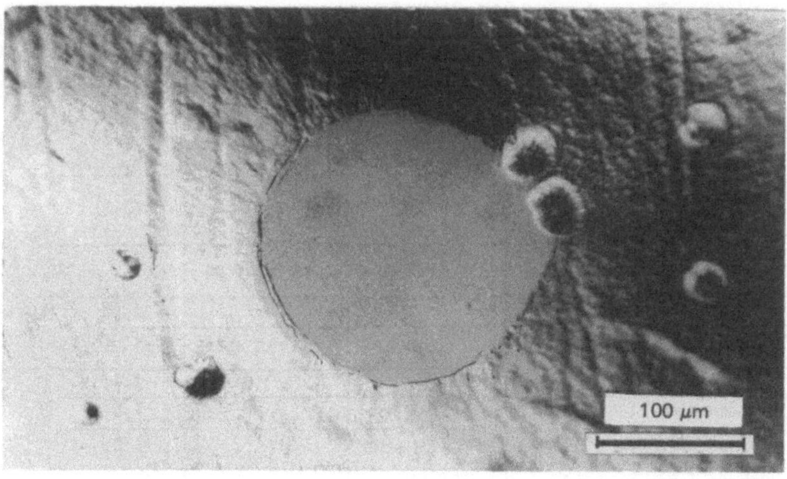


Fig. 3. (111) face on a germanium film grown without dope on a spherical substrate.

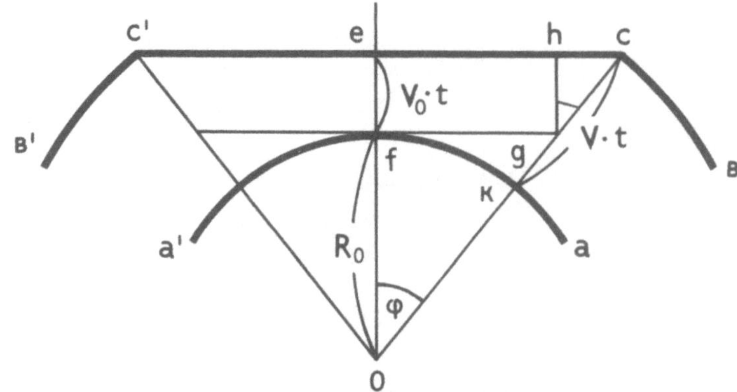


Fig. 4. Scheme for determining the relationship between growth rates V and $V_0$.

If the initial radius is known together with the face size, and if one of the growth rates (V or $V_0$) can be determined independently, one can find the other and hence the ratio $V/V_0$.

In Fig. 4, the circle a'a is the surface of the initial substrate, while b'c'cb is the growth surface after a time t, and c'c is a smooth face. The triangles ghc and 0fg give us that

$$\frac{V}{V_0}=\frac{Vt}{V_0t}=\frac{1}{\cos\varphi}+\frac{R_0}{V_0t}\left(\frac{1}{\cos\varphi}-1\right), \quad (1)$$

where V relates to the orientation to the right of c (and to the left of c'), which is the point of face propagation. If $V_0$ is known from independent evidence, one can measure the face size after a given time and use (1) to determine $V(\varphi)$. If $V_0$ is unknown, the relationship can still be utilized in relative terms. Here we used V to determine $V_0$ and hence $V/V_0$.

As V we used the growth rate measured with a flat substrate deviating slightly from the exact face orientation (Fig. 2); the range in $\varphi$ for which we determined $V/V_0$ was 0.1-1.1°, and the minimum angle was determined by the minimum face size that could be observed. The upper limit was set by the distortion that set in at the boundary of the spherical surface after a period of growth, where a diffuse macrostep was produced, so it became uncertain what orientation then corresponded to V.

We found that $V/V_0$ remained constant within this range of angles to within 1-3%, which means that V is independent of $\varphi$ within this range in the latter, i.e., the minimum growth rate extends down to deviations $\varphi$ ~0.1°. Table 1 gives the mean $V/V_0$ for the various faces.

TABLE 1

| Specimen | Growth conditions | (111) | (110) | (100) | (311) |
|---|---|---|---|---|---|
| 1 | Without dope | | | Face not seen | 1.07 |
| | With dope | | | 1.15 | 1.08 |
| 2 | Without dope | | | Face not seen | 1.07 |
| | With dope | | | 1.19 | 1.10 |
| 3 | Without dope | 1.10 | 1.04 | | 1.07 |
| | With dope | 1.17 | Face not seen | | 1.08 |
| 4 | Without dope | 1.14 | 1.14 | | 1.08 |
| | With dope | 1.17 | Face not seen | | 1.06 |

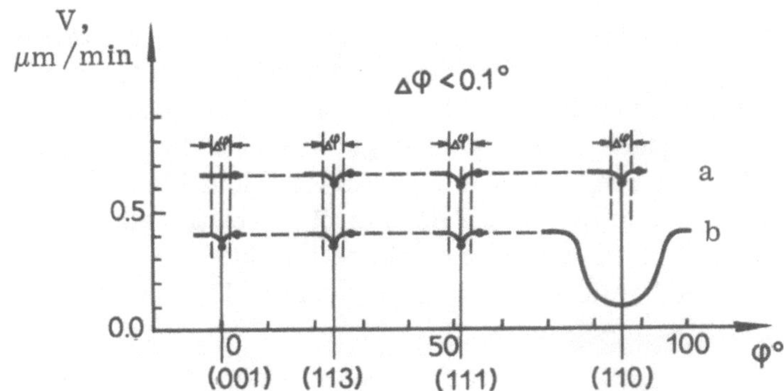


Fig. 5. Growth rate of germanium as a function of orientation for growth: (a) without dope and (b) with $AsCl_3$ dope.

It is clear from Table 1 that $V/V_0$ is 1.1-1.2 for all the faces appearing on the growth surface, no matter what the dope level, so the growth rates are the same for all orientations in the absence of dope, with only minor minima in the growth rate for the (111), (110), and (311) faces (Fig. 5a).

In the presence of $AsCl_3$, a (100) face shows a minimum growth rate (Table 1), while the growth rate of a (110) face is substantially reduced (Fig. 1). However, it proved impossible to measure the growth rate as a function of orientation near a (110) face in this way because this face did not produce a plateau in the presence of $AsCl_3$.

## 3. Measurement of Growth Rate as a Function of Orientation Near a (110) Face

A detailed study of $V(\varphi)$ near (110) was made as follows; a spherical surface containing the (110) orientation was used in the presence of $AsCl_3$ ($M_{AsCl_3/GeCl_4} = 3.5 \cdot 10^{-2}$), and the surface afterwards clearly showed the position of the (110) face to the unaided eye, along with the major crystallographic directions. A Talysurf profile recorder was used to measure the relief along the [112] direction (Fig. 6b). A circle on the sphere corresponds to a horizontal straight line on the chart, i.e., the instrument records the deviation of the relief from a regular circle. For

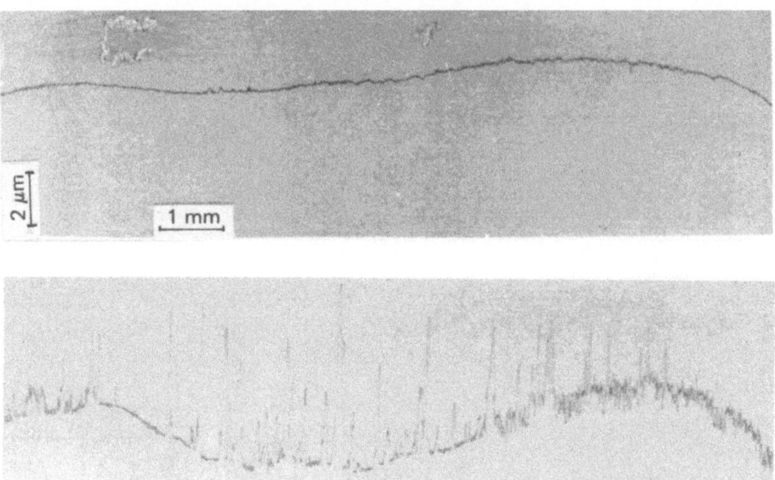


Fig. 6. Surface relief of film growth on a spherical substrate: (a) without dope and (b) with $AsCl_3$ dope.

comparison, Fig. 6a shows the relief after growth in the absence of dope on an analogous spherical surface. It is clear that a (110) face has a minimum growth rate, and this minimum extends over a large angular range (~20°), which is a difference from other faces. The small peaks on the recording correspond to growth pyramids, which occur in large numbers on a (110) face grown in the presence of $AsCl_3$. The growth-rate diagram takes the form shown in Fig. 5b in that case. The corresponding morphological changes have previously been described [2]. The more marked effect of the dope on the growth rate for (110) is ascribed to stronger adsorption on this face. Indirect evidence for this comes from the deposition of arsenic from supersaturated solid solution in germanium [3]. Deposits oriented on (110) planes of germanium are the most stable.

## Conclusions

1. $AsCl_3$ and $PCl_3$ affect the growth rates of germanium faces in similar ways, but $PCl_3$ exerts its effects at lower temperatures.

2. The growth rates of all faces are reduced at high dope concentrations, and there is a tendency for growth to give way to etching.

3. Above a certain dope concentration, a (110) face grows much more slowly than the others.

4. In the absence of dope, (111), (110) and (311) faces have minimal growth rates ($V_0/V$ = 1.1-1.2; $\Delta\varphi \leqslant 0.1°$), while in the presence of $AsCl_3$ similar minima occur for (111), (100) and (311) faces.

5. In the presence of $AsCl_3$, the minimum in the growth rate for (110) becomes much deeper and wider ($V/V_0 = 4$; $\Delta\varphi \simeq 20°$ for $M_{AsCl_3/GeCl_4} = 3.5 \cdot 10^{-2}$).

## Literature Cited

1. E. I. Givargizov, Fiz. Tverd. Tela, 6:1804 (1964).
2. A. N. Stepanova and N. N. Sheftal', Kristall u. Technik, 7:133 (1972).
3. V. N. Rozhanskii and N. D. Zakharov, Fiz. Tverd. Tela, 13:440 (1971).

# EFFECTS OF CRYSTALLIZATION CONDITIONS ON GROWTH AND DOPING ANISOTROPY FOR EPITAXIAL GERMANIUM

**L. G. Lavrent'eva, I. S. Zakharov, I. V. Ivonin, and S. E. Toropov**

*Kuznetsov Siberian Technical Physics Institute*

Our previous studies have shown [1, 2] that the growth-rate anisotropy for the Ge – HI – $H_2$ system has some major features, namely the growth rate falls when the orientation deviates from the (111) and (100) singular faces. Additional studies on the anisotropy in the orientation range (111)-(110) have confirmed that the growth rate is maximal at the (111) point (Fig. 1). The general applicability of this result was examined by means of additional experiments with a Ge – $I_2$ system, the general working conditions being as for the iodide – hydrogen system, with the source and substrate temperatures 600 and 400°C respectively. The iodine concentration was 0.6 mg/cm$^3$, while the source was doped with gallium ($2 \cdot 10^{17}$ cm$^{-3}$). Figure 1 shows the growth-rate anisotropy for this system. Here the singular faces correspond to minima in the growth rate, with an initial rise on deviation followed by a fall to a deep level minimum in the middle of the orientation range. Although the two systems differ considerably in the shapes of the growth-rate anisotropy curves, they do have some features in common, namely that there is a certain angular range in which there is difficulty in step motion due to deviation from the singular faces, with the result that the step growth rate falls. One possible reason is [1, 2] that the kink density is dependent on the step curvature. Another is change in step height. If the step movement is due to a surface diffusion flux, then the step speed should fall as the height increases [3-5]. High steps should produce faceted forms, and such a step should move mainly by emitting small steps, which consumes a certain amount of energy.

A check on this mechanism was made by electron microscopy of films grown in both systems.

We used platinum-shadowed carbon replicas; the resolution in step height (as corrected for possible oxidation in air and coalescence of the platinum during shadowing) was estimated as 30 Å. Several replicas were prepared in succession from each specimen, which differed in platinum shadowing angle. This provided more precise determination of the distance between steps and of the step shape, while also enabling us to avoid effects due to accidental contamination of the specimen.

We found that the main trends in the relief were the same for both systems for the ranges (111)-(311) and (100)-(311), so we subsequently used only the range (111)-(311).

First we consider the morphology of films produced in a closed Ge – $I_2$ system (Fig. 2). A (111) face produces virtually no steps, apart from areas corresponding to the flanks of vici-

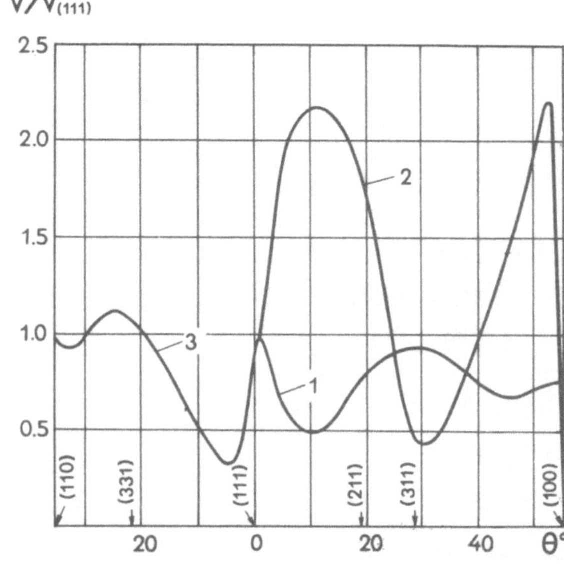


**Fig. 1.** Growth-rate anisotropy of germanium in the systems: (1, 3) Ge−HI−$H_2$; (2) Ge−$J_2$.

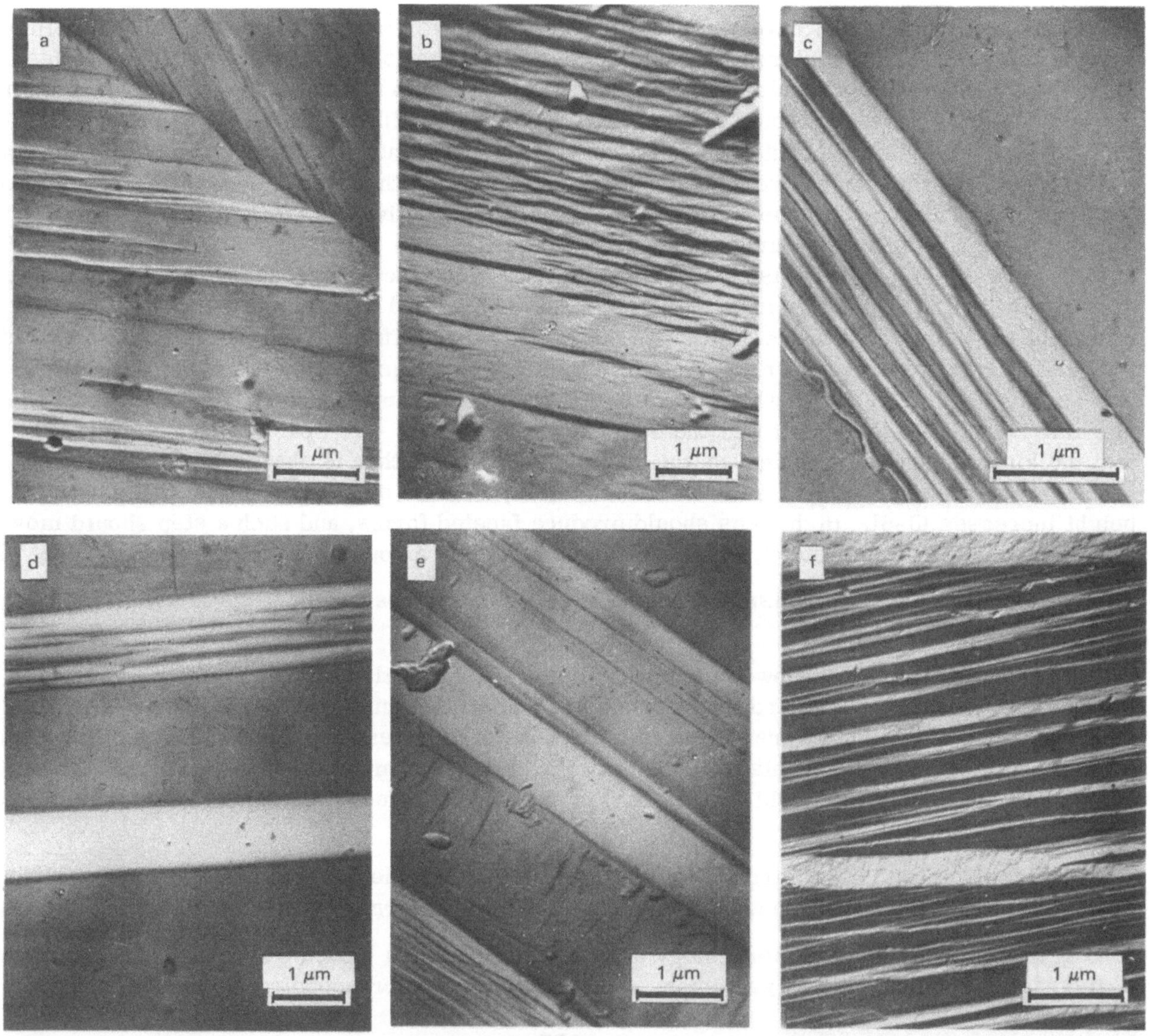

**Fig. 2.** Micromorphology of germanium in the Ge−$I_2$ system for the following cases with or without deviation: (a) (111); (b) 2°; (c) 10°; (d) (211); (e) (311); (f) (211).

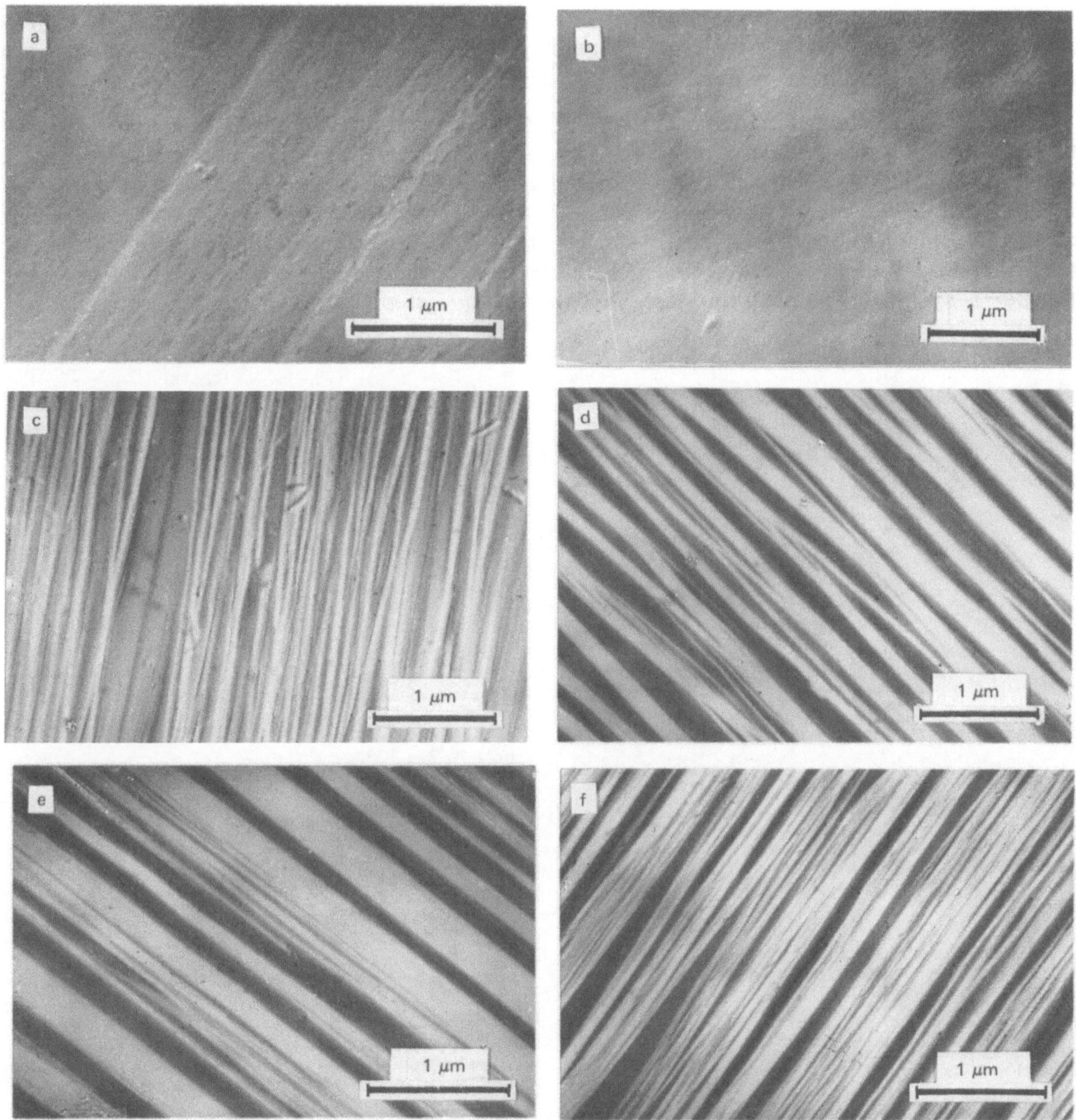


Fig. 3. Micromorphology of germanium in the Ge−HI−$H_2$ system in relation to orientation with or without deviation: (a) (111); (b) 2°; (c) 10°; (d) (211); (e) (311); (f) (511).

nal figures, on which one usually finds clumps of steps. Surfaces deviating from (111) by small angles showed relief as very thin parallel steps, which frequently formed clumps. Any further increases in the deviation from (111) caused the clumps to become more extensive, and these bunches sometimes gave rise to facets (Fig. 2c). On (211) and (311), the numbers of facetted steps and the heights increased, while the individual steps on (311) had heights up to 3 μm and prominent facets representing (111) and (100) faces. Deviations exceeding 10°, where there is a sharp fall in growth rate, result in the most marked change in the surface structure, and large facetted steps appear, which indicates that the minimum growth rate arises from a fall in step performance because the steps become linked and give rise to facets. The main driving force for the rearrangement is the change in the surface energy. The latter is indicated also by the following feature. Parts of the specimen deviating from a given orientation, i.e., towards (010), showed the large steps splitting up (in that case there was no crystallographic facetting), as is clear from Fig. 2f for a (211) plane.

Figure 3 shows films grown in the Ge−HI−$H_2$ system; a (111) plane is virtually structureless, apart from the flanks of vicinal figures, where there are clumps of small steps. Surfaces deviating by 2 and 4° show parts without visible steps and also parts covered with submicroscopic low-contrast steps. The 10° specimens show fairly large parallel steps of height up to 150 Å and numerous very small steps. At 30° there is a further increase in the step height, and facetted steps are seen, but there is none of the marked change in relief seen in the Ge−$I_2$ system. The maximum height of the facetted steps on (311) is 800 Å.

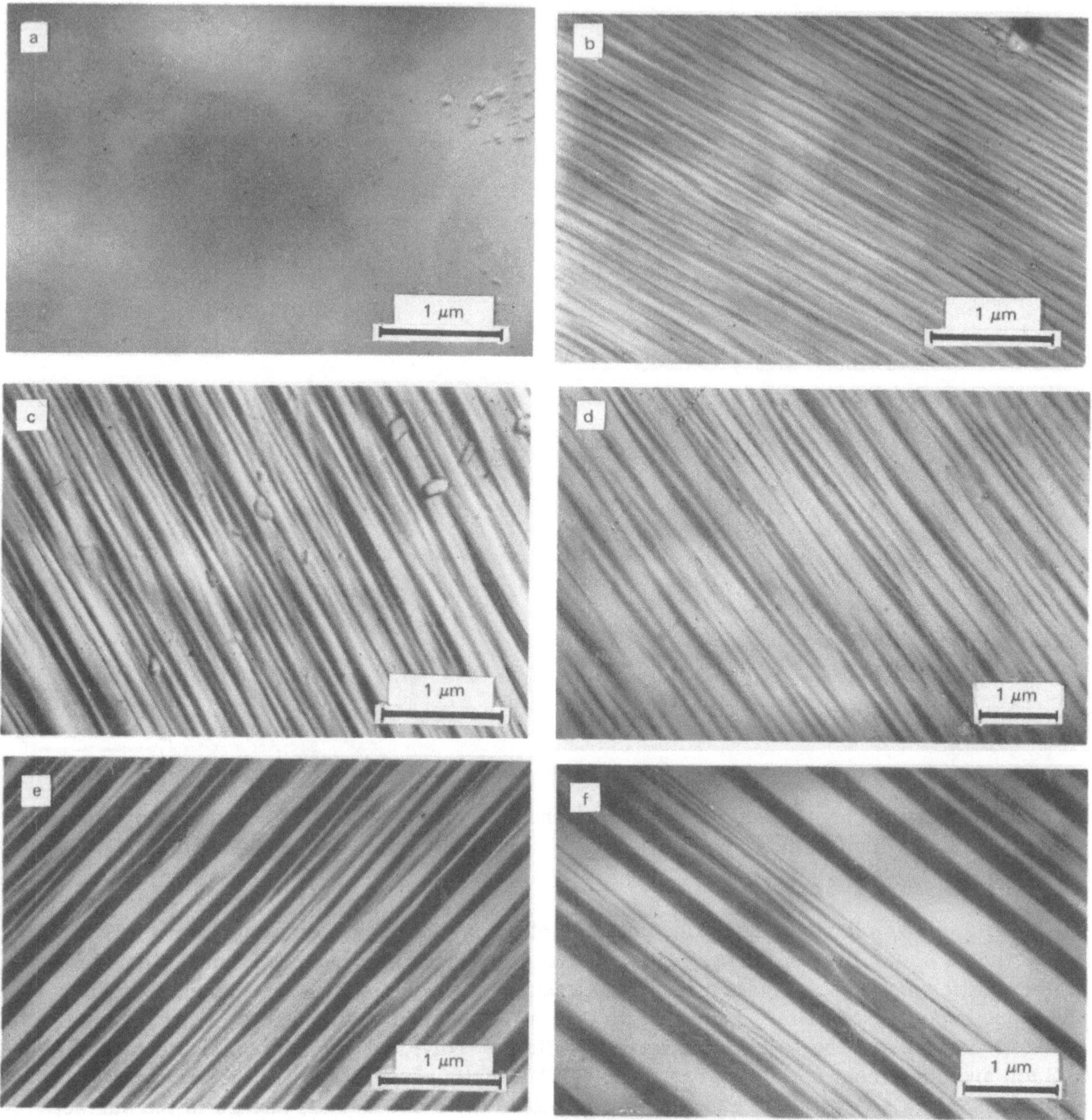


Fig. 4. Change in the morphology of germanium with time in a Ge−HI−$H_2$ system, (311) orientation, growth times in min: (a) 0; (b) 1; (c) 15; (d) 30; (e) 60; (f) 240.

The morphology indicates that the main trends are the same for the two systems; the mean step height increases with the deviation from a singular face, and facetted steps develop. This process is less pronounced in the $Ge-HI-H_2$ system, so the (311) growth rate in that case should be somewhat higher than for the $Ge-I_2$ system, as is observed. However, these results do not explain the fall in growth rate at small deviations from the singular faces observed for the $Ge-HI-H_2$ system, since there is little change in the relief in this range. It would seem that here there is some additional mechanism that reduces the speed.

We now examine how the steps tend to coalesce and how rapidly the relief develops. We examined this in terms of the morphology of a function of time for the $Ge-HI-H_2$ system. The changes were most marked on (311), as would be expected. Figure 4 shows a surface of this orientation at various instants ranging from 1 min to 4 hr, as well as the morphology of the initial substrate after gas etching. The original substrate was almost structureless, so our method did not reveal any edge steps. However, only 1 min of growth on (311) resulted in a clear-cut system of parallel steps of height 60-70 Å. The step system altered in structure as the growth time increased, in particular by step enlargement. Growth for 1 hr resulted in clear facets, but the size of these continued to increase, and at 4 hr it was still increasing, although more slowly. Therefore, the rate of enlargement of the steps is dependent on the orientation, and the anisotropy in the growth rate should show some change with time, perhaps substantial.

These results thus confirm the conclusion [5] that a growing surface has steps of two types, which pass one into the other under certain conditions.

We have shown [6] that the change in surface relief has a marked effect on the rejection of impurities by the growing crystal. It would seem that the change in morphology is one of the major reasons why transitional layers occur in epitaxial germanium [7], in much the same way as for gallium arsenide [6].

The differences in the growth-rate anisotropy between the $Ge-I_2$ and $Ge-HI-H_2$ systems may be due to various factors, including the hydrogen. The initial material in the $Ge-I_2$ system was doped with gallium, whereas that in the $Ge-HI-H_2$ system was doped with antimony. Figure 5 shows the carrier concentration as a function of orientation for each system. The curves are similar in shape, but the relationship to the growth-rate anisotropy indicates that the open system (with antimony as dope) shows the electron concentration falling as the growth

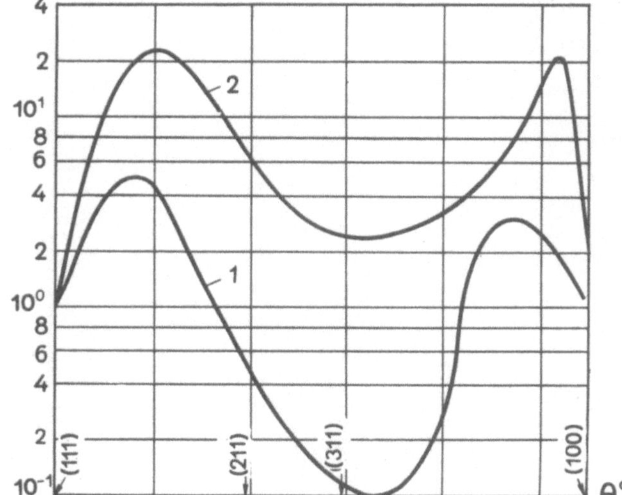


Fig. 5. Effects of orientation deviation from (111) on the relative carrier concentration in films grown in the systems $Ge-HI-H_2$ (1, electrons) and $Ge-I_2$ (2, holes).

rate increases, while the closed system (gallium dope) has the doping level running parallel to the growth rate. This means that the donor dope interacts with the growing surface in a way different from the acceptor dope, so there are major differences in the effects on the growth kinetics. However, this aspect requires more detailed research.

We are indebted to A. A. Chernov for a valuable discussion and to L. M. Krasil'nikova for advice in performing the electron microscope studies.

## Literature Cited


1. L. G. Lavrent'eva, I. S. Zakharov, and Yu. M. Rumyantsev, Kristallografiya, 15:854 (1970); 16:423 (1971).
2. L. G. Lavrent'eva, I. S. Zakharov, and Yu. M. Rumyantsev, Izv. VUZov, Fizika, No. 4, 121 (1970).
3. G. G. Lemmlein and E. D. Dukova, Kristallografiya, 1:112 (1956).
4. A. A. Chernov, Kristallografiya, 1:119 (1956).
5. A. A. Chernov, Usp. Fiz. Nauch., 73:277 (1961).
6. L. G. Lavrent'eva, M. D. Vilisova, I. V. Kronin, L. M. Krasil'nikova, and Yu. M. Rumyantsev, Izv. VUZov, Fizika, No. 2 (1973).
7. L. G. Lavrent'eva and I. S. Zakharov, Proceedings of the International Conference on the Physics of Chemical Semiconductor Heterojunction Layer Structures, Budapest, Vol. 1 (1970), p. 253.

# EFFECTS OF FOREIGN PARTICLES ON A MACROSCOPICALLY SMOOTH SURFACE ON THE MOVEMENT OF STEPS DUE TO EVAPORATION AND CONDENSATION

Ya. E. Geguzin, V. V. Kalinin, and Yu. S. Kaganovskii

*Kharkov University*

## Introduction

The details of the high-temperature relief on a macroscopically smooth surface should move systematically during evaporation and condensation because they act as sources or sinks for the atoms involved.

If these steps move freely, the mass transport is determined by the equilibrium value of the corresponding kinetic coefficient. The situation is entirely different if obstacles are present on the surface, e.g., particles of some other phase. The obstacles of course cannot halt the motion completely if the driving chemical-potential gradient is appropriate. However, they do influence the speed, i.e., they ultimately influence the mass-transport rate and the resulting surface geometry.

This situation is essentially similar to that arising during dispersion hardening in a crystal containing dislocations and a dispersed phase [1]. The surface steps are then analogous to dislocations, while the particles of the secondary phase are similar to the particles in the bulk lying along the dislocations.

Here we report experimental evidence on the interaction between moving steps and macroscopic particles.

## 1. Experiment: Qualitative Observations

The experiments were as follows. The smooth natural cleavage surface of an LiF crystal (macroscopically smooth surface) was coated with a layer of gold by evaporation at $5 \cdot 10^{-5}$ mm Hg, and this on subsequent annealing split up into individual particles, which lay on the smooth parts and in the reentrant parts of steps. These specimens were annealed under vacuum ($10^{-2}$ mm Hg) under conditions such that mass transport occurred by evaporation, which caused the steps to move, and thus to interact with the gold particles. After preset intervals at 600°C, the surface was examined in the electron microscope by means of a carbon-replica technique, with gold shadowing at about 15°. This revealed the steps, whose heights were nearly monotonic, and also any changes in the shape of the particles and effects on the geometry of the steps.

Further experiments were done in which two crystals were placed one above the other at a small distance apart in a temperature gradient. Under these conditions, there is systematic transport to the cooler crystal, and therefore the steps move in the opposite direction.

The following main observations were made. Figures 1 and 2 show that the gold particles serve to block steps moving in evaporation foci, as is clear from the retarded position of the steps. These moving steps can also displace the particles, and the moving particles collide and may become larger [2].

A hindered step of macroscopic height can detach from a particle, as is clear from the cascade of smaller steps that results (Fig. 3). A step retarded by particles breaks up into smaller steps, which surround the latter.

At the late stages, one finds numerous particles lying on pinnacles, which may be thinner than the particle size; this is clear from the shape of the shadows arising on the peak shadowing. The shadow may be thinner at the base of the particle (Fig. 3). If the surface is coated by vapor deposition, the steps can drive the particles (Fig. 4), and one sometimes finds that the particles become covered, which is the opposite to the formation of particles on pinnacles during evaporation.

## 2. Interaction of Steps with Particles of Second Phase

We now discuss the interaction between a moving step and particles on a macroscopically smooth surface, whose orientation deviates slightly from that represented by the equilibrium faces of the crystal. We assume that the mean free path for an adsorbed atom on the surface

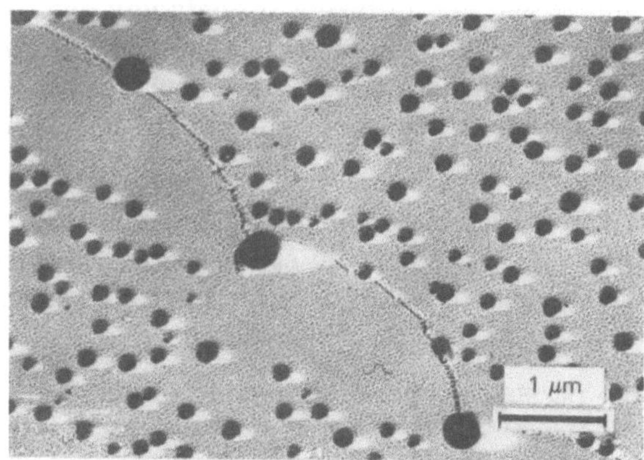


Fig. 1. Gold particles blocking the motion of a cleavage step during evaporation at 600°C, t = 30 min.

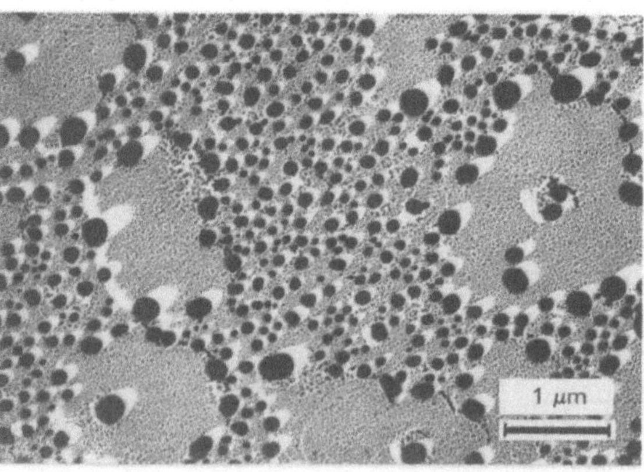


Fig. 2. Evaporation foci on a (100) surface of LiF, with moving steps also displacing gold particles, 600°C, t = 30 min.

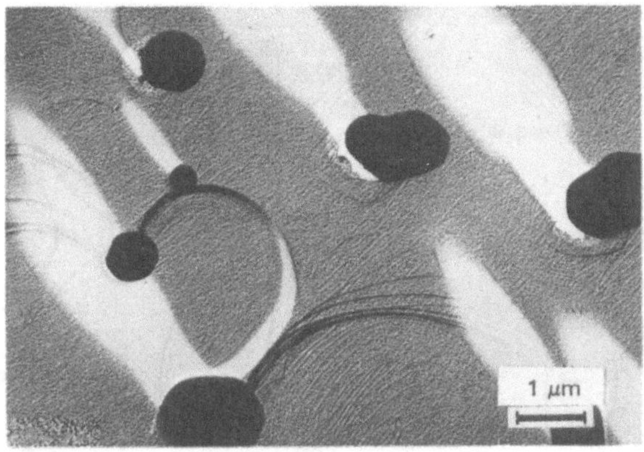


Fig. 3. Steps surrounding gold particles and detaching in layers from particles. Particles remaining on LiF surface. T = 600°C, t = 1.5 hr.

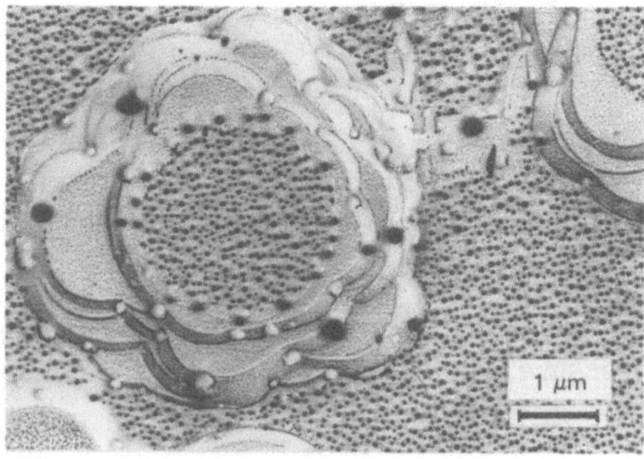


Fig. 4. Enlargement of gold particles displaced by moving steps during condensation of LiF on a (100) surface.

$\lambda_s$ is much less than the distance d between steps, and therefore the interaction between steps can be neglected.

We first discuss the situation corresponding to evaporation [3]. The particles on a real surface may lie either on smooth parts or in reentrant angles in steps, where the radius of curvature of a particle is $R \gg h$ (h is the height of a step), and therefore the contact between the particle and the step may be of complicated geometry. For simplicity, we consider two main types of contact between particle and step (Fig. 5a, b), which will now be discussed.

Under evaporation conditions, the free steps move along a surface with some speed $v_c$, which is governed by the difference $\Delta\mu$ in the chemical potential for the crystal and the vapor [4]. If particles lie along the step and are bound to the latter because the interfacial energy is $\gamma_{12} < \gamma_1 + \gamma_2$ (subscript 1 relates to the crystal and subscript 2 to the material of the particle), then the moving step will be deflected.

We first assume that the particles cannot move. The particles are also assumed to be identical and uniformly distributed along the step at distances of $l$ (the linear particle density is $n_p = 1/l_0$). During the motion of a step, the latter is deflected, and the angle $\varphi$ (Fig. 5a) should remain unchanged, which is possible only if the line of contact between the step and particle slides along the surface of the particle. During this process, the radius of curvature $\rho$ of the step is reduced, finally obtaining the minimum value $\rho_{min}$. If $R \ll l$, then we have $\rho_{min} \simeq l/2$, and it is clear that if $|\Delta\mu| < \gamma_1\omega/\rho_{min}$ (where $\omega$ is the atomic volume), the step motion will cease when the value $\rho^* = \gamma_1\omega/\Delta\mu$ is attained. If on the other hand $|\Delta\mu| > \gamma_1\omega/\rho$, the motion will continue, and the radius of curvature will increase. This process will be completed by contact between the parts of the steps moving from opposite sides of the particle, which is followed by detachment from the latter (Fig. 5a).

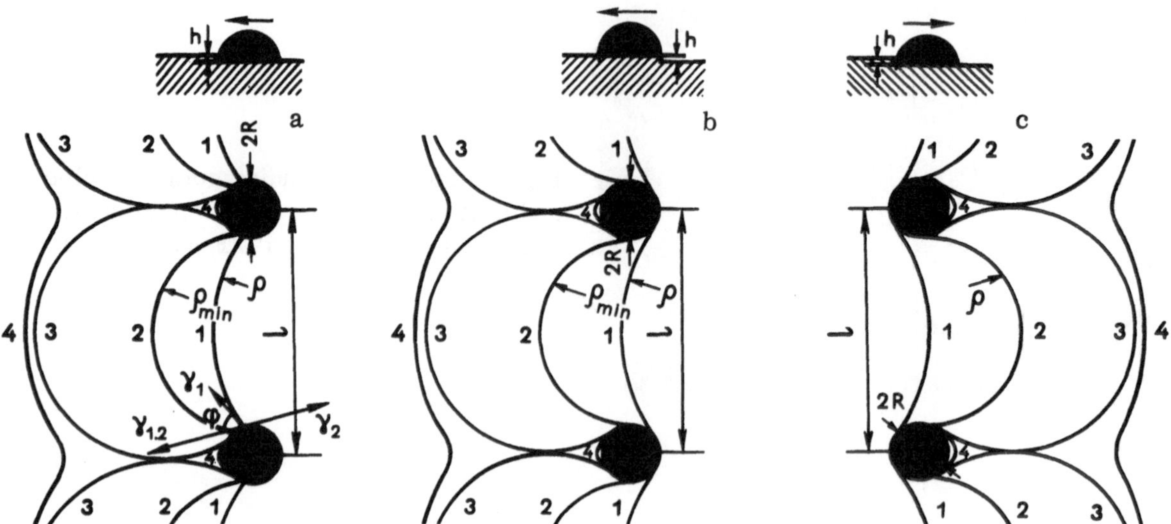


**Fig. 5.** Schematic representation of successive stages in step motion between particles: (a) Particles in a reentrant angle in a step during evaporation; (b, c) particles in the path of a moving step during evaporation and condensation respectively.

The force $F_p$ applied to a particle can be put in the form $F_p = P_c S$, where $P_c = \gamma_1/\rho$ is the pressure that deflects the step and $S = hl$ is the surface area of a step. Then

$$F_r = \frac{\gamma_1 h l}{\rho} = \frac{h l}{\omega}\,\Delta\mu. \qquad (1)$$

We now consider this process under conditions such that the particles can move as a whole with the steps, on account of material transport around the step in the direction of the effective force.

If $\Delta\mu$ is insufficient to deflect a step and such that $\rho_{min}$ reaches the threshold value $\rho^*$, the steps can move only together with the particles at some self-consistent speed $\tilde{v} = v_p = v_c$; the resultant mass transport within each particle may involve a mechanism different from that responsible for the movement of the step, since the mobilities of the two are controlled in different ways.

We calculate $\tilde{v}$ in terms of the step mobility ($[\zeta_c]$ = cm · sec/g) and the particle mobility ($[\zeta_p]$ = sec/g) as phenomenological constants whose relationship to the parameters of the transport mechanism will be established later on.

Equality of $v_p$ and $v_c$ means that

$$\tilde{v} = \zeta_c F^*_c = \zeta_p F_p, \qquad (2)$$

$$F_c^* = F_c - n_p \cdot F_p, \qquad (3)$$

where $F_c$ is the force applied to a step of unit length free from particles; it follows from (2) and (3) that

$$\tilde{v} = F_c \tilde{\zeta} = F_c \zeta_c \frac{1}{1+\psi}. \qquad (4)$$

where

$$\tilde{\zeta}=\frac{\zeta_p\zeta_c}{\zeta_p+n_p\zeta_c},\quad \psi=\frac{\zeta_c}{\zeta_p}n_p.$$

From (4) we have the following possible limiting cases: (a) $\psi \ll 1$; then $\tilde{v} \sim \zeta_c$, which can occur only for $n_p \to 0$, i.e., when the particle density is low, or else when $\zeta_p \gg n_p\zeta_c$, i.e., when the particles are small and correspondingly their mobility is high; (b) $\psi \gg 1$; in that case $\tilde{v} \sim \zeta_p/n_p$, which can occur only if the linear density of the particles is high, or else the mobility is low.

We now discuss the quantities appearing in (4). The force driving a step of unit length is $F = -d\Phi/dx$, where $d\Phi = \Delta\mu dN = -\Delta\mu(h/\omega)dx$ is the change in the energy of the system when the step is displaced by dx, while $dN = (h/\omega)dx$ is the number of atoms that leave the step and become adsorbed atoms. Therefore,

$$F_c=\frac{h}{\omega}\Delta\mu. \tag{5}$$

By definition

$$\zeta_c\simeq\frac{v_c}{F_c}=\frac{v_c\omega}{h\Delta\mu}=\frac{b\omega}{h}. \tag{6}$$

Equation (4) can also be derived directly from the circumstances encountered on the part of the path hindered by two particles (Fig. 5); this part of a step moves [5] with the speed $v(\rho)=v_c(1-\rho/\rho^*)$, where $v_c = b\Delta\mu$ and $\rho^*=\gamma_1\omega/\Delta\mu$. The value of b is dependent on the step motion mechanism [4]. Then these equations and (1) give the following expression for the speed of the step-particle assembly: $v=\zeta_p b\Delta\mu(\zeta_p+b\omega n_p/h)^{-1}$, and it is readily seen that this with (5) and (6) gives an expression coincident with (4).

If material is removed from a step by surface diffusion, then $b_s=2D_a n_a\omega/kTh\lambda_s$; where $D_a$ is a diffusion coefficient for adsorbed atoms, whose surface density is $n_a$, and in that case

$$\zeta_{cs}=\frac{2D_s}{kT}\frac{a^4}{\lambda_s h^2}, \tag{7}$$

where $D_s=D_a\xi_a$, $\xi_a\simeq a^2 n_a$, where $a$ is the lattice parameter of the substrate. In (7) we have thus incorporated the feature that the atoms can leave the step along the two surfaces bounding it.

If the step motion is restricted by the diffusion rate of the evaporating atoms in the gas, then $b_g=D_g p_0\omega d/kTh\delta$, in the steady state, where $\delta$ is the characteristic linear dimension, which is dependent on the geometry, which itself determines the evaporation rate. In that case

$$\zeta_{cg}\simeq\frac{D_g}{h^2}\left(\frac{\omega}{kT}\right)^2\frac{d}{\delta}, \tag{8}$$

where $D_g$ is the diffusion coefficient in the gas and $P_0$ is the equilibrium vapor pressure. The result $\zeta_c \sim d$ follows from (8) and is reasonable, since it reflects the feature that the speed of the steps increases with the separation (for a given evaporation rate).

The following formulas [1] define $\zeta_p$ when the transport is by surface or bulk diffusion:

$$\zeta_{ps}\simeq\frac{D_s^p}{kT}\left(\frac{a_p}{R}\right)^4,\quad \zeta_{pv}\simeq\frac{D_v^p}{kT}\left(\frac{a_p}{R}\right)^3, \tag{9}$$

where $D_s^p$ and $D_v^p$ are the diffusion coefficients for those cases respectively for the particle material, while $a_p$ is the lattice parameter of the particle.

As $F_c \sim h$, while $\zeta_c \simeq b/h^2$, the condition $\partial v/\partial h = 0$ implies that there is an optimum step height $h^*$ at which the motion of the step and particles together occurs at the maximum rate:

$$h^* = (n_p b/\zeta_p)^2, \tag{10}$$

where b takes the value $b_s$ or $b_g$ in accordance with the conditions. Estimates show that $h^* \lessgtr R$.

If the particles lie on smooth parts of the surface, they can also influence the step motion, because motion under a particle must be accompanied by a change in particle position, i.e., by a change in energy proportional to $\Delta\gamma = \gamma_1 + \gamma_2 - \gamma_{12} > 0$; the following condition is then that for retention of the coupling between a deflected step and the set of particles above it:

$$\Delta\mu < \frac{2R\omega}{hl}\,\Delta\gamma. \tag{11}$$

In any solution, (11) must be supplemented by the condition $\Delta\mu < 2\gamma_1\omega/l$, which implies that a step cannot be completely suppressed by a particle. In that case, the motion will be with the speed described by (4). If $R \ll l$, the force applied to the particle will be as in the previous case, as (1) shows.

The step will encircle the particle if the condition $\Delta\mu > 2\gamma_1\omega/l$. is met at the same time as (11); after encirclement, the particle should remain on a pedestal of height h made up of the crystal material, as is illustrated by Fig. 3. Clearly, subsequent steps arriving at this pedestal will increase the height of the latter by their own height, and the column under the particle will resemble a dislocation ring arising from encirclement by dislocations. At the start, we have $R \gg h$, and then the pedestal will be essentially cylindrical, i.e., the negative curvature of the surface will be slight, but at later stages the surface will be described by $1/r_1 - 1/r_2 = \Delta\mu/\gamma_1\omega$, where $r_1$ and $r_2$ are the principal radii of curvature. At a late stage, the pedestal should vanish (break off) because the opposite sides come into contact (Fig. 6).

In the case of condensation, particles may be displaced because they are repelled by the moving steps; it is clear that in this case the form of a step near the particle will be as shown in Fig. 5. The joint motion of step and particles will be described by (4). If on the other hand $\Delta\mu > \gamma_1\omega/\rho_{min}$, the steps will be deflected around the particles, and this will result in the observed trapping.

## 3. Effects of Particles on Evaporation and Condensation Rates

Particles on a crystal should affect these rates, and we assume for simplicity that the surface is vicinal, with an inclination to the singular plane $\alpha = h/d \ll 1$, in which case we can

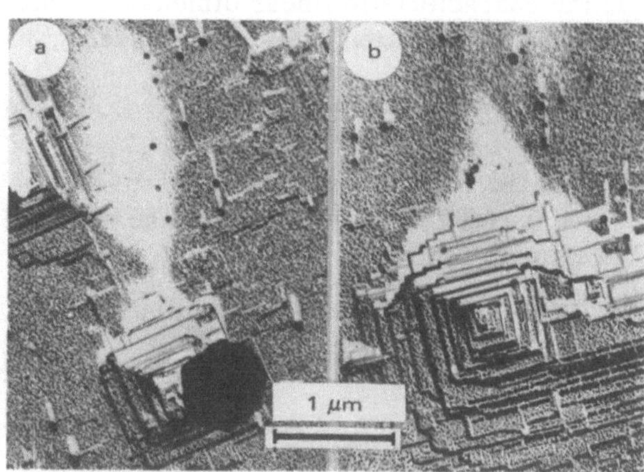


Fig. 6. Effects under large particles at a late stage in evaporation: (a) high pedestal bearing particle; (b) fractured pedestal without particle.

estimate the effects on the evaporation and condensation rates. The flux of atoms from unit surface in unit time (or to unit surface) is defined by

$$I = \tilde{v} h/\omega d, \tag{12}$$

where $\tilde{v}$ is the step speed, which in general is defined by (4) and (5). Then on the basis that $\Delta\mu = kT\Delta p/p_0$ we get

$$I = \tilde{\xi}\left(\frac{h}{\omega}\right)^2 \frac{kT}{d} \frac{\Delta p}{p_0}, \tag{13}$$

where $\tilde{\xi}$ can be calculated by the formulas given above for the possible mechanisms.

A quantitative measure of the particle effect is given by

$$æ = \frac{\tilde{\xi}}{\xi_c} = \frac{1}{1+\varphi}. \tag{14}$$

The retardation will be substantial for $\psi \gg 1$; we now estimate æ for the case where the motion of step and particle is determined by surface diffusion. From (8) and (9) we have

$$\psi = (D_s/D_s^p)(R^4 n_p/\lambda_s h^2)(a/a_p)^4.$$

For $(D_S/D_S^P) \simeq 10^2$, $R \simeq 10^{-6}$ cm, $h \simeq 10^{-7}$, $\lambda_s \simeq 5 \cdot 10^{-5}$ cm, $n_p \simeq 10^5$ cm$^{-1}$, $(a/a_p)^4 \simeq 1$, we find $\psi \simeq 10$, and consequently æ $\simeq 1/10$; the estimates for the other cases are readily made by the reader.

This retardation will occur only if $\Delta\mu < \gamma\omega/\rho$, and then for $\rho \simeq l = 1/n_p$ we have a restriction on the supersaturation:

$$\left|\frac{\Delta p}{p_0}\right| < \frac{\gamma\omega n_p}{kT}$$

Under our conditions ($\gamma \simeq 5 \cdot 10^2$ dyn/cm, $n_p \simeq 10^5$ cm$^{-1}$), we find $\Delta p/p_0| < 10^{-2}$, which is reasonable.

## 4. Complicating Circumstances

The idealized step-particle geometry is in fact complicated by various features. The following are the main ones. Firstly, particles coalesce during high-temperature annealing, so $n_p$ alters, as does the mean particle size. There are three different coalescence mechanisms, whose relative importance varies with the particle size and the rate of removal of material. One of these is diffusion of material from small particles to larger ones in the self-consistent diffusion field, namely two-dimensional diffusion coalescence [6]. Another mechanism (one-dimensional diffusion coalescence) occurs when the material is exchanged between particles by a diffusion flux along the steps. A specific mechanism is the third one, when the moving particles fuse on account of collision. The experiments described above indicate that the third mechanism is the main one, which also governs the changes in $l = 1/n_p$.

Secondly, we had previously supposed that a step interacting with particles is a stable formation, but in fact it exists in a dynamic state, since a low step approaching a particle will be retarded, and such steps may accumulate, interact, and fuse. On the other hand, a step becomes detached from an obstacle layer by layer. Therefore, steps are continually encountering and leaving obstacles, and the maximum height occurs for a step linked to an obstacle. It may be that this height is given by (10).

## Literature Cited

1. Ya. E. Geguzin and M. A. Krivoglaz, Movement of Macroscopic Inclusions in Solids [in Russian], Metallurgiya, Moscow (1971).
2. Ya. E. Geguzin, V. V. Kalinin, and Yu. S. Kaganovskii, Dokl. Akad. Nauk SSSR, 206:917 (1972).
3. Ya. E. Geguzin, Yu. S. Kaganovskii, and V. V. Kalinin, Proceedings of the Fourth All-Union Conference on Crystal Growth: Mechanism and Kinetics of Crystallization, Part 2 [in Russian], Izd. AN Arm. SSR, Erevan (1972), p. 58.
4. A. A. Chernov, Usp. Fiz. Nauk, 73:277 (1961).
5. B. Barton, N. Cabrera, and F. Frank, In: Elementary Crystal Growth Processes [Russian translation], IL (1959), p. 11.
6. Ya. E. Geguzin, Yu. S. Kaganovskii, and V. V. Slyozov, J. Phys. Chem. Solids, 30:1173 (1969).

# ESTIMATION OF THE MEAN DISPLACEMENT OF ADSORBED MOLECULES FROM THE GROWTH OF LOCHKEIMS


## H. Krohn

*Central Institute for Solid State Physics and Materials Science, Institute for Solid State Physics and Electron Microscopy, Halle*


In 1960 Bethge and Keller [1] using the gold decoration technique were able to reveal the surface structures that are formed by free evaporation of alkali halide crystals in vacuum. They observed that the first structures becoming visible after a slight evaporation of the surface are numerous small holes in the surface. A typical example is presented in Fig. 1. In the gold decoration micrographs these holes appear as nearly circular patterns. They are the places where the decrystallization begins. Since in general a nucleation is necessary for such processes, these holes were called "lochkeims," i.e., hollow or negative nuclei. By the methods for the determination of step height described by Bethge and Keller [2] it was proved that lochkeims of the kind shown in Fig. 1 have a depth of only one lattice distance, i.e., $a/2$.

The diameter of the lochkeims increases with increasing time of evaporation, and it is the main purpose of this paper to determine their growth rate. The essential experimental difficulty lies in the fact that the density of lochkeims is not reproducible: On one and the same crystal surface, i.e., for identical experimental conditions, densities of lochkeims $\rho$ between $10^8$ and $10^{10}$ $cm^{-2}$ can be observed. Since the mean radius of the lochkeims $\bar{R}$ depends on the density of lochkeims, the values of $\bar{R}$ differ in a similar way as $\rho$.

The difficulties arising from the fluctuations of $\rho$ can be met as follows: One does not determine the mean radius of lochkeims $\bar{R}$, but one measures the radii of some of the largest lochkeims of the micrograph examined and calculates their mean value $R_m$. It is not important how many of the largest lochkeims are used for determining the mean value, since the radii of the largest lochkeims visible in a single micrograph differ remarkably little. The largest lochkeims meet two requirements important for the measurement of the growth rate. On the one hand, it might be supposed that they have hardly been influenced by the presence of other lochkeims during growth, because the distance between largest lochkeims and their nearest neighbors is relatively large. On the other hand, the largest lochkeims evidently originated at the beginning of the evaporation process, i.e., at time $t = 0$, and have grown all the time during the experiment, while smaller lochkeims were probably produced later on.

In Fig. 2 the values for $R_m$ measured at the temperature $T = 285°C$ are plotted against the time of evaporation t. The growth velocity decreases with increasing size of the lochkeims. It can be demonstreted that this behavior is in accordance with the molecular kinetic theory of Burton, Cabrera, and Frank (BCF theory) [3]. However, instead of using the complicated

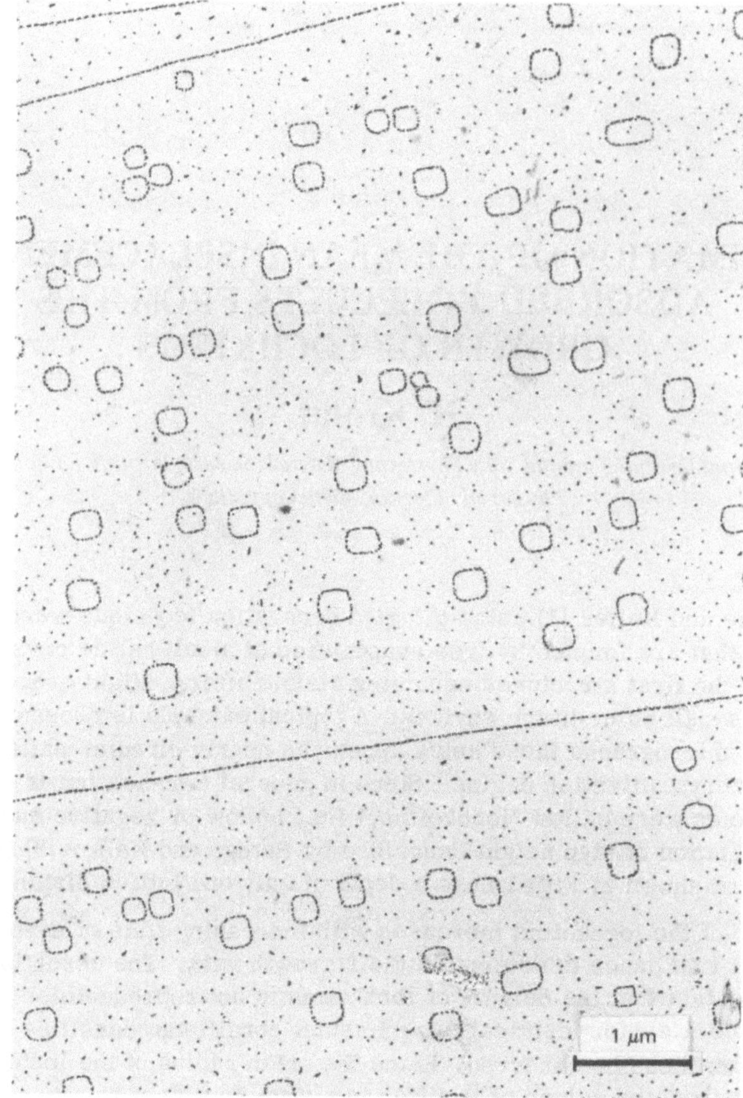


Lochkeims on a (100) surface of NaCl produced by evaporation in high vacuum for 10 min at 325°C. Gold decoration; 30,000 ×.

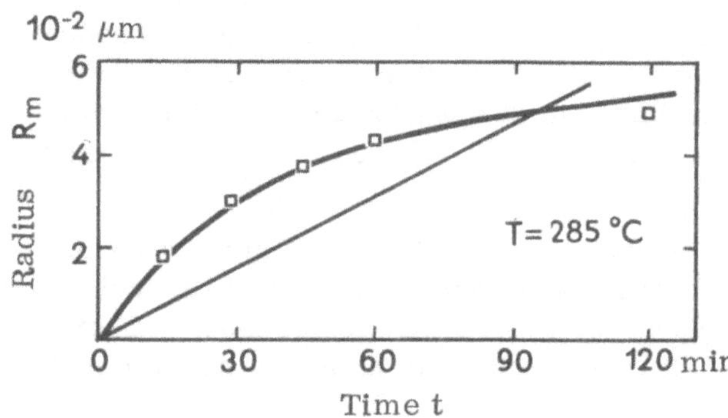


Fig. 2. Increase of the radius of lochkeims on NaCl surfaces with time of evaporation.

expressions for the displacement of an annular step we will here interpret our results in a different and simplified manner.

For comparing growth velocities of the lochkeims at different temperatures one has to define a characteristic velocity. The value given by the tangent at any point of the curve $R_m = R_m(t)$ is not accurate enough because of the small number and relatively large fluctuations of the measuring points. Therefore the tangent was not used at all, and the velocity was defined by the mean velocity v in the range between $R_m = 0$ and $R_m = 0.05\ \mu m$ for the following reason: During the evaporation process, the straight cleavage steps are displaced by a certain distance d. By means of the double decoration technique this displacement d was observed to be approximately equal to the radius of lochkeims $R_m$ if $R_m = 0.05\ \mu m$. Consequently, the velocity v defined above also represents the velocity of movement of a straight step with a step height of $a/2$.

In Fig. 3 the velocities v obtained at different temperatures are plotted in a logarithmic scale against the reciprocal temperature. According to this Arrhenius plot the growth velocity can be written as follows:

$$v = 2.67 \cdot 10^7 \exp(-1.80\ \text{eV}/kT)\ \text{cm/sec.} \tag{1}$$

For the discussion of our experimental data it is assumed that an extended straight step moves according to the following model: The molecules separated from the step diffuse on the flat surface. During their residence time $\tau$ on the surface they move away from the step a mean distence $\lambda$ before their desorption into vacuum. Near the step the concentration of admolecules is that at equilibrium $n_{so}$, and far from the step – since we have free evaporation – the concentration of admolecules is zero. With this model there is a simple solution of the diffusion problem given by BCF theory. The local concentration of admolecules is found to be

$$n_s = n_{so} \exp(-y\sqrt{2}/\lambda), \tag{2}$$

y being the distance from the step. The flux of molecules directed on one side of the step is

$$I_0 = -D\,\text{grad}\, n_s|_{y=0} = \frac{\sqrt{2}\, D n_{so}}{\lambda}. \tag{3}$$

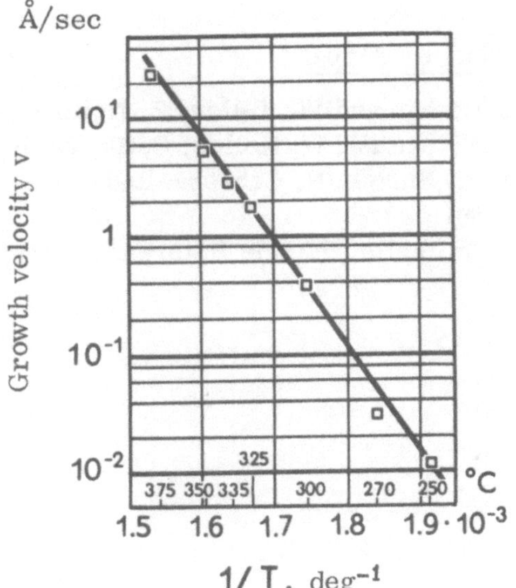


Fig. 3. Growth velocity of lochkeims in dependence on temperature.

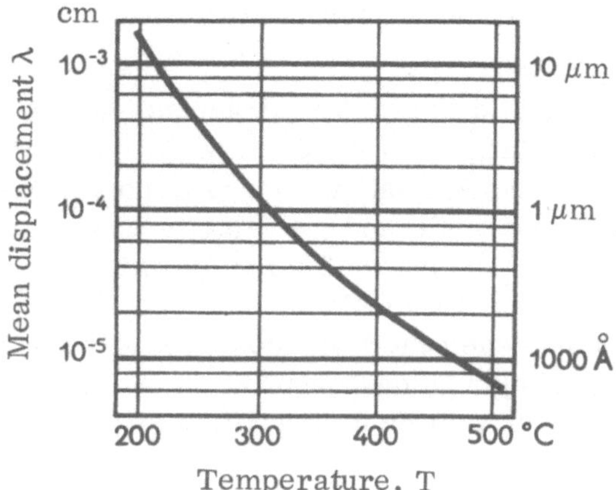


Fig. 4. Temperature dependence of the mean displacement of admolecules.

From the Einstein relation $\lambda = (2D\tau)^{1/2}$ it follows that

$$I_0 = \frac{\lambda n_{so}}{\sqrt{2}\,\tau}. \tag{4}$$

The quotient of the two unknown quantities, $n_{so}$ and $\tau$, can be replaced by one single measureable quantity, the impinging rate at equilibrium $N_e\downarrow$, since at equilibrium with the saturated vapor we have $n_{so}/\tau = N_e\downarrow$. Thus the mean displacement becomes

$$\lambda = \sqrt{2}\; I_0/N_e. \tag{5}$$

The flux of particles $I_0$ is given by the measured velocity v: $I_0 = \frac{1}{2}vn_0$, where $n_0$ is the surface concentration of lattice sites. The impinging rate $N_e\downarrow$ follows from the vapor pressure p: $N_e\downarrow = p/(2\pi mkT)^{1/2}$, if we assume that there is no reflection at the surface. If the values of vapor pressure listed by Smithells [4] are taken, one obtains the temperature dependence of the mean displacement λ as presented in Fig. 4.

The assumption of equilibrium concentration of admolecules near the step has not yet been proved. It may be mentioned, however, that the values obtained for λ correspond very well to the values obtained by other methods.

## Literature Cited

1. H. Bethge and W. Keller, Z. Naturforsch., 15a:271 (1960).
2. H. Bethge and W. Keller, Optik, 23:462 (1965/66).
3. W. K. Burton, N. Cabrera, and F. C. Frank, Phil. Trans. Roy. Soc. London, A243:299 (1950).
4. J. Smithells, Metals Reference Book, Vol. 2, Plenum Press, New York (1967).

# OBSERVATIONS OF THE KINEMATIC INTERACTION OF SURFACE STEPS DURING EVAPORATION OF NaCl

**K. W. Keller**

*Central Institute for Solid State Physics and Materials Science, Institute for Solid State Physics and Electron Microscopy, Halle*

In the Burton–Cabrera–Frank theory [1], the speed of a surface step is dependent on the step separation on account of the interaction by surface diffusion (molecules leaving one step diffuse to another). This is essentially the kinematic theory of crystallization, and the interaction is kinematic. The theory is usually applied to macroscopic effects, e.g., etch-pit shape. Here we describe direct electron-microscope observations on kinematic interaction between elementary steps with gold decoration [2, 3]. During evaporation from a NaCl crystal under vacuum, pits are produced around dislocations on $\{100\}$ surfaces, which have a characteristic step structure, step height $d_0/2 = 2.81$ Å or $d_0 = 5.62$ Å [4]. The steps are equidistant in steady-state evaporation, and the separation is determined by the conditions at the center. Kinematic interaction is most prominent in unsteady states, e.g., when the sequences are not indefinitely long ones with equal step heights.

Mullins and Hirth's calculations [7] show that a finite step sequence does not remain equidistant during kinematic interaction. The separations increase steadily behind the leading edge, while near the end the steps tend to be paired (alternating large and small separations). All of these effects have been observed.

The step trains in a pit may migrate to an area free from steps (as in the simple circular and double spirals in Figs. 1 and 2), in which case the separations increase at the edge. The same occurs for the rectangular spiral in Fig. 3. A region free from steps is then produced within the pit because the double spiral (step height $d_0/2$) becomes a rectangular spiral (height $d_0$) [6] and the speed is dependent on the height. Figure 4 shows the increase in distance between turns for all types of spiral. These results are similar to those of Surek et al. [8] from computer simulation of a step train at the edge of a crystal. The numbers on the curves are the total numbers of turns in the spirals and the numbers persisting after some of the leading turns have been destroyed by steps from other sources. Many turns are lost by large-diameter spirals, so the curves become less steep. The separations are constant in the inner parts of these spirals. Although these distances for simple circular spirals are different from those for double spirals, the curves are very similar in shape. However, the increase in terrace width at the pit edges is larger for the rectangular spirals, whose step heights are twice those of circular ones. Surek has shown [8] that this must be due to speed differences correlated with step height. On the other hand, the stability of steps of height $d_0$ indicates (see below) that such steps interact more weakly than do $d_0/2$ steps, evidently because molecule capture by a $d_0$ step is more difficult.

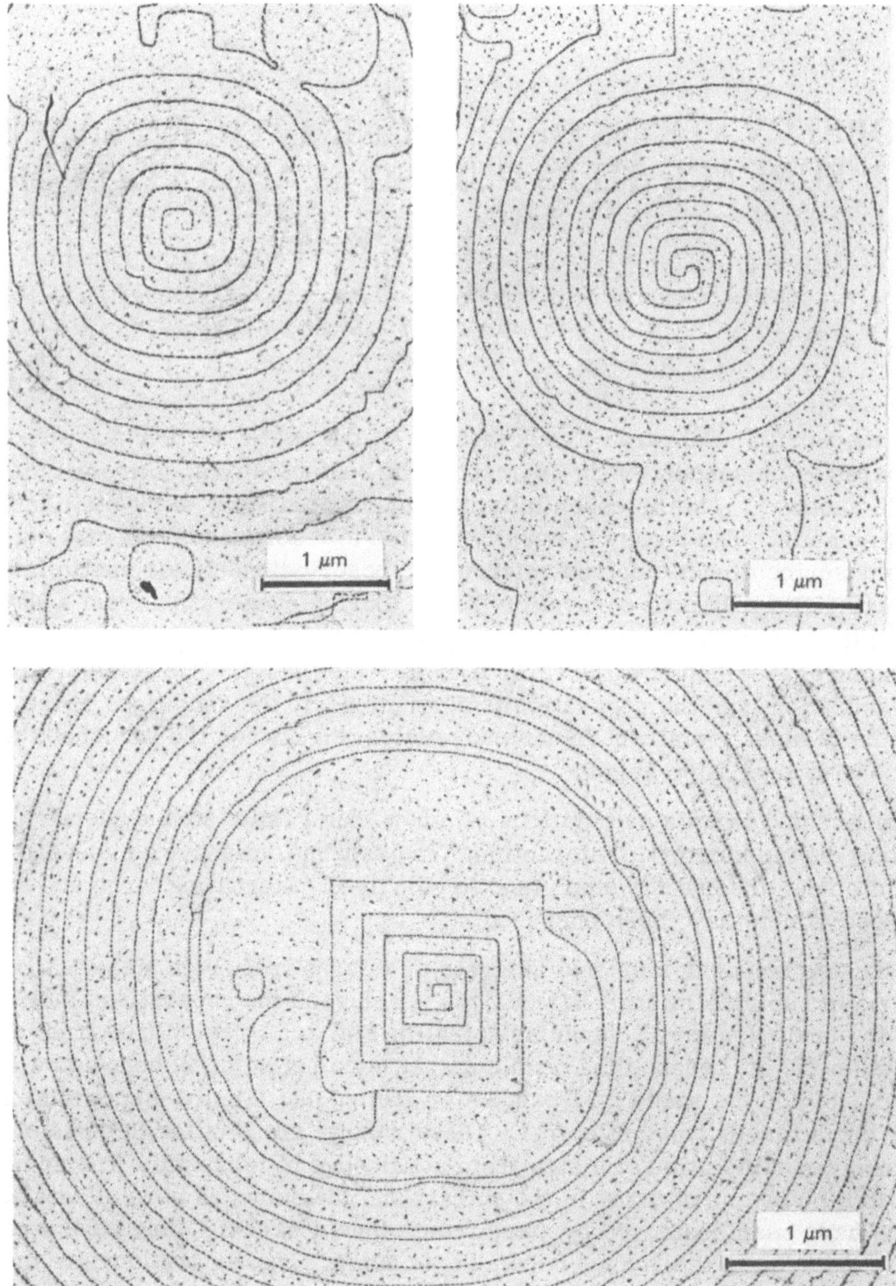


Fig. 1-3. Evaporation spirals on rocksalt: simple circular spiral (1), double spiral (2), and rectangular spiral (3) with increasing distances between turns. Electron-microscope specimens made by gold decoration.

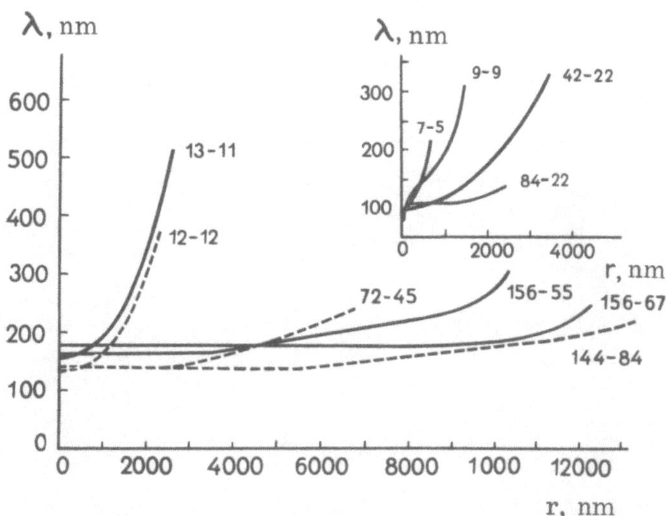


Fig. 4. Increase in the distance λ between turns with the distance r from the center of the spiral. Solid lines for elementary circular spirals, broken lines for double spirals; top right, square spirals.

Figure 2 also shows that the last two steps at the trailing edge in the outer sequence form a pair; Fig. 5 shows this pairing more clearly (arrow), and there is an area free from steps for a time. The bottom right corner shows the alternating large and small separations. Point A in Fig. 6 shows that the pairing may end in formation of steps of height $d_0$, where the following points may be noted:

1. Pairing is pronounced only in ⟨100⟩ directions, while it is hardly seen along ⟨110⟩ so fusion of $d_0/2$ steps to give $d_0$ ones is slower in such directions (Fig. 6). The interaction along ⟨110⟩ must therefore be less, which may be due to anisotropy on surface diffusion or variation in step microstructure from point to point.

2. Paired steps are not distributed so evenly as equidistant steps, which may be due to increased speed fluctuations, particularly when the distances between adjacent steps are unequal. The poles along the ⟨110⟩ directions (B in Fig. 6) show very great shape instability; a facetted structure tends to be produced.

These variations in interaction explain the stability of the $d_0$ steps in a rectangular spiral; the work of detachment of the first molecule from a $d_0$ step (two different modes are possible) indicates scope for step splitting, which actually occurs only if there is a sufficiently large step-free area ahead of the front (Fig. 3). Then the interaction does not prevent $d_0/2$ steps from forming. The $d_0$ steps can move in a rectangular spiral, but $d_0/2$ steps cannot be produced, because these interact more strongly. The splitting starts in ⟨110⟩ directions because the $d_0/2$ steps interact more weakly along such directions. The results for $d_0$ step stability lead us to examine the splitting of larger steps into $d_0$ ones (rather than $d_0/2$) in the same way. A small step separates from a large one more readily if its height is $d_0$ rather than $d_0/2$.

A particular consequence of kinematic interaction is that stellate rosettes (center of Fig. 6) are formed by evaporation. If a dislocation has its Burgers vector parallel to the surface, the periodic preferential nucleation results in $d_0/2$ circular steps or $d_0$ rectangular ones (Fig. 7) [6]. These types are formed at random. A rosette is produced when a sequence of circular steps follows a rectangular step (pit A in Fig. 7). The shape is due on the one hand to the initial pincushion form but also (particularly) to the larger number

Fig. 5. Step pairing within an evaporation step.

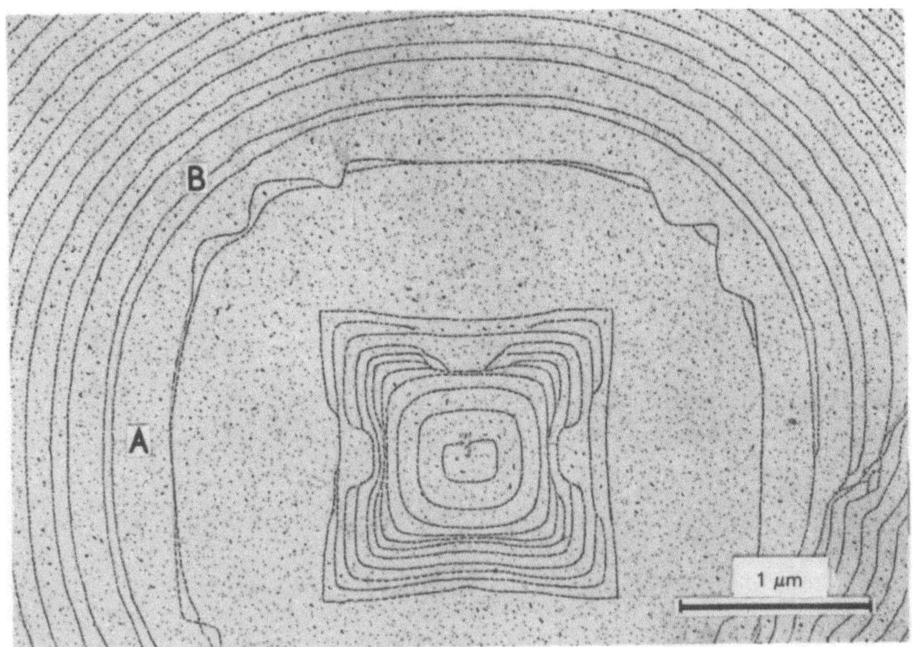


Fig. 6. Etch rosette on rocksalt crystal.

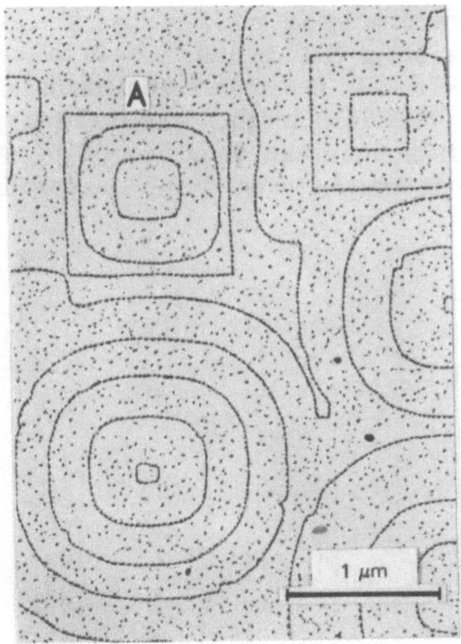


Fig. 7. Circular and rectangular spiral steps at points of emergence of edge dislocations on rocksalt crystals.

of steps along ⟨110⟩, because the steps do not fuse along these directions, in contrast to ⟨100⟩. At about 450°C, an outer step has a height of $d_0$ along ⟨110⟩, whereas the subsequent steps are of height $d_0/2$. At 300-400°C the ⟨110⟩, steps also partially fuse, and the outer step height in some pits becomes $3d_0/2$, while subsequent steps are of height $d_0$ over part of their length. Consequently, the interaction along ⟨100⟩ is different from that along ⟨110⟩. A fast $d_0/2$ step moving along ⟨100⟩ may overtake a $d_0$ step, in which case the motion is retarded (Fig. 8), and the $d_0$ step acquires a pincushion form because depressions are produced at the sides of the

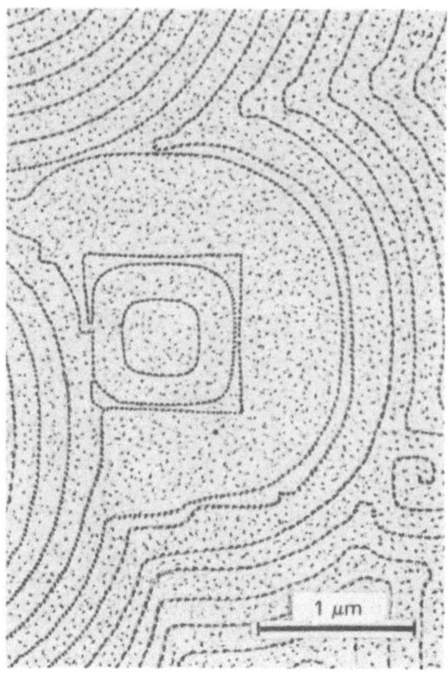


Fig. 8. Initial phase of rosette formation.

rectangle. Then the $d_0/2$ steps collide with a $d_0$ one (left side of Fig. 6) or slow down so far as to coalesce (lower part of pit), as has already occurred on three sides. This results in high steps, which move more slowly, which itself accentuates the coalescence. The steps along ⟨100⟩ retard the other steps, but the interaction along ⟨110⟩ is unilateral in the sense that each step retards the one behind it but not the one before it. It may be that a special configuration occurs (corners on steps), and one expects an elevated molecular concentration within a corner, so there should be relatively little effect from any fresh molecule. At the same time this elevated concentration may impede the motion of the subsequent step. The result is stellate

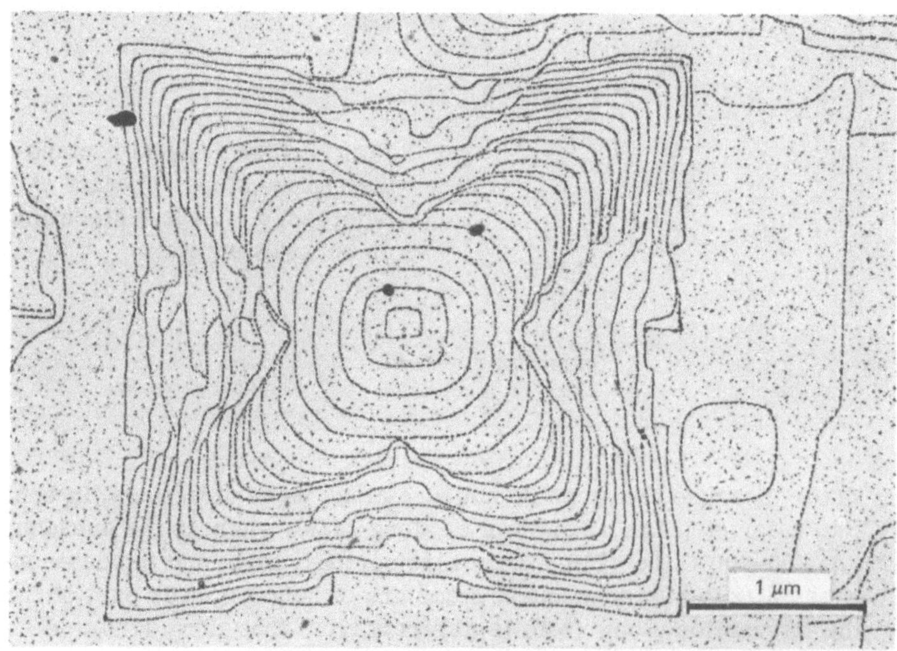


Fig. 9. Rosette showing unequal interaction between steps along ⟨100⟩ and ⟨110⟩.

etch pits (Figs. 6 and 9). There is some relationship between these rosettes and the deviations from polyhedral form in crystals discussed by Chernov [9], but the comparison must be restricted to steps at the edges of the pits, where the molecule concentration is affected by any approaching circular step. In other respects the shape of a rosette is governed by the kinematic interaction. The same should apply to growth hummocks on silicon carbide [9], which closely resemble etch rosettes.

## Literature Cited

1. W. K. Burton, N. Cabrera, and F. C. Frank, Phil. Trans. Roy. Soc., London, A243:299 (1950).
2. G. A. Bassett, Phil. Mag., 3:1042 (1958).
3. H. Bethge, Phys. Status. Solidi, 2:775 (1962).
4. H. Bethge and W. Keller, Z. Naturforsch., 15a:271 (1960).
5. K. W. Keller, Growth of Crystals, Vol. 7, Consultants Bureau, New York (1969), p. 145.
6. K. W. Keller, Phys. Status Solidi, 36:557 (1969).
7. W. W. Mullins and J. P. Hirth, J. Phys. Chem. Solids, 24:1391 (1963).
8. T. Surek, G. M. Pound, and J. P. Hirth, J. Chem. Phys., 55:5157 (1971).
9. A. A. Chernov, Kristallografiya, 8:87 (1963).

# ON THE SHAPE OF GROWTH AND EVAPORATION SPIRALS

**T. Surek, J. P. Hirth, and G. M. Pound**

*Department of Metallurgical Engineering, McGill University, Montreal, Quebec. Canada*
*Metallurgical Engineering Department, Ohio State University, Columbus, Ohio*
*Department of Materials Science, Stanford University, Stanford, California*

As conceived by Frank [1], a spiral dislocation emergent at a point on the crystal surface provides a ledge on the surface with a height equal to the component of the Burgers' vector of the dislocation normal to the surface. Crystal evaporation of low-index surfaces then proceeds by the terrace–ledge–kink (TLK) mechanism [2-4], whereby atoms dissociate from kink positions in ledges, diffuse along the low-index terraces, and finally desorb into the vapor phase. Crystal growth from the vapor takes place by the reverse of the above mechanism. Under these conditions, evaporation or growth spirals are composed of straight, low-index ledge segments [3]. At higher temperatures ($T \gtrsim 0.5T_m$, where $T_m$ is the melting temperature, for molecular or covalent crystals and $T \gtrsim 0.05T_m$ for monatomic metals), the TLK mechanism leads to a smoothly curved spiral-shaped ledge front, the case of interest here, which rotates at a constant angular velocity when steady state is reached. The shape of the curved spiral is determined by kinetic considerations for the movement of the curved ledge fronts and also by capillarity considerations; the limiting radius of curvature at the center of the spiral corresponds to the radius of a disc-shaped hole (or monatomic disc in the case of growth) which is in unstable equilibrium with the local undersaturation (or supersaturation for growth) existing at the point of emergence of the dislocation on the surface [5].

Although numerous observations [6-8] of growth and evaporation or dissolution spirals have been made in the past, the exact mathematical description of the steady-state spiral shape under a range of experimental conditions is still unattained. Also, the problem of the transient development of the spiral shape has thus far escaped analytical treatment; an approximate solution to this latter problem was obtained recently by replacing the spiral ledge front with concentric, circular ledge fronts [5]. The most often quoted result relating to the spiral mechanism is the solution by Cabrera and Levine [9]: for cases where the ledge motions are governed by the surface diffusion of adatoms, and when the overlap of the adatom diffusion fields from adjacent spiral turns is negligible, they find for the steady-state spacing between successive turns

$$\lambda_{CL} = 19\,r^* = -\frac{19\,\gamma}{(kT/V)\ln(p/p_e)}, \tag{1}$$

where $\gamma$ is the specific solid–vapor interfacial free energy of the spiral ledge surface, V is the atomic volume, T is the temperature, and k is Boltzmann's constant. $p/p_e < 1$ is the undersaturation ratio (for crystal growth, $p/p_e > 1$ and the sign in equation (1) is reversed); therefore, $r^*$ is the critical disc size which in unstable equilibrium with the undersaturated vapor phase.

Equation (1) is likely to be satisfied near equilibrium (i.e., $p/p_e \approx 1$) for either growth or evaporation spirals. The shape of the spiral in this case can be rationalized on physical grounds as follows: Since the initial $\lambda$-spacing is large, i.e., proportional to $r^*$ if not precisely $19r^*$ as in equation (1), there will be no diffusion-induced acceleration (cf. Hirth and Pound [4]) of the ledge fronts as they move away from the spiral center. There will be a small acceleration caused by curvature, however, which will be nearly negligible because of the small curvature of the spiral ledge. Therefore, $\lambda$ will stay essentially constant (i.e., a simple Archimedian spiral) over most of the spiral length with $\partial\lambda/\partial r \gtrless 0$ near the center of the spiral. For large undersaturations, there will be overlap of adatom diffusion fields from adjacent spiral turns and equation (1) will not apply; from a physical point of view, however, it could still be argued that $\partial\lambda/\partial r$ for the steady-state spiral should be positive near the spiral center, partly because of the diffusion-induced acceleration of the ledges and partly because of the effect of decreasing capillarity away from the spiral center. A transient-to-steady-state solution of the spiral ledge dynamics, using a concentric, circular-disc model to approximate the spiral shape, indeed resulted in $\partial\lambda/\partial r > 0$ near the spiral center, in agreement with the above physical expectation [5].

Thus far we have considered only the adatom diffusion-controlled movement of ledges away from the spiral center, but, clearly, other factors may also influence the spiral ledge dynamics and the resulting spiral morphologies during the growth or vaporization of some substances. The contribution of surface energy considerations, the attachment kinetics of atoms to kink sites, the effects of impurities in the crystal or adsorbed from the vapor, and the effects of diffusion in the vapor phase are only some of the factors which can influence the kinetics of the spiral ledge fronts. Some of these factors could lead to an attractive interaction between adjacent spiral turns, compared to the repulsive interaction that exists in the case of the adatom diffusion-controlled kinetics which causes the "spreading" of the spiral, i.e., $\partial\lambda/\partial r > 0$. It is conceivable, therefore, that these other factors would tend to give a steady-state spiral with $\partial\lambda/\partial r < 0$ near the spiral center. This raises the fundamental question of whether it is possible, for purely geometric reasons, to have $\partial\lambda/\partial r < 0$ as well as $\partial\lambda/\partial r > 0$ for a spiral in a stationary state, i.e., one that rotates at a constant angular velocity without any change in shape. In the remainder of this paper, we present an affirmative demonstration of this concept.

Consider any spiral $r = r(\Theta)$ with $\partial r/\partial\Theta > 0$ (i.e., r increases monotonically with $\Theta$). When necessary for simplicity or concreteness, we consider spirals of the explicit form

$$r = a\Theta^n \tag{2}$$

with the constant parameters $a$ and n being restricted to the physically interesting cases $a > 0$ and $n > 0$. Figure 1 defines the variables used in the following calculations. The angular velocity, $\omega$ = constant $= -\partial\Theta/\partial t$.

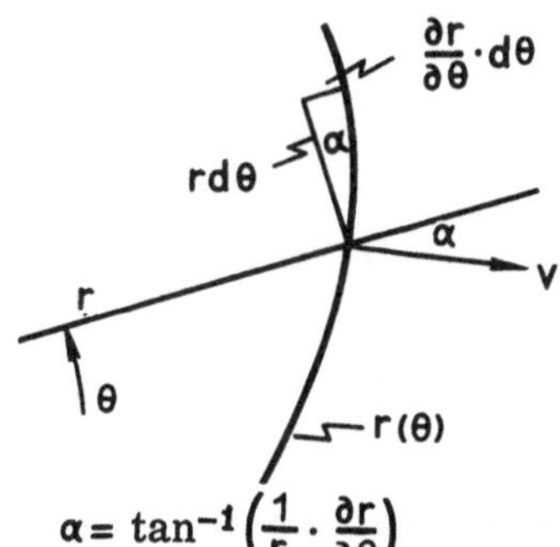


Fig. 1. Definition of polar coordinate variables.

The Sign of $\partial v/\partial\Theta$. The normal velocity of the spiral ledge front at $(r, \Theta)$ in Fig. 1 is given by

$$v=\frac{\partial r}{\partial t}\cos\alpha=\frac{\partial r}{\partial\Theta}\cdot\frac{\partial\Theta}{\partial t}\cos\alpha=-\omega\frac{\partial r}{\partial\Theta}\Bigg/\left[1+\left(\frac{1}{r}\cdot\frac{\partial r}{\partial\Theta}\right)^2\right]^{1/2}. \tag{2a}$$

Therefore,

$$\frac{\partial v}{\partial\Theta}=-\omega\left\{\frac{\dfrac{\partial^2 r}{\partial\Theta^2}}{\left[1+\left(\dfrac{1}{r}\cdot\dfrac{\partial r}{\partial\Theta}\right)^2\right]^{1/2}}-\frac{\dfrac{1}{r}\cdot\left(\dfrac{\partial r}{\partial\Theta}\right)^2\left[\dfrac{1}{r}\cdot\dfrac{\partial^2 r}{\partial\Theta^2}-\dfrac{1}{r^2}\cdot\left(\dfrac{\partial r}{\partial\Theta}\right)^2\right]}{\left[1+\left(\dfrac{1}{r}\cdot\dfrac{\partial r}{\partial\Theta}\right)^2\right]^{3/2}}\right\}. \tag{2b}$$

Hence, since $\omega < 0$, $\partial v/\partial\Theta > 0$ if $\{\partial^2 r/\partial\Theta^2 + (1/r^3)(\partial r/\partial\Theta)^4\} > 0$. Substituting equation (2) in this inequality, we obtain

$$an(n-1)\Theta^{n-2}+an^4\Theta^{n-4}>0. \tag{3}$$

Equation (3) is satisfied always for $n \geq 1$. However, for $0 < n < 1$, we can have a region for which $\partial v/\partial\Theta < 0$. In fact, $\partial v/\partial\Theta > 0$ for small $\Theta$, and $\partial v/\partial\Theta < 0$ for large $\Theta$ in this case, with a critical value, $\Theta_{crit}$, given by

$$\Theta^2_{crit}=\frac{n^4}{n(1-n)}. \tag{4}$$

The Sign of $\partial\alpha/\partial\Theta$. From the definition of $\alpha$ in Figure 1, we have

$$\frac{\partial\alpha}{\partial\Theta}=\left[\frac{1}{r}\frac{\partial^2 r}{\partial\Theta^2}-\frac{1}{r^2}\left(\frac{\partial r}{\partial\Theta}\right)^2\right]\Bigg/\left[1+\left(\frac{1}{r}\frac{\partial r}{\partial\Theta}\right)^2\right]. \tag{4a}$$

Therefore,

$$\partial\alpha/\partial\Theta<0 \quad \text{if} \quad (\partial r/\partial\Theta)^2>r(\partial^2 r/\partial\Theta^2). \tag{5}$$

Substituting equation (2) in (5), we get $n > n-1$, which is always true. In other words, for any continuously curved spiral of the form in equation (2), the angle between the radius vector and the normal to the spiral decreases with increasing $\Theta$ (or r).

The Sign of $\partial\lambda/\partial\Theta$. For the spiral in equation (2), we define the spacing $\lambda$ between successive spiral turns by

$$\lambda=a\{(2\pi+\Theta)^n-\Theta^n\}. \tag{5a}$$

Therefore,

$$\frac{\partial\lambda}{\partial\Theta}=an\{(2\pi+\Theta)^{n-1}-\Theta^{n-1}\}. \tag{5b}$$

Thus, $\partial\lambda/\partial\Theta > 0$ if $n > 1$ and $\partial\lambda/\partial\Theta < 0$ if $n < 1$. For an Archimedean spiral (i.e., $r = a\Theta$), of course, $\partial\lambda/\partial\Theta = 0$.

The Sign of $\partial l/\partial\Theta$. In Fig. 2, we define the normal spacing $l$ between successive spiral turns as that measured along the normal direction to the spiral line. The significance of the

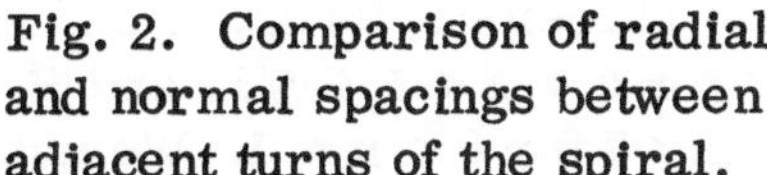
Fig. 2. Comparison of radial and normal spacings between adjacent turns of the spiral.

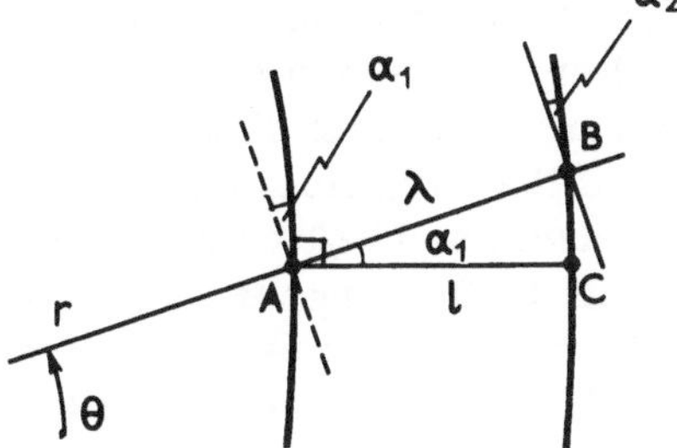


spacing $l$ is that in problems where adatom diffusion influences the ledge kinetics, the normal velocity v is determined by the spacing $l$ rather than $\lambda$ between the spiral turns. Not too close to the spiral center, we can approximate ABC in Fig. 2 by a triangle; therefore,

$$\frac{l}{\sin(\pi/2-\alpha_2)}=\frac{\lambda}{\sin(\pi/2-\alpha_1+\alpha_2)}, \tag{5b}$$

or

$$l=\lambda\frac{\cos\alpha_2}{\cos(\alpha_1-\alpha_2)}. \tag{6}$$

Using the definition of $\alpha$ in Fig. 1 and substituting equation (2), we obtain, after considerable simplification,

$$l=\lambda\,\frac{\left(1+\dfrac{n^2}{\Theta^2}\right)^{1/2}}{1+\dfrac{n^2}{\Theta^2(1+2\pi/\Theta)}}. \tag{7}$$

Therefore, at least for large $\Theta$, we find $l < \lambda$. More rigorously, it can be shown from equation (7) that $l < \lambda$ for $\Theta > 2\pi(1-n^2/4\pi^2)^{1/2}$ or $\Theta \gtrsim 2\pi$, while for $\Theta < 2\pi$ the assumption leading to equation (6) is probably not valid. Moreover, since $\partial\alpha/\partial\Theta < 0$, in the limit of large $\Theta$ we have $\alpha \to 0$ and $l \to \lambda$. In other words, for very large $\Theta$, the spiral can be replaced, to a very good approximation, by concentric, circular ledge fronts.

The sign of $\partial l/\partial\Theta$ can be obtained, in principle, from equation (6) and substituting equation (2). This procedure is much too cumbersome; for our present purposes, we note that for large $\Theta$, $\alpha_2 \approx \alpha_1 = \alpha$ and, therefore,

$$l \approx \lambda\cdot\cos\alpha \tag{7a}$$

and

$$\frac{\partial l}{\partial\Theta}=\cos\alpha\,\frac{\partial\lambda}{\partial\Theta}-\lambda\cdot\sin\alpha\,\frac{\partial\alpha}{\partial\Theta}. \tag{8}$$

Since $\partial\alpha/\partial\Theta < 0$ always, $\partial l/\partial\Theta > 0$ when $\partial\lambda/\partial\Theta > 0$, i.e., for $n \geq 1$. For $0 < n < 1$, $\partial\lambda/\partial\Theta < 0$, and it is expected that $\partial l/\partial\Theta$ will behave qualitatively in the same manner as $\partial v/\partial\Theta$, i.e., $\partial l/\partial\Theta > 0$ for small $\Theta$ and $\partial l/\partial\Theta < 0$ for large $\Theta$ with some $\Theta_{crit}$ [cf. equation (4)] where $\partial l/\partial\Theta = 0$. These same conclusions can be reached quantitatively by substituting for the various terms in equation (8) from above and examining the large $\Theta$ limit.

In conclusion, we have shown that for the families of spiral shapes $r = a\Theta^n$ both $\partial\lambda/\partial\Theta < 0$ and $\partial\lambda/\partial\Theta > 0$ are consistent with the concept of a spiral which rotates at a constant angular velocity without any change in shape. Since $\partial/\partial r = (\partial\Theta/\partial r)\cdot\partial/\partial\Theta$, our findings also apply to the derivatives of the variables with respect to r. Therefore, for $n > 1$ and $n = 1$ (Archimedian spiral), $\partial v/\partial r$, $\partial\lambda/\partial r$, and $\partial l/\partial r$ are all $\geq 0$; while for $0 < n < 1$, $\partial v/\partial r$ and $\partial l/\partial r$ are $>0$ for small $\Theta$ (or r) but $<0$ for large $\Theta$. $\partial\lambda/\partial r$ is always $<0$ for $0 < n < 1$. We note that many observations of growth and evaporation spirals [6-8] give $n \approx 1$ or $n \gtrsim 1$, which, for reasons explained earlier, is the physically expected behavior when the spiral ledge kinetics are at least partly adatom diffusion-controlled.

## Literature Cited


1. F. C. Frank, Disc. Faraday Soc., 5:48, 67 (1949).
2. W. Kossel, Nachr. Ges. Wiss. Göttingen, 135 (1927).
3. W. K. Burton, N. Cabrera, and F. C. Frank, Phil. Trans. Roy. Soc. London A243:299 (1950).
4. J. P. Hirth and G. M. Pound, J. Chem. Phys., 26:1216 (1957).
5. T. Surek, G. M. Pound, and J. P. Hirth, J. Chem. Phys., 55:5157 (1971); T. Surek, Ph. D. Thesis, Stanford University (1971).
6. E. D. Dukova, Dokl. Akad. Nauk SSSR, 121:288 (1958) [Sov. Phys. – Dokl., 3:703 (1958)].
7. A. R. Verma, Crystal Growth and Dislocations, Butterworths, London (1953).
8. A. J. Forty, Adv. Phys., 3:1 (1954).
9. N. Cabrera and M. M. Levine, Phil. Mag., 1:450 (1956).

# HOLOGRAPHIC TECHNIQUES IN CRYSTAL GROWTH

**R. J. Schaefer, J. A. Blodgett, and M. E. Glicksman**

*U.S. Naval Research Laboratory, Washington, D. C.*

## Introduction

A hologram is a recording, on some photosensitive medium, of the amplitude and phase of coherent light scattered from an object. When the hologram is illuminated by monochromatic light, it can reproduce the amplitude and phase of this scattered light, and thus reconstruct an image of the original object. This reconstructed image has unique properties which make it especially useful for quantitative observations of the shapes of objects. Objects everywhere within the field of view of the hologram are reconstructed in sharp focus, irrespective of their distance from the plane of the hologram. Because the hologram reconstructs the phase, as well as the amplitude, of the light scattered from the object, several interferometric techniques can be used to make precision measurements of the object. Holography is more difficult and costly than conventional photography, and its application is usually advantageous only when makes use of these special depth of focus and phase information properties.

In studies of crystal growth, holography uniquely makes possible the recording of transient events occurring at unpredictable locations. Crystallization events throughout the volume of a relatively large study chamber can be observed microscopically in the reconstructed image. Moreover, the transient shapes of the crystals can be measured in detail, and interferometric techniques can be used to map out the growth rates at all points on the crystal surface.

## Holographic Principles

To form a hologram, an object (such as a crystallization test chamber) is illuminated by coherent monochromatic light from a laser. The light scattered from the object is allowed to fall upon a high-resolution photosensitive emulsion (film or plate). Simultaneously, a portion of the laser beam is split off before striking the object, and is directed onto the emulsion by a path which bypasses the object. The second, or reference, beam forms an interference pattern when combined with the first, or object, beam, and this interference pattern is recorded on the emulsion to form the hologram. Because the reference beam must be coherent with (i.e., have a fixed phase relationship to) the object beam, the path lengths of the object and reference beams must be approximately equal, with their path difference less than the "coherence length" of the laser. A typical arrangement for making holograms of transparent objects is shown in Fig. 1.

After the hologram has been developed and fixed, it is illuminated by a reconstruction beam identical to the reference beam used during exposure. The interference pattern recorded on the hologram acts as a highly complex grating, diffracting light to reproduce the

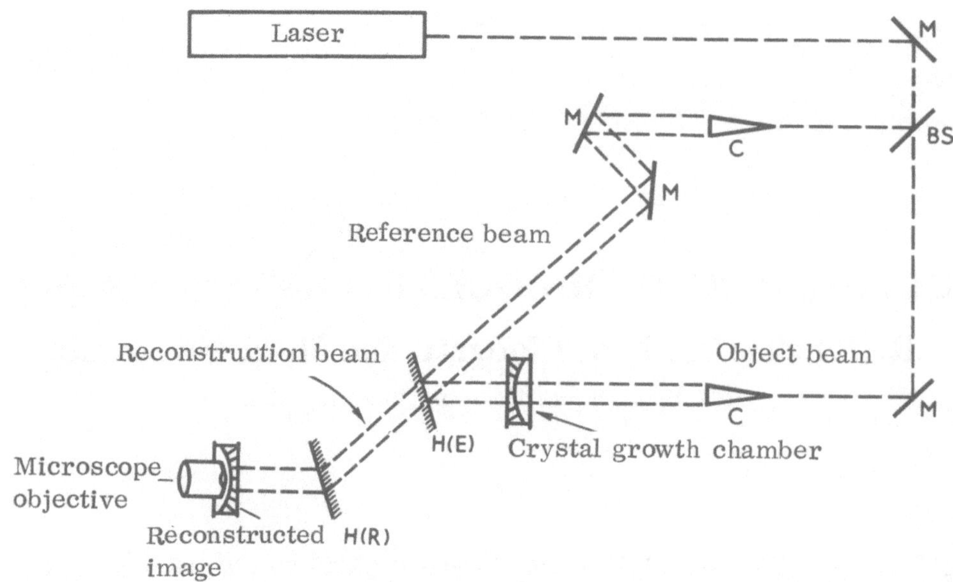


Fig. 1. Schematic representation of a typical arrangement for making and reconstructing holograms of transparent objects. M, mirror; BS, beam splitter; C, collimator; H(E), hologram during exposure; H(R), hologram during reconstruction. The same beam is used as the reference beam during exposure and the reconstruction beam during reconstruction.

relative amplitude and phase of the light originally scattered from the object. This diffracted light reconstructs an image of the object, and all points within this image are reconstructed in sharp focus, irrespective of their distance from the plane of the hologram.

## Quality of the Reconstructed Image

The quality of the image reconstructed from the hologram is influenced by many factors. Although holograms can be produced when many of these factors are significantly compromised, the quality of the reconstructed image necessarily suffers. Acceptable limits on all of these factors are decided by the resolution and image quality required for any particular project. Resolution of 200-300 cycles per millimeter can be obtained without great care, whereas resolution of 500 or more cycles per millimeter requires careful attention to all details. In practice, the higher resolutions can be obtained only by using emulsions on flat glass plates. For holograms of microscopic objects, such as growth features on crystal surfaces, the following are important considerations.

(a) Film Quality, Including Resolution and Linearity. Films should be capable of recording 2000 or more lines per millimeter, so that the interference patterns of light waves incident at large angles can be recorded. The amplitude transmission of the film should be a linear function of the exposure, so that the hologram will not produce higher-order diffracted beams during reconstruction, which would add confusing optical "noise" to the reconstructed image. These requirements demand the use of special high-resolution (thus low-speed) emulsions, with careful processing.

(b) Emulsion Flatness. If the hologram does not have exactly the same shape during reconstruction as it had during exposure, aberrations will be introduced into the reconstructed image. The use of emulsions on glass plates can minimize shape changes in processing. When holograms must be exposed in rapid succession, however, the use of glass plates is awkward, and it may be more desirable to make holograms on 35-mm film with a motorized film transport. The film transport mechanism should hold the film as flat as possible, and reconstruction of the individual holograms can be improved by sandwiching each hologram between two pieces of flat glass with a fluid that matches the index of refraction between the hologram and the glass.

(c) Hologram Alignment. Optimum resolution is obtained only when the reconstructing beam exactly duplicates the reference beam used during exposure of the hologram. The hologram should thus be mounted in a tilting mechanism with micrometer adjustments, and the tilt adjusted for optimum resolution. It is useful to make a hologram of a known object with a simple pattern, such as a high-resolution test target, to assist in the alignment of the holograms for reconstruction. Glass plates can often be supported in mounts on the optical bench which will allow them to be replaced, after development, exactly in their original position. Glass chambers are now available commercially which allow photosensitive plates to be exposed, developed, fixed, and reconstructed without ever being removed from the optical bench.

(d) Stability. To prevent any change in the interference pattern on the emulsion during exposure, all parts of the optical system, including the object (specimen) must remain stationary within a fraction of a wavelength during the exposure time. This generally requires that the components must be mounted rigidly, and may require the use of a high-quality vibration isolation system. To achieve easily the shortest possible exposure time, it is advantageous to use a relatively high-powered laser, such as an argon ion laser. If sufficient monochromatic power is available to make exposures of a few hundred microseconds or less, then the vibration isolation system may not be needed during exposure. However, vibration isolation is needed for those interference techniques described below which demand good stability during reconstruction.

(e) Aperture. The ultimate resolution possible in a reconstructed holographic image is limited by the numerical aperture of the hologram. In this respect the hologram acts like any other diffraction-limiting aperture in an optical system. Although resolution can be improved by simply increasing the area of the hologram, a large hologram requires longer exposure time (for a given total laser power), is more difficult to hold flat, and is more costly.

(f) Viewing Technique. The finest details of the reconstructed image can be seen only by examining the image through a microscope with large numerical aperture. If the hologram is rotated 180° about a vertical axis, it reconstructs a real image of the crystallization chamber, into which the microscope objective can move for high-resolution viewing (Fig. 1). The reconstruction beam must be identical to the reference beam, but pass through the hologram in the opposite direction, a situation which is readily attainable if both beams are collimated. Other beam arrangements can generate a real image, but this image will contain aberrations and will not have unit magnification. The unrotated hologram would reconstruct a virtual image, appearing behind the hologram. This is the type of reconstruction most commonly seen in holograms of macroscopic objects, but it proves not useful for microscopic observation because a high-power microscope objective (with its short working distance) cannot be focussed on the virtual image.

## Techniques for Crystal Growth Studies

(a) Transmitted Light. For studies of crystal growth in transparent systems, crystals can be grown in a temperature-controlled glass-walled observation chamber, and illuminated by laser light which passes through the chamber and onto the emulsion. The crystal interfaces are rendered visible by the difference between the refractive indices of the crystal and the surrounding medium (liquid or vapor).

If the specimen is illuminated by an expanded, collimated laser beam, then the phase of the light emerging from any point on the specimen will depend on the total optical path length through which it passed, the local optical path length being equal to the actual path length times the local refractive index. For a crystal with refractive index $n_c$ and thickness h, growing in a medium with refractive index $n_m$, the optical path length of light passing through a specimen chamber of thickness d is $L = hn_c + (d-h)n_m$. If the thickness of the crystal changes by $\Delta h$, the change in optical path length is $\Delta L = \Delta h(n_c - n_m)$, and the path length will thus change by one wavelength $\lambda$ when $\Delta h = \lambda/(n_c - n_m)$. Because these phase relationships are preserved in the reconstructed image, the thickness of a growing crystal can be measured by interference between the reconstructed

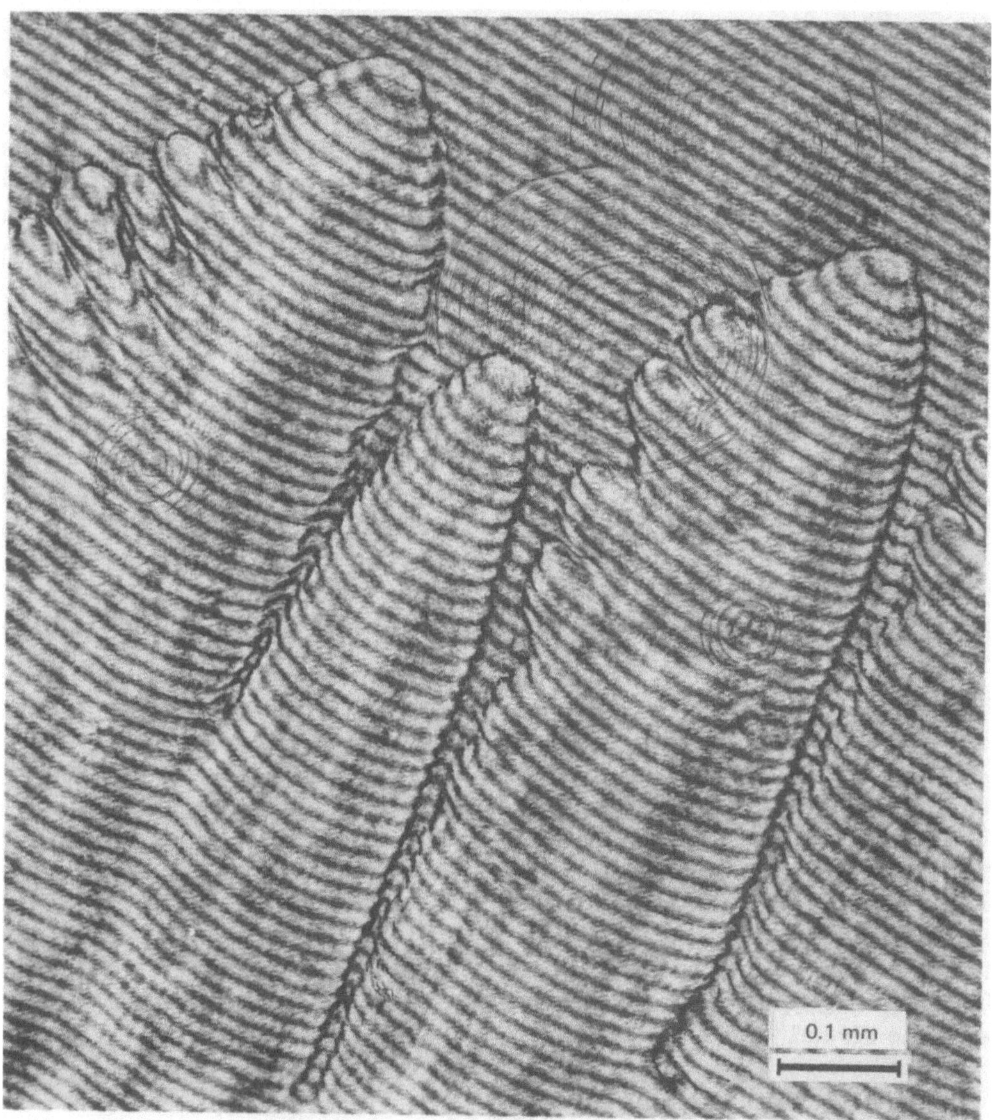


Fig. 2. Fringes in reconstructed image of succinonitrile dendrites, showing shape of tips and initiation of branches. Note complete suppression of branching on the shortest dendrite, and solidification of interdendritic material to form single crystal.

image and a plane wave, introduced via a beam splitter positioned between the hologram and the reconstructed image. The interference fringes so formed then delineate contours of constant crystal thickness relative to a reference plane determined by the alignment of the beam splitter. By adjusting the beam splitter, the fringes can be aligned with features of interest on the crystal surface, thus greatly facilitating the measurement of important dimensions. As with most interferometric techniques, good mechanical stability is needed to prevent fringe motion during observation. The contour interval, $\lambda/(n_c - n_m)$, being a function of the refractive indices of the crystal and the surrounding medium, will be much smaller at a crystal–vapor interface (typically about 1 μm) than at a crystal–liquid interface (typically 15-40 μm). The depth of focus and phase information preserved in the hologram make this interference microscopy technique especially suitable for the study of rapidly growing crystals, where the location of significant events and the alignment of the fringes would be extremely difficult by con-

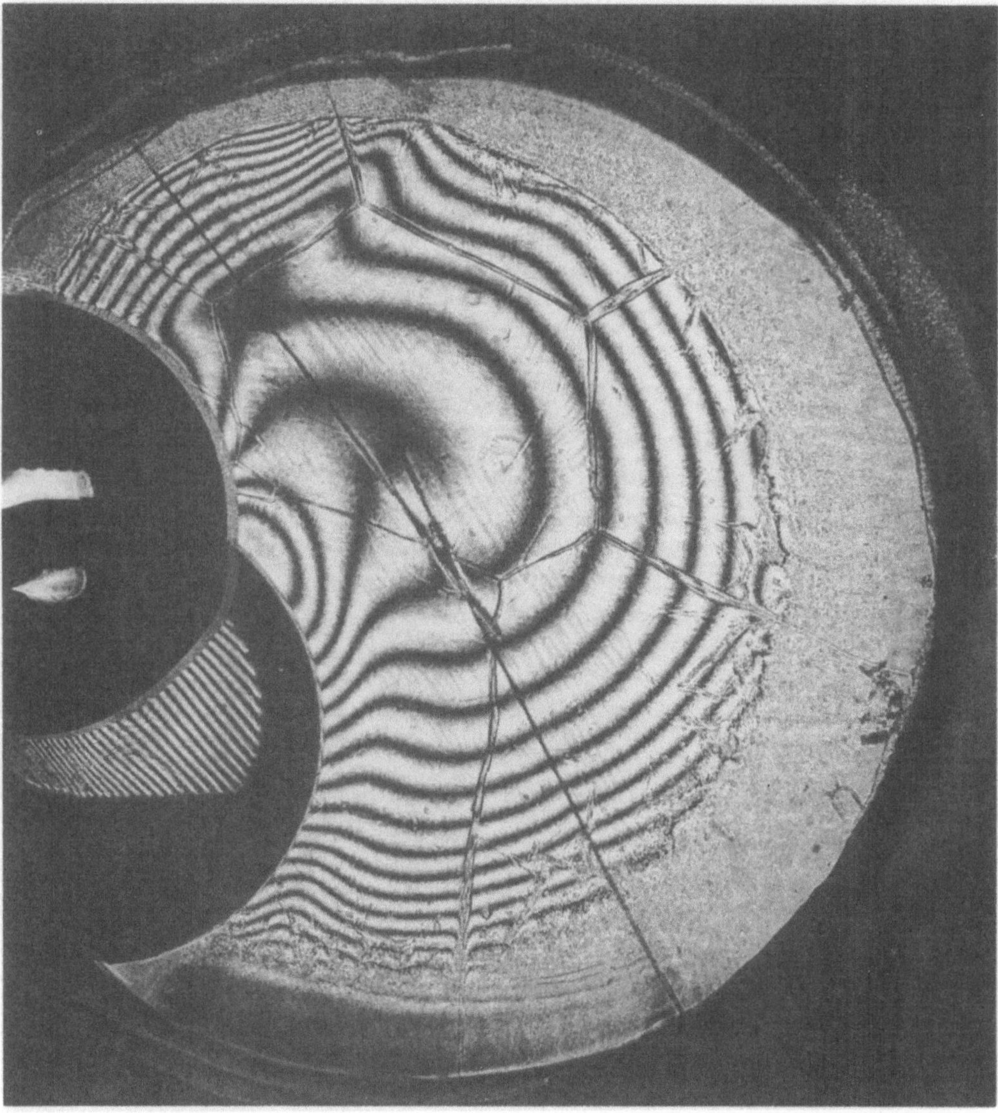

Fig. 3. Reconstruction of double-exposure hologram of solidigying succinonitrile. The fringes indicate the crystal growth occurring in the 1-min interval between exposures. Two large bubbles have clearly distorted the shape of the solid–liquid interface. A fine thermocouple crosses the center of the 19-mm-diameter cell.

ventional interferometric techniques. Figure 2 shows an example of the use of this technique to produce fringes on the reconstructed image of succinonitrile dendrites growing at a small supercooling.

Some unique double-exposure techniques are possible with holography. If a holographic emulsion is exposed twice, using two different objects, then the hologram will reconstruct images of both objects. Thus, if two exposures of a crystal are recorded at slightly different stages of growth, the resulting hologram will reconstruct two slightly different images of the crystal, with interference fringes showing the changes in crystal shape. In this way, the rate of crystal growth at all points on the crystal surface can be mapped out. Figure 3 shows a reconstructed hologram with interference fringes delineating the growth of succinonitrile crystals in a small test chamber; the fringes indicate a greater growth rate near the sides of the

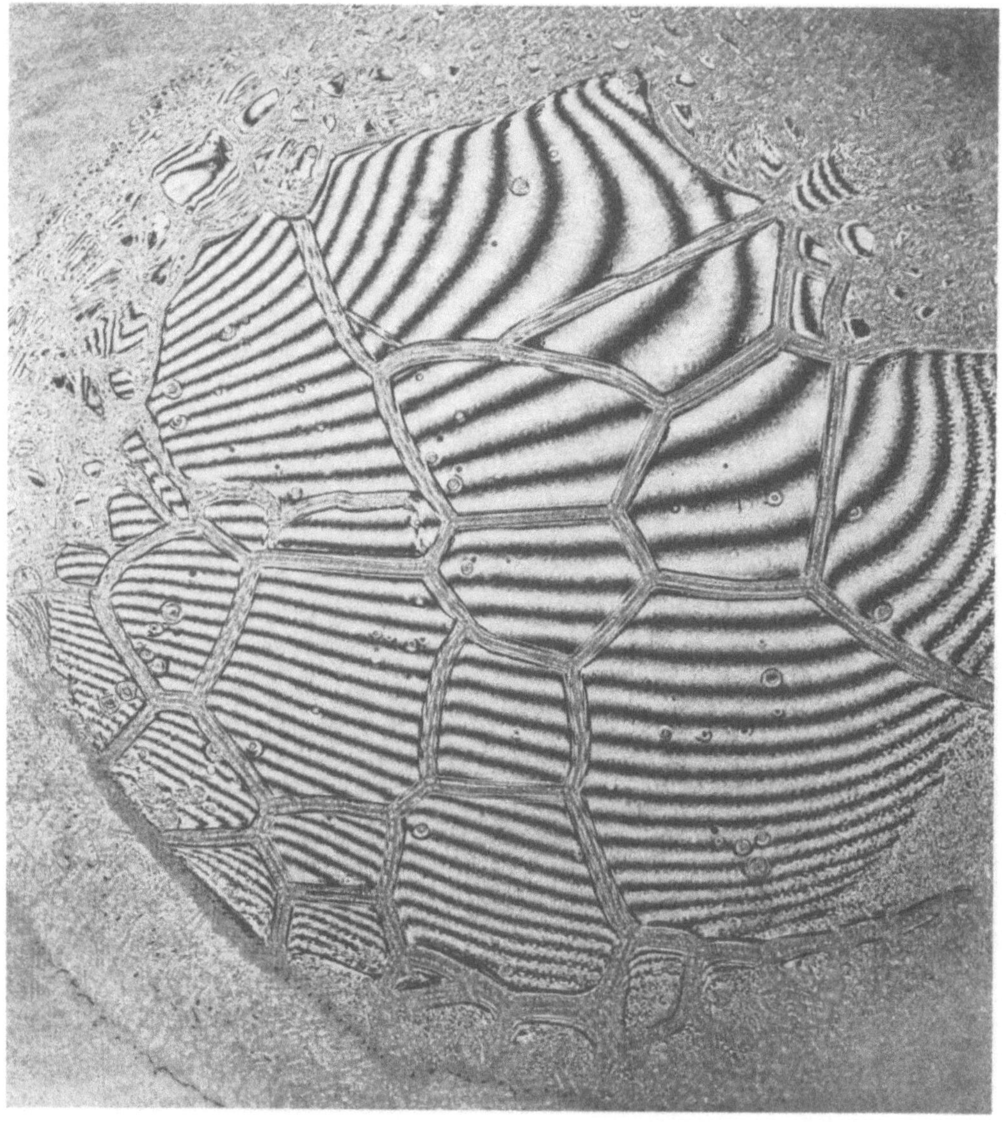

Fig. 4. Reconstruction of double-exposure hologram of solidifying camphene. Since one exposure was of the object beam without any specimen, the fringes delineate contours of constant crystal thickness. Fringe interval 18.5 $\mu$m.

chamber, where cooling was most rapid. In this case, each contour represents 35 $\mu$m of growth.

Double exposures may also be made by making one exposure of the specimen, followed by a second exposure of the object beam itself, with the specimen removed from the system. The reconstructed image will then show fringes indicating the thickness of the crystal, similar to those seen with the interference microscopy method. Although these fringes cannot be aligned with features of interest, they do have the advantage of being "frozen" into the hologram so that they are not influenced by vibrations during reconstruction, and can be readily observed and then photographed without vibration isolation equipment. Figure 4 is a reconstruction from a double-exposure hologram made by this technique, with contours showing the total thickness of crystals in an observation chamber. Figure 5 is a higher-magnification view of the detailed structure near a crystal trijunction, reconstructed from four different double-exposure holograms exposed at intervals of 15 sec.

(b) Reflected Light. Holography can also be used to observe the growth of crystals which are not transparent, if they are growing in a transparent medium. Examples would

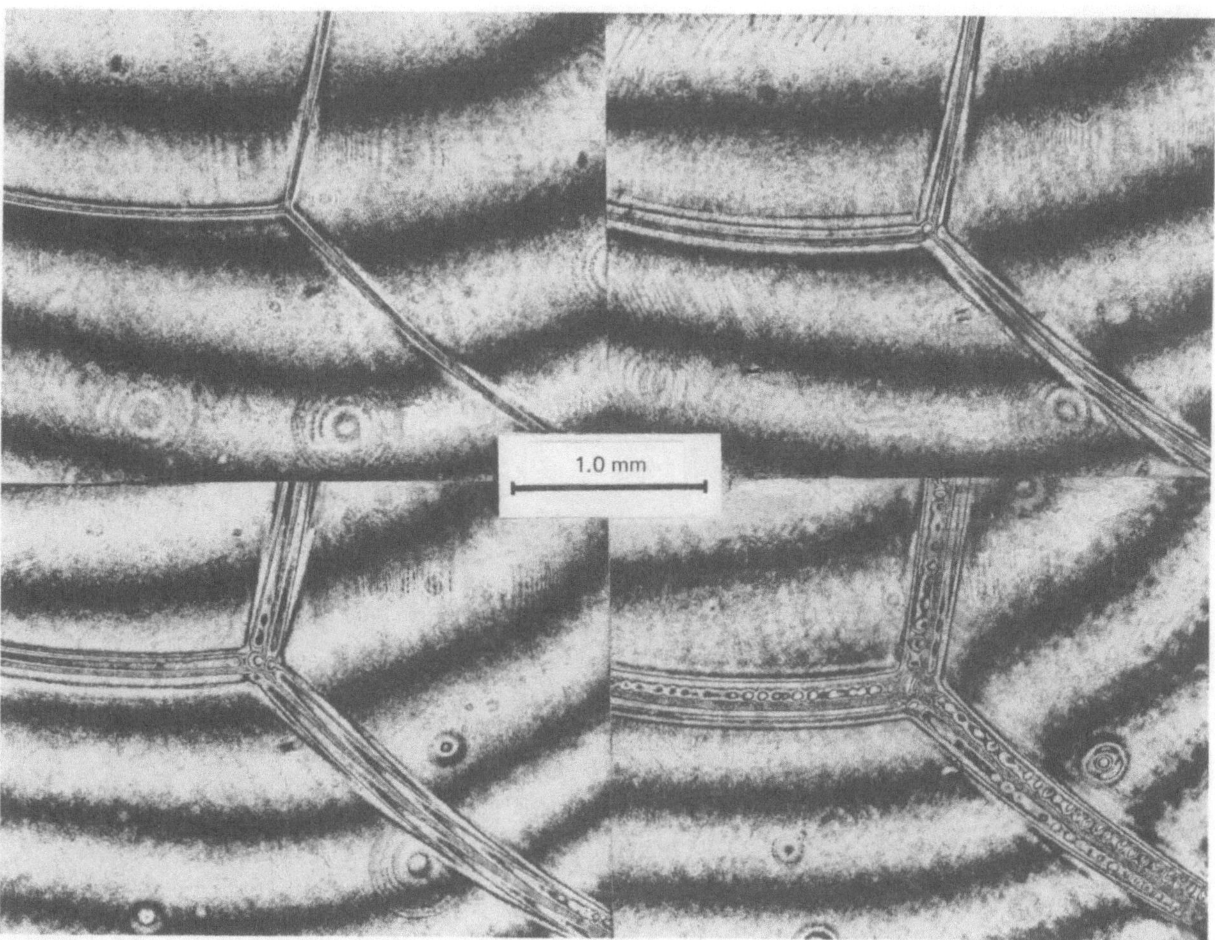


Fig. 5. High-magnification views from four hologram similar to that shown in Fig. 4, showing development of features on a camphene solid—liquid interface near a crystal trijunction. Grain boundaries and impurity particles are seen to generate distortions of the interface. Time interval between successive holograms about 15 sec.

be metal crystals grown by electrodeposition or vapor deposition. An object beam incident normal to the metal surface will be reflected straight back, and can be directed onto the emulsion by means of a beam splitter. Interference techniques similar to those used for transparent systems can also be used for reflecting systems, and the sensitivity is greatly increased. For reflection from a growing crystal surface advancing a distance $\Delta h$, the optical path length change is $\Delta L = 2n_m \Delta h$. Each fringe represents a $\Delta L$ of one wavelength $\lambda$, corresponding to $\Delta h = \lambda/2n_m$, typically about 0.2 $\mu$m. With double-exposure methods, it is possible to measure shape changes on crystal surfaces which are originally very complex or irregular.

## Summary

Holography can contribute to the study of crystal growth by recording images with unlimited depth of focus, on which several interferometric techniques can be applied. This paper has described only a few of the holographic methods available to the investigator. Additional variations can be devised readily to meet the needs of specific experiments. Several books [1-3] describing the basic principles and techniques used in holography in detail have been published, and these books will be useful to investigators who initiate holography projects.

## Literature Cited

1. R. J. Collier, C. B. Burckhardt, and L. H. Lin, Optical Holography, Academic Press, New York (1971).
2. H. M. Smith, Principles of Holography, Wiley, New York (1969).
3. J. W. Goodman, Introduction to Fourier Optics, McGraw-Hill, New York (1968).

# CONCENTRATION IN HOMOGENEITY IN A SOLUTION DURING CRYSTAL GROWTH AND DISSOLUTION

I. M. Guseva, V. M. Ginzburg and V. A. Kramarenko

*All-Union Physical Optics Research Institute, Moscow*

It has been shown [1-9] that the state of a solution [1-6] or melt [7-9] may be monitored during the growth of a macroscopic crystal by holographic methods, which can also be applied to examine the crystal morphology; particular interest attaches to holographic interferometry, which has various advantages over classical interferometry [2, 7, 10]. The difficulties encountered in classical interferometry have meant that until recently it has been used only with microscopic objects [11, 12].

Here we report results on the concentration nonuniformity in a solution around a $KH_2PO_4$ crystal (~1 cm × 1 cm × 2 cm) during growth under static conditions. The measurements were made by holographic interferometry in two ways: by real-time observation, which involves superimposing the real object with the virtual image [2], and the double-exposure method [7, 10]. The real-time method allows one to observe the interference pattern directly during the growth, which is of considerable interest. The double-exposure method requires *a posteriori* processing, and it provides patterns that indicate the changes in solution uniformity between exposures; it also provides a three-dimensional image of the crystal, as well as of the convection flux.

The experiments with $KH_2PO_4$ crystals were performed on z-cut seeds in a crystallizer of volume over 1 $dm^3$.

Previous real-time experiments [4, 5] have shown that concentration nonuniformities appear when the pyramid faces have been produced and the temperature is about 3° below the saturation point; horizontal layers with clear-cut boundaries are visible. There is a complicated concentration pattern within each such layer (Fig. 1). Subsequent experiments have been performed by both methods and have confirmed this phenomenon during dissolution as well as during growth.

The double-exposure method was used in recording a series of patterns, including dissolution in an unsaturated solution, metastable equilibrium with a supersaturated solution, and crystal growth. The results for the unsaturated range were recorded at intervals of 5 min as the temperature was steadily reduced, and the nonuniformity in the solution was dependent on the deviation from saturation. At a certain stage, the convection is replaced by stratification of the solution (Fig. 2a). Any further fall in the temperature causes the thickness of the layer to increase, but at the same time the boundary between the layer and the rest of the solution becomes diffuse, while the convection is very much attenuated. In the metastable range, there

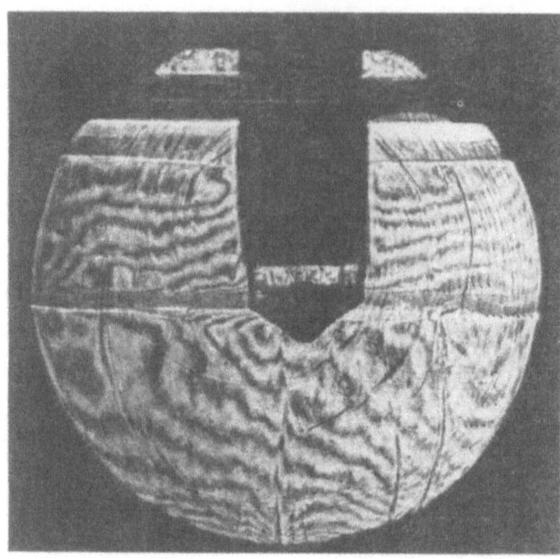

Fig. 1. Interferogram of solution recorded in real time.

is no longer any boundary at the bottom of the crystallizer, and convection is absent (Fig. 2b). Figure 2c shows the labile region. The convection flux is clearly seen, which passes into the horizontal layer in the upper part of the crystal. In the main volume of solution in contact with the crystal, the fringes are almost invisible, which indicates that there is no change in solution parameters between exposures.

Subsequently, we examined the dissolution in relation to saturation temperature in real time. The solutions were saturated at temperatures between 20 and 60°C by steps of 5-10°. If the saturation temperature did not exceed 40°C, the interference patterns during growth differed from those during dissolution; the dissolution pattern under these conditions is composed of a layer of concentric rings of more or less regular shape around the seed. This layer arises at the level of the seed (Fig. 3a) at low subsaturations (about 0.01), and it then descends without changing structure (Fig. 3b). Such patterns are extremely stable, no matter what the changes in subsaturation, mixing, or rotation of the seed.

Experiments with hotter solutions (from 40 to 60°C) showed that the patterns in the growth and dissolution stages were extremely similar (Fig. 3c). Under these conditions, the fringes were less deformed by the crystal than for the case of dissolution in a cooled solution.

These results show that there is some form of reversibility in growth and dissolution [13, 14], but this exists for any given crystal only in a certain temperature range. That range is above 40°C for $KH_2PO_4$.

The concentration gradients occurring at low subsaturation are the larger the lower the temperature, which may be due to the reduced diffusion rate at low temperatures [13].

The close relationship between crystal quality and solution state indicates that this stratification may be one of the causes of defects.

This stratification was predicted by Zemyatchenskii in 1914 [15], when he observed the behavior of alum crystals at various depths in a vessel containing a supersaturated solution under static conditions, when he found that a seed at a low level grew, whereas one at a higher level dissolved. On this basis he supposed that the solution may be stratified.

We are indebted to I. V. Gavrilova, N. P. Shakol'skaya, A. A. Chernov, and Yu. O. Punin for a discussion.

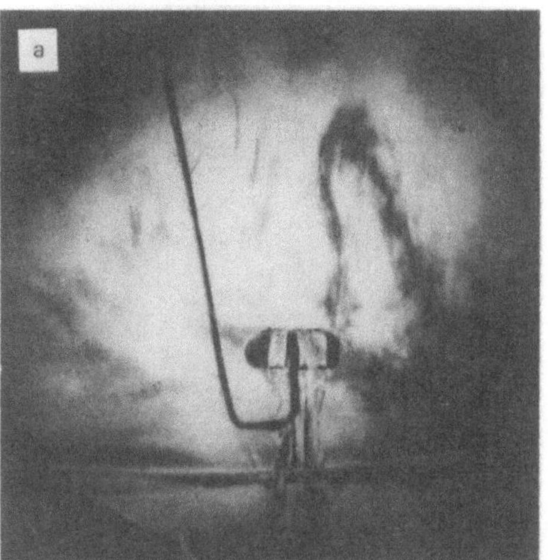


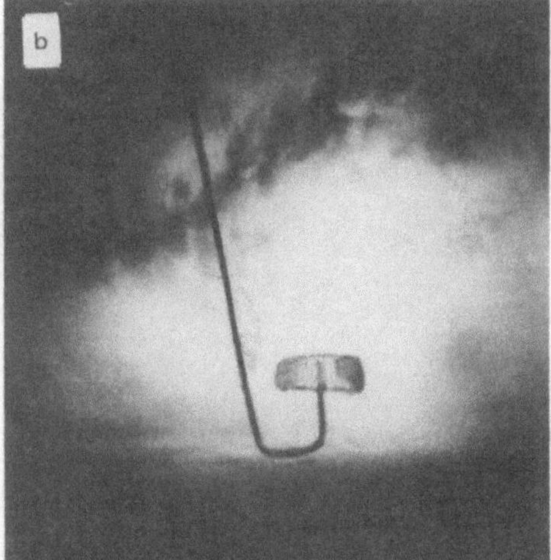


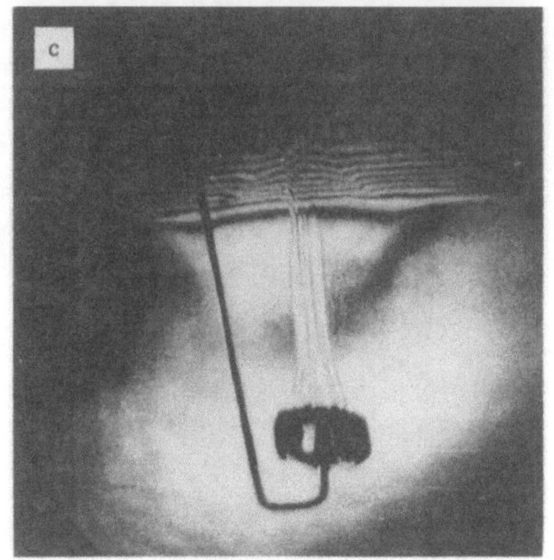


Fig. 2. Interference patterns recorded by the double-exposure method for the following ranges: (a) subsaturation, time between exposures 5 min, solution temperature falling 1.4°; at the start the temperature difference $\Delta T$ from the saturation point was 3°; (b) metastable state, time between exposures 5 min, temperature falling 0.2°C, and $\Delta T$ = 0.6° at start; (c) labile state, time between exposures 1 hr, temperature fall 0.4°, and $\Delta T$ = 6.0° at start.

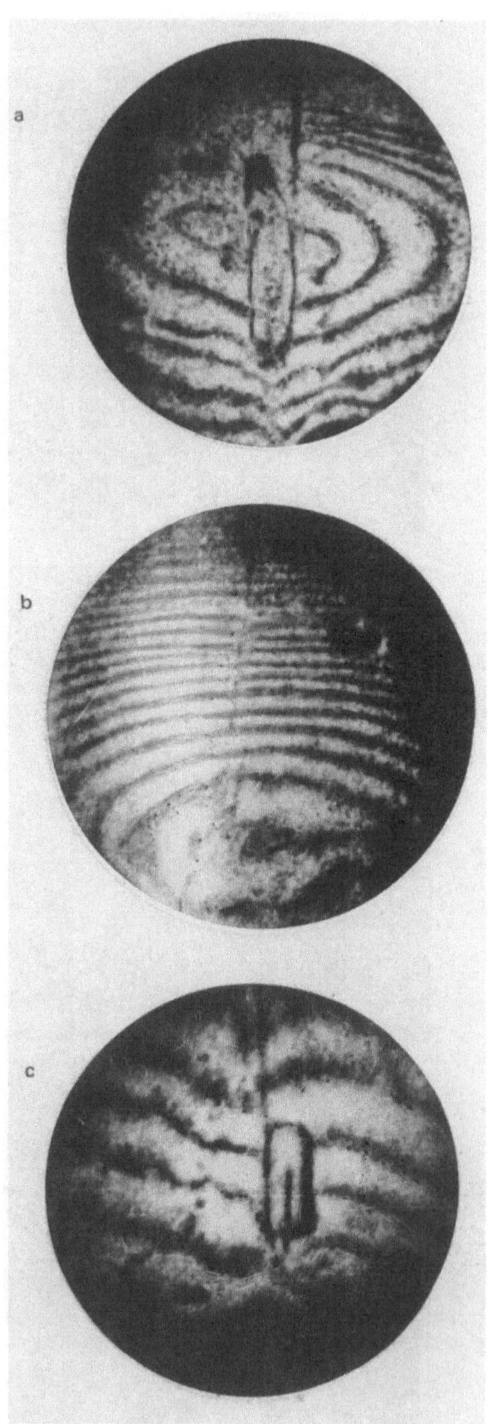


Fig. 3. Interference patterns recorded in real time during dissociation of a seed: (a, b) saturation temperature 25°C; (c) saturation temperature 53°C.

## Literature Cited

1. I. N. Guseva, V. M. Ginzburg, V. A. Kramarenko, E. G. Semenov, A. S. Sonin, and B. M. Stepanov, Abstracts for the Fourth All-Union Conference on Crystal Growth, Erevan, 1972, Growth and Perfection in Crystals, Part 2 [in Russian], p. 247.
2. V. M. Ginzburg, I. N. Guseva, V. A. Kramarenko, E. G. Semenov, A. S. Sonin, and B. M. Stepanov, Kristallografiya, 17:1012 (1972).
3. V. M. Ginzburg, I. N. Guseva, E. N. Lekhtsier, E. G. Semenov, A. S. Sonin, and B. M. Stepanov, Metrologiya, 9:11 (1971).
4. V. M. Ginzburg, I. N. Guseva, E. N. Lekhtsier, V. A. Kramarenko, E. G. Semenov, A. S. Sonin, and B. M. Stepanov, Trudy VNIIOFI, Holography [in Russian], Series B, No. 2, 112 (1972).
5. V. M. Ginzburg, I. N. Guseva, V. A. Kramarenko, E. G. Semenov, A. S. Sonin, and B. M. Stepanov, Proceedings of the Fourth All-Union Conference on Crystallization Mechanisms and Kinetics [in Russian], Minsk (1971).
6. V. M. Ginzburg, I. N. Guseva, V. A. Kramarenko, E. G. Semenov, A. S. Sonin, and B. M. Stepanov, Abstracts for the All-Union Conference on the State of the Art and Development Prospects in High-Speed Photography, Cinematography, and Fast-Process Metrology [in Russian], Moscow (1972), p. 130.
7. R. J. Schaefer, J. A. Blodgett, and M. E. Glicksman, this volume, p. 207.
8. R. J. Schaefer, J. A. Blodgett, and M. E. Glicksman, J. Crystal Growth, 13/14:68 (1972).
9. R. H. McFee, Laser Focus, 5:21, 76 (1969).
10. Yu. I. Ostrovskii, Holography [in Russian], Nauka, Leningrad (1970).
11. V. N. Varikash, Z. Scholtz, and I. Myl, Izv. AN BSSR, Ser. Fiz.-Mat., 3:124 (1967).
12. S. Goldstaub, In: Growth of Single Crystals [Russian translation], Metallurgiya, Moscow (1970), p. 31.
13. T. G. Petrov, E. B. Treivus, and A. P. Kasatkin, Growth of Crystals from Solution [in Russian], Nedra, Leningrad (1967), p. 49.
14. H. Buckley, Crystal Growth [Russian translation], IL, Moscow (1954).
15. A. P. Zemyatchenskii, Zapiski Akad. Nauk, Vol. 33, No. 5 (1914).

Part III

# GROWTH SHAPE STABILITY AND TRANSPORT PROCESSES

# STABILITY OF A PLANAR GROWTH FRONT FOR ANISOTROPIC SURFACE KINETICS


## A. A. Chernov

*Institute of Crystallography, Academy of Sciences of the USSR, Moscow*


## 1. Introduction and Physical Formulation

The rate of growth of a crystal from a melt is often dependent on the orientation only to a slight extent, with the growth front reproducing the crystallization isotherm. The stability of the growth front under such isotropic conditions has repeatedly been examined, and it has been found that a planar front is almost always unstable if the melt is supercooled (more precisely, when the generalized temperature gradient at the front given by (27) in this paper is negative). High surface energy tends to increase the stability, but the increase is not particularly marked for typical numerical values [1]. Only the rate of increase in the instability is dependent on the rates of the isotropic surface processes.

However, growth rates are often highly anisotropic in growth from vapors, most solutions, and many melts, as is clear from the facets on the crystals. Such facetted crystals are far more stable than spherical ones against degeneration into skeletal or dendritic forms. There is a critical size above which such growth cannot occur, and this corresponds for a polyhedron with typical values of the kinetic coefficient to a size larger by about two orders of magnitude than the critical radius for absolute stability in a sphere. One naturally expects [2] that anisotropy in the growth rate will extend substantially the stability range for a planar front. Here we derive stability criteria for a planar front for a particular type of perturbation in the anisotropic case.

The anisotropy in the normal growth rate is particularly pronounced near singular faces that grow by a layer mechanism, whose stability is the primary object of study here, although we impose no restrictions on the extent of the anisotropy. The growth rate of a singular face at moderate supersaturation (supercooling) is determined by the performance of the most active layer sources (dislocations, block boundaries, packing defects, etc.). Near the points of emergence of defects there are vicinal humps, and the numbers of steps emitted by the vertices of such humps per unit time go with the tangential speeds of such steps to determine $p_0$, namely the tangent of the angle formed by the flank of the hump with the singular face. As the step is proportional to the supersaturation, the growth rate of any vicinal surface along the normal to the singular plane is put as

$$\Phi = \Omega b(p)(c - c_e), \tag{1}$$

where $\Omega$ is the specific volume of a particle* in the crystal, $c_e$ is the density of the substance in the saturated solution, c is the actual concentration at the growth front, and b(p) is the kine-

* By particle we mean atom, molecule, or complex attached to the growing crystal.

tic crystallization coefficient in cm/sec. We have $p = p_0$ if the surface is covered with vicinal cones. As $p_0$ is an increasing function of the supersaturation, (1) defines a fairly general increase in the growth rate with supersaturation, not merely a linear one. Here by p we mean the local face orientation, not the mean orientation. The rate of growth from a melt is related to the supersaturation at the front in a similar way:

$$\Phi = b^T(p)(\overset{\circ}{T} - T), \quad (2)$$

where $\overset{\circ}{T}$ is the melting point, T is the actual temperature, and $b^T$ is the kinetic coefficient in cm/sec · deg; if $p = 0$ corresponds to a singular face, then b(p) and $b^T(p)$ have singular minima for $p = 0$, with $b(0) = 0$ and $b^T(0) = 0$; $b(p) \sim p$ near this minimum, so $\theta \equiv (1/b)(\partial b/\partial p)$ is a measure of the anisotropy and is equal to 1/p. As the characteristic value is $p_0 \sim 10^{-2}$ for vicinal cones, we have that $\theta \sim 10^2$. We must emphasize here again that by p we mean the local orientation, not the mean over the face, while the supersaturation is less than the critical value needed to produce nuclei on the whole face. If a supersaturation is sufficiently high to produce multiple nuclei or steep vicinal cones of high step density, then any change in slope will not affect the growth rate substantially, i.e., the anisotropy in b(p) will be slight ($\theta \sim 1$ or even $\theta \ll 1$, which in fact means transition to a normal growth). See [2] for some data on kinetic coefficients.

We now turn to the stability, restricting ourselves at first to the physical significance of growth from solution. In the initial stationary state, there is a front flat on average with vicinal cones, which grows at a rate $\Phi$ governed by the supersaturation and $p_0$, i.e., by the output of the layer sources. This initial front is approximated by a plane $p = 0$ having the kinetic coefficient $b(p_0)$. Now let the front become distorted, e.g., one of the vicinal cones becomes sharper and steeper than the adjacent ones.* This perturbation in this model represents a sharp-ended projection on the planar front; on the steeper flanks we have $p > p_0$, and therefore $b(p) > b(p_0)$, i.e., the absorbing capacity of the perturbed region is higher than that of the rest of the surface. Therefore, the concentration over the perturbed part, including the vertex, becomes less, which will tend to suppress the perturbation. On the other hand, the vertex tends to enter a more highly supersaturated solution, since it rises above the initial front, which corresponds to instability for the isotropic case.† In the present (anisotropic case) the result (increase in the deviation or suppression) is dependent on the balance between the above two opposing factors. If instability dominates, then multiheaded growth tends to occur, which is the anisotropic analog of cellular growth.

A similar argument can be carried through readily for growth from a melt by considering the temperature distribution instead of the concentration. We now turn to detailed calculations.

## 2. Stability during Growth from Solution

We assume that the concentration c satisfies the steady-state diffusion equation

$$D\nabla^2 c + \Phi \frac{\partial c}{\partial z} = 0. \quad (3)$$

Here the z axis is perpendicular to the face, while the x and y axes lie in the plane, $\nabla^2$ is the

* In general, there may be other perturbations, for instance, simultaneous sharpening in all cones; the criterion is affected by the type of perturbation, which is an aspect not examined here. Unfortunately, it becomes difficult to use harmonic perturbations if b(p) shows singular anisotropy or even strong anisotropy.

† Sears [3] considers this factor responsible for the onset of whisker growth.

three-dimensional Laplace operator, and D is the diffusion coefficient. The conservation equation should apply at the growth front:

$$D\frac{\partial c}{\partial z}+c\Phi=\Omega^{-1}\Phi=b(c-c_e), \tag{4}$$

whence we get

$$D\frac{\partial c}{\partial z}=b(c-c_e)(1-\Omega c). \tag{5}$$

This condition differs from the one normally employed since Bertout's time [4],

$$D\frac{\partial c}{\partial z}=b(c-c_e), \tag{6}$$

in that it contains the factor $(1 - \Omega c)$, which is important only for concentrated solutions, but which converts the boundary condition into a nonlinear one. Nevertheless, the form taken by (4) and (5) is essentially correct because the left side of (4) is the total flux of material arriving at the growth front by diffusion and by collision of the front with particles of the solute (the $c\Phi$ term). The steady-state solution to (3)-(5) for an unperturbed surface exists only if the diffusion occurs in a layer near the surface of arbitrary but finite thickness. Therefore, the analysis below remains meaningful only while the size of the perturbed region is less than the thickness of the diffusion layer, or else the characteristic times for surface reconstruction and diffusion are much less than the time for substantial change in the speed of the unperturbed front. The means clearly that the quasistationary approximation can be used. In practice, we can assume that the initial flux is $q = \partial c/\partial z$, i.e., the growth rate and c are given at the unperturbed surface.

We examined the behavior of a small perturbation of the form

$$\tilde{z}(\rho)=\zeta_0 e^{-\rho/\lambda}, \tag{7}$$

where $\rho = (x^2 + y^2)^{1/2}$ is the radius vector in the (x, y) plane (Fig. 1). The concentration over the perturbed surface is given in the form

$$c(\rho, z)=\bar{c}(z)+\tilde{c}(\rho, z), \tag{8}$$

where $\bar{c}(z)$ is the concentration distribution over the unperturbed front and $\tilde{c}$ is the concentration change due to the perturbation. We write (5) for a perturbed system and retain only the terms of the first order in the small quantities $\tilde{z}$ and $\tilde{c}$ to get for $z = 0$ that

$$D\frac{\partial \tilde{c}}{\partial z}=b^*\tilde{c}+(b^*+\Phi+D\theta/\lambda)q\tilde{z}, \tag{9}$$

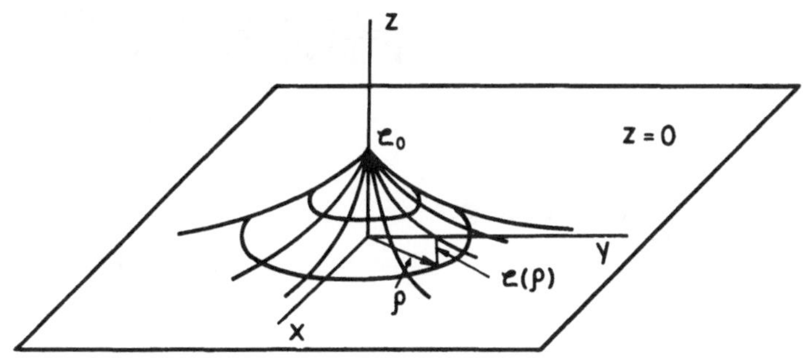


Fig. 1. Form of perturbation in a growth front.

where

$$b^* = b(p_0)[1 - \Omega(2\bar{c} - c_e)].$$

We have used the following expansions of the concentration and the derivative at the perturbed surface in deriving (9):

$$c\Big|_{\tilde{z}} = \bar{c}\Big|_0 + \frac{\partial \bar{c}}{\partial z}\Big|_0 \tilde{z} + \tilde{c}\Big|_0,$$

$$\frac{\partial c}{\partial z}\Big|_{\tilde{z}} = \frac{\partial \bar{c}}{\partial z}\Big|_0 + \frac{\partial^2 \bar{c}}{\partial z^2}\Big|_0 \tilde{z} + \frac{\partial \tilde{c}}{\partial z}\Big|_0,$$

where the subscript $\tilde{z}$ denotes the value at the perturbed surface, while subscript 0 denotes that at the unperturbed one. The $D\theta/\lambda$ term represents the anisotropy in the surface kinetics and is positive throughout the perturbed surface (i.e., for p positive and negative) on account of the minima in b(p) at p = 0; we neglect the anisotropy in the x, y plane.

The solution to (3) and (9) takes the form

$$c(\rho, z) = \frac{1}{2\pi}\iint_{-\infty}^{\infty} e^{-\chi_z z - i\vec{\chi}\vec{\rho}} A(\chi) d\vec{\chi},$$

$$\vec{\rho} = \{x, y\}, \quad \vec{\chi} = \{\chi_x, \chi_y\}, \quad \chi = \sqrt{\chi_x^2 + \chi_y^2}, \tag{10}$$

$$\chi_z = \frac{\Phi}{2D} + \frac{\Phi}{2D}\sqrt{1 + \frac{4D^2\chi^2}{\Phi^2}}; \quad A(\chi) = -q\varepsilon(\chi)\frac{b^* + \Phi + D\theta/\lambda}{b^* + D\chi_z},$$

while the Fourier transform of the perturbation is

$$\varepsilon(\chi) = \frac{1}{2\pi}\iint_{-\infty}^{\infty} \tilde{z}(\rho') e^{i\vec{\chi}\vec{\rho}'} d\vec{\rho}' = \frac{\zeta_0 \lambda^2}{(1 + \chi^2\lambda^2)^{3/2}}. \tag{11}$$

The perturbation cone vanishes and the front is stable if the supersaturation at the vertex is less than that on the unperturbed surface, i.e., when

$$\delta c \equiv q\zeta_0 + \tilde{c}(0, 0) < 0. \tag{12}$$

We calculate c(0,0) from (10) and (11) to get

$$\frac{\delta c}{q\zeta_0} = 1 - \left(1 + s + \frac{\theta}{\mu}\right)\int_0^{\infty} \frac{\chi d\chi}{(1 + \chi^2\lambda^2)^{3/2}(1 + D\chi_z/b^*)} = 1 - \frac{\theta + \mu(1+s)}{1 + \mu^2(1+s)}\Bigg\{1 + \mu -$$

$$- \frac{\mu(1 + s/2)}{\sqrt{1 + \mu^2(1+s)}} \ln \frac{\mu^2 s(1+s) - 1 - \sqrt{1 + \mu^2(1+s)}}{\mu(1+s)[\mu(1+s/2) - \sqrt{1 + \mu^2(1+s)}]}\Bigg\}, \tag{13}$$

$$\mu = \frac{b^*\lambda}{D}, \quad s = \frac{\Phi}{b^*}.$$

We use (4) and (5) to get

$$s = \frac{(\bar{c} - c_e)\Omega}{1 - (2\bar{c} - c_e)\Omega}. \tag{14}$$

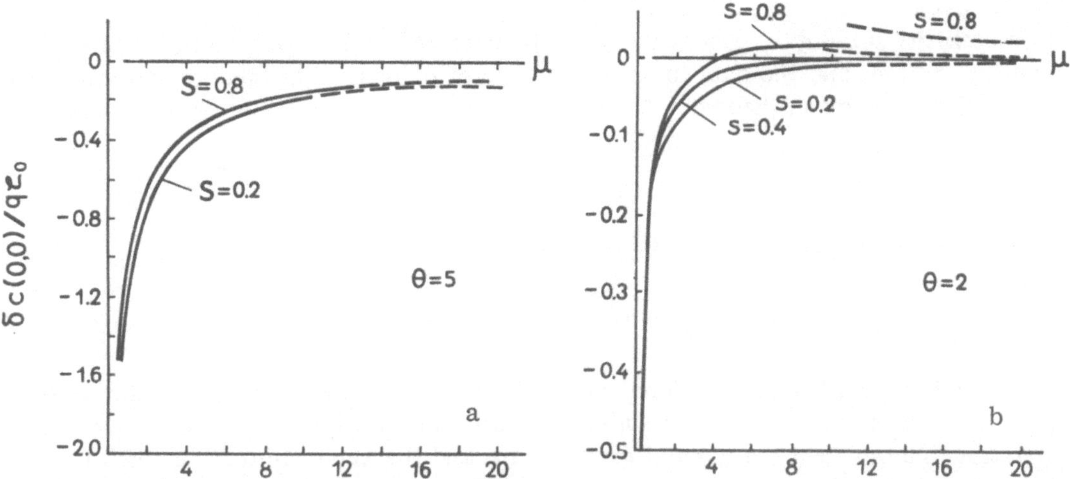


Fig. 2. Concentration change at the vertex of a perturbation having the characteristic dimension $\lambda = (D/b^*)\mu$ in the plane of the front: (a) slight anisotropy ($\theta = 2$); (b) considerable anisotropy ($\theta = 5$).

If the concentration at the front is $\bar{c} \sim 1.5\ c_e$ (i.e., about 50% supersaturation), while the concentration of the saturated solution is $c_e \sim 0.2\Omega^{-1}$ (i.e., about 20% if the density of the solvent is roughly equal to that of the crystal), then $s \simeq 0.2$; if the supersaturation is about 10% ($\bar{c} \simeq 1.1c_e$) and the concentration is about 10% ($c_e\Omega \simeq 0.1$), then we have $s \simeq 10^{-2}$, and therefore $s \ll 1$ in all important cases.

We now examine how $\delta c/q\zeta_0$ varies with $\mu$, $\theta$, and s; curves a and b of Fig. 2 show the general dependence on $\mu$ for several values of $\theta$ and s, which are derived from (13). If $\mu \ll 1$, we have for s variable that

$$\frac{\delta c}{q\zeta_0} = 1 - \theta \tag{15}$$

no matter what s may be, which means that the front is stable against perturbations that cover only a small fraction ($\lambda \ll D/b^*$) of the surface in the neighborhood of the vertex of a vicinal cone in the (x, y) plane provided that $\theta > 1$. Perturbations of this type arise, for example, on account of brief accidental rise or fall in the supersaturation at the vertex of a vicinal cone, and as the time required for such an effect is short, such perturbations would appear to be the most frequent and hazardous. Perturbations with $\mu \gg 1$, which extend over large areas, should be rarer, but they should also cause instability. In practice, $2\lambda$ should be considered as the limit to the mean distance between the vertices of adjacent cones or to the face size. If $\mu \gg 2/s \gg 1$, then

$$\frac{\delta c}{q\zeta_0} = -\frac{1}{\mu}\left(\frac{\theta}{1+s} + 1 - \ln\frac{4}{s}\right). \tag{16}$$

Parts a and b of Fig. 2 (broken lines) show curves corresponding to (16), which correspond to the same s as the solid lines. It is clear that (16) represents the better approximation the smaller s and the larger $\theta$.

It follows from (16) that for

$$1 \gg s > s_k = 4e^{-\left(1+\frac{\theta}{1+s}\right)} \simeq 4e^{-(1+\theta)} \tag{17}$$

we have that $\delta c/q\zeta_0$ is always negative, and as the right side of (13) is monotone for $s > s_k$ we have absolute stability for any $\lambda$. Usually, the inclination of a stationary vicinal cone increases with the supersaturation, and the data indicate [5, 6] that a linear relationship applies, which is also implied by the dislocation theory:

$$p_0 = Ms, \tag{18}$$

while for a cone around a single screw dislocation we have

$$M = \frac{a}{19R_c s} = \frac{kTa[1-(2\bar{c}-c_e)\Omega]}{19\alpha c_e}, \tag{18'}$$

where $R_c$ is the radius of a two-dimensional critical nucleus, $a$ is step height, and $\alpha$ is the surface energy of the end of the step. If the supersaturation is so slight as to correspond to a quadratic relationship between the normal growth rate and the supersaturation, then $\theta = 1/p_0$, and (17) gives an equation for $s_k$:

$$s_k = 4e^{\frac{1}{Ms_k}-1}. \tag{19}$$

The equation (19) has two roots if $M > e^2/4$; it has one root $s_k = 4/e^2$ if $M = e^2/4$ and has no roots for $M < e^2/4$. If we take $kT \simeq 5 \cdot 10^{-14}$ erg, $\Omega \simeq 3 \cdot 10^{-23}$ cm$^3$, $a \simeq 10^2$ ergs/cm$^2$, $\Omega c_e \simeq 0.1$, $a \simeq 3 \cdot 10^{-8}$ cm, we get that M is about 0.25, so the front is always stable in a range of supersaturations in which the growth rate has a quadratic dependence on the supersaturation and $\theta = 1/p_0$, since (17) is met. However, in the linear range of $\Phi(s)$, where the anisotropy may become small, we have $\theta \not\ll 1$, so the system will be unstable. It follows from (13) that the loss of stability occurs first at high $\mu = b^*\lambda/D$ as $\theta$ decreases. Therefore, the first sign of loss of stability is the occurrence of large-cell multiheaded growth, which occurs fairly slowly by virtue of the smallness of $\delta c/q\zeta_0$ for large $\mu$. The sizes $\lambda_k$ of the cell will fall as follows as the supersaturation increases:

$$\lambda_k = \frac{D}{b^*}\mu_k(\theta, s), \tag{20}$$

where $\mu_k$ is the root of $\delta c = 0$; we cannot find the result for $\mu_k(\theta, s)$ in general form analytically from (13), while it is undesirable to tabulate $\mu_k$ numerically from (13) because the form of the latter reflects not only the trends of interest to us but also the detailed form of the perturbation of (7). It is therefore sufficient to perform an approximate examination of $\mu_k(\theta, s)$, and we expand the right side of (13) as a series to show that for $s = 0$ and $\mu_k \gg 1$ we have

$$\mu_k(\theta, 0) = 0.5e^{(\theta+1)}. \tag{21}$$

Then $\mu_k$ increases with s and becomes infinite at $s = s_k$, so we have $\mu_k \simeq 10^3$ for $\theta \simeq 6$ even if the anisotropy is quite weak, i.e., a perturbation results in instability only when its radius in the xy plane is $(xy)\lambda_k \gtrsim 10^2 D/b^*$, while for $D \simeq 10^{-5}$ cm$^2$/sec and $b \simeq 10^{-3}$ cm/sec we get that $\lambda_k \approx 10$ cm, which in practice denotes stability.

It is of some significance to establish the physical meaning of absolute stability for $s > 0$ ($s > s_k$) and the absence of such stability for $s = 0$; the following illustrates the cause. The concentration is minimal at the surface and increases linearly away from it for $s = 0$ and more slowly for $s > 0$; a projection due to a perturbation causes the concentration to fall [$\tilde{c} < 0$, as (10) and (13) show], which applies not only because the projection has grown in the solution but also because the absorbing capacity of the projection has increased in the anisotropic case.

The slower rise in the concentration at greater distances from the front for $s > 0$ means that the fall in c in this case is more pronounced, i.e., the growth is stable. Formally speaking, this is expressed by the term containing s in the parentheses in front of the integral in (13), although the integral itself tends to fall somewhat as s increases. Finally, it is clear that instability occurs at high $\mu$ primarily on account of the decrease in the inclination of the perturbation as it spreads over a large area $\sim\pi\lambda^2$, since the increase in area $\pi\lambda^2$ is balanced by the weak reduction in the kinetic coefficient only for $s > s_k$.

## 3. Stability of Growth from the Melt in the Presence of Impurities

We now consider growth from a melt containing an impurity with a partition coefficient k; we assume that the temperature in the crystal $T_1$ and in the melt $T_2$ will satisfy (5), as will the impurity concentration C, while the boundary conditions at the front take the form

$$\begin{gathered} T_1=T_2, \\ a_1\frac{\partial T_1}{\partial z}-a_2\frac{\partial T_2}{\partial z}=T_Q b^\tau(p)(\mathring{T}-mC-T)=T_Q\Phi, \\ -D\frac{\partial C}{\partial z}=(1-k)\,C\Phi. \end{gathered} \tag{22}$$

Here $a_1$ and $a_2$ ($cm^2/sec$) are the thermal diffusivities of melt and crystal (it is assumed that the specific heats and densities are identical), while D is the diffusion coefficient for the impurity, $\mathring{T}$ is the melting point of the pure substance, $-m$ is the slope of the liquidus line in the phase diagram, and $T_Q$ (degrees) is the ratio of the heat of crystallization to the specific heat. We take the perturbation in the form of (7). We specify that (22) shall apply on the perturbed surface and get the corrections $\tilde{T}_1$, $\tilde{T}_2$, and $\tilde{C}$ to the initial temperature and concentration distributions in terms of the following boundary conditions at $z = 0$:

$$\begin{gathered} \tilde{T}_1-\tilde{T}_2=(G_2-G_1)\tilde{z};\quad G_1\equiv\left.\frac{\partial\bar{T}_1}{\partial z}\right|_0,\quad G_2\equiv\left.\frac{\partial\bar{T}_2}{\partial z}\right|_0; \\ a_1\frac{\partial\tilde{T}_1}{\partial z}-a_2\frac{\partial\tilde{T}_2}{\partial z}+T_Q b^\tau\tilde{T}_1+T_Q b^\tau m\tilde{C}=T_Q b^\tau\left[\frac{(1-k)m\bar{C}\Phi}{D}-G_1+\frac{\theta\Phi}{\lambda}\right]\tilde{z}; \\ D\frac{\partial\tilde{C}}{\partial z}+(1-k)(\Phi-b^\tau m\bar{C})\tilde{C}-(1-k)b^\tau\bar{C}\tilde{T}_1=-(1-k)\left[\frac{\bar{C}\Phi}{D}(b^\tau m\bar{C}+k(\Phi-b^\tau m\bar{C}))-b\bar{C}G_1+\frac{\theta\Phi\bar{C}}{\lambda}\right]. \end{gathered} \tag{23}$$

Here, as before, $\theta=\partial\ln b^\tau(p)/\partial p|_{p_0}$, while the values of b and $\theta$ are taken for $p=p_0$, and a bar above a symbol denotes the unperturbed value. The solution is sought in the form

$$\begin{gathered} \tilde{T}_1=\frac{1}{2\pi}\iint_{-\infty}^{\infty}A_1(\chi)e^{\chi_1 z-i\vec{\chi}\vec{\rho}}\,d\vec{\chi};\quad \chi_1=-\frac{\Phi}{2a_1}+\frac{\Phi}{2a_1}\sqrt{1+\frac{4\chi^2D^2}{\Phi^2}}, \\ \tilde{T}_2=\frac{1}{2\pi}\iint_{-\infty}^{\infty}A_2(\chi)e^{-\chi_2 z-i\vec{\chi}\vec{\rho}}\,d\vec{\chi};\quad \chi_2=+\frac{\Phi}{2a_2}+\frac{\Phi}{2a_2}\sqrt{1+\frac{4\chi^2D^2}{\Phi^2}}, \\ \tilde{C}=\frac{1}{2\pi}\iint_{-\infty}^{\infty}B(\chi)e^{-\chi_z z-i\vec{\chi}\vec{\rho}}\,d\vec{\chi}, \end{gathered} \tag{24}$$

where the Fourier transforms of $A_1$, $A_2$, and B satisfy a linear system of equations derived by substituting (24) into (23) and expressing the perturbations $\varepsilon(\chi)$ from (11) in terms of the Fourier transforms. The stability criterion is derived from the requirement that there is a fall $\delta\Delta T$ in the supercooling $\Delta T = \overset{\circ}{T} - mC - T$ at the crest of the perturbation:

$$-\delta\Delta T = \left[ m\tilde{C} + m\frac{\partial \bar{C}}{\partial z}\tilde{z} + \tilde{T}_1 + G_1\tilde{z} \right]_{\substack{\rho=0\\ z=0}} = \int_0^\infty [mB + A_1]\,\chi d\chi + \left( G_1 - \frac{m(1-k)\bar{C}\Phi}{D} \right)\zeta_0 > 0. \tag{25}$$

As the expressions for A and B are cumbrous, the integration has been carried through only for the limiting case of local perturbations extending over small areas around the initial vertices: $\lambda - 0$. The corresponding expressions give the stability expression for such perturbations:

$$\frac{G}{\Phi} > \frac{m(1-k)C_\infty}{Dk}(1-\theta) - \frac{\theta T_Q}{a_1 + a_2}, \tag{26}$$

where

$$G = \frac{a_1 G_1 + a_2 G_2}{a_1 + a_2}, \tag{27}$$

and C is the impurity concentration far from the front. Condition (26) becomes the usual criterion for concentration supercooling for $\theta = 0$. In the case of a pure melt ($C_\infty = 0$), we get the stability condition as

$$\frac{G}{\Phi} > -\frac{\theta T_Q}{a_1 + a_2}. \tag{28}$$

If $m \simeq 3°C/\%$, $C \simeq 10^{-2}\%$, $D \simeq 10^{-5}$ cm$^2$/sec, and $k \simeq 0.1$, then we must have for stability that $G/\Phi > 3 \cdot 10^4(1-\theta) - 10^5\theta$, or for $\Phi \simeq 10^{-2}$ cm/sec, $G > 3 \cdot 10^2(1-\theta) - 10^3\theta$ deg/cm; (26) implies that the growth front is stable in a supercooled melt (more precisely for $G < 0$) always if

$$\theta > \left( 1 + \frac{T_Q D k}{(a_1 + a_2)(1-k)mC_\infty} \right),$$

i.e., for $\theta > 0.25$ if the above figures are used, while extremely high supercooling can be maintained if there is more anisotropy in the growth rate. It is sufficient to have $\theta > 0$ to obtain stability for a pure melt. We emphasize again that (26) and (27) relate only to localized perturbations. Large-scale perturbations can cause instability at lower $G/\Phi$, and any estimate of the characteristic dimension corresponding to onset of instability, i.e., of the characteristic size of the cellular structure, requires a detailed analysis of the integral in (25), or, more precisely, a solution for the simultaneous perturbation of all growth cones for an arbitrary type of perturbation.

Stability of a planar growth front in a supersaturated solution or supercooled melt can be the cause of spontaneous banding or zoning in crystals; in fact, let the melt contain a certain amount of impurity having $k < 1$ at the start. As the crystal grows, the impurity concentration ahead of the front increases, so the concentration supercooling does the same. When the critical value is reached, instability sets in, and the resulting unstable cones penetrate the region rich in impurity, while the impurity uptake factor for the now-unstable front approaches 1, so the impurity is absorbed and the front flattens out again. Then the process starts again. This mechanism is clearly responsible for single crystals containing bands of impurity, while it differs from the mechanism in which fresh nuclei are produced in the region of maximum supercooling ahead of the front [7, 8].

## Literature Cited


1. R. F. Sekerka, J. Crystal Growth, No. 3/4, 71 (1971).
2. A. A. Chernov, Kristallografiya, 16:842 (1971).
3. R. L. Schwoebel, J. Appl. Phys., Vol. 38 (1967).
4. H. Buckley, Crystal Growth [Russian translation], IL, Moscow (1954), p. 113.
5. A. A. Chernov and E. D. Dukova, Kristallografiya, 14:169-170 (1969).
6. R. Kaischev and E. Budevski, Contemporary Physics, 8:489 (1967); 8:489 (1967).
7. A. I. Landau, Fiz. Met. Metalloved., 6:148 (1958).
8. B. N. Aleksandrov, B. I. Verkin, N. M. Lifshits, and G. M. Stepanova, Fiz. Met. Metalloved., 6:167 (1958).

# MORPHOLOGICAL STABILITY NEAR A GRAIN BOUNDARY GROOVE IN A SOLID–LIQUID INTERFACE DURING SOLIDIFICATION OF A PURE SUBSTANCE

**S. R. Coriell and R. F. Sekerka**

*National Bureau of Standards, Washington D. C.*
*Carnegie-Mellon University, Pittsburgh, Pennsylvania*

## 1. Introduction

The theory of morphological stability [1-3] deals with the stability of the shape of the interface that separates two phases during a phase transformation. Analysis for stability consists in carrying out the following steps: (1) solve the phase transformation problem for the smooth two-phase interface to be tested for stability; (2) allow a small arbitrary perturbation in the shape of the otherwise smooth interface; (3) solve the problem for the perturbed interface to determine if the perturbation will grow or decay. If the perturbation grows, the shape is deemed to be morphologically unstable; whereas if the perturbation decays, the shape is said to be stable. It is also possible that certain special perturbations will neither grow nor decay; in this case, the new interface shape is possibly a stable shape which, however, must itself be tested for stability.

In linear theory the results of morphological stability calculations can be stated in terms of the Fourier components of the interface shape, viz., the dependence of a Fourier component of frequency $\omega$ on time, t, is proportional to $\exp[f(\omega)t]$, where $f(\omega)$ is a function of $\omega$ and depends on the particular crystallization conditions. The interface is morphologically unstable if $f(\omega) > 0$ for any value of $\omega$. A review of experimental tests of stability theory is given in ref. 3.

In a number of experiments [4-6] instabilities are apparently first observed near defects, an important class of such defects being grain boundaries. Possible reasons for this that immediately come to mind are: (1) a grain boundary gives rise to an initial perturbation which is large compared to other perturbations on the interface; as the perturbations are amplified, the perturbations near the grain boundary are the first to reach observable size; (2) the stability criterion is different in the presence of a grain boundary, e.g., $f(\omega)$ is not the same when a grain boundary is present; (3) the initial perturbation resulting from a grain boundary is sufficiently large that linear theory is inapplicable. Alternatively, the morphological features observed near grain boundaries are not truly indicative of instability at all but, instead, are steady-state shapes that resemble the shapes that are observed when instability is really present.

In this article, we seek to investigate the role of defects by using linear stability theory to study the time dependent shape of a nearly planar interface between a pure liquid and a bicrystal, oriented such that the grain boundary is perpendicular to the solid–liquid interface.

The grain boundary produces a groove in the solid–liquid interface, and we take the slope of the planar interface to be zero; the slope, s, of the groove is then given by $(\gamma_B/2\gamma)/\{1-(\gamma_B/2\gamma)^2\}^{1/2}$, where $\gamma_B$ and $\gamma$ are the grain boundary and the solid-liquid surface free energies, respectively.* The use of linear stability theory restricts our treatment to boundaries for which $(\gamma_B/2\gamma) \ll 1$. For simplicity we consider a situation such that in the absence of a grain boundary, the temperature gradients, $G_S$ and $G_L$, in the solid and liquid, respectively, are constant. Depending on the relative magnitudes of these gradients, the solid–liquid interface advances, recedes, or remains stationary. We make the usual assumptions of time-independent morphological stability theory, e.g., Laplace's equation is used for the temperature fields.†

The important difference between the present calculation and previous stability calculations is that the interface is required to have a fixed non-zero slope at the grain boundary. When $G_L = G_S = 0$ (isothermal system), the time evolution of an interface containing a grain boundary groove has been treated previously [7]. In the present calculation we employ a method that combines the methods of previous morphological stability theory with that of [7] to allow for nonzero and unequal temperature gradients.

For $G_L = G_S = 0$, Mullins showed that the groove profile has a fixed shape and linear dimensions that are proportional to $t^{1/3}$, thus the depth of the groove is an increasing function of time. This indicates the rather subtle influence of a grain boundary on the time evolution of an interface; since, for this case, $f(\omega)$ is never positive and hence the interface is morphologically stable according to the previous definition of stability. We note also that the stationary shape of a grain boundary groove in a temperature gradient has recently been calculated by Nash and Glicksman [8] without any restrictions on the magnitude of the groove slope.

## 2. Theory

We consider a liquid in constact with its solid, which contains a grain boundary. Sufficiently far from the grain boundary, the solid–liquid interface is planar and we assume that the grain boundary is perpendicular to this planar interface. As illustrated in Fig. 1, our coordinate system is chosen such that the grain boundary lies in the plane $x = 0$ and the position of the solid–liquid interface is described by the equation $y = W(x, t)$, where $t$ is the time and $W(x, 0) \to 0$ as $|x| \to \infty$. All quantities are taken as independent of the z-coordinate, i.e., we are treating a two-dimensional problem. Because of the presence of the grain boundary, a further condition on $W(x, t)$ is that $\partial W/\partial |x| = s$ at $x = 0$, where $s \ll 1$. We also assume that $|\partial W/\partial x| \ll 1$ and $W(x, t)$ is small since, in subsequent calculations, we neglect all powers of W greater than unity. In addition, we take $W(x, 0)$ to be a symmetric function of x; it follows that $W(x, t)$ is a symmetric function of x and thus we need only consider positive values of x.

We assume that the temperature fields in the solid and liquid are given by Laplace's equation. Thus, we write

$$T_L(x, y) = T_M + G_L y + \overline{T}_L(x, y), \tag{1a}$$

* Although this expression for s corresponds to the case where $\gamma_B$ and $\gamma$ are isotropic, there exist corresponding expressions for the case where these surface tensions depend on orientation. However, our subsequent development only requires that s be some specified constant throughout the calculation. The possibility that s might depend upon growth velocity or might change as a function of time because of departure from local equilibrium has never been fully explored.

† The validity of the use of time-independent morphological stability theory is considered in the Discussion.

$$T_S(x, y) = T_M + G_S y + \bar{T}_S(x, y), \tag{1b}$$

where $T_L$ and $T_S$ are the temperatures in the liquid and solid, respectively, $T_M$ is the melting point of solid at a flat interface, $G_L$ and $G_S$ are the temperature gradients in the liquid and solid, respectively, sufficiently far from the interface, and $\bar{T}_L(x, y)$ and $\bar{T}_S(x, y)$ satisfy Laplace's equation. $\bar{T}_L(x, y)$ and $\bar{T}_S(x, y)$ vanish if $W(x, t) = 0$ and for small $W(x, t)$, $\bar{T}_L$ and $\bar{T}_S$ will be correspondingly small. We consider $G_s$ as positive (otherwise the solid would be superheated), but $G_L$ may be either positive or negative (negative $G_L$ corresponds to a supercooled liquid).

The temperature at the solid–liquid interface is given by the Gibbs–Thomson equation

$$T_{LI} = T_{SI} = T_M - T_M \Gamma K(x, t). \tag{2}$$

where the subscript I indicates evaluation at the solid–liquid interface, $\Gamma$ is a capillary constant, and the curvature $K(x, t) = -(\partial^2 W/\partial x^2)$ since we have assumed that $|\partial W/\partial x| \ll 1$. Here

$$\Gamma = (1/L_V)(\gamma + \gamma''),$$

where $L_V$ is the latent heat per unit volume, $\gamma$ is the solid–liquid surface tension, and $\gamma''$ is the second derivative of $\gamma$ with respect to the angle made by the normal to the surface and a fixed reference line.* $\gamma$ and $\gamma''$ are evaluated at the orientation of the planar interface; for isotropic $\gamma$, $\gamma'' = 0$.

The interface velocity v is evaluated from the heat conservation condition

$$L_V v = -k_L(\partial T_L/\partial y)_I + k_S(\partial T_S/\partial y)_I, \tag{3}$$

where $k_L$ and k are the thermal conductivities of the liquid and solid, respectively. It is convenient to write the interface velocity as

$$v = V + \partial W/\partial t, \tag{4}$$

where V is independent of x and is given by

$$V = (k_S G_S - k_L G_L)/L_V = \bar{k}(\mathcal{G}_S - \mathcal{G}_L)/L_V,$$

where $2\bar{k} = k_L + k_S$, $\mathcal{G}_L = k_L G_L/\bar{k}$, and $\mathcal{G}_S = k_S G_S/\bar{k}$.

We solve Laplace's equation for $\bar{T}_L$ and $\bar{T}_S$; the boundary condition (2) may be rewritten as

$$\bar{T}_{LI} = -T_M \Gamma K - G_L W, \tag{5a}$$

$$\bar{T}_{SI} = -T_M \Gamma K - G_S W. \tag{5b}$$

In addition, $\bar{T}_L \to 0$ as $y \to \infty$ and $\bar{T}_S \to 0$ as $y \to -\infty$. The part of the interface velocity that depends on x is given by

$$\frac{\partial W}{\partial t} = -\frac{k_L}{L_V}\left(\frac{\partial \bar{T}_L}{\partial y}\right)_I + \frac{k_S}{L_V}\left(\frac{\partial \bar{T}_S}{\partial y}\right)_I. \tag{6}$$

* Although this formally allows for a dependence of $\gamma$ on orientation, we have confined ourselves to the case where $\Gamma$ can be treated as a constant. See [9] for details.

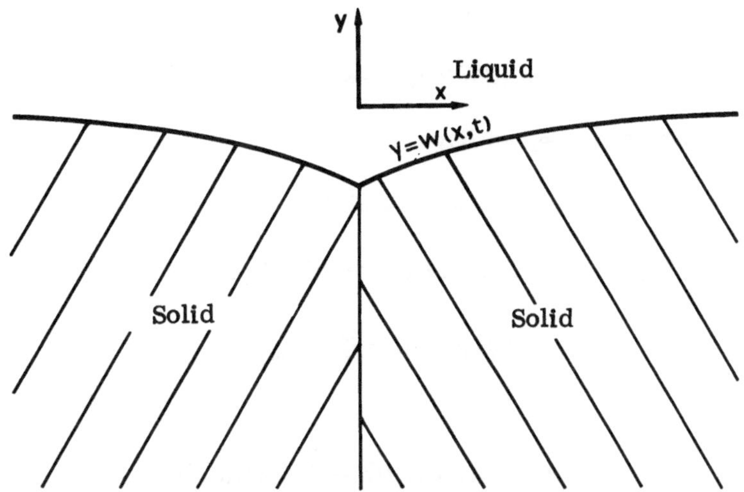


Fig. 1. Sketch of solid–liquid interface containing a grain boundary groove. The position of the interface is described by the equation $y = W(x, t)$. The plane $y = 0$ is chosen such that $W(x, t) \to 0$ as $|x| \to \infty$.

We solve for $\overline{T}_L$ and $\overline{T}_S$ by carrying out a Fourier cosine transform to replace the dependence on x by a dependence on the transform variable $\omega$. Given a function $F(x, y, t)$, we define $\widetilde{F}(\omega, y, t)$ as

$$\widetilde{F}(\omega, y, t) = \int_0^\infty dx \cos \omega x \, F(x, y, t). \tag{7}$$

The inverse transform is

$$F(x, y, t) = \frac{2}{\pi} \int_0^\infty d\omega \cos \omega x \, \widetilde{F}(\omega, y, t). \tag{8}$$

We apply the transform to the differential equations (Laplace's equations) for $\overline{T}_L(x, y)$ and $\overline{T}_S(x, y)$, to the boundary conditions, and to (6). The differential equations for $\widetilde{\overline{T}}_L(\omega, y)$ and $\widetilde{\overline{T}}_S(\omega, y)$ become

$$[-\omega^2 + (\partial^2/\partial y^2)] \, \widetilde{\overline{T}}_L = 0, \tag{9a}$$

$$[-\omega^2 + (\partial^2/\partial y^2)] \, \widetilde{\overline{T}}_S = 0. \tag{9b}$$

The solutions satisfying the boundary conditions are

$$\widetilde{\overline{T}}_L = (-T_M \Gamma \widetilde{K} - G_L \widetilde{W}) e^{-\omega y}, \tag{10a}$$

$$\widetilde{\overline{T}}_S = (-T_M \Gamma \widetilde{K} - G_S \widetilde{W}) \, e^{\omega y}. \tag{10b}$$

Upon evaluating the transform of (6) using (10), we obtain

$$(\partial \widetilde{W}/\partial t) = (-\omega \bar{k}/L_V) \, [(G_L + G_S) \, \widetilde{W}(\omega, t) + 2 T_M \Gamma \widetilde{K}(\omega, t)]. \tag{11}$$

Since $K = -(\partial^2 W/\partial x^2)$, two integrations by parts yield

$$\tilde{K} = -\cos \omega x \, (\partial W/\partial x)|_0^\infty - \omega \sin \omega x \, W|_0^\infty + \omega^2 \tilde{W}. \tag{12a}$$

Since W and $(\partial W/\partial x)$ vanish as $x \to \infty$, we have

$$\tilde{K} = s + \omega^2 \tilde{W}, \tag{12b}$$

where s is the slope, $(\partial W/\partial x)$, at $x = 0$. Substitution of this relationship into (11) yields

$$(\partial \tilde{W}/\partial t) = f(\omega)\tilde{W} + g(\omega), \tag{13}$$

where

$$f(\omega) = -\omega(\bar{k}/L_V)\,(G_L + G_S + 2T_M\Gamma\omega^2),$$

$$g(\omega) = -\omega(\bar{k}/L_V)\,(2T_M\Gamma s).$$

The solution of (13) is

$$\tilde{W}(\omega, t) = \{\tilde{W}(\omega, 0) + [g(\omega)/f(\omega)]\} \exp[f(\omega)t] - [g(\omega)/f(\omega)], \tag{14}$$

where $\tilde{W}(\omega, 0)$ is the Fourier transform of the interface shape at $t = 0$. We now obtain W(x, t) by taking the inverse transform of (14), i.e.,

$$W(x, t) = \frac{2}{\pi}\int_0^\infty d\omega \cos \omega x \, \tilde{W}(\omega, t). \tag{15}$$

It should be noted that $\tilde{W}(\omega, t)$ is a well behaved function, e.g., $\tilde{W}(\omega, t)$ is not singular when $f(\omega) = 0$. The important difference, however, between this calculation and previous morphological stability calculations is the constraint that W(x, t) has a finite fixed slope at $x = 0$. This constraint enters into the calculation through (12). If this constraint was not present, and we could take $s = 0$, then $g(\omega) = 0$, and we would obtain the result of previous theory [1].

The interface is stable with respect to perturbation if $f(\omega) < 0$ for all $\omega$ ($\omega \geq 0$), and is unstable if $f(\omega) > 0$ for some values of $\omega$. It is interesting that $f(\omega)$ does not depend on s, and thus the grain boundary groove does not change the stability criterion. However, a grain boundary groove certainly does provide an initial perturbation; this can be appreciated by consideration of (14) for $\tilde{W}(\omega, 0) = 0$. Indeed, the fact that (14) leads to a finite result even when $\tilde{W}(\omega, 0) = 0$ makes it appear as if one gets something from nothing. However, (11) shows that the time rate of change of $\tilde{W}$ depends not only on $\tilde{W}$ itself but also on the transform of the curvature. This can be understood in terms of a limiting process with respect to some variable b on which W is dependent. Indeed, there are functions W(x, b) which tend to zero as $b \to \infty$ but whose derivatives with respect to x do not. For instance, if

$$W = -(s/b) \exp(-b|x|)$$

then

$$\lim_{b\to\infty} W = 0,$$

whereas

$$\lim_{b\to\infty} [-(d^2W/dx^2)] = 2s\,\delta(x),$$

where $\delta(x)$ is the Dirac delta function.

The time evolution of the interface shape W(x, t) depends strongly on the behavior of f(ω), which depends on whether $(G_L + G_S)$ is positive, zero, or negative (we take $\Gamma > 0$). Since it appears necessary to discuss each of these possibilities separately, we denote $(G_L + G_S)$ positive as case (1), $(G_L + G_S) = 0$ as case (2), and $(G_L + G_S) < 0$ as case (3). We first consider the long time behavior.

Case (1) $(G_L + G_S) > 0$

For this case $f(\omega) \leq 0$ for all values of ω (the equality holds only for $\omega = 0$). Since g(0)/f(0) is finite, it is clear from (14) and (15) that for sufficiently large times $(e^{f(\omega)t} \rightarrow 0)$

$$W(x, t) \cong -\frac{2}{\pi}\int_0^\infty d\omega \cos \omega x \, [g(\omega)/f(\omega)] = -(s/a) \exp(-ax), \tag{16}$$

where $a^2 = |G_L + G_S|/2T_M\Gamma$. Note that the velocity $V = (\bar{k}/L_V)(G_S - G_L)$ is proportional to the difference of the weighted temperature gradients, so $(G_L + G_S)$ can be varied independently of V. Therefore if $G_L = G_S$, so that $V = 0$, but $(G_L + G_S) \neq 0$, equation (16) describes the shape of a grain boundary groove for a stationary interface. Furthermore, we note that for a moving interface the groove shape is given by the exact same function of $(G_S + G_L)$, although $G_S \neq G_L$ and $V \neq 0$.

Case (2) $(G_L + G_S) = 0$

The case $G_L = G_S = 0$, has been treated by Mullins $\tilde{W}(\omega, 0) = 0$ and is formally identical to the case $(G_L + G_S) = 0$ for that initial condition. Although $f(\omega) \leq 0$ for all ω and the equality holds only for $\omega = 0$, $g(\omega)/f(\omega) = s/\omega^2$, which diverges as $\omega \rightarrow 0$, so the long-time behavior is not obvious. However, for a reasonable choice of W(x, 0), i.e., such that $\tilde{W}(\omega, 0)$ is not singular, the factor $\tilde{W}(\omega, 0) \exp[f(\omega)t]$ in (14) makes no contribution to W(x, t) at large times; therefore, for large times W(x, t) does not depend on our choice of W(x, 0). Thus, the results of Mullins can be used for the long-time behavior; the groove profile has a fixed shape and the - linear dimensions are proportional to $t^{1/3}$ [see equation (8) of [7]].

Case (3) $(G_L + G_S) < 0$

The interface is morphologically unstable because f(ω) is positive for a range of values of ω. The long-time behavior without the grain boundary constraint has been previously described [10]. We strongly suspect that for long times the grain boundary constraint will have little effect and the interface shape will be similar to that described previously. Further discussion will be given subsequently.

## 3. Analytical Forms for Three Cases

In this section, we obtain expressions for each case that are suitable for numerical evaluation. It is therefore necessary to specify W(x, 0). We have already shown that when $(G_L + G_S) \geq 0$, the long-time behavior of W(x, t) is independent of W(x, 0). Furthermore, when $(G_L + G_S) < 0$ [and hence $f(\omega) > 0$ for some range of values of ω], the long-time behavior of W(x, t) depends on W(x, 0) in a simple manner [10]; in particular, the Fourier transform of W(x, 0), evaluated at the maximum of f(ω), appears as an overall amplitude factor for W(ω, t) but the other details of W(x, 0) are irrelevant. Hence, we are reasonably free to choose W(x, 0) rather arbitrarily and a reasonable choice appears to be

$$W(x, 0) = -(s/b) \exp(-bx). \tag{17}$$

This choice satisfies the conditions that $\partial W/\partial x) = s$ at $x = 0$ and $W(x, 0) \to 0$ as $x \to \infty$; in addition it corresponds to a member of the family of steady state shapes of a grain boundary groove.* We carry out the calculations for various values of b. In the limit $b \to \infty$, $W(x, 0) \to 0$ and our initial condition is a flat interface containing a grain boundary groove. From (17), it follows that

$$\tilde{W}(\omega, 0) = -s/(\omega^2 + b^2), \tag{18}$$

We can then write $\tilde{W}(\omega, t)$ as

$$\tilde{W}(\omega, t) = \left\{\frac{s}{\omega^2 + (G_L + G_S)/2T_M\Gamma}\right\}\left\{\frac{[b^2 - (G_L + G_S)/2T_M\Gamma]\exp[f(\omega)t]}{\omega^2 + b^2} - 1\right\}. \tag{19}$$

We now consider separately the three possibilities, viz., $(G_L + G_S)$ is (1) positive, (2) zero, (3) negative.

Case (1) $(G_L + G_S) > 0$

Defining

$$a^2 \equiv |G_L + G_S|/2T_M\Gamma,$$

and

$$(1/\tau) \equiv (\bar{k}/L_V)(2a/3\sqrt{3})|G_L + G_S|,$$

and introducing the change of variable $\eta = \omega/a$, we have

$$(\pi a/2s)W(x, t) = \int_0^\infty d\eta \frac{\cos(ax\eta)}{\eta^2 + 1}\left\{\frac{1 - (a/b)^2}{1 + (a\eta/b)^2}\exp[-(3\sqrt{3}/2)(t/\tau)\eta(1 + \eta^2)] - 1\right\}, \tag{20a}$$

or

$$(\pi a/2s)W(x, t) = -[\pi/2\exp(-ax) + \int_0^\infty d\eta \frac{\cos(ax\eta)}{\eta^2 + 1}\left\{\frac{1 - (a/b)^2}{1 + (a\eta/b)^2}\exp[-(3\sqrt{3}/2)(t/\tau)\eta(1 + \eta^2)]\right\}. \tag{20b}$$

For long times, the integral on the right-hand side of (20b) approaches zero and [see (16)] we obtain $(\pi a/2s)W(x, t) = -(\pi/2)\exp(-ax)$.

Case (2) $(G_L + G_S) = 0$

Defining $A = [\bar{k}/L_V)(2T_m\Gamma)$ and introducing the change of variable $\eta = (At)^{1/3}\omega$, we obtain

$$(\pi/2s)W(x, t) = (At)^{\frac{1}{3}}\int_0^8 d\eta \frac{\cos[\eta x/(At)^{\frac{1}{3}}]}{\eta^2}\left\{\frac{\exp(-\eta^3)}{1 + [\eta/b(At)^{\frac{1}{3}}]^2} - 1\right\}. \tag{21}$$

For b infinite, (21) reduces to the situation treated by Mullins, and $W(x, t)/(At)^{1/3}$ is a function of $x/(At)^{1/3}$.

---

* Only in the case where $b^2 = (G_L + G_S)/2T_M\Gamma > 0$ will (17) represent the steady state shape for the actual growth conditions [see (16)].

Case (3) $(G_L + G_S) < 0$

Using the previous definitions of $a$, $\tau$, and $\eta$, noting that

$$a^2 \equiv |G_L + G_S|/2T_M\Gamma = -(G_L + G_S)/2T_M\Gamma$$

and that

$$(1/\tau) \equiv (\bar{k}/L_V)(2a/3\sqrt{3})|G_L + G_S| = -(\bar{k}/L_V)(2a/3\sqrt{3})(G_L + G_S),$$

we can write

$$(\pi a/2s)W(x, t) = \int_0^\infty d\eta \frac{\cos(ax\eta)}{\eta^2 - 1}\left\{\frac{[1+(a/b)^2]}{[1+(a\eta/b)^2]} \exp[(3\sqrt{3}/2)(t/\tau)\ \eta(1-\eta^2)] - 1\right\}. \tag{22}$$

In the appendix, we show that the preceding expressions (20), (21), and (22) for W(x, t) satisfy the condition that $(\partial W/\partial x) = s$ at the grain boundary groove.

The exponential in (22) has a maximum at $\eta = 3^{-1/2}$. Following [10], we can obtain an approximate evaluation of (22) for $(t/\tau)$ sufficiently large. We expand the exponential about $\eta = (3)^{-1/2}$, neglecting higher than quadratic terms in $\eta$. In the terms in the denominator, we set $\eta = (3)^{-1/2}$; we also neglect the time-independent term in (22). For a more detailed discussion set [10]; basically we are using equation (50) of [10] to evaluate W(x, t). Thus we have

$$(\pi a/2s)W(x, t) \cong -\frac{3[1+a^2/b^2]}{2[1+a^2/3b^2]} \exp(t/\tau) \int_0^\infty d\eta \cos(ax\eta) \exp[(-9/2)(t/\tau)(\eta - 3^{-\frac{1}{2}})^2]. \tag{23a}$$

Recognizing that integrating from $-\infty$ to $\infty$ rather than from 0 to $\infty$ does not appreciably change the value of the integral on the right-hand side of (23a), we can evaluate this integral to obtain

$$(\pi a/2s)W_G(x, t) = -(\pi/2)^{\frac{1}{2}} (\tau/t)^{\frac{1}{2}}\{[1+a^2/b^2]/[1+a^2/3b^2]\}\cos(ax/3^{\frac{1}{2}}) \exp(t/\tau) \exp\{-(ax)^2/18(t/\tau)\}, \tag{23b}$$

where the subscript G on W(x, t) is used to denote the approximate W(x, t) obtained from the right-hand side of (23b). For large $(t/\tau)$, we expect $W_G(x, t)$ to be an excellent approximation to W(x, t); note however, that $W_G(x, t)$ has zero slope at x = 0, i.e., this approximation does not satisfy $(\partial W/\partial x) = s$ at x = 0. Note also that the parameter b, which describes the initial interface shape, appears only in the factor $[1 + a^2/b^2]/[1 + a^2/3b^2]$ in (23b); for an initially flat interface $(b \to \infty)$, this factor is unity; it increases monotonically as b decreases and approaches a value of three as $b \to 0$.

Before proceeding to the results of numerical evaluation of (20), (21), and (22), we note that there exists an alternative method of evaluating W(x, t) for the initial condition of a flat interface. This method yields a tractable expression for W(0, t) and thus provides a useful check on the numerical calculations. Differentiation of (15) with respect to t yields

$$\partial W(x, t)/\partial t = (2/\pi) \int_0^\infty d\omega \cos \omega x\ [\partial \widetilde{W}(\omega, t)/\partial t], \tag{24a}$$

where [from (14) with $W(\omega, 0) = 0$]

$$\partial \widetilde{W}/\partial t = g(\omega)\exp[f(\omega)t]. \tag{24b}$$

In obtaining (24a), we have interchanged the order of integration with respect to $\omega$ and differentiation with respect to t. This is valid [11] since the integral on the right-hand side of (24a) is uniformly convergent with respect to t for $t > 0$. It is possible to express the right-hand side of (24a) in terms of Lommel functions [12]. Then using series representations of the Lommel functions, one can obtain series representations of $\partial W/\partial t$. However, we proceed as follows: introducing a new variable of integration $\eta = At\omega^3$ and taking $x = 0$ we have

$$\frac{\partial W(0,t)}{\partial t} = -(2sA/3\pi)(At)^{-2/3} \sum_{k=0}^{\infty} \{[\pm a^2(At)^{2/3}]^k/k!\} \int_0^{\infty} d\eta \, \eta^{\frac{1}{3}(k-1)} e^{-\eta}. \tag{25}$$

where the identity

$$e^z = \sum_{k=0}^{\infty} z^k/k!$$

has been used and we have interchanged the order of summation and integration. With regard to $\pm a^2$, the plus sign is to be used if $(G_L + G_S) < 0$ and the minus sign if $(G_L + G_S) > 0$. The integrals in (25) can be evaluated in terms of the gamma function $\Gamma(Z)$; we then integrate (25) with respect to t using the initial condition $W(0, 0) = 0$. This gives

$$W(0,t) = -(s/\pi)(At)^{\frac{1}{3}} \left\{ 2\Gamma(2/3) + \sum_{k=1}^{\infty} \left[ (\pm 1)^k 3^k \, \Gamma\left(\frac{k+2}{3}\right) (t/2\tau)^{2/3k} \right] \Big/ \left[ (k+\tfrac{1}{2})k! \right] \right\}, \tag{26}$$

where we have used the identity $2Aa^3\tau = 3^{3/2}$. If $(G_L + G_S) = 0$, $1/\tau = 0$ and the series reduces to one term, i.e., $W(0, t) = -(2s/\pi)(At)^{1/3}\Gamma(2/3)$, which agrees with Mullins [7]. For small $(t/\tau)$ we have used (26) to provide a check on the values of $W(0, t)$ obtained by numerical integration.

## 4. Numerical Results

Equations (20)-(22) have been evaluated numerically (trapezoidal rule) for various values of the parameters; results are given graphically in Figs. 2-12. The content of the figures is summarized in Table 1. The parameters $a$, b, and $\tau$ that appear in the Table have been previously defined; for cases 1 and 3 a particular calculation of $(\pi a/2s)W(x, t)$ as a function of $(ax)$ requires specification of a dimensionless time $(t/\tau)$ and the ratio $a/b$, which determines the initial profile.

TABLE 1. Summary of Figures Giving Numerical Results

| Fig. | Case* | Ordinate | Abscissa | Initial condition $(a/b)$ | Time $(t/\tau)$ | Additional inclusions |
|---|---|---|---|---|---|---|
| 2 | 1,2,3 | $(-\pi a_0/2s)W(0, t)$ | $(t/\tau_0)$ | 0 | | |
| 3 | 1,2,3 | $ln[(-\pi a_0/2s)W(0, t)]$ | $(t/\tau_0)$ | 0 | | |
| 4 | 1 | $(-\pi a/2s)W(0, t)$ | $(t/\tau)$ | 0, 1/2, 1, 2, 3 | | |
| 5 | 3 | $ln[(-\pi a/2s)W(0, t)]$ | $(t/\tau)$ | 0, 1, 2, 3 | | |
| 6 | 3 | $(\pi a/2s)W(x, t)$ | $ax$ | $10^{-3}$ | 0; 0.05; 0.2; 0.8 | |
| 7 | 3 | $(\pi a/2s)W(x, t)$ | $ax$ | $10^{-3}$ | 0.8; 1.6; 3.2 | $W_G(x, 3.2\tau)$ |
| 8 | 1 | $(\pi a/2s)W(x, t)$ | $ax$ | $10^{-3}$ | 0; 0.05; 0.2; 0.8; ∞ | |
| 9 | 3 | $(\pi a/2s)W(x, t)$ | $ax$ | 3 | 0; 0.2; 0.8 | |
| 10 | 3 | $(\pi a/2s)W(x, t)$ | $ax$ | 3 | 0.8; 1.6; 3.2 | $W_G(x, 3.2\tau)$ |
| 11 | 1 | $(\pi a/2s)W(x, t)$ | $ax$ | 3 | 0; 0.2; 0.8; 1.6, ∞ | |
| 12 | 2 | $(\pi b_1/2s)W(x, t)$ | $b_1x$ | $b=\infty$, $b_1$ | $b_1(At)^{1/3}=0, \frac{1}{2}, 1.2$ | |

* Case 1, $(G_S + G_L) > 0$; Case 2, $(G_S + G_L) = 0$; Case 3, $(G_S + G_L) < 0$.

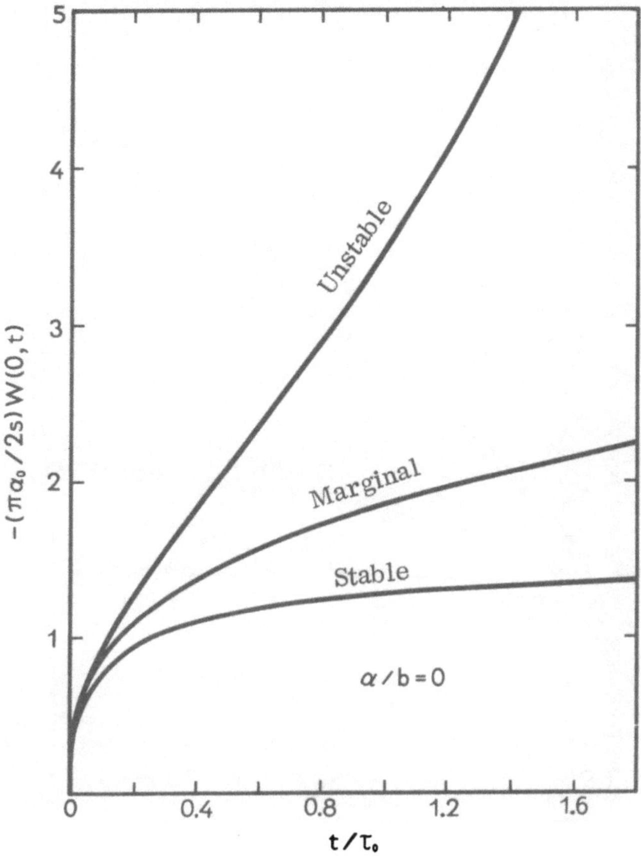


Fig. 2. Dimensionless groove depth $[-(\pi a_0/2s)W(0, t)]$, as a function of dimensionless time, $t/\tau_0$, for a morphologically stable, marginal, and unstable interface. At $t/\tau_0 = 0$, the interface is planar.

For case 2, however, $a = 0$ so there exists no parameter $a^{-1}$ for the scaling of distances. Therefore, to directly compare case 2 with cases 1 and 3, we rewrite (21) in the form

$$(\pi b_1/2s)W(x, t) = b_1(At)^{\frac{1}{3}} \int_0^\infty d\eta \, \frac{\cos[\eta(b_1x)/b_1(At)^{\frac{1}{3}}]}{\eta^2} \left\{ \frac{\exp(-\eta^3)}{1 + [\eta(b_1/b)/b_1(At)^{\frac{1}{3}}]^2} - 1 \right\}, \tag{27}$$

where $b_1$ is an arbitrary inverse length. If, for finite b, we choose $b_1 = b$, it is clear that $(\pi b_1/2s)W(x, t)$ as a function of $b_1x$ depends on only the single parameter $b_1(At)^{1/3}$. Alternatively, we can choose $b_1$ in case 2 to be numerically equal to $a$ for cases 1 or 3; the quantity $(\pi a/2s)W(x, t)$ can then be compared directly in all three cases.

When we compare the three different cases, we are really considering the results of three different experiments on the same substance. In the first experiment, $G_L + G_S = |G_{L0} + G_{S0}|$ (we use the subscript zero to denote specific values of a quantity), in the second experiment $G_L + G_S = 0$, and in the third experiment $G_L + G_S = -|G_{L0} + G_{S0}|$. Thus experiments 1, 2, and 3 correspond to cases 1, 2, and 3, respectively. The values of parameters $a_0$ and $\tau_0$ are identical in cases 1 and 3; if for case 2 we take $b_1 = a_0$ and use the relation $2Aa_0^3\tau_0 = 3^{3/2}$ to determine $\tau_0$, then we can directly compare all three cases.

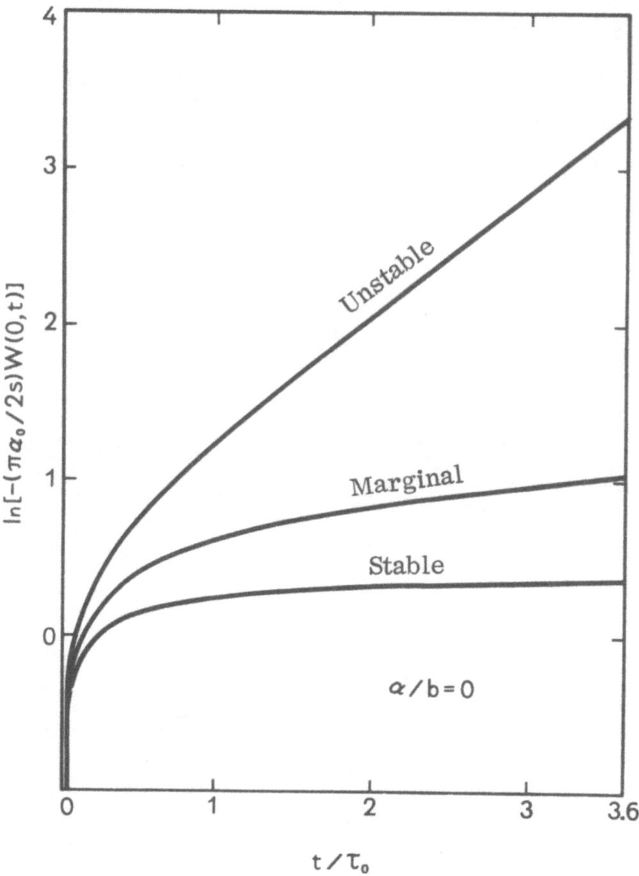


Fig. 3. Natural logarithm of the dimensionless groove depth as a function of dimensionless time; parameters identical to those in Fig. 2.

For an initially flat interface, Fig. 2 shows the dimensionless groove depth, $[-(\pi a_0/2s)\cdot W(0,t)]$ as a function of $(t/\tau_0)$ for all three cases. For small times the behavior of the groove depth is given by the first few terms in (26). Since the leading term is independent of $(G_L + G_S)$, all three cases have the same dependence on time for very small times, i.e., $[-(\pi a_0/2s)W(0,t)] \cong \sqrt{3}\,\Gamma(2/3)(t/2\tau_0)^{\frac{1}{3}} \cong 2.3\,(t/2\tau_0)^{\frac{1}{3}}$. (For case 2, this equation is exact for all times.) The sign of the k = 1 term in (26) is the negative of the sign of $(G_L + G_S)$; thus the groove depth is greater when $(G_L + G_S)$ is negative (case 3). In fact, as can be seen from Fig. 2, this is correct not only for small times but for all times. The long time behavior is quite different in the three cases. For case 1, $-(\pi a_0/2s)W(0,t)$ approaches $^1/_2\pi$ as t approaches infinity. In fact, for large $(t/\tau_0)$ it can be shown* from (20b) that $[-(\pi a_0/2s)W(0,t)] \cong \pi/2 - [(3^{3/2}/2)(t/\tau_0)]^{-1}$. For case 2, the groove depth increases as $t^{1/3}$. For case 3, the groove depth increases exponentially. This is evident from Fig. 3 in which the natural logarithm of the dimensionless groove depth is against $(t/\tau_0)$; for $(t/\tau_0) \gtrsim 1.4$, the plot is linear for case 3.

For case 1, Fig. 4 shows the groove depth as a function of time for various initial conditions. For long times, it follows from (20b) that $[-(\pi a/2s)W(0,t)] \cong \pi/2 - [1-(a/b)^2]/[(3^{3/2}/2)(t/\tau)]$,

---

* For large values of $(t/\tau_0)$, the integrand of (20b) vanishes except for very small $\eta$; thus we can approximate the integral by $\int_0^\infty d\eta \exp[(-3^{3/2}/2)(t/\tau)\eta]$.

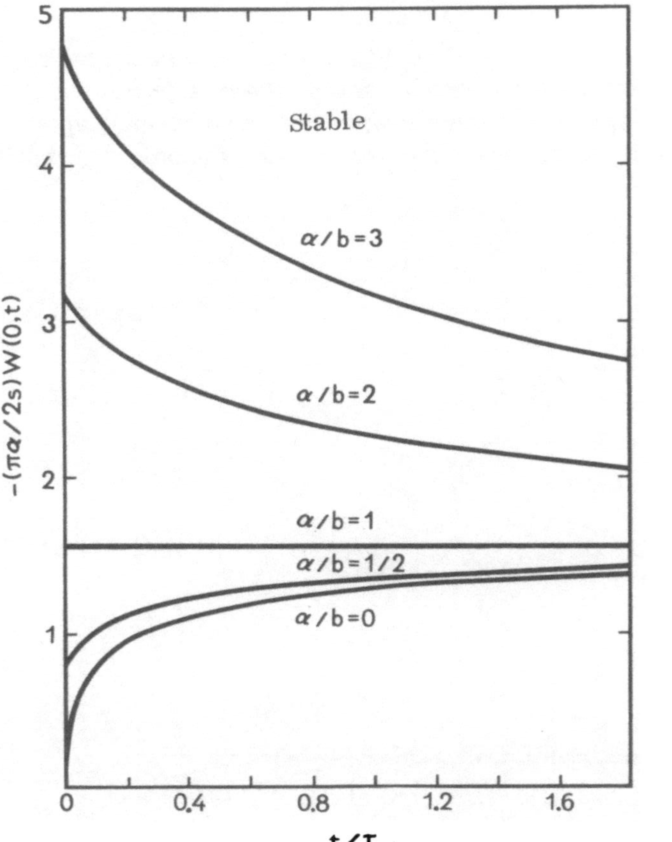


Fig. 4. Dimensionless groove depth, $[-(\pi a/2s)W(0, t)]$, as a function of dimensionless time, $t/\tau$, for a morphologically stable interface and various initial conditions. At $t/\tau = 0$, the dimensionless groove shape, $-(\pi a/2s)W(x, 0) = (\pi/2)(a/b)\exp[-(b/a)(ax)]$.

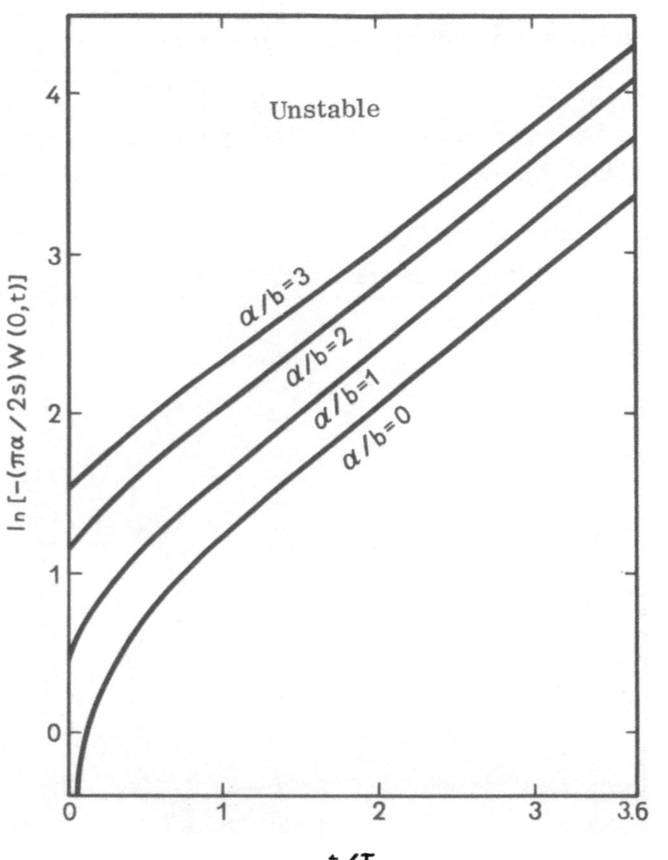


Fig. 5. Natural logarithm of dimensionless groove depth as a function of dimensionless time for a morphologically unstable interface and various initial conditions.

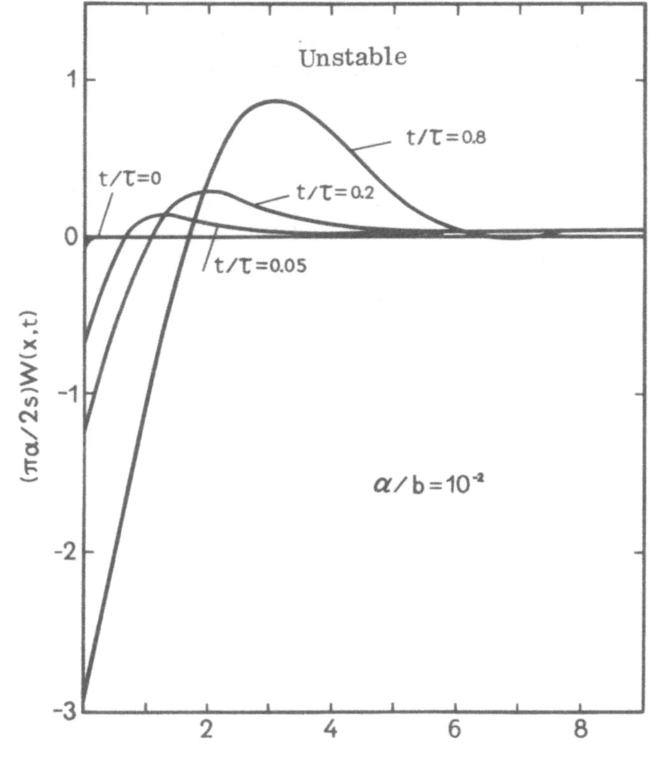


Fig. 6. Groove shape for various times for an unstable interface (dimensionless units). $a/b = 10^{-2}$, so the initial shape is almost planar. Shapes for longer times are given in Fig. 7.

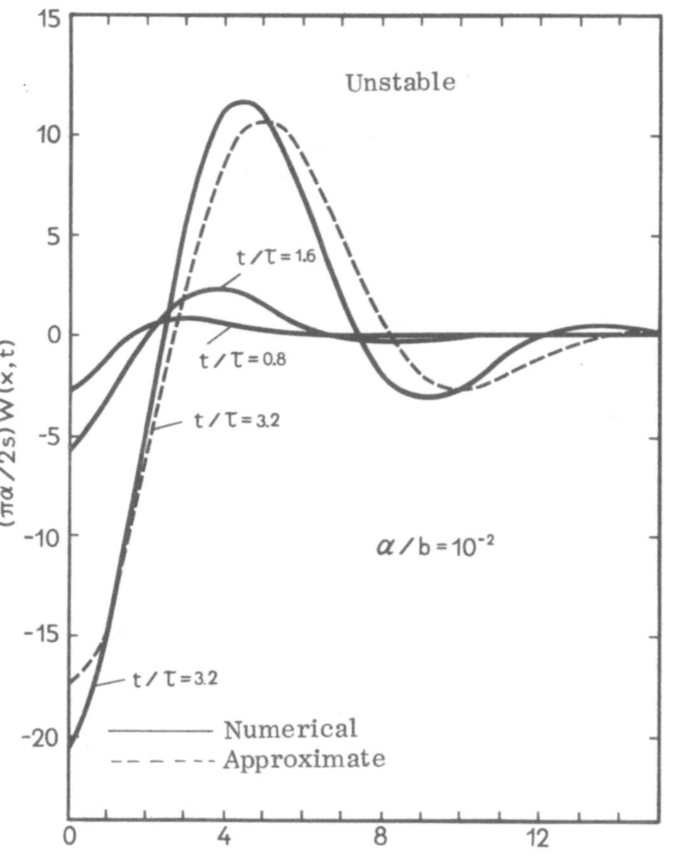


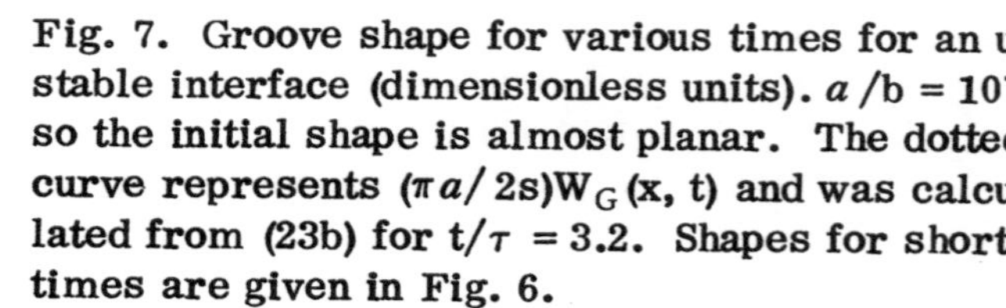

Fig. 7. Groove shape for various times for an unstable interface (dimensionless units). $a/b = 10^{-2}$, so the initial shape is almost planar. The dotted curve represents $(\pi a/2s)W_G(x, t)$ and was calculated from (23b) for $t/\tau = 3.2$. Shapes for shorter times are given in Fig. 6.

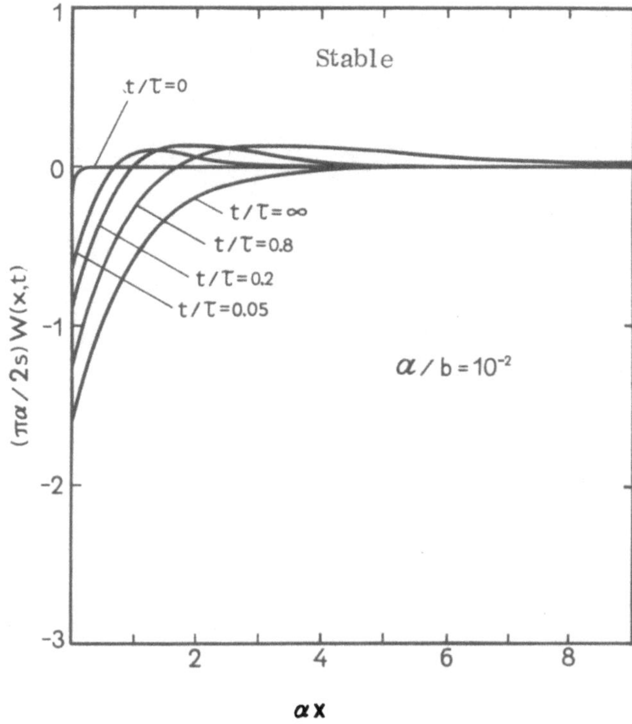


Fig. 8. Groove shape for various times for a stable interface (dimensionless units). The initial shape is the same as in Figs. 6 and 7. The final shape is given by $(\pi a/2s)W(x,\infty) = -(\pi/2)\exp(-ax)$.

when $[(3^{3/2})(t/\tau)]$, $(b/a) \gg 1$. Thus, for large values of $(a/b)$, the approach to the steady state value, $^1/_2\pi$, is slower than for small values of $(a/b)$. For $a/b = 1$, it is evident from (20b) that $W(x, t)$ [and hence $W(0, t)$] is independent of time. For $(a/b) > 1$ the groove depth decreases with time while for $(a/b) < 1$ it increases with time.

For case 3, Fig. 5 gives $\ln\{(-\pi a/2s)W(0, t)\}$ as a function of $(t/\tau)$ for various initial conditions. As previously remarked, we see that the long time behavior is exponential and that the curves tend to become parallel, i.e., the groove depths differ by a multiplicative factor.

Figures. 6, 7, 8, and 12 show the time evolution of the interface shape for an essentially flat initial interface.* In terms of $\tau_0$, the times in Fig. 12 are $(t/\tau_0) = 0.048$, 0.385, and 3.08 corresponding to $b_1^3At = {}^1/_8$, 1, and 8. For small times, e.g., $(t/\tau_0) \cong 0.05$, the curves for the three different cases are remarkably similar; whereas for long times the curves are very distinct for the three cases.

For case 1, the interface approaches the stationary shape $W(x, t) = -(s/a_0)\exp(-a_0x)$. The maximum which occurs in $W(x, t)$ (as a function of x) follows from the conservation of solid material. The behavior for case 2 has previously been described by Mullins; the linear dimensions increase as $t^{1/3}$. For case 3, which is the morphologically unstable case, the long-time behavior is essentially the same as for any morphologically unstable interface, i.e., the grain

*It can be seen from (20) and (22) that except for extremely small values of $(t/\tau)$ there is no difference between $(a/b) = 10^{-2}$ and $(a/b) = 0$ (i.e., b infinite), which corresponds to an initially flat interface.

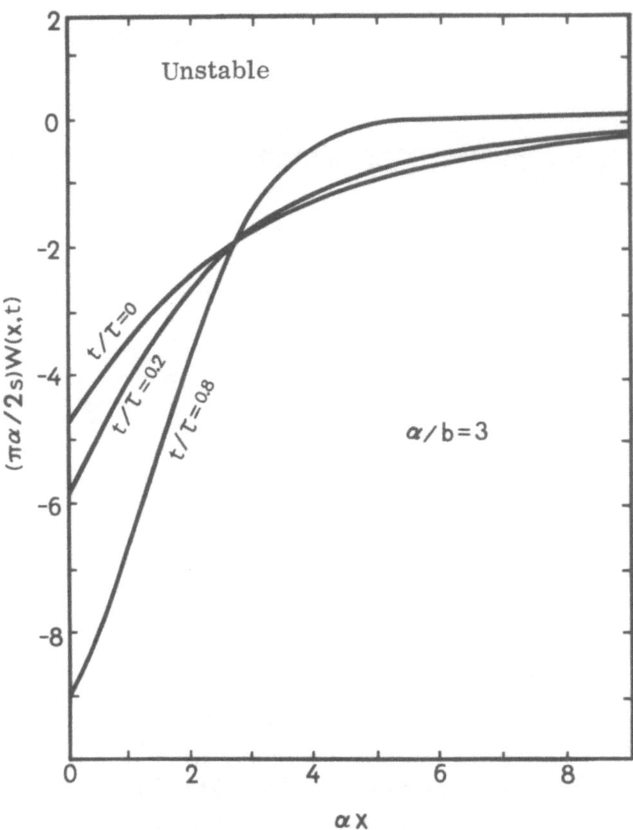


Fig. 9. Groove shape for various times for an unstable interface (dimensionless units). $a/b = 3$, so the initial shape (the $t/\tau = 0$ curve) is given by $(\pi a/2s)W(x, 0) = -(3\pi/2)\exp(-ax/3)$. Shapes for longer times are given in Fig. 10.

boundary simply provides an initial perturbation. The rather good agreement between $W_G(x, t)$ and $W(x, t)$ for $(t/\tau_0) = 3.2$ in Fig. 7 confirms this, since $W_G(x, t)$ neglects the grain boundary constraint $(\partial W/\partial x) = s$ at $x = 0$. Although not illustrated, for very long times (assuming linear theory) the interface becomes sinusoidal with wavelength $(2\pi 3^{1/2}/a_0)$. [See (23b) and [10]]

We now consider (Figs. 9-12) the profiles when the initial interface shape is not flat. For $(G_L + G_S) < 0$, comparing Fig. 10 ($a/b = 3$) with Fig. 7 ($a/b = 10^{-2}$), we see that longer times are required to develop a pronounced maximum in the profiles for $a/b = 3$ than for $a/b = 10^{-2}$. However, from Fig. 10, it is clear that for $t/\tau = 3.2$, $W_G(x, t)$ is a good approximation to $W(x, t)$. From (23b), the initial profile influences $W_G(x, t)$ through the factor $(1 + a^2/b^2)/(1 + a^2/3b^2)$; thus when $W(x, t) \cong W_G(x, t)$, the only effect of the initial profile is to provide different initial amplitudes for the fastest growing Fourier component. In Fig. 11 for $(a/b) = 3$ and $(G_L + G_S) > 0$, the initial profile is deeper than the stationary profile and $W(0, t)$ increases with time. Conservation of solid material requires that the curves for finite times cross the $t = 0$ curve. Finally, in Fig. 12 ($G_L + G_S = 0$) we illustrate our earlier remark that for long times $W(x, t)$ does does not depend on $W(x, 0)$. Although initially the curves are quite far apart, it is apparent that when $b^3_1 At = 8$ the two curves have approached each other. Thus, for long times the initial profile is unimportant, and we can use the results of Mullins.

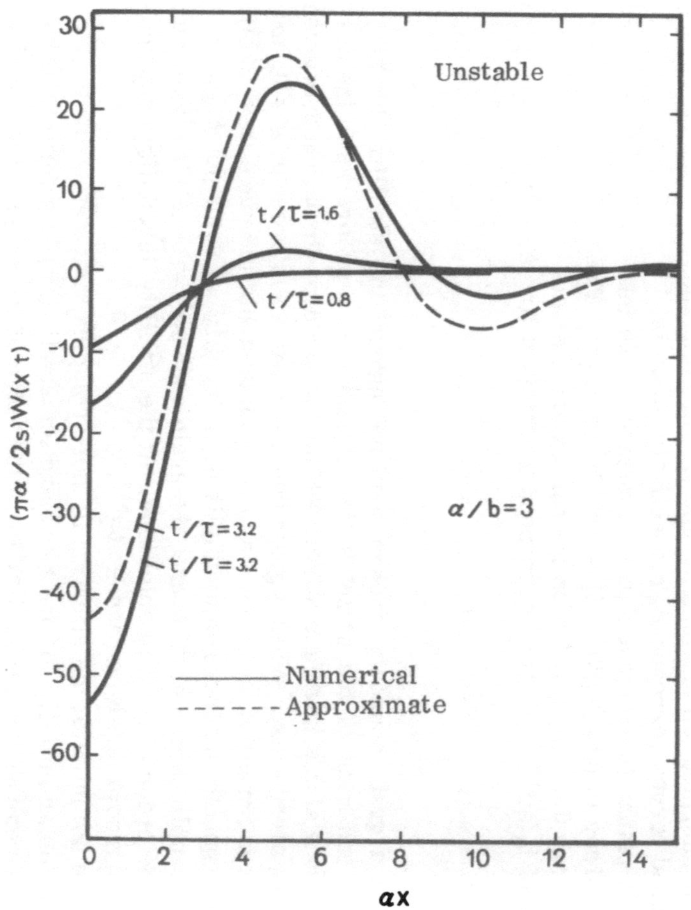

Fig. 10. Groove shape for various times for an unstable interface (dimensionless units). $a/b = 3$, so the initial shape (the $t/\tau = 0$ curve) is given by $(\pi a/2s)W(x, 0) = -(3\pi/2)\exp(-ax/3)$. The dashed curve represents $(\pi a/2s)W_G(x, t)$ and was calculated from (23b) for $(t/\tau) = 3.2$. Shapes for shorter times are given in Fig. 9.

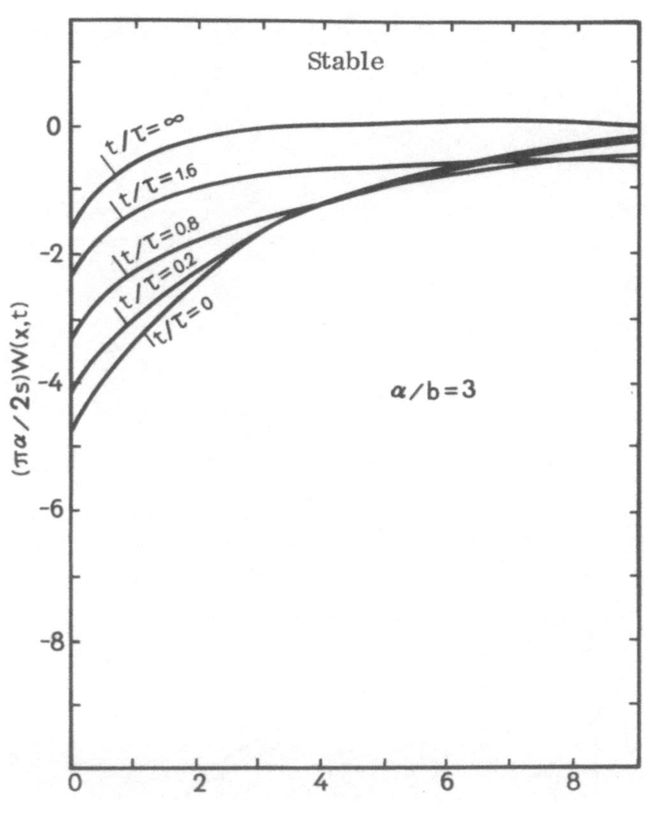


Fig. 11. Groove shape for various times for a stable interface (dimensionless units). The initial shape is the same as in Figs. 9 and 10. The final shape is given by $(\pi a/2s)W(x, \infty) = -(\pi/2)\exp(-ax)$.

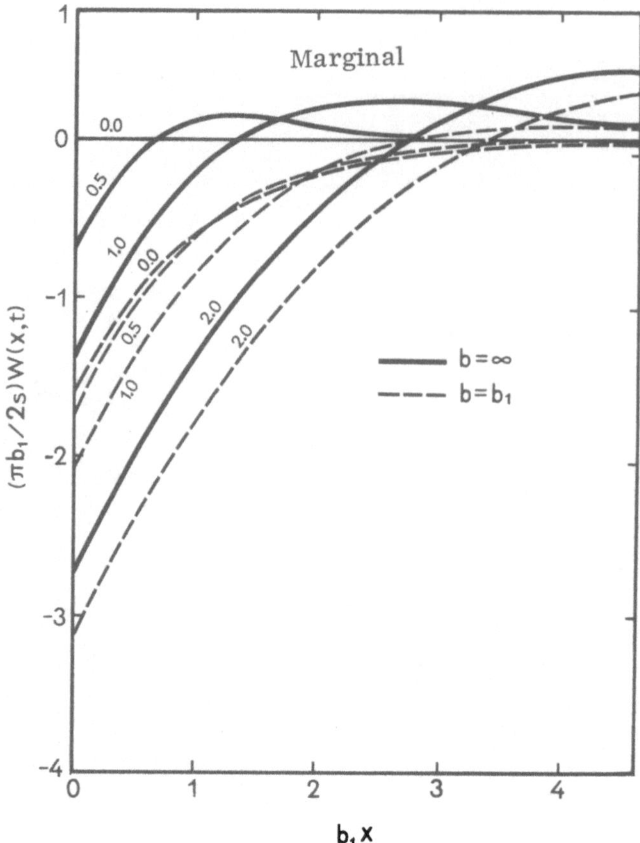


Fig. 12. Groove shape for various times for a marginal interface (dimensionless units). The solid and dashed curves correspond, respectively, to an initially planar interface and to an interface whose initial shape is $(\pi b_1/2s)W(x, 0) = -(\pi/2)\exp(-b_1x)$. The numbers 0.0, 0.5, 1.0, 2.0 on the curves are the values of $b_1(at)^{1/3}$ and thus correspond to different times.

## 5. Discussion

We have shown that a grain boundary groove does not change the stability–instability criterion, i.e., the criterion is the same in the presence of a groove as in the absence of a groove. As $t \to \infty$ a stable interface attains a time-independent shape, which is given by eq. (16), viz., $W(x, t) = -(s/a)\exp(-ax)$. When a grain boundary groove is present, $s \neq 0$, and the time-independent shape is not a flat interface. As $t \to \infty$, an unstable interface [see (23b)], becomes sinusoidal with exponentially increasing amplitude (in reality, at some point linear stability theory becomes inapplicable). In this case, the grain boundary groove simply provides an initial perturbation of the order of $(s/a)$ in amplitude. At the stability-instability demarcation, $(G_L + G_S) = 0$, and the groove depth increases at $t^{1/3}$.

In practice, since t is not infinite, the situation is more complex than stated above. We have previously noted that for an initially flat interface and for small $(t/\tau)$ [e.g., $(t/\tau) = 0.05$] the time-dependent interface shape is quite similar for either sign of $(G_L + G_S)$. Thus, in these circumstances, it would be impossible to determine whether the interface is morphologically stable or unstable from observations at small values of $(t/\tau)$. In addition, $a^2$ is proportional to $|G_L + G_S|$ and $\tau$ is proportional to $|G_L + G_S|^{-3/2}$; thus as the stability–instability demarcation is approached, i.e., $|G_L + G_S| \to 0$, $a^2 \to 0$, and $\tau \to \infty$. Hence, from qualitative observations of interface shape, it is exceedingly difficult to precisely delineate the stability–instability criterion. The theory quantitatively predicts the interface shape as a function of time; however, near the instability is the appearance of oscillations in the interpredictions are very similar whether the interface is morphologically stable or morphologically unstable.

A possible observational criterion for morphological instability is the appearance of oscillations in the interface shape. It is clear from Fig. 8 (stable interface) that the observation of a single maximum does not indicate an unstable interface. For the case treated here (a single component), it appears that the observation of minima (other than at the grain boundary) would indicate an unstable interface. However, further work for the case of a dilute binary alloy (to be published subsequently) shows that this is not a universally valid criterion.

Indeed, the only universally meaningful criterion for instability seems to be the exponential growth of perturbations; while stability, conversely, entails their exponential decay. Situations such as the marginal state (case 2, $G_L + G_S = 0$) of grain boundary grooving are characterized by a profile with algebraically increasing amplitude in order to satisfy certain constraints (e.g., $\partial W/\partial x = s$ at $x = 0$); in the absence of such constraints, these situations do not exist.

We discuss briefly the validity of the small slope approximation. As previously noted for $V = 0$ and $(G_L + G_S) > 0$, the time-independent shape of a grain boundary groove has been calculated exactly by Nash and Glicksman [8]. With the additional condition that $k_L = k_S$, the following analytic formula has been obtained by Bolling and Tiller [13]

$$-2^{-\frac{1}{2}} a\, W^*(x) = \{1 - [1 + (\partial W^*/\partial x)^2]^{-\frac{1}{2}}\}^{\frac{1}{2}}. \tag{28}$$

Here, the asterisk on W indicates that the small slope approximation has not been made. If $|\partial W^*/\partial x| \ll 1$, eq. (28) yields $(\partial W/\partial x) = -aW$ from which we obtain $W = -(s/a)\exp(-ax)$. However, for any value of s, the groove depth, $[-W^*(0)]$, is given exactly by

$$-W^*(0) = \{2/a^2\}^{\frac{1}{2}}\{1 - [1 + s^2]^{-\frac{1}{2}}\}^{\frac{1}{2}},$$

which can be compared with the small slope expression $[-W(0)] = s/a$. The ratio $W(0)/W^*(0)$ is $\{{}^1/_2[1 + s^2 + (1 + s^2)^{1/2}]\}^{1/2}$, which is equal to unity for $s = 0$ and is a monotonic increasing function of s. For $s = 0.1$, $0.5$, and $1.0$, $W(0)/W^*(0)$ is equal to 1.004, 1.088, and 1.307, respectively. Thus, the error in $W(0)$ is less than 10% when $s \le 0.5$. To further substantiate our approximation, we have made a comparison of $W(0)$ with the exact solutions of Nash and Glicksman [8, 14]) for $s = 0.1$ and $1.0$ and $k_S/k_L = 4$; agreement similar to the $k_S = k_L$ case is found.

We also wish to point out that the present two-dimensional (no z-dependence) results are rather easily extended to three dimensions. Briefly, the position of the solid–liquid interface is described by the equation $y = W(x, z, t)$ and without loss of generality we write $W(x, z, t) = W(x, t) + W_0(x, z, t)$, where $W(x, t)$ is the solution of the two-dimensional problem [(15)] and $W_0$ is defined by the above equation. At $t = 0$, any arbitrary $W_0(x, z, 0)$ can be decomposed into Fourier components of the form $\cos\omega x\,(C_1\cos\omega' z + C_2\sin\omega' z)$. The derivative with respect to x of such terms vanishes at $x = 0$; hence, at $x = 0$ the constraint

$$\partial W(x, z, t)/\partial x = \partial W(x, t)/\partial x = s$$

is satisfied at the grain boundary. The time evolution of the z-dependent Fourier components is exponential i.e., $W_0(x, z, t)$ can be decomposed into Fourier components of the form

$$\cos\omega x\,(C_1\cos\omega' z + C_2\sin\omega' z)\ \exp\{f([\omega^2 + \omega'^2]^{\frac{1}{2}})\,t\},$$

where $f([\omega^2 + \omega'^2]^{1/2})$ is idential to the previously defined $f(\omega)$ except that $[\omega^2 + \omega'^2]^{1/2}$ replaces $\omega$. Note that such a replacement does not change the stability–instability demarcation. An important physical consequence of the above discussion is that the observation of oscillations along a direction parallel to the grain boundary (i.e., of the form $\cos \omega' z$) indicates morphological instability. This is in contrast to oscillations perpendicular to the grain boundary which do not necessarily imply morphological instability. It is to be emphasized, however, that these conclusions apply only to a single isolated grain boundary. The interaction among perturbations when several grain boundaries or other defects are present has not yet been explored.

We also remark that our results should be applicable to the freezing of pure materials in crucibles with perfectly insulating walls where, now, the wall–crystal interface plays the role of the grain boundary. For indeed, an equivalent constant slope of the interface will be maintained at the wall by a balance of surface tensions and the plane of the wall will be a plane of zero heat flux.

We shall now discuss the degree to which our use of Laplace's equation [satisfied by the expressions in (1)] affects our results. General aspects of the use of Laplace's equation in morphological stability theory have been discussed in morphological stability theory have been discussed in several places [10, 15, 16]) so only a brief account will be given here.

The complete time-dependent equation for T in a frame of reference which moves uniformly with the solid–liquid interface is

$$\nabla^2 T + (V/\alpha)(\partial T/\partial y) = \alpha^{-1}(\partial T/\partial t), \tag{29}$$

where $\alpha$ is the thermal diffusivity. This equation differs from Laplace's equation by two terms. The first term, $\alpha^{-1}(\partial T/\partial t)$, is necessary for the description of those conditions brought about because of changes in the interface shape as well as transient effects which arise because the temperature must adjust to initial conditions. The second term, $(V/\alpha)(\partial T/\partial y)$, on the other hand, accounts for the use of a moving frame of reference.

We shall proceed to show that the effects of the term $\alpha^{-1}(\partial T/\partial t)$ can be made negligibly small provided that we confine ourselves to a suitable spatial and temporal domain; whereas the effects of the term $(V/\alpha)(\partial T/\partial y)$ will be shown to be negligible except very near to the stability–instability demarcation.

For the effects caused by changes in interface shape to be unimportant, it is necessary for changes in the interface groove depth (which take place in a time $\tau_{th} \sim \lambda_0^2/4\pi^2\alpha$) to be small compared to the interface groove depth itself. Thus,

$$(\partial W/\partial t)\tau_{th} \ll W$$

or

$$\lambda_0^2/(4\pi^2\alpha) \ll \tau, \tag{30}$$

where $\tau$ is given just prior to (20a) and $\lambda_0$ characterizes the spatial extent of the interface feature of interest. If we take

$$\lambda_0^2 = a^{-2} = 2T_M\Gamma/|G_L + G_S|,$$

(30) becomes

$$(\bar{k}/L_V\alpha)(3\pi^2 6^{\frac{1}{2}})^{-1}(T_M\Gamma|G_L + G_S|)^{\frac{1}{2}} \ll 1. \tag{31}$$

Except for enormous temperature gradients, (31) is well satisfied and is therefore not a basic limitation.

In order to examine the other effects described by the term $\alpha^{-1}(\partial T/\partial t)$, we must acknowledge the fact that we are really dealing with an initial value problem. If we are interested in the shape of the interface at a distance x from the grain boundary, we must wait a time of the order of $t \sim x^2/4\pi^2\alpha$ in order for transient effects arising from imposed initial conditions of the problem to die away. Thus, for fixed x, one must wait a time $x^2/4\pi^2\alpha$ for our results to apply; while for any fixed t, our results will not apply at distances greater than $(4\pi^2\alpha t)^{1/2}$.

We now turn to examine the somewhat more restrictive conditions under which the term $(V/\alpha)(\partial T/\partial y)$ can be neglected. Its effects will be negligible provided that the distance traveled by the interface in a time necessary for heat to be conducted over a distance $\lambda_0$ is negligible with respect to $\lambda_0$. The corresponding time is $\tau_{th}$ and the distance advanced by the interface in this time is $V\tau_{th}$; consequently, we need

$$V\tau_{th} \ll \lambda_0$$

or

$$V\lambda_0/4\pi^2\alpha \ll 1. \tag{32}$$

Equation (32) is analogous to the criterion given in [16] which pertains to an isothermal diffusion-controlled phase transformation. Again letting $\lambda_0 = a^{-1}$, we obtain

$$(V/4\pi^2\alpha)\,(2T_M\Gamma/|G_L + G_S|)^{\frac{1}{2}} \ll 1. \tag{33}$$

The quantity on the left-hand side of (33) is recognized to be essentially that on the left of (31) multiplied by the factor

$$(L_V V/\bar{k})/|G_L + G_S| = (G_S - G_L)/|G_S + G_L|. \tag{34}$$

Thus (33) will be satisfied except very near the stability–instability demarcation where $|G_S + G_L| \to 0$; however, in this case, $a^{-1} \to \infty$ and is no longer a good estimate of $\lambda_0$. Thus, in the vicinity of the stability–instability demarcation, the time constant for transient phenomena becomes very large and a more rigorous analysis is needed.

We conclude, therefore, that after a time $t \sim x^2/4\pi^2\alpha$ has elapsed, the use of Laplace's equation to calculate the temperature field at a given x is an excellent approximation (except very near the stability–instability demarcation) for the growth velocities and temperature gradients typical of crystal growth processes.

## Appendix

We wish to show that the integral expression (15) for W(x, t) satisfies the condition $(\partial W/\partial x) = s$ at $x = 0$. We recall [11] that given a function $\varphi(x, \eta)$, the validity of the equation

$$\frac{\partial}{\partial x}\int_0^\infty \varphi(x, \eta)\, d\eta = \int_0^\infty (\partial\varphi/\partial x)\, d\eta$$

requires that the integral on the right be uniformly convergent. For example if $\varphi(x, \eta) = [\cos(\eta x)]/(\eta^2 + 1)$ the above equation is not valid for $x = 0$. Thus we cannot calculate $(\partial W/\partial x)$ from (15) by differentiation under the integral sign. In calculating $(\partial W/\partial x)$ at $x = 0$, it is convenient to consider the three cases separately.

Case (1), $(G_S + G_L) > 0$

The result follows immediately from differentiation of (20b).

Case (2), $(G_S + G_L) = 0$

Following [7], we write (21) in the form $W(x, t) = (2s/\pi)(At)^{1/3}(I_1 + I_2)$, where

$$I_1 = \int_0^\infty d\eta\,\eta^{-2}\left\{\frac{\cos[\eta x/(At)^{\frac{1}{3}}]\exp[-\eta^3]}{1+[\eta/b(At)^{\frac{1}{3}}]^2} - 1\right\}$$

and

$$I_2 = \int_0^\infty d\eta\,\eta^{-2}\{1-\cos[\eta x/(At)^{\frac{1}{3}}]\}.$$

Differentiation of $I_1$ yields $(\partial I_1/\partial x) = 0$ at $x = 0$ and evaluation of $I_2$ gives $I_2 = x\pi/2(At)^{1/3}$. Thus $(\partial I_2/\partial x) = x\pi/2(At)^{1/3}$ and $(\partial W/\partial x) = s$ at $x = 0$.

Case (3), $(G_L + G_S) < 0$

We rewrite (22) in the form* $W(x, t) = (2s/\pi a)(I_3 + I_4)$,

where

$$I_3 = \int_0^\infty d\eta\,\frac{\cos(ax\eta)}{\eta^2-1}\left\{\frac{[1+(a/b)^2]}{[1+(a\eta/b)^2]}\exp[(3^{3/2}/2)(t/\tau)\eta(1-\eta^2)] - \exp[1-\eta]\right\}$$

and

$$I_4 = \int_0^\infty d\eta\,\frac{\cos(ax\eta)}{\eta^2-1}\{\exp[1-\eta]-1\}.$$

We may differentiate $I_3$ under the integral sign; it follows that $(\partial I_3/\partial x) = 0$ at $x = 0$. To evaluate $(\partial I_4/\partial x)$, we carry out a contour integration in the complex $\eta$ plane. With $\cos(ax\eta) = \mathrm{Re}[\exp(iax\eta)]$, it follows that the integral $I_4$ along the positive real axis is equal to the same integral along the positive imaginary axis (close the contour with a infinite quarter circle and note that there are no singularities inside this contour and the integrand vanishes faster than $|1/\eta|$ as $|\eta| \to \infty$). Thus

$$I_4 = -\int_0^\infty d\eta\,\frac{\sin\eta}{\eta^2+1}\exp[1-ax\eta].$$

In this form, it is now permissible to differentiate under the integral sign; we find $(\partial I_4/\partial x) = (\pi a/2)$ at $x = 0$. Thus $(\partial W/\partial x) = s$ at $x = 0$.

Acknowledgments

We thank M. E. Glicksman, R. D. Mountain, F. W. J. Olver, and J. A. Simmons for helpful discussions.

*We thank F. W. J. Olver for suggesting this method of proof.

Literature Cited

1. W. W. Mullins and R. F. Sekerka, J. Appl. Phys., 34:323 (1963).
2. R. F. Sekerka, J. Crystal Growth, 3/4:71 (1968).
3. R. F. Sekerka, Four Lectures on Morphological Stability, P. Hartman, ed., North-Holland, Amsterdam, to be published.
4. L. R. Morris and W. C. Winegard, J. Crystal Growth, 5:361 (1969).
5. R. J. Schaefer and M. E. Glicksman, Metal. Trans., 1:1973 (1970).
6. T. Sato and G. Ohira, Trans. Japan Inst. Metals, 12:285 (1971).
7. W. W. Mullins, Trans. AIME, 218:354 (1960).
8. G. E. Nash and M. E. Glicksman, Phil. Mag., 24:577 (1971).
9. W. W. Mullins, Metal Surfaces, American Society for Metals, Metals Park, Ohio (1963).
10. R. F. Sekerka, J. Crystal Growth, 10:239 (1971).
11. E. T. Whittaker and G. N. Watson, A Course of Modern Analysis, Cambridge (1958).
12. W. Magnus, F. Oberhettinger, and R. P. Soni, Formulas and Theorems for the Special Functions of Mathematical Physics, Springer-Verlag, New York (1966).
13. G. F. Bolling and W. A. Tiller, J. Appl. Phys., 31:1345 (1960).
14. G. E. Nash and M. E. Glicksman, private communication.
15. R. F. Sekerka, In: Crystal Growth, H. S. Peiser, ed., Pergamon, Oxford (1969) p. 691.
16. R. F. Sekerka, J. Phys. Chem. Solids, 28:983 (1967).

# MORPHOLOGICAL STABILITY NEAR A GRAIN BOUNDARY GROOVE IN A SOLID–LIQUID INTERFACE DURING SOLIDIFICATION OF A BINARY ALLOY

**S. R. Coriell and R. F. Sekerka**

*National Bureau of Standards, Washington, D.C.*
*Carnegie-Mellon University, Pittsburgh, Pennsylvania*

## 1. Introduction

Morphological stability theory [1–4]) describes the stability of the shape of a solid–liquid interface during a crystallization process. In a recent article [5], we determined the shape and the morphological stability of a grain boundary groove in a solid–liquid interface during unidirectional solidification of a pure substance at constant velocity. In this paper, we extend our previous treatment to the solidification of a dilute binary alloy. In part, this work was motivated by experimental observations [6–8] that during the solidification of binary alloys instabilities are first observed near defects such as grain boundaries.

We will use linear stability theory to calculate the time dependent shape of a nearly planar interface, intersected perpendicularly by a grain boundary (the physical properties and interface shape are assumed symmetric with respect to reflection through the grain boundary plane). The grain boundary produces a groove in the interface; the slope, s, of this groove is determined by the balancing of grain boundary and solid-liquid surface tensions (there is a discontinuity in the slope of the interface at the grain boundary; the slope on the left being $-s$, the slope on the right $+s$). Linear stability theory restricts us to small slopes which in turn requires that the ratio of grain boundary to solid–liquid surface energy be small. Thus, the grain boundary causes a constraint on the interface, viz., the interface slope (and the discontinuity in slope) at the grain boundary is a constant independent of time.

The presence of the grain boundary constraint is the important difference between the present calculation and previous treatments of morphological stability [2, 9] during solidification of a binary alloy. Except for this distinction, the methods and approximations used are identical to those of [2], e.g., local equilibrium at the interface is assumed.

In the next section, we develop the theory which results in an integral representation of the time-dependent interface shape. In Section 3, we obtain approximate analytic results for the long time behavior of the interface for the three possibilities, i.e., stable, marginal, and unstable behavior. An integral expression for the solute concentration at the interface is also given. In Section 4, we present numerical calculations of the interface shape and the solute distribution.

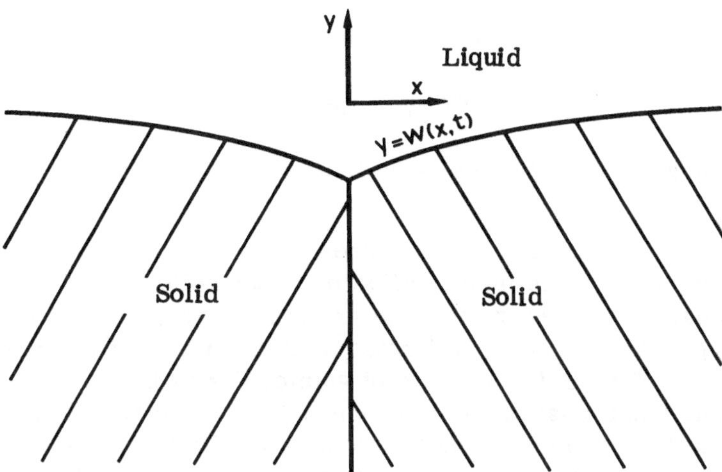


Fig. 1. Sketch of solid–liquid interface containing a grain boundary groove. The position of the interface is described by the equation $y = W(x, t)$. The plane $y = 0$ is chosen such that $W(x, t) \to 0$ as $|x| \to \infty$.

## 2. Theory

We wish to determine the effect of a grain boundary on the shape of a nearly planar solid–liquid interface during unidirectional solidification of a dilute binary alloy at constant velocity. As illustrated in Fig. 1, an (x, y, z) coordinate system is chosen such that the grain boundary lies in the plane $x = 0$ and the position of the solid–liquid interface is described by $y = W(x, t)$, where t is the time and $W(x, 0) \to 0$ as $|x| \to \infty$. All quantities are assumed independent of the z-coordinate. Further we take $W(x, 0)$ as a symmetric function of x; consequently $W(x, t)$ is a symmetric function of x and we consider only non-negative values of x. The presence of the grain boundary imposes the additional condition that $(\partial W/\partial x) = s \neq 0$ at $x = 0$. In addition we make essentially the same assumptions as in [5], e.g., $W(x, t)$ and $(\partial W/\partial x)$ are assumed small and all powers of W greater than unity are neglected.

In detail, we wish to solve the equations*

$$\nabla^2 T_L = 0, \tag{1a}$$

$$\nabla^2 T_S = 0, \tag{1b}$$

$$\nabla^2 C + (V/D)(\partial C/\partial y) = 0 \tag{1c}$$

subject to the conditions that

$$T_{LI} = T_{SI} = T_M - T_M \Gamma\, K(x,t) + mC_i, \tag{2a}$$

$$(\partial T_L/\partial y)_{y\to\infty} = G_L, \tag{2b}$$

* The validity of using these equations in place of the fully time-dependent equations has been previously discussed [5, 11, 13, 14].

$$(\partial T_S/\partial y)_{y\to-\infty}=G_S, \tag{2c}$$

$$C(x,y\to\infty)=C_\infty, \tag{2d}$$

$$v=V+(\partial W/\partial t)=(k_S/L_V)(\partial T_S/\partial y)_I-(k_L/L_V)(\partial T_L/\partial y)_I=D[C_I(k-1)]^{-1}(\partial C/\partial y)_I. \tag{2e}$$

In these equations, $T_L$ and $T_S$ are the temperatures in the liquid and solid, respectively, C is the concentration of impurity in the liquid (diffusion in the solid is neglected), V (a constant) is the velocity of the planar interface far from the grain boundary, D is the diffusion coefficient of the impurity in the liquid (the subscript I denotes evaluation at the interface), $T_M$ is the melting temperature of a flat interface in the absence of solute, $\Gamma = \gamma/L_V$ is a capillary constant, $\gamma$ is the solid-liquid surface tension (assumed isotropic), $L_V$ is the latent heat per unit volume, $K(x, t) = -(\partial^2 W/\partial x^2)$ is the curvature, m is the slope of the liquidus line on the phase diagram, $G_L$ and $G_S$ (constants) are the temperature gradients in liquid and solid, respectively, far from the interface, $C_\infty$ is the solute concentration in the liquid far from the interface, v is the interface velocity, $k_L$ and $k_S$ are the thermal conductivities of liquid and solid, respectively, and k is the partition coefficient (ratio of solid to liquid solute concentration at the interface). In preparation for the application of perturbation theory, we write the temperature and solute fields in the form

$$T_L(x, y)=T_M+(mC_\infty/k)+G_Ly+\bar{T}_L(x, y), \tag{3a}$$

$$T_S(x, y)=T_M+(mC_\infty/k)+G_Sy+\bar{T}_S(x, y), \tag{3b}$$

$$C(x, y)=(C_\infty/k)+(DG_C/V)[1-\exp(-Vy/D)]+\mathcal{C}(x, y), \tag{3c}$$

where $G_C = VC_\infty(k-1)/Dk$. The quantities $\bar{T}_L(x, y)$, $\bar{T}_S(x, y)$, and $\mathcal{C}(x, y)$ are defined by the above equations. If $W(x, t) = 0$, $\bar{T}_L$, $\bar{T}_S$, and $\mathcal{C}$ vanish and for small W, they are proportional to W. $\bar{T}_L$, $\bar{T}_S$, and $\mathcal{C}$ satisfy the same differential equations at $T_L$, $T_S$, and C, respectively. The boundary conditions (2), together with (3), imply that

$$G_LW+\bar{T}_L(x, 0)=G_SW+\bar{T}_S(x, 0)=-T_M\Gamma K+mG_CW+m\mathcal{C}(x, 0), \tag{4a}$$

$$(\partial\bar{T}_L/\partial y)_{y\to\infty}=0, \tag{4b}$$

$$(\partial\bar{T}_S/\partial y)_{y\to-\infty}=0, \tag{4c}$$

$$\mathcal{C}(x, y\to\infty)=0; \tag{4d}$$

$$(\partial W/\partial t)=(k_S/L_V)(\partial\bar{T}_S/\partial y)_{y=0}-(k_L/L_V)(\partial\bar{T}_L/\partial y)_{y=0}=$$
$$=(V/G_C)\{(-k/C_\infty)[G_C^2W+G_C\mathcal{C}(x, 0)]-G_C(V/D)W+(\partial\mathcal{C}/\partial y)_{y=0}\}, \tag{4e}$$

where we have used

$$V=(k_S/L_V)G_S-(k_L/L_V)G_L=DG_Ck/[C_\infty(k-1)].$$

In order to find $\bar{T}_L$, $\bar{T}_S$, and $\mathcal{C}$ subject to the preceding conditions, we use the Fourier cosine transform to replace their dependence on x by a dependence on a transform variable, $\omega$. Given a function $F(x, y, t)$, we define its Fourier cosine transform, $\tilde{F}(\omega, y, t)$, by

$$\tilde{F}(\omega, y, t) = \int_0^\infty dx \cos(\omega x)F(x, y, t), \tag{5}$$

whose inverse is

$$F(x, y, t)=(2/\pi) \int_0^\infty d\omega \cos(\omega x) \tilde{F}(\omega, y, t). \tag{6}$$

Then the transformed temperature and solute fields, which satisfy the transformed differential equations and boundary conditions at infinity, are

$$\tilde{\tilde{T}}_L(\omega, y)=a_L \exp(-\omega y), \tag{7a}$$

$$\tilde{\tilde{T}}_S(\omega, y)=a_S \exp(\omega y), \tag{7b}$$

$$\tilde{C}(\omega, y)=a_C \exp(-\omega^* y), \tag{7c}$$

where $\omega^*=(V/2D)+[(V/2D)^2+\omega^2]^{1/2}$, and the constants $a_L$, $a_S$, $a_C$ are chosen to satisfy the interface boundary conditions (4a) and (4e). Upon solving for $a_L$, $a_S$, and $a_C$ and then using (4e) to calculate $(\partial W/\partial t)$, we obtain

$$(\partial\tilde{W}/\partial t)=h(\omega)[-\mathcal{G}_L+\mathcal{G}_S)\tilde{W}-2T_M\Gamma\tilde{K}+2mG_C\tilde{W}(\omega^*-V/D)/(\omega^*-Vp/D)], \tag{8}$$

where

$$h(\omega)=(V\omega)/[(\mathcal{G}_S-\mathcal{G}_L)+2mG_C\,\omega/(\omega^*-Vp/D)],$$

$$\mathcal{G}_L=k_L G_L/\bar{k},\ \mathcal{G}_S=k_S G_S/\bar{k},\ 2\bar{k}=k_L+k_S,\ p=1-k.$$

A straightforward calculation [5] shows that

$$\tilde{K}(\omega, t)=s+\omega^2\tilde{W}(\omega, t).$$

Eliminating K from (8) and solving for $\tilde{W}(\omega, t)$, we have

$$\tilde{W}(x,t)=[\tilde{W}(x,0)-T_M\Gamma s/S(\omega)]\exp[f(\omega)t]+T_M\Gamma s/S(\omega), \tag{9}$$

where

$$2S(\omega)=-(\mathcal{G}_L+\mathcal{G}_S)-2T_M\Gamma\,\omega^2+2mG_C(\omega^*-v/D)/(\omega^*-vp/D),$$

$f(\omega)=2S(\omega)\,h(\omega)$, and $\tilde{W}(\omega, 0)$ is the Fourier cosine transform of W(x, 0). We assume W(x, 0) is finite and approaches zero sufficiently rapidly as $x\to\infty$ that $\tilde{W}(\omega, 0)$ is finite for all $\omega$. Wherever possible we have used notations similar to those previously used; in particular $f(\omega)$ is identical to $f(\omega)$ of [3, 4] and $S(\omega)$ is identical to $S(\omega)$ of [2]. When s = 0 (grain boundary not present), (9) reduces to previous results [2].

## 3. Analytical Results

### 3.1. General

As is evident from (9), the time evolution of the interface shape is strongly dependent on the sign of $f(\omega)$. If $f(\omega)$ is positive for some range of $\omega$, the interface is morphologically unstable. If $f(\omega) < 0$ for all positive values of $\omega$, then the interface is morphologically stable. The stability-instability demarcation is thus determined by $f(\omega)$. Since $f(\omega)$ does not depend on s, the presence of a grain boundary has no effect on the stability-instability criterion.

The properties of $f(\omega)$ have been extensively discussed [2, 10]. In brief, since $h(\omega) > 0$ for all positive $\omega$, the sign of $f(\omega)$ is the same as the sign of $S(\omega)$. As $\omega \to \infty$, $2S(\omega) \to -2T_M\Gamma\omega^2 < 0$; also, $2S(0) = -(G_L + G_S) < 0$ provided that we restrict ourselves to the usual case where $G_L > 0$. Further it has been shown that $S(\omega)$ has at most one maximum as $\omega$ ranges from zero to infinity. If there is no maximum, it is clear that $S(\omega) < 0$ for all $\omega$ $(0 \le \omega < \infty)$. In the case of a maximum, let us denote the value of $\omega$ corresponding to the maximum by $\omega_m$. If $S(\omega_m) < 0$, then $S(\omega) < 0$ for all $\omega$; if $S(\omega_m) = 0$, $S(\omega) < 0$ for all $\omega$ except $\omega_m$; if $S(\omega_m) > 0$, then $S(\omega) > 0$ for some range of $\omega$. Thus, for $S(\omega_m) > 0$, the interface is morphologically unstable. When there is no maximum or $S(\omega_m) < 0$, the interface is morphologically stable. $S(\omega_m) = 0$ is a marginal case and will be discussed subsequently.

The three cases just delineated can be discussed conveniently with the aid of the following dimensionless variables:

$$I \equiv 2mG_C/(G_S - G_L) = 2\bar{k}(k-1)mC_\infty/(L_V D k), \tag{10a}$$

$$J \equiv 3IT_M\Gamma V^2/(4mG_C D^2) = 3\bar{k}T_M\Gamma V/(2L_V D^2), \tag{10b}$$

$$\Sigma \equiv (G_L + G_S)/2mG_C, \tag{10c}$$

$$(\chi_0')^2 \equiv (I/J)\,|\,1-\Sigma\,|\,. \tag{10d}$$

Furthermore, to treat all possible cases formally in a uniform manner, we define a parameter

$$\chi_0 \equiv \begin{cases} \chi_0' & \text{if} \quad \chi_0' \neq 0 \\ \chi_0'' & \text{if} \quad \chi_0' = 0, \end{cases} \tag{10e}$$

where we are free to choose any convenient positive value for $\chi_0''$. Except for the somewhat more general definition of $\chi_0$, the above notation is identical to that of [11]. We next introduce the length

$$L = [\sqrt{3}\,\chi_0(V/2D)]^{-1}$$

and the dimensionless time

$$t^* = (2\bar{k}T_M\Gamma/L_V L^3)t.$$

Using these definitions and introducing a new variable of integration $\eta = L\omega$, we write

$$(\pi/2Ls)\,W(x,\,t) = \int_0^\infty d\eta\,\cos\,(\eta x/L)\{[\widetilde{W}(\omega,0)/sL^2 - T_M\Gamma/L^2S(\omega)]\exp[f(\omega)t] + T_M\Gamma/L^2S(\omega)\}, \tag{11}$$

where

$$T^M\Gamma/L^2S(\omega) = \{U - \eta^2 + [(I/J\chi_0^2)2(p-1)]/[1-2p+(1+3\chi_0^2\eta^2)^{\frac{1}{2}}\}^{-1},$$

$$U = \begin{cases} 1 & \text{if} \quad \Sigma < 1 \\ 0 & \text{if} \quad \Sigma = 1 \\ -1 & \text{if} \quad \Sigma > 1, \end{cases}$$

and

$$f(\omega)t = t^*\eta[L^2S(\omega)/T_M\Gamma]/\{1+(\sqrt{3}\,\chi_o I\eta)/[1-2p+(1+3\chi_o{}^2\eta^2)^{\frac{1}{2}}]\}.$$

If we specify the initial shape [in particular, $\tilde{W}(\omega, 0)/sL^2$], the dimensionless time $t^*$, and the four dimensionless quantities p, I, J, and $\Sigma$, then (11) allows us to calculate $(\pi/2Ls)W(x, t)$ as a function of $(x/L)$. For an alloy of a specified composition p and I are fixed whereas J and $\Sigma$ depend on the solidification conditions. For instance, if V and $G_S$ are regarded as independent variables and $G_S$ is eliminated, then J is proportional to V [(10b)] and

$$\Sigma = [I + (2G_L\bar{k}/VL_V)]/I.$$

## 3.2. Long-Time Behavior of W(x, t)

3.2.1. Stable Case. If the interface is morphologically stable, then $S(\omega) < 0$ for all $\omega$ and $f(\omega) < 0$ for all positive $\omega$ [$f(0) = 0$]. Thus, for sufficiently long times, $\exp[f(\omega)t] \to 0$ for all positive $\omega$ and $\exp[f(0)t] = 1$. Since the factor multiplying $\exp[f(\omega)t]$ in (11) is finite for $\omega = 0$ [assuming that $W(x, 0)$ is a reasonable function and that $S(0) = -(G_L + G_S) \neq 0$], it is clear that for long times the shape of a stable interface, which is intersected by a grain boundary, is given by

$$(\pi/2Ls)W(x, t\to\infty) = \int_0^\infty d\eta \cos(\eta x/L)[T_M\Gamma/L^2S(\omega)]. \tag{12}$$

For $V = 0$, $L^2 = 2T_m\Gamma/(G_L + G_S)$ and the integral in (12) can be evaluated to give

$$(\pi/2Ls)W(x, t\to\infty) = -(\pi/2)\exp(-x/L). \tag{13}$$

This is identical to the shape previously obtained for all V for a one component system. Since the solute field is uniform for $V = 0$ and $t \to \infty$ the above result is not surprising.

When $V \neq 0$, we have not been able to evaluate (12) analytically. An approximate analytic evaluation is possible, however, when $S(\omega_m) \to 0$ [and $S(\omega_m) < 0$], i.e., the interface is stable but near the stability–instability demarcation. Under the circumstances, the major contribution to the integral in (12) is from values of the integrand in the vicinity of $\omega_m$. For $\omega$ near $\omega_m$, we can approximate $S(\omega)$ by

$$S(\omega) \cong S(\omega_m) + 1/2S''(\omega)(\omega - \omega_m)^2,$$

where $S''(\omega)$ is the second derivative of $S(\omega)$. Since values of $S(\omega)$ for $\omega$ not near $\omega_m$ make only a minor contribution to the integral, we can use the above approximation for all values of $\omega$. For similar reasons (provided $\omega_m \neq 0$), we can extend the range of integration from $-\infty$ to $\infty$. Thus for $S(\omega_m)$ near zero, we have

$$(\pi/2Ls)W(x, t\to\infty) \simeq (T_M\Gamma/L^2)\int_{-\infty}^{\infty} d\eta \cos(\eta x/L)[S(\omega_m) + \tfrac{1}{2}S''(\omega_m)(\omega-\omega_m)^2]^{-1},$$

or

$$W(x, t\to\infty) = \{-2T_M\Gamma s/[\tfrac{1}{2}S(\omega_m)S''(\omega_m)]^{\frac{1}{2}}\}\cos(\omega_m x)\exp\{-[2S(\omega_m)/S''(\omega_m)]^{\frac{1}{2}}x\}. \tag{14}$$

Hence the interface shape is sinusoidal with an exponential damping factor. For many experimental conditions, $S''(\omega_m)$ is approximately constant as $S(\omega_m) \to 0$ and the damping decreases as $S(\omega_m) \to 0$. Thus, very near the stability–instability demarcation, the interface shape is nearly sinusoidal. Numerical evaluation of (12) leads to the same result; the numerical calculations will be discussed subsequently.

An alternative method of understanding this behavior is to consider $\omega$ (or $\eta = L\omega$) as a complex variable. The integration in (12) has branch points at $\eta = \pm i/(3\chi_0^2)^{1/2}$ and poles at the zeroes of $S(\omega)$. By contour integration we can write $W(x, t \to \infty)$ as the sum of a branch cut integral along the positive imaginary axis from $i/(3\chi_0^2)^{1/2}$ to $\infty$ and a sum of residues from poles in the upper half plane. Although the branch cut integral can not be evaluated, it has the form of a Laplace transform of a nonnegative function. Thus, the branch cut integral is a nonnegative monotonic decreasing function of x. Denoting the zeroes of $S(\omega)$ by $\omega_j^*$, a given residue contributes a term proportional to $\exp(i\omega_j^* x)$ to $W(x, t \to \infty)$. When $V = 0$, the branch cut integral vanishes, $\omega_j^*$ is pure imaginary and the interface shape is exponential. As V increases, $\omega_j^*$ has both real and imaginary parts. Finally, at the stability–instability demarcation $\omega_j^*$ is real. Thus, as the stability–instability demarcation is approached, $\exp(i\omega_j^* x)$ becomes increasingly oscillatory in character in agreement with (14).

3.2.2. Marginal Case. We now discuss the behavior of the interface shape at the stability–instability demarcation, i.e., $S(\omega_m) = 0$. Since $f(\omega) \leq 0$ for all $\underset{\sim}{\omega}$ and the equality holds only for $\omega = 0$ and $\omega = \omega_m$, it is clear that for sufficiently long times $\tilde{W}(\omega, 0) \exp[f(\omega)t]$ does not contribute to the integral in (11), i.e., $W(x, t \to \infty)$ does not depend on the initial interface shape. However, since $[T_M\Gamma/L^2S(\omega)]$ is infinite at $\omega = \omega_m$, the term $[T_M\Gamma/L^2S(\omega)] \times \exp[f(\omega)t]$ does contribute to the integral. Thus

$$(\pi/2Ls)W(x, t\to\infty) = \int_0^\infty d\eta \cos(\eta x/L)[T_M\Gamma/L^2S(\omega)]\{1-\exp[f(\omega)t]\}. \tag{15}$$

Although it does not appear possible to evaluate (15) analytically, the following procedure leads readily to an approximate formula for $\partial W/\partial t$. Differentiation of $W(x, t)$ with respect to time yields

$$(\pi/2L)(\partial W/\partial t) = -(2T_M\Gamma/L^2)\int_0^\infty d\eta \cos(\eta x/L)\, h(\omega)\exp[f(\omega)t]. \tag{16a}$$

For large times, the only significant contribution to the integral in Eq.(16a) occurs near $\omega = \omega_m$ where $f(\omega) = 0$.* We approximate $f(\omega)$ by

$$f(\omega) \simeq \tfrac{1}{2} f''(\omega_m)(\omega-\omega_m)^2 \equiv -D(\omega-\omega_m)^2,$$

Furthermore, we can extend the range of integration from $-\infty$ to $\infty$ and approximate $h(\omega)$ by $h(\omega_m)$. The integral can then be evaluated and we obtain

$$(\partial W/\partial t) \cong -[4sT_M\Gamma h(\omega_m)\cos(\omega_m x)\exp(-x^2/4Dt)]/[\pi Dt]^{\frac{1}{2}}. \tag{16b}$$

Thus, the groove depth increases as $t^{1/2}$; the lateral spreading and the oscillatory behavior of the interface shape follow from (16b).

---

* Although $f(0) = 0$, $h(0) = 0$ and there is no significant contribution near $\omega = 0$.

3.2.3. Unstable Case. When the interface is morphologically unstable, $f(\omega) > 0$ for some range of $\omega$. Consequently for long times, the exponential term in (11) makes the dominant contribution to the integral. Denoting by $\omega_0$ the value of $\omega$ at which $f(\omega)$ has a maximum, it is also clear that only values of $f(\omega)$ for $\omega$ near $\omega_0$ are important. Thus, we can use the Green's function technique, previously developed [11] for interfaces not containing grain boundary grooves, to approximate the interface shape. Briefly, we approximate $f(\omega)$ by

$$f(\omega) \cong f(\omega_0) + \tfrac{1}{2} f''(\omega_0)(\omega - \omega_0)^2 \equiv (1/\tau) - D(\omega - \omega_0)^2,$$

where $f''(\omega_0)$ is the second derivative of $f(\omega)$ with respect to $\omega$ evaluated at $\omega_0$. We also make the approximation

$$[\tilde{W}(\omega, 0)/sL^2 - T_M\Gamma/L^2S(\omega)] \cong [\tilde{W}(\omega_0, 0)/sL^2 - T_M\Gamma/L^2S(\omega_0)],$$

Further, we neglect the time-independent term and we extend the range of integration from $-\infty$ to $\infty$. Thus,

$$(\pi/2Ls)W(x, t\to\infty) \cong \int_{-\infty}^{\infty} L d\omega \cos\omega x[\tilde{W}(\omega_0, 0)/sL^2 - T_M\Gamma/L^2S(\omega_0)] \exp[\Gamma(t/\tau) - Dt(\omega - \omega_0)^2], \tag{17a}$$

or

$$W_G(x, t) = [2\tilde{W}(\omega, 0) - 2T_M\Gamma s/S(\omega_0)][\pi D t]^{-\frac{1}{2}} \cos(\omega_0 x) \exp(t/\tau) \exp(-x^2/4Dt), \tag{17b}$$

where we have added the subscript G to $W(x, t)$ to denote this particular approximation. When $S = 0$, (17b) agrees with the results of [11]. Actually, (17b) is valid for intermediate times, i.e., times sufficiently large that $\exp[f(\omega)t]$ is large but sufficiently small that linear theory is valid (see discussion in [11]). From (17b), we see that the main effect of a grain boundary groove is to provide an initial perturbation of the interface.

### 3.3. Interface Concentration

Since for opaque materials it is usually the concentration distribution (rather than the interface shape) which is measured, we state the appropriate equations for calculating the interface concentration. Using (3c), (6), and (7c), we have

$$C_I = (C_\infty/k) + G_C W + (2/\pi)\int_0^\infty d\omega\, a_C \cos(\omega x). \tag{18}$$

Calculation of the quantity $a_c$, defined by (7c), yields

$$a_C = -(G_C/V\{\dot{\tilde{W}} + (V^2k/D)\tilde{W}\}/(\omega^* - vp/D), \tag{19}$$

where $\dot{\tilde{W}}$ and $\tilde{W}$ are given by (8) and (9), respectively. Equation (18) will be examined numerically in the following section.

## 4. Numerical Results

As previously remarked, calculation of the interface shape as a function of time requires specification of the four dimensionless parameters p, I, J, and $\Sigma$. It is clear that it is not feasible to explore all possible values of each of these parameters. In fact, we will report

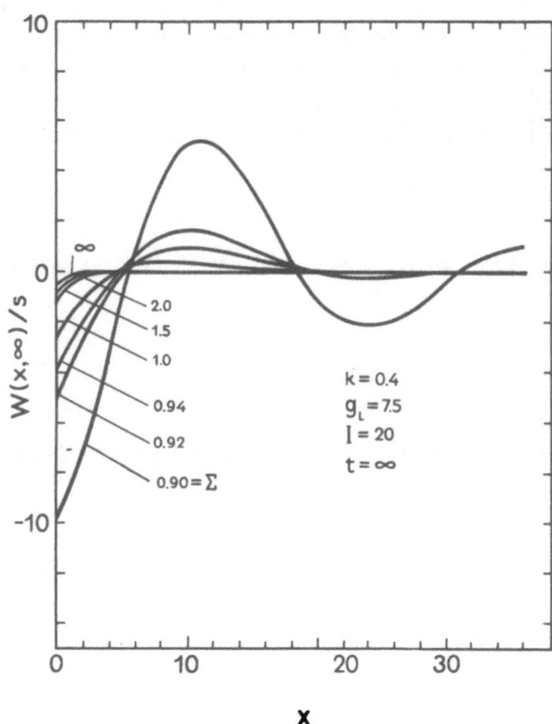


Fig. 2. Long time groove shape for a stable interface. As $\Sigma$ decreases, instability is approached (and would occur for $\Sigma = 0.893$). $\Sigma = \infty$ corresponds to zero interface velocity.

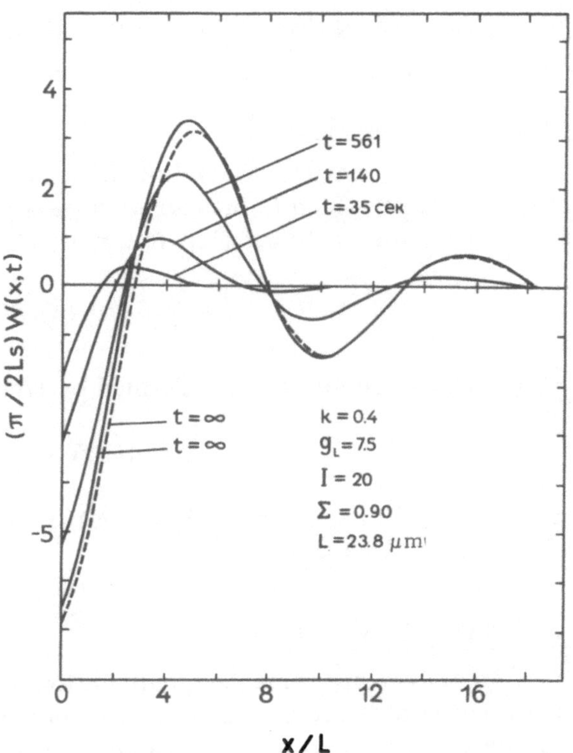


Fig. 3. Groove shape for various times for a stable interface. The solid t = $\infty$ curve is the same as the $\Sigma = 0.90$ curve of Fig. 2. The dashed t = $\infty$ curve was calculated from the approximate analytic expression given by (14).

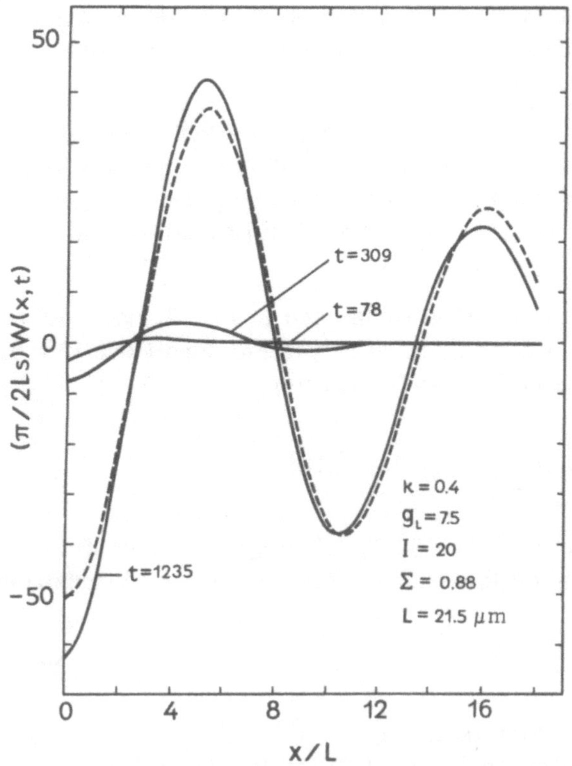


Fig. 4. Groove shape for various times for an unstable interface. The dashed curve for t = 1235 sec was calculated from the approximate analytic expression given by (17).

only a very limited number of calculated results. We choose these parameters so that they lie in the range used by Morris and Winegard [6] in their experiments on dilute alloys of Sb in Pb. Thus, we take p = 1 − k = 0.6 and I = 20, which corresponds to 0.02 wt.% Sb.* Further we vary J and Σ in such a manner that $G_L$ remains constant and we take $G_L$ = 7.5 K/cm ($G_L \cong$ 10.6 K/cm). Thus $J = 3G_L T_M \Gamma(\bar{k}/L_v D)^2/(I\Sigma - 1)$ and we carry out calculations for various values of Σ. For these values of p, I, and $G_L$, the stability−instability demarcation occurs at $\Sigma \cong$ 0.893.

It is also necessary to specify the initial shape. From [5] and the previous discussion of the long-time behavior of the interface shape, we know that the long-time behavior is only trivially influenced by the initial shape. Therefore we are free to choose W(x, 0) and a convenient choice is to take an essentially flat interface as our initial shape.

We have evaluated (12) numerically to obtain the shape of a stable interface for large times. The results for Σ = ∞, 2.0, 1.5, 1.0, 0.94, 0.92, and 0.90 are shown in Fig. 2. Σ = ∞ corresponds to V = 0 and the interface shape is exponential [(13)]. Although not evident from the figure, all the curves for finite Σ have at least one maximum. As Σ decreases, the groove depth increases, e.g., when V = 0, Σ = ∞ and −W(0, ∞)/s ≅ 8 μm whereas when V = 0.88 · $10^{-3}$ cm/sec, Σ = 0.9 and −W(0, ∞)/s ≅ 100 μm. It is also clear that as Σ decreases and the stability−instability demarcation is approached, the interface shape becomes more and more oscillatory.

Figure 3 shows the time evolution of a stable interface, whose final shape corresponds to the Σ = 0.90 curve of Fig. 2. The curves were calculated using (11). For t = 561 sec, the groove depth has attained an appreciable fraction of its final value whereas the first minimum

*We use the following approximate values for the physical properties: $\bar{k}/L_v = 10^{-3}$ cm²/sec K, $2T_M\Gamma = 10^{-5}$ cm K, D = 2 · $10^{-5}$ cm²/sec, m = −5 K/wt.%. Thus $C_\infty \cong 10^{-3}$ I(wt.%).

has attained less than half its final value. The dashed curve was calculated from the approximate analytical formula given by (14); for $\Sigma = 0.90$ this approximation is in good agreement with the numerical results.

The time evolution of an unstable interface is shown in Fig. 4 for t = 78, 309, and 1235 sec. The dashed curve corresponds to $(\pi/2Ls)\, W_G(x, t)$ for $t = 3.35\ \tau = 1235$ sec and was calculated from the approximate analytic formula given by (17b). As can be seen, there is rather good agreement between $W_G(x, t)$ and $W(x, t)$ for $t/\tau = 3.35$. Even better agreement would be obtained for longer times.

We have calculated the solute distribution under the conditions used in Fig. 2, i.e., the long-time solute distribution for a stable interface. Under these steady-state conditions $\tilde{W} = 0$, and (18) and (19) for the solute distribution at the interface may be written

$$C_I = (C_\infty/k) + G_C(2/\pi)\int_0^\infty d\omega \cos\omega x\, \tilde{W}\{(\omega^* - V/D)/(\omega^* - Vp/D)\}. \tag{20}$$

When $\omega \gg (V/2D)$, the quantity $[(\omega^* - V/D)/(\omega^* - Vp/D)] \cong 1$. Thus, when the major contribution to the integral in (20) occurs for values of $\omega$ such that $\omega \gg (V/2D)$, the solute distribution is given by

$$C_I \cong (C_\infty/k) + G_C W. \tag{21}$$

It may also be shown* that for x = 0 the above expression is an upper bound (if $G_C < 0$) for the solute concentration. The solute distribution was calculated from (20) for the parameters used in Fig. 2. For these parameters, it was found that (21) represents a reasonable approximation to the solute distribution. For example, the incremental solute concentration at x = 0 calculated from (21) is at most in error by 8%. Thus, the solute distribution is approximately the same as the interface shape shown in Fig. 2, and there is no need to include a separate figure. The maximum solute concentration occurs at x = 0; for $\Sigma = 2.0, 1.5, 1.0, 0.94, 0.92, 0.90$, the quantity $\{[C_I(x=0) - C_\infty/k]/s(C_\infty/k)\}$ is equal to 0.013, 0.020, 0.059, 0.093, 0.124, and 0.248, respectively. For $\Sigma = \infty$, $G_C = 0$ and $C_I = C_\infty/k$. Although the dependence of the solute distribution on x is correctly given by Fig. 2, the actual magnitude of the incremental solute concentration, i.e., $(C_I - C_\infty/k) \cong G_C W$, is given by multiplying W by $G_C$. Since $G_C$ depends on $\Sigma$, each curve in Fig. 2 will be multiplied by a different factor. The concentration in the solid at the interface is given by $kC_I$; since for these calculations $\tilde{W} = 0$, the concentration in the solid is independent of y and the concentration as a function of x is also given by $kC_I$.

It is clear from these results that solute segregation (and even an oscillatory solute distribution) does not indicate morphological instability.

## 5. Discussion

As in the case of a pure substance [5], the preceding two-dimensional calculations are easily extended to three dimensions. Briefly, the position of the solid–liquid interface is now described by

$$y = W(x, z, t) = W_I(x, t) + W_0(x, z, t),$$

where $W(x, t)$ is the solution of the two-dimensional problem [eq. (9)] and $W_0(x, z, t)$ is the

* This conclusion follows because $\tilde{W} < 0$ and $(\omega^* - V/D)/[\omega^* - (V/D)p] \le 1$ for all $\omega$.

interface shape in the absence of a grain boundary [2]. Thus, $W_0(x, z, t)$ is given by [11]

$$W_0(x,z,t) = \pi^{-2}\int_0^\infty d\omega \int_{-\infty}^{\infty} d\omega' \cos(\omega x)\exp(i\omega' z)\exp\{[f(\omega^2+\omega'^2)^{\frac{1}{2}}]\,t\}\,\tilde{\tilde{W}}_0(\omega,\omega',0), \quad (22)$$

where $\tilde{\tilde{W}}_0(\omega, \omega', 0)$ is defined by the above equation with $t = 0$, i.e., $\tilde{\tilde{W}}_0(\omega, \omega', 0)$ is a two-dimensional Fourier transform of $W_0(x, z, 0)$; $f[(\omega^2 + \omega'^2)^{1/2}]$ is identical to $f(\omega)$ except that $\omega$ has been replaced by $(\omega^2 + \omega'^2)^{1/2}$. As a consequence of the above equation (and linear stability theory), instabilities having z-dependence, i.e., of the form $\cos(\omega x)[C_1 \sin(\omega' z) + C_2 \cos(\omega' z)]$, are not influenced by an isolated grain boundary lying in the $x = 0$ plane.

As noted in Section 3, the presence of a grain boundary groove does not change $f(\omega)$, and hence does not change the stability−instability criterion, i.e., the demarcation between stability and instability is the same whether or not a grain boundary groove is present. Under conditions of instability the main effect of a grain boundary groove on the long time behavior of the interface is to provide an initial perturbation of the interface of the order of

$$[2T_M\Gamma s/S(\omega_0)(\pi D\tau)^{1/2}]$$

in amplitude. If $\omega_0 \gg (V/2D)$,

$$[2T_M\Gamma s/S(\omega_0)(\pi D\tau)^{1/2}]$$

is of the order of $s/\omega_0$. For example, using the same parameters as in Fig. 4, $s/\omega_0 \cong 40$ sec μm. If the perturbation due to the grain boundary is large compared to other perturbations on the interface, then instabilities would first be observed at grain boundaries.

Under conditions of stability, after initial transients have decayed, the interface attains a shape which is independent of time. If a grain boundary was not present, this would be a flat interface. In the presence of a grain boundary, as illustrated in Fig. 2, the shape depends strongly on the proximity to the stability−instability demarcation. Near instability, the stable shape is oscillatory and exhibits the same features as an unstable shape. In fact, it is only in the long-time behavior that one can distinguish between stability and instability, i.e., perturbations on an unstable interface increase exponentially with time whereas the shape of a stable interface becomes independent of time. As can be seen from a comparison of Fig. 3 (stable interface) and Fig. 4 (unstable interface), for small times it is not possible to distinguish between stability and instability.

We now discuss the experimental observations [6-8] of instabilities near grain boundaries. In a study of the solidification of dilute alloys of Sb in Pb, Morris and Winegard [6, 12] studied interface stability by observing the Sb microsegregation by an etching technique. They found that segregation occurred at defects, such as grain boundaries and the container walls, at lower growth velocities than the velocity, $V^*$, required for segregation in a defect-free region of crystal. Segregation near high-angle boundaries and the container wall occurred at velocities of about $0.8\ V^*$, whereas segregation near low-angle boundaries (striations) occurred at about $0.98\ V^*$. Using the relation $V = 2(\bar{k}/L_V)G_L/(I\Sigma - 1)$ and the same parameters as in Figs. 2-4, we find that $V^*$, $0.98\ V^*$, and $0.80\ V^*$ correspond to $\Sigma = 0.893$, $0.910$, and $1.104$, respectively. From the calculations of Section 4, segregation always occurs at grain boundaries, the amount increasing as $\Sigma$ decreases and s increases. From the discussion in the preceding section and Fig. 2, it is clear that for $\Sigma = 0.91$ there is an oscillating solute distribution, whereas for $\Sigma = 1.1$, there is a maximum in solute concentration at the grain boundary followed by a weak minimum at some finite distance from the boundary. A quantitative test of the theory, however, would require measurement of the solute distribution. The calculated wavelength (distance between the grain boundary groove and the first minimum in the interface shape) of approximately 240 $\mu$m is in good agreement with the measured elongated cell spacings.

Schaefer and Glicksman [7] observed the interface shape during solidification of succinonitrile, which contained small amounts of impurities. Oscillatory behavior was first observed at grain boundaries. At later times, dendritic growth was observed indicating that the interface was morphologically unstable. Since it appears extremely likely that the amplitude of the perturbation due to the grain boundary is large compared to the amplitude of other perturbations of the interface, the observation of instability at grain boundaries prior to general interface instability is to be expected.

Both Morris and Winegard [6] and Schaefer and Glicksman [7] observed oscillatory behavior along a direction parallel to the grain boundary (i.e., of the form $\cos \omega' z$). In the case of succinonitrile, these oscillations developed after the oscillatory behavior perpendicular to the grain boundary (i.e., of the form $\cos \omega x$) had developed. The present linear stability theory for an isolated grain boundary does not explain these observations since perturbations of the form $\cos \omega' z$ do not interact with the grain boundary. Thus, the growth of $\cos \omega' z$ perturbations should be the same whether they are near or far away from the grain boundary, in contradiction to the experimental observations. A proper explanation would appear to require consideration of nonlinear effects and the interaction of defects. For example, Schaefer and Glicksman observed that trijunctions were more effective than grain boundaries in initiating interface instability. Finally, it is evident from the work of Morris and Winegard that crystalline anisotropy should be accounted for in any detailed analysis of interface morphology.

Sato and Ohira [8] investigated the transition from a planar to cellular structure in alloys of Cu in Al. Several of their findings are in agreement with the present theory. Grain boundary segregation occurred prior to the breakdown of the interface. As instability was approached, the grain boundary grooves deepended (see Fig. 2).

## Literature Cited

1. W. W. Mullins and R. F. Sekerka, J. Appl. Phys., 34:323 (1963).
2. W. W. Mullins and R. F. Sekerka, J Appl. Phys., 35:444 (1964).
3. R. F. Sekerka, J. Crystal Growth, 3/4:71 (1968).
4. R. F. Sekerka, Four Lectures on Morphological Stability, P. Hartman, ed., North-Holland, Amsterdam, to be published.
5. S. R. Coriell and R. F. Sekerka, J. Crystal Growth, 19:90 (1973).
6. L. R. Morris and W. C. Winegard, J. Crystal Growth, 5:361 (1969).
7. R. J. Schaefer and M. E. Glicksman, Metal. Trans., 1: 1973 (1970).
8. T. Sato and G. Ohira, Trans. Japan Inst. Metals, 12:285 (1971).
9. V. V. Voronkov, Soviet Phys. – Solid State 6:2378 (1965).
10. R. F. Sekerka, J. Appl. Phys., 36:264 (1965).
11. R. F. Sekerka, J. Crystal Growth, 10:239 (1971).
12. L. R. Morris, Ph. D. thesis, University of Toronto (1967).
13. R. F. Sekerka, In: Crystal Growth, H. S. Peiser, ed., Pergamon, Oxford (1967) p. 691.
14. R. F. Sekerka, J. Phys. Chem. Solids, 28:983 (1967).

# SIZE BEHAVIOR OF CRYSTAL INTERACTING BY DIFFUSION DURING PHASE TRANSITION


## B. Ya. Lyubov and V. V. Shevelev

*Central Ferrous Metallurgy Research Institute, Moscow*


Crystals interact by diffusion during phase transitions, which substantially complicates the size as a function of time. Special mathematical techniques are needed to deduce the time dependence, and previously such problems have usually been handled by computing methods or else by neglecting the effects of phase-boundary movements on the concentration patterns [1-3].

Recently a method has been developed [4-6] for defining the solution as a series in terms of a small parameter. Here this method is extended, with the zeroth approximation as the solution for a fixed phase boundary. Two applications are given.

Consider the growth of two crystals differing in physical nature, e.g., austenite and cementite, which interact by diffusion in a liquid solution supersaturated with respect to both phases. The concentrations are equal to the equilibrium values at the surfaces, which grow in opposite directions; we neglect the diffusion within the crystals. A major point is that the process is nonstationary, which is usually neglected [7]. The equations are formulated as follows:

$$\frac{\partial^2 C}{\partial x^2}=\frac{\partial C}{\partial \tau}, \tag{1}$$

$$C(x,\ 0)=C_0, \tag{2}$$

$$C(y_\alpha,\ \tau)=C_{\alpha e} \tag{3}$$

$$C(y_\beta,\ \tau)=C_{\beta e}, \tag{4}$$

$$(C_{\alpha e}-C_\alpha)\ \frac{dy_\alpha}{d\tau}=-\left(\frac{\partial C}{\partial x}\right)_{x=y_\alpha}, \tag{5}$$

$$(C_{\beta e}-C_\beta)\frac{dy_\beta}{d\tau}=-\left(\frac{\partial C}{\partial x}\right)_{x=y_\beta}. \tag{6}$$

Here $x = x'/l_0$ is a dimensionless coordinate, $l_0$ is half the distance between the center of the crystals of the $\alpha$ and $\beta$ phases, $\tau = Dt/l_0^2$ is the dimensionless time, D is the diffusion coefficient of a solute in the liquid solution, $C_\alpha$ and $C_\beta$ are the compositions of the $\alpha$ and $\beta$ crystals respectively, $C_{\alpha e}$ and $C_{\beta e}$ are the equilibrium concentrations at the surfaces of the $\alpha$ and $\beta$ crystals respectively, $y_\alpha(\tau)$ and $y_\beta(\tau)$ are variable dimensions of the corresponding $\alpha$ and $\beta$

crystals, and conditions (5) and (6) represent mass balance at the moving surfaces (phase interfaces). The functions $y_\alpha(\tau)$ and $y_\beta(\tau)$ satisfy the following initial conditions:

$$y_\alpha(0)=0; \quad y_\beta(0)=2. \tag{7}$$

The solution to (2)-(6) is sought in the form

$$C(x,\tau)=C_0+\sum_{\kappa=0}^{\infty}\frac{(x-1)^{2\kappa}}{(2k)!}\frac{d^\kappa}{d\tau^\kappa}A(\tau)+\sum_{\kappa=0}^{\infty}\frac{(x-1)^{2\kappa+1}}{(2k+1)!}\frac{d^\kappa}{d\tau^\kappa}B(\tau). \tag{8}$$

The functions $A(\tau)$ and $B(\tau)$ satisfy the following initial conditions:

$$\lim_{\tau\to 0}\frac{d^\kappa}{d\tau^\kappa}A(\tau)=0; \quad \lim_{\tau\to 0}\frac{d^\kappa}{d\tau^\kappa}B(\tau)=0; \quad \kappa=0,\ 1,\ 2,\ldots \tag{9}$$

which follow from (2); when (3)-(6) are met, we get the following system of equations for the unknown functions $A(\tau)$, $B(\tau)$, $y_\alpha(\tau)$, $y_\beta(\tau)$:

$$\sum_{\kappa=0}^{\infty}\frac{(y_\alpha\lambda-1)^{2\kappa}}{(2k)!}\frac{d^\kappa}{d\tau^\kappa}A(\tau)+\sum_{\kappa=0}^{\infty}\frac{(y_\alpha\lambda-1)^{2\kappa+1}}{(2k+1)!}\frac{d_\kappa}{d\tau_\kappa}B(\tau)=(C_{\alpha e}-C_0), \tag{10}$$

$$\sum_{\kappa=0}^{\infty}\frac{(y_\beta-1)^{2\kappa}}{(2k)!}\frac{d^\kappa}{d\tau^\kappa}A(\tau)+\sum_{\kappa=0}^{\infty}\frac{(y_\beta-1)^{2\kappa+1}}{(2k+1)!}\frac{d^\kappa}{d\tau^\kappa}B(\tau)=(C_{\beta e}-C_0), \tag{11}$$

$$(C_{\alpha e}-C_\alpha)\frac{dy_\alpha}{d\tau}=-\lambda\sum_{\kappa=1}^{\infty}\frac{(y_\alpha\lambda-1)^{2\kappa-1}}{(2k-1)!}\frac{d^\kappa}{d\tau^\kappa}A(\tau)-\lambda\sum_{\kappa=0}^{\infty}\frac{(y_\alpha\lambda-1)^{2\kappa}}{(2k)!}\frac{d^\kappa}{d\tau^\kappa}B(\tau), \tag{12}$$

$$(C_{\beta e}-C_\beta)\frac{dy_\beta}{d\tau}=-\lambda\sum_{\kappa=1}^{\infty}\frac{(y_\beta-1)^{2\kappa-1}}{(2k-1)!}\frac{d^\kappa}{d\tau^\kappa}A(\tau)-\lambda\sum_{\kappa=0}^{\infty}\frac{(y_\beta-1)^{2\kappa}}{(2k)!}\frac{d^\kappa}{d\tau^\kappa}B(\tau), \tag{13}$$

where $\lambda$ is a small parameter. The functions $A(\tau)$, $B(\tau)$, $y_\alpha(\tau)$, and $y_\beta(\tau)$ are defined as power series in $\lambda$:

$$A(\tau)=A_0(\tau)+\lambda A_1(\tau)+\lambda^2A_2(\tau)+\ldots+\lambda^nA_n(\tau)+\ldots, \tag{14}$$

$$B(\tau)=B_0(\tau)+\lambda B_1(\tau)+\lambda^2B_2(\tau)+\ldots+\lambda^nB_n(\tau)+\ldots, \tag{15}$$

$$y_\alpha(\tau)=\varphi_0(\tau)+\lambda\varphi_1(\tau)+\lambda^2\varphi_2(\tau)+\ldots+\lambda^n\varphi_n(\tau)+\ldots, \tag{16}$$

$$y_\beta(\tau)=2-\lambda\psi_0(\tau)-\lambda^2\psi_1(\tau)-\lambda^3\psi_2(\tau)-\ldots-\lambda^{n+1}\psi_n(\tau)+\ldots \tag{17}$$

The zeroth approximation corresponds to the assumption that $y_\alpha \ll 1$ and $(2-y_\beta) \ll 1$; we substitute the series of (14)-(17) into (10)-(13) and equate the coefficients to identical powers of $\lambda$ to get equations defining $A(\tau)$, $B(\tau)$, $y_\alpha(\tau)$, and $y_\beta(\tau)$ in the zeroth, first, and higher approximations; in the zeroth approximation we have

$$\sum_{\kappa=0}^{\infty}\frac{1}{(2k)!}\frac{d^\kappa}{d\tau^\kappa}A_0(\tau)-\sum_{\kappa=0}^{\infty}\frac{1}{(2k+1)!}\frac{d^\kappa}{d\tau^\kappa}B_0(\tau)=(C_{\alpha e}-C_0), \tag{18}$$

$$\sum_{\kappa=0}^{\infty}\frac{1}{(2k)!}\frac{d^\kappa}{d\tau^\kappa}A_0(\tau)+\sum_{\kappa=0}^{\infty}\frac{1}{(2k+1)!}\frac{d^\kappa}{d\tau^\kappa}B_0(\tau)=(C_{\beta e}-C_0), \tag{19}$$

$$(C_{\alpha e}-C_\alpha)\frac{d\varphi_0}{d\tau}=\sum_{\kappa=1}^{\infty}\frac{1}{(2k-1)!}\frac{d^\kappa}{d\tau^\kappa}A_0(\tau)-\sum_{\kappa=0}^{\infty}\frac{1}{(2k)!}\frac{d^\kappa}{d\tau^\kappa}B_0(\tau), \tag{20}$$

$$(C_{\beta e}-C_\beta)\frac{d\psi_0}{d\tau}=\sum_{\kappa=1}^{\infty}\frac{1}{(2k-1)!}\frac{d^\kappa}{d\tau^\kappa}A_0(\tau)+\sum_{\kappa=0}^{\infty}\frac{1}{(2k)!}\frac{d^k}{d\tau^k}B_0(\tau). \tag{21}$$

We apply a Laplace–Carson transformation to (18)-(21) to get in transform space that

$$\bar{A}_0(P)=\frac{\{(C_{\alpha e}-C_0)+(C_{\beta e}-C_0)\}}{2\cosh\sqrt{P}}, \tag{22}$$

$$\bar{B}_0(P)=\frac{(C_{\beta e}-C_{\alpha e})\sqrt{P}}{2\sinh\sqrt{P}}, \tag{23}$$

$$\bar{\varphi}_0(P)=\frac{\{(C_{\alpha e}-C_0)+(C_{\beta e}-C_0)\}}{2(C_{\alpha e}-C_\alpha)}\frac{\tanh\sqrt{P}}{\sqrt{P}}-\frac{(C_{\beta e}-C_{\alpha e})}{2(C_{\alpha e}-C_\alpha)}\frac{\coth\sqrt{P}}{\sqrt{P}}, \tag{24}$$

$$\bar{\psi}_0(P)=\frac{(C_{\alpha e}-C_0)+(C_{\beta e}-C_0)}{2(C_{\beta e}-C_\beta)}\frac{\tanh\sqrt{P}}{\sqrt{P}}+\frac{(C_{\beta e}-C_{\alpha e})}{2(C_{\beta e}-C_\beta)}\frac{\coth\sqrt{P}}{\sqrt{P}}. \tag{25}$$

We convert to the original via the second expansion theorem to get expressions for $A_0(\tau)$, $B_0(\tau)$, $\varphi_0(\tau)$, and $\psi_0(\tau)$; as we are subsequently interested only in the behavior of the size, we give the results only for the functions $y_\alpha(\tau)$ and $y_\beta(\tau)$, and in the zeroth approximation we have

$$\varphi_0(\tau)=\frac{1}{2}(\lambda_\alpha+\lambda_\beta\lambda_{\beta\alpha})\omega_1(\tau)-\frac{1}{2}(\lambda_\beta\lambda_{\beta\alpha}-\lambda_\alpha)\omega_2(\tau), \tag{26}$$

$$\psi_0(\tau)=\frac{1}{2}(\lambda_\beta+\lambda_\alpha/\lambda_{\beta\alpha})\omega_1(\tau)+\frac{1}{2}\left(\lambda_\beta-\frac{\lambda^\alpha}{\lambda_{\beta\alpha}}\right)\omega_2(\tau), \tag{27}$$

where

$$\omega_1=1-\frac{8}{\pi^2}\sum_{\kappa=0}^{\infty}\frac{1}{(2k+1)^2}e^{-\frac{(2k+1)^2\pi^2\tau}{4}}, \tag{28}$$

$$\omega_2=\tau+{}^1/_3-\frac{2}{\pi^2}\sum_{\kappa=1}^{\infty}\frac{1}{k^2}e^{-\pi^2\kappa^2\tau}, \tag{29}$$

and

$$\lambda_\alpha=\frac{(C_{\alpha e}-C_0)}{(C_{\alpha e}-C_\alpha)};\quad \lambda_\beta=\frac{(C_{\beta e}-C_0)}{(C_{\beta e}-C_\beta)};\quad \lambda_{\beta\alpha}=\frac{(C_{\beta e}-C_\beta)}{(C_{\alpha e}-C_\alpha)}. \tag{30}$$

Parameters $\lambda_\alpha$ and $\lambda_\beta$ characterize the growth rates of the $\alpha$ and $\beta$ crystals in the absence of diffusion interaction, while $\lambda_{\beta\alpha}$ characterizes in a sense the ratio of the growth rates. Steps similar to those used in the zeroth approximation give us, for example, the following expression for $\varphi_1(\tau)$:

$$\varphi_1(\tau)=\frac{1}{4}\frac{d}{d\tau}\int_0^\tau\{\varphi_0^2(\eta)+\lambda_{\beta\alpha}\psi_0^2(\eta)\}\,\vartheta_2[0,(\tau-\eta)]d\eta+\frac{1}{4}\frac{d}{d\tau}\int_0^\tau\{\varphi_0^2(\eta)-\lambda_{\beta\alpha}\psi_0^2(\eta)\}\,\vartheta_3[0,(\tau-\eta)]d\eta, \tag{31}$$

where

$$\vartheta_2(0,x)=2\sum_{\kappa=0}^{\infty}\exp\left[-\frac{\pi^2(2k+1)^2x}{4}\right],\qquad \vartheta_3(0,x)=1+2\sum_{\kappa=1}^{\infty}\exp(-\pi^2k^2x).$$

Figure 1 shows graphs of $y_\alpha(\tau)$ and $y_\beta(\tau)$ for the crystallization of an Fe–C eutectic at about 1126°C; the abscissa (bottom scale) is logarithmic for $\tau$ large, whereas for $\tau$ small (upper scale) it is linear. The nonstationary feature of the process is particularly important at the early stages. The steady-state solution indicates that there is slower growth in the earlier and middle stages but more rapid growth at the end. The diffusion interaction makes itself felt at the later stages, where the growth rate is substantially higher than that for a single

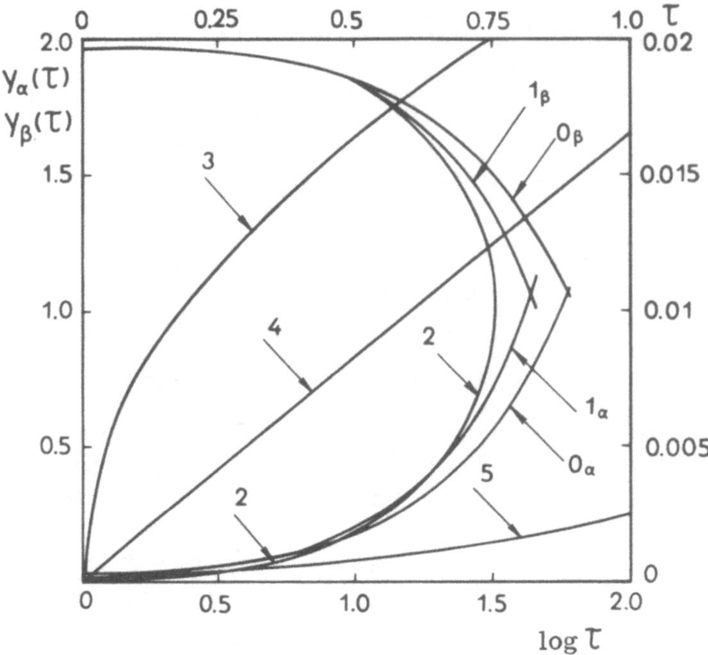


Fig. 1. Curves for $y_\alpha(\tau)$ and $y_\beta(\tau)$, where $0_\alpha$, $1_\alpha$, $0_\beta$, and $1_\beta$ are the zeroth and first approximations for large t for the $\alpha$ and $\beta$ phases respectively; 2 represents the stationary solution; 3 and 4 represent the solutions for small $\beta$ of crystal of the $\alpha$ phase in the nonstationary and stationary approximations respectively; and 5 represents the growth of a single crystal of the $\alpha$ phase.

crystal. The point of encounter $y_{\alpha c}$ is readily found from the law of conservation of matter:

$$y_{\alpha c}=2\,\frac{(C_\beta-C_0)}{(C_\beta-C_\alpha)}. \tag{32}$$

The steady-state solution gives

$$y_{\alpha st}=2\,\frac{(C_{\beta e}-C_\beta)}{\{(C_{\beta e}-C_{\alpha e})+(C_\alpha-C_\beta)\}}. \tag{33}$$

It follows from (32) and (33) that the steady-state solution is applicable if $C_{\beta e}\approx C_{\alpha e}\approx C_0$, i.e., when the concentration differences in the liquid and the initial supersaturations with respect to the two phases are small. Also, the solute diffuses only small distances from the completed crystal to the enriched one at the end of the process, and the concentration pattern between the crystals is then closely represented by a straight line, which corresponds to the steady-state growth condition.

As the second example we consider the nonstationary dissolution of planar crystals interacting by diffusion. The interaction is represented by zero flux through a surface passing halfway between the crystals. The mathematical formulation is

$$\frac{\partial^2 C_\beta}{\partial x^2}=\frac{1}{\gamma}\,\frac{\partial C_\beta}{\partial \tau}, \tag{34}$$

$$C_\beta(x,\ 0)=C_{\beta 0}, \tag{35}$$

$$C_\beta(y,\ \tau)=C_{\beta\alpha}, \tag{36}$$

$$\left(\frac{\partial C_\beta}{\partial x}\right)_{x=0}=0, \tag{37}$$

$$\frac{\partial^2 C_\alpha}{\partial x^2}=\frac{\partial C_\alpha}{\partial \tau}, \tag{38}$$

$$C_\alpha(x,\ 0)=C_{\alpha 0}, \tag{39}$$

$$C_\alpha(y,\ \tau)=C_{\alpha\beta}, \tag{40}$$

$$(C_{\beta\alpha}-C_{\alpha\beta})\frac{dy}{d\tau}=\left(\frac{\partial C_\alpha}{\partial x}\right)_{x=y^+}-\gamma\left(\frac{\partial C_\beta}{\partial x}\right)_{x=y^-}, \tag{41}$$

$$\left(\frac{\partial C_\alpha}{\partial x}\right)_{x=1}=0. \tag{42}$$

Here $x = x'/l_0$ is the dimensionless coordinate, and $l_0$ is half the distance between the middles of the crystals, $\tau = D_\alpha t/l_0^2$ is dimensionless time, $D_\alpha$ is the diffusion coefficient of the solute in the matrix (the $\alpha$ phase), $\gamma = D_\beta/D_\alpha$, $D_\beta$ is the diffusion coefficient in the crystal, $C_{\alpha 0}$ and $C_{\beta 0}$ are the initial concentrations in the matrix and in the crystal respectively, $C_{\beta\alpha}$ and $C_{\alpha\beta}$ are the equilibrium concentrations at the interface in the crystal and matrix respectively, and $y(\tau)$ is the coordinate of the phase interface.

The solution is sought in the form

$$C_\alpha(x,\ \tau)=C_{\alpha 0}+\sum_{\kappa=0}^{\infty}\frac{(1-x)^{2\kappa}}{(2k)!}\frac{d^\kappa}{d\tau^\kappa}A(\tau), \tag{43}$$

$$C_\beta(x,\ \tau)=C_{\beta 0}+\sum_{\kappa=0}^{\infty}\frac{x^{2\kappa}}{(2k)!\gamma^\kappa}\frac{d^\kappa}{d\tau^\kappa}B(\tau). \tag{44}$$

It is readily seen that conditions (37) and (42) are satisfied; by virtue of (35) and (39), the functions $A(\tau)$ and $B(\tau)$ satisfy condition (9), while (36), (40), and (41) can be satisfied, as in the first case, by a system of equations for $A(\tau)$, $B(\tau)$, and $y(\tau)$:

$$(C_{\beta\alpha}-C_{\beta 0})=\sum_{\kappa=0}^{\infty}\frac{y^{2\kappa}}{(2k)!\gamma^\kappa}\frac{d^\kappa}{d\tau^\kappa}B(\tau), \tag{45}$$

$$(C_{\alpha\beta}-C_{\alpha 0})=\sum_{\kappa=0}^{\infty}\frac{(1-y)^{2\kappa}}{(2k)!}\frac{d^\kappa}{d\tau^\kappa}A(\tau), \tag{46}$$

$$(C_{\alpha\beta}-C_{\beta\alpha})\frac{dy}{d\tau}=\lambda\sum_{\kappa=1}^{\infty}\frac{(1-y)^{2\kappa-1}}{(2k-1)!}\frac{d^\kappa}{d\tau^\kappa}A(\tau)+\gamma\lambda\sum_{\kappa=1}^{\infty}\frac{y^{2\kappa-1}}{(2k-1)!\gamma^\kappa}\frac{d^\kappa}{d\tau^\kappa}B(\tau), \tag{47}$$

where $\lambda$ is a small parameter, as in the previous case; $A(\tau)$ and $B(\tau)$ are sought in the form of (14) and (15), while $y(\tau)$ is sought as a series:

$$y(\tau)=y_0+\lambda\psi_0(\tau)+\lambda^2\psi_1(\tau)+\ldots+\lambda^{n+1}\psi_n(\tau)+\ldots \tag{48}$$

We equate coefficients for identical powers of $\lambda$ and perform operations similar to those in the first case to find $A(\tau)$, $B(\tau)$, and $y(\tau)$ in the zeroth, first, and higher approximations; in the

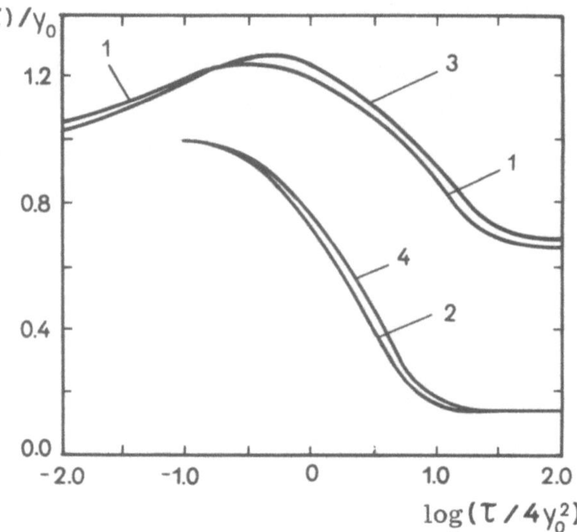


Fig. 2. Graphs for $y(\tau)$: (1) $y_0 = 0.094$; $C_{\beta 0} = 1$; $C_{\beta\alpha} = 0.75$; $C_{\alpha 0} = 0$, $C_{\alpha\beta} = 0.05$; $\gamma = 1$; (3) computer calculations for the same values of the parameters [3]; (2) $y_0 = 0.167$; $C_{\beta 0} = 1$; $C_{\beta\alpha} = 0.85$; $C_{\alpha 0} = 0$; $C_{\alpha\beta} = 0.15$; $\gamma = 1$; (4) computer calculation [3].

zeroth approximation, the motion of the phase boundary is described by

$$y(\tau)=y_0+\frac{(C_{\alpha\beta}-C_{\alpha 0})(1-y_0)+y_0(C_{\beta\alpha}-C_{\beta 0})}{C_{\alpha\beta}-C_{\beta\alpha}}-\frac{C_{\alpha\beta}-C_{\alpha 0}}{C_{\alpha\beta}-C_{\beta\alpha}}\frac{8(1-y_0)}{\pi^2}\sum_{n=0}^{\infty}\frac{1}{(2n+1)^2}\exp\left[-\frac{(2n+1)^2\pi^2\tau}{4(1-y_0)^2}\right]-$$
$$-\frac{C_{\beta\alpha}-C_{\beta 0}}{C_{\alpha\beta}-C_{\beta\alpha}}\frac{8y_0}{\pi^2}\sum_{n=0}^{\infty}\frac{1}{(2n+1)^2}\exp\left[-\frac{(2n+1)^2\pi^2\tau\gamma}{4y_0^2}\right], \tag{49}$$

where $y_0$ is half the initial size of the crystal. We apply this result to dissolution of crystals in the two-phase region, i.e., when the crystals dissolve partially. Figure 2 shows $y(\tau)$ curves constructed from (49). The data for the purpose were taken from [3]. The zeroth approximation gives good agreement with the result from finite-difference methods operated by computer [3]. We have to incorporate into (48) the higher terms $\psi_1(\tau), \psi_2(\tau), \ldots, \psi_n(\tau), \ldots$ in order to discuss the behavior in the case of complete dissolution; further, there is an effect from diffusion interaction even at the early stage of dissolution, as has been observed by experiment [8, 9]. In order to predict the conditions under which such growth may occur at the start of dissolution, we write (49) in a different form, which is convenient for analyzing the kinetics in the initial stages:

$$y(\tau)=y_0+2\sqrt{\frac{\tau}{\pi}}\left\{\frac{(C_{\alpha\beta}-C_{\alpha 0})}{(C_{\alpha\beta}-C_{\beta\alpha})}+\sqrt{\gamma}\,\frac{(C_{\beta\alpha}-C_{\beta 0})}{(C_{\alpha\beta}-C_{\beta\alpha})}\right\}-$$
$$-2\frac{C_{\alpha\beta}-C_{\alpha 0}}{C_{\alpha\beta}-C_{\beta\alpha}}\sum_{\kappa=0}^{\infty}(-1)^{\kappa}\left\{2\sqrt{\frac{\tau}{\pi}}\exp\left[-\frac{(1-y_0)(k+1)^2}{\tau\gamma}\right]-2(1-y_0)(k+1)\times\right.$$
$$\left.\times\operatorname{erfc}\left[\frac{(1-y_0)(k+1)}{\sqrt{\tau\gamma}}\right]\right\}-2\sqrt{\gamma}\frac{C_{\beta\alpha}-C_{\beta 0}}{C_{\alpha\beta}-C_{\beta\alpha}}\sum_{\kappa=0}^{\infty}(-1)^{\kappa}\left\{2\sqrt{\frac{\tau}{\pi}}\exp\left[-\right.\right.$$
$$\left.\left.-\frac{y_0^2(k+1)^2}{\tau\gamma}\right]-\frac{2y_0(k+1)}{\sqrt{\gamma}}\operatorname{erfc}\left[\frac{y_0(k+1)}{\sqrt{\tau\gamma}}\right]\right\}. \tag{50}$$

If $\alpha$ is very small we have

$$y(\tau)\approx y_0+\frac{\tau}{(C_{\beta\alpha}-C_{\alpha\beta})}2\sqrt{\frac{\tau}{\pi}}\{\sqrt{\gamma}(C_{\beta 0}-C_{\beta\alpha})-(C_{\alpha\beta}-C_{\alpha 0})\}, \tag{51}$$

which means that if $(C_{\alpha\beta}-C_{\alpha 0})\geqslant\sqrt{\gamma}\,(C_{\beta 0}-C_{\beta\alpha})$, there will be no growth at the start of dissolution. Therefore, any growth corresponds to a large difference between the initial equilibrium concentrations in the crystal and a small one in the matrix, and also to cases where $D_\beta > D_\alpha$ for large t for the $\alpha$ and $\beta$ phases respectively; 2 represents the stationary solution; 3 and 4 represents the solution for small $\tau$ of crystal of the $\alpha$ phase in the nonstationary and stationary approximations respectively; and 5 represents the growth of a single crystal of the $\alpha$ phase.

This method of calculating the kinetic behavior of the size for interacting crystals can be used in many such aspects of phase-transition theory.

## Literature Cited


1. F. V. Nolti, P. G. Shewmoh, and J. S. Foster, Trans. Met. Soc., AIME, 245:1427 (1969).
2. R. D. Lanam and R. W. Heckel, Metallurg. Trans., 2:2255 (1971).
3. R. A. Tanzilli and R. W. Heckel, Trans. Met. Soc., AIME, 242:2313 (1968).
4. B. Ya. Lyubov and N. I. Yalovoi, Izv. AN SSSR, Metally, 2:152 (1970).
5. I. P. Arkusha and B. Ya. Lyubov, Fiz. Met. Metalloved., 29:449 (1970).
6. V. V. Ios'kov and B. Ya. Lyubov, In: Abstracts for the Conference on Mechanisms and Kinetics of Crystallization [in Russian], Minsk (1971), p. 111.
7. B. Ya. Pines, Zh. Tekh. Fiz., 18:831 (1948).
8. J. R. Eifert, D. A. Chatfield, G. W. Powel, and J. Speretnak, Trans. Met. Soc., AIME, 242:66 (1968).
9. J. B. Clark, H. I. Aaronson, and M. A. Domian, Trans. Amer. Soc. Metals, 56:774 (1963).
10. B. Ya. Lyubov, The Kinetic Theory of Phase Transitions [in Russian], Metallurgiya, Moscow (1969).

# BULK SUPERCOOLING FOR GROWTH FRONT MOTION LIMITED BY THE HEAT-TRANSFER RATE


N. A. Avdonin

*Latvian State University, Riga*


We consider the motion of a phase boundary for a crystal growing from a melt on the assumption that the crystallization rate is determined only by the heat-transfer conditions in the bulk from the interface, and that the phase transition occurs at the temperature $T_e$ for phase equilibrium (Stefan problem). In the simplest formulation [1], the problem amounts to determining the temperature distribution T(x, t) and the position of the phase interface x = y(t) in a coordinate system x, t linked to the heater on the basis of the following conditions:

$$\lambda \frac{\partial^2 T}{\partial x^2} - \alpha(T - T_1(x)) = c\rho\left(\frac{\partial T}{\partial t} + v_0 \frac{\partial T}{\partial x}\right), \tag{1}$$

$$-\infty < x < \infty, \quad t > 0;$$

$$\gamma\rho(y'(t) - v_0) = \lambda\left(\left.\frac{\partial T}{\partial x}\right|_{x=y(t)+} - \left.\frac{\partial T}{\partial x}\right|_{x=y(t)-}\right), \tag{2}$$

$$T|_{x=y(t)_+} = T|_{x=y(t)_-} = T_e, \tag{3}$$

$$\lim_{x\to\pm\infty} \frac{\partial}{\partial x}(T - T_1(x)) = 0, \tag{4}$$

$$T(x, 0) = T_1(x). \tag{5}$$

Here $v_0$ is the speed of the billet, $\lambda$ is the thermal conductivity of the material,* c is specific heat, $\rho$ density, $\gamma$ latent heat of the phase transition, $\alpha$ heat-transfer coefficient, and $T_1(x)$ temperature of the resistance furnace. For simplicity we specify $T_1(x)$ as

$$T_1(x) = T_e - kx. \tag{6}$$

We assume that the temperature distribution in the billet has reached a steady state, i.e., $\lim_{t\to\infty} T(x, t) = T(x)$, $\lim_{t\to\infty} y(t) = x_1$, $\lim_{t\to\infty} y'(t) = 0$, and we solve (1)-(5) as a quasistationary problem. The solution is readily derived [1]:

$$T(x) = T_1(x) + \frac{2\beta k}{\alpha} + Lv_0 \exp(\beta(x - x_1) - \sqrt{\beta^2 + \alpha}|x - x_1|), \tag{7}$$

---

* To simplify the subsequent expressions it has been assumed that the solid and liquid phases are equal in thermal conductivity.

$$x_1 = 2\beta\alpha^{-1} + L\frac{v_0}{k}, \quad L = \frac{\gamma\rho}{2\lambda\sqrt{\beta^2+\alpha}}, \quad \beta = c\rho v_0(2\lambda)^{-1}. \tag{8}$$

We see that T(x) does not show a monotone decrease along the billet for all values of the parameter; if

$$v_0 > v_k, \quad v_k = \frac{2\lambda k}{\gamma\rho}\frac{\sqrt{\beta^2+\alpha}}{\sqrt{\beta^2+\alpha}+\beta}, \tag{9}$$

the temperature distribution takes the form shown by curve 2 of Fig. 1; point $x_2$ with temperature $T = T_e$ is not the phase-transition point, so the solution is not a solution to the Stefan problem in the classical formulation, which requires that all points having $T = T_e$ should be phase-transition points.

It can be shown that a classical solution to the problem does not exist in this case for $v_0 > v_k$; it is true that Oleinik [2] has shown that a generalized solution exists in this case, but the structure of the phase boundary for that solution has not been examined, although it would appear to be complicated. In the present instance, it is more natural to seek a possible solution in a class of smooth functions such as the solution to (7)-(8), namely one that allows supercooling. The part $x_2x_1$ should be considered as a supercooled zone in the bulk of the liquid. However, one can be certain that a supercooled zone exists only if the solution to (1)-(5) tends in the limit to the solution for the stationary case. The solution for (1)-(5) has been examined neglecting the convective term in (1). It is found that the solution in that case tends asymptotically to a limit, and therefore a supercooled zone arises at the time $t_0$ after the start in the case $v_0 > v_k$ [3]. Therefore, we have the following pattern: if the crystallization rate is slow, the crystal grows without supercooling within the bulk of the melt. If the growth rate exceeds the critical value, a supercooled zone arises in the liquid ahead of the front, and the extent of the supercooling increases with the growth rate.

This picture is not substantially altered if we assume [4] that condition (3) at the front is replaced by the kinetic equation

$$v = m\Delta T, \tag{10}$$

where m is a kinetic coefficient and $\Delta T = T_e - T$ is the supercooling.

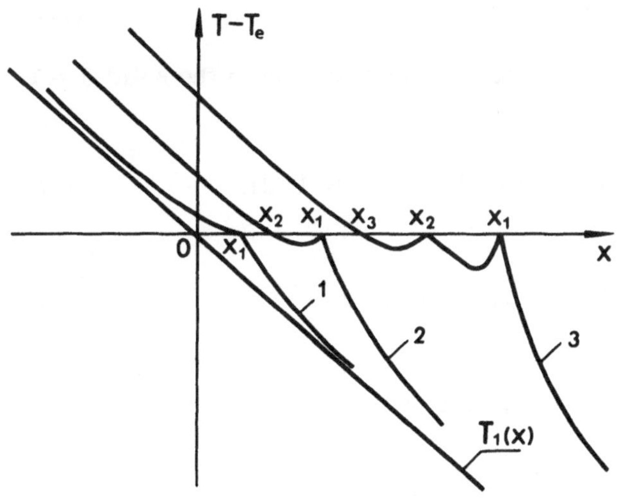


Fig. 1. Axial temperature distribution in a billet: $v_0 < v_k$; 2) $v_0 > v_k$; 3) $v_0 > v_k$, with partial crystallization at point $x_2$.

In fact, in the quasistationary case we should have for $t \to \infty$ that $v \to v_0$, and so

$$\Delta T = v_0 m^{-1}, \text{ i.e., } T|_{x=y(t)} = T_e - \Delta T = T_e - v_0 m^{-1}. \quad (3')$$

However, the solution remains as before for a given temperature at the phase interface, as (7) shows; the condition for supercooling of (9) also is unaltered. The only difference is that the temperature at the phase interface is then defined by (3'), while $x_1$ is given by

$$x_1 = v_0 m^{-1} + 2\beta\alpha^{-1} + L v_0 \cdot k^{-1}. \quad (8')$$

Normal engineering speeds of $v_0 \simeq \cdot 10^{-2}$ cm/sec (m = 10 cm/sec·deg) result in the supercooling at the front being negligibly small ($\Delta T \simeq 0.001°C$), whereas there is considerable supercooling ahead of the front.

Supercooling in the bulk of the melt disturbs the normal growth; the onset of dendritic growth has been ascribed [5] to attainment of a supercooling gradient grad $\Delta T$ ahead of the crystallization front exceeding grad $\Delta T)_{cr}$ , while the onset of bulk spontaneous crystallization is associated with a critical maximal supercooling $(\Delta T_{max})_{cr}$ in the bulk. In the case of directional crystallization, the solution of (7)-(8) defines the maximum supercooling and the temperature gradient ahead of the front:

$$\Delta T_{max} = k\left[Lv_0 - \frac{1}{\beta_2}\,(1 + \ln(L\beta_2 v_0))\right]. \quad (11)$$

$$\operatorname{grad} T = k(L\beta_2 v_0 - 1), \quad (12)$$

where

$$L = \gamma\rho(2\lambda\sqrt{\alpha + \beta^2})^{-1}; \quad \beta_{1,2} = \beta \mp \sqrt{\beta^2 + \alpha}. \quad (13)$$

The critical gradient is related to stability loss in a planar crystallization front [6, 7]. We examined the stability of a planar front by the approximate method of [6]. The quasistationary temperature distribution is described by an equation of the form of (1), which is written for the two-dimensional case. The initial state for a planar front is taken as the solution of (7)-(8). The deviation from planar form is sought as

$$x = y = x_1 + \varphi = x_1 + \delta(t) \cdot \sin\omega z, \quad (14)$$

and the initial deviation corresponds to $\delta(0) = \delta_0$; the temperature distribution in the solid ($i = 1$) and in the liquid ($i = 2$) is given by

$$T_i = T_1(x) + \frac{2\beta k}{\alpha} + Lv_0 e^{\beta_i(x - x_1)} + C_i\delta(t)\sin\omega z\, e^{-\omega_i(x - x_1)}; \quad (i = 1,\ 2). \quad (15)$$

Here

$$\omega_{1,2} = -\beta \pm \sqrt{\beta^2 + \alpha + \omega^2}; \quad C_i = k - \Gamma T_e \omega^2 - \beta_i L v_0; \quad (i = 1, 2), \quad (16)$$

where $\Gamma$ is the surface tension.

We substitute from (15) and (14) into (2) to get up to terms of the first order in $\delta$ that

$$\frac{\dot\delta}{\delta} = -\frac{2\lambda}{\gamma\rho}\sqrt{\beta^2 + \alpha + \omega^2}\left[k + \Gamma T_e \omega^2 - L\beta v_0\left(1 - \frac{\sqrt{\beta^2 + \alpha}}{\sqrt{\beta^2 + \alpha + \omega^2}}\right)\right] \quad (17)$$

The stability is governed by the sign of the expression within the brackets; stability applies for any $\omega$ if $k - L\beta v_0 > 0$, whereas the spectral range where $\omega$ is small will be unstable for $k - L\beta v_0 < 0$, so the limiting speed $v_i$ at which the instability occurs for a planar front is given by

$$v_i = \frac{k}{L\beta} = \frac{2\lambda k}{\gamma \cdot \rho} \frac{\sqrt{\beta^2 + \alpha}}{\beta}. \tag{18}$$

We compare (18) with the critical velocity of (9) to get

$$v_i = v_k \frac{\sqrt{\beta^2+\alpha}+\beta}{\beta} = v_k(1+\sqrt{1+\alpha/\beta^2}), \tag{18a}$$

so the crystallization rate corresponding to loss of stability exceeds $v_k$ by at least a factor of 2. The critical gradient is found from (12) with $v_0 = v_i$:

$$(\mathrm{grad}T)_{cr} = k\sqrt{1+\alpha/\beta^2}. \tag{19}$$

Calculations performed for gallium arsenide with a gradient 5 deg/cm in the oven gave $v_k = 0.4$ mm/min, $v_i = 1.6$ mm/min.

The critical maximal supercooling in the melt is governed by the conditions for the onset of bulk crystallization; a full discussion of the scope for bulk spontaneous crystallization would require a discussion of nucleation in the bulk of the melt, and here we merely note that a solution exists in the above formulation that involves partial crystallization at additional boundaries as in curve 3 of Fig. 1 if there is a restriction on the maximal supercooling. In that case, part of the material crystallizes at point $x_2$, and the rest crystallizes at point $x_1$ [1]. The existence of such a solution confirms that bulk crystallization can occur.

## Literature Cited

1. N. A. Avdonin, Fiz. Khim. Obrab. Mat., No. 4, 22 (1972).
2. O. A. Oleinik, Dokl. Akad. Nauk SSSR, 135:1054 (1960).
3. N. A. Avdonin, Proceedings of the Fourth All-Union Conference on Heat and Mass Transfer [in Russian], Minsk (1972), p. 156.
4. B. Ya. Lyubov and B. L. Sepozhnikov, In: Growth and Defects of Metallic Crystals [in Russian], Naukova Dumka, Kiev (1972), p. 23.
5. A. I. Landau, Fiz. Met. Metalloved., 6(1):148 (1958).
6. W. Mullins and R. Sekerka, In: Problems in Crystal Growth [Russian translation], Mir (1958), p. 106.
7. A. A. Chernov, Kristallografiya, 16:842 (1971).

# A THEORETICAL STUDY OF THE EFFECTS OF VARIOUS FACTORS ON IMPURITY ZONING IN CRYSTALS

**B. I. Birman**

*All-Union Single Crystals Research Institute, Kharkov*

One of the commoner forms of inhomogeneity in crystals is impurity banding, namely bands parallel to the crystallization front with elevated and reduced impurity contents. There are many papers [1-9] on this effect in semiconductor single crystals and in refractory oxides; recently there have also been several theoretical studies [10-12].

It has been shown [13] that impurity banding has much in common with the morphological instability of a crystallization front; this study is a continuation of [13] and deals with the theoretical effects of various factors: the temperature gradient in the melt, the degree of melt mixing (in the boundary-layer approximation), the latent heat of crystallization, and the frequency of the temperature fluctuations in the melt as regards the relative amplitude of the impurity banding $Y = m\Delta C/k\Delta T$, where $\Delta T$ is the amplitude of the temperature variation in the melt, $\Delta c$ is the amplitude of the impurity concentration nonuniformity in the crystal, and m and k are the slope of the liquidus line and the partition coefficient.

Mathematically, the problem is formulated as follows. Let the position of a planar front correspond to the plane z = 0 in a coordinate system moving with a constant speed V in the positive direction of the z axis in the steady state, while the result for periodic variation in the growth conditions is put as

$$z=\varphi(t)=\varepsilon \exp(i\omega t), \tag{1}$$

Here $\varepsilon$ is an infinitely small amplitude of position deviation in the front, $\omega = 2\pi/\tau_0$ is the frequency of the temperature fluctuation in the melt, and $\tau_0$ is the period of oscillation; the actual growth rate is given by

$$v(t)=V+i\omega\varepsilon \exp(i\omega t), \tag{2}$$

The temperatures in the crystal and melt, T'(z, t) and T(z, t), and the impurity concentration in the melt c(z, t), vary periodically with the same frequency and are given by the solution to the following system of equations:

$$\frac{\partial^2 T'}{\partial z^2}+\frac{V}{\chi_S}\frac{\partial T'}{\partial z}=\frac{1}{\chi_S}\frac{\partial T'}{\partial t} \quad \text{for} \quad -h<z<\varphi \quad \text{(crystal)}, \tag{3a}$$

$$\frac{\partial^2 T}{\partial z^2}+\frac{V}{\chi_L}\frac{\partial T}{\partial z}=\frac{1}{\chi_L}\frac{\partial T}{\partial t} \quad \text{for} \quad \varphi<z<\delta_1 \quad \text{(thermal boundary layer)}, \tag{3b}$$

$$\frac{\partial^2 c}{\partial z^2}+\frac{V}{D}\frac{\partial c}{\partial z}=\frac{1}{D}\frac{\partial c}{\partial t} \quad \text{for } \varphi<z<\delta \quad \text{(diffusion boundary layer).} \tag{3c}$$

The boundary conditions at the crystallization front take the form

$$T'|_{z=\varphi}=T|_{z=\varphi}, \tag{4a}$$

$$\varkappa_s \left.\frac{\partial T'}{\partial z}\right|_{z=\varphi}=\varkappa_l \left.\frac{\partial T}{\partial z}\right|_{z=\varphi}+Q\rho v(t), \tag{4b}$$

$$(1-k)c|_{z=\varphi}\cdot v(t)+D\left.\frac{\partial c}{\partial z}\right|_{z=\varphi}=0, \tag{4c}$$

$$T|_{z=\varphi}=T_M+mc|_{z=\varphi}. \tag{4d}$$

The boundary conditions are as follows at the outer surfaces of these regions:

$$\left.\frac{\partial T'}{\partial z}\right|_{z=-h}=\frac{\alpha_s}{\varkappa_s}[T'|_{z=-h}-T_c], \tag{5a}$$

$$T|_{z=\delta_T}=\bar{T}_0+\Delta T\exp(i\omega t), \tag{5b}$$

$$c|_{z=\delta}=\bar{c}_0. \tag{5c}$$

Here $h$, $\delta_T$, and $\delta$ are the thicknesses of the crystal, the thermal boundary layer, and the diffusion boundary layer, while $\varkappa_S$, $\varkappa_l$, $\chi_S$, and $\chi_l$ are the thermal conductivity and thermal diffusivity for the crystal and liquid respectively, D is diffusion coefficient of the impurity in the melt, $T_M$, m, $\rho$, and Q are the melting point of the pure substance, the slope of the liquidus line, the density, and the latent heat of crystallization, $\alpha_s$ and $T_c$ are the heat-transfer coefficient and the temperature of the cooling medium, $T_0$ and $c_0$ are the mean temperature and concentration in the bulk of the melt, and $\Delta T$ is the amplitude of the temperature variation in the bulk of the melt.

We solve system (3) subject to (4) and (5) to get the temperature and concentration at the front as

$$T|_{z=\varphi}=\bar{T}_\varphi+[\Delta T N_L(\xi)+\varepsilon A(\xi)]\exp(i\omega t), \tag{6}$$

$$c|_{z=\varphi}=\bar{c}_\varphi+\varepsilon G_c B(\eta,\gamma,k)\exp(i\omega t), \tag{7}$$

where the function

$$N_L(\xi)=\frac{1}{\cosh\xi\sqrt{i}+W\sinh\xi\sqrt{i}} \tag{8}$$

describes the damping and the phase shift in the thermal wave on passage through the thermal boundary layer, while the function

$$A(\xi,\theta_0)=G_L+\theta_0\frac{(-mc_0)}{D}\sqrt{\omega\chi_2}\,\Phi(\xi)+\frac{W}{\sqrt{i}}(G_S-G_L)\Phi(\xi) \tag{9}$$

describes the effects of the thermal factors on the amplitude of the temperature variation at the front arising from the displacement, and the function

$$B(\eta,\gamma,k)=\frac{\sqrt{\eta^2+i\gamma}-\eta(1+i\gamma/2\eta^2)\tanh\sqrt{\eta^2+i\gamma}}{\sqrt{\eta^2+i\gamma}-\eta(1-2k)\tanh\sqrt{\eta^2+i\gamma}} \tag{10}$$

describes the effect of impurity segregation and transport by diffusion on the amplitude of the concentration fluctuation at the front.

The following symbols have been used in (6)–(10):

$$\Phi(\xi)=\frac{\sqrt{i}\,\sinh\xi\sqrt{i}}{\cosh\xi\sqrt{i}+W\sinh\xi\sqrt{i}}, \tag{11a}$$

$$W=\sqrt{\frac{\varkappa_s^2\chi_l}{\varkappa_l^2\chi_s}}\,\frac{\tanh\xi^*\sqrt{i}\quad\xi^*\sqrt{i}+\alpha_s h/\varkappa_s}{\xi^*\sqrt{i}+(\alpha_s h/\varkappa_s)\tanh\xi^*\sqrt{i}}, \tag{11b}$$

where $G_c$, $G_s$, and $G_L$ are respectively gradients in the impurity concentration in the melt and the temperature gradients in the crystal and melt at the front, $\eta = V\delta/2D$ is the dimensionless thickness of the diffusion boundary layer, which characterizes the degree of mixing in the melt, $\gamma = \omega\delta^2/D$ is the dimensionless oscillation frequency, and $\xi=\sqrt{\frac{\omega\delta_T^2}{\chi_e}}$, $\xi^*=\sqrt{\frac{\omega h^2}{\chi_s}}$, $\theta_0=\frac{Q\rho D}{\varkappa_e(-mc_0)}$ is a parameter representing the ratio of the latent heat of crystallization to the impurity concentration in the bulk of the melt.

If the thermophysical parameters of the crystal and melt are equal, while the crystal is very thick ($\omega h^2/\chi_s \gg 1$), then W tends to 1, while the formulas for $N_L$ and $\Phi$ simplify:

$$N_L(\xi)=\exp(-\xi\sqrt{i}), \tag{12a}$$

$$\Phi(\xi)=\sqrt{i}\,\frac{1-\exp(-2\xi\sqrt{i})}{2}. \tag{12b}$$

We substitute the periodic parts of (6) and (7) into (4d) to get the expression for the amplitude:

$$\varepsilon=\frac{N_L(\xi)\Delta T}{mG_cB(\eta,\gamma,k)-A(G_L,\gamma,\theta_0)}, \tag{13}$$

and substitute (13) into (7) to get the following expression for the impurity concentration at the crystallization front on the melt side:

$$c_{z=\varphi}=c_\varphi+\frac{BN_LG_c\Delta T}{mG_cB-A}\exp(i\omega t). \tag{14}$$

We isolate the real and imaginary parts in these quantities and get the following expression for the relative amplitude of the impurity band in the crystal after various simple steps:

$$Y=\frac{m\tilde{\Delta}c}{k\Delta T}=\frac{|B|\,|N_L|}{\sqrt{\left[(1-X)-\left(1-B_1+\frac{\sqrt{\gamma}}{2\eta}\theta_0k^*\Phi_1\right)\right]^2+\left[\frac{\sqrt{\gamma}}{2\eta}\theta_0k^*\Phi_2-B_2\right]^2}}, \tag{15}$$

where

$$\Phi_1=\mathrm{Re}\,\Phi,\quad \Phi_2=\mathrm{Im}\,\Phi;\quad B_1=\mathrm{Re}\,B;\quad B_2=\mathrm{Im}\,B;$$

$$k^*=\frac{k+(1-k)\exp(-2\eta)}{1-k};\qquad |B|=\sqrt{B_1^2+B_2^2};\quad |N_L|=e^{-\xi/\sqrt{2}},$$

and $X = G_L/mG_0$ denotes the relative temperature gradient in the melt, which determines the degree of deviation from the condition for concentration-dependent supercooling to arise.

We see from (15) that the X dependence of Y is of resonant type; Y takes its maximum value when $X=X^*=B_1-\frac{\sqrt{\gamma}}{2\eta}\theta_0k^*\Phi_1$, and the value of X may be positive or negative, but it is always less than 1, i.e., the resonant conditions always correspond to concentration supercooling. The comparatively small temperature fluctuations in the melt may result in a comparatively large impurity nonuniformity in the crystal. The terms in $\frac{\sqrt{\gamma}}{2\eta}\,\theta_0\Phi_2 - B_2$ characterize the width in $\gamma$ for the resonant region; the algebraic sum of these is always positive and increases with $\gamma$.

Any detailed analysis of the effects of the various factors on the banding requires numerical calculation of Y from (15), which was performed with a Promin-2 computer; some results are shown in the figures.

Parts a and b of Fig. 1 show the relative banding amplitude $Y = m\Delta c/k\Delta T$ as a function of the relative temperature gradient in the melt $X = G_L/mG_c$ for various dimensionless temperature fluctuation frequencies for the melt $\gamma = \omega\delta^2/D$: 0.1(1); 0.2(2); 0.5(3); 1(4); 4(5); 16(6), and 64(7) for $k = 0.1$, $\theta_0 = 0$ and comparatively strong mixing ($\eta = V\delta/2D = 0.1$) and moderate mixing ($\eta = V\delta/2D = 1$); if the dimensionless frequency is small, the curve has a peak, which always lies in the region where there is concentration supercooling ($X \ll 1$). The height of this peak is largest at medium mixing rates. Oscillations at comparatively high frequencies ($\gamma \gg 1$) result in monotonic curves, with little banding. Branches of the Y(X) curves extend also to the region where there is no concentration supercooling, where any increase in temperature gradient in the melt always reduces the relative amplitude of the banding. Similar curves for the same mixing rate, but for small k, show that in the latter case the peak shifts to higher X; if $k > 1$, i.e., the impurity tends to accumulate in the crystal, the peak is very much displaced into the region of concentration supercooling, while the relative amplitude of the banding is comparatively small in the region where there is no such supercooling.

Parts a and b of Fig. 2 show $Y = m\Delta c/k\Delta T$ as a function of the dimensionless frequency $\gamma = \omega\delta^2/D$ for the following values of $\eta = V\delta/2D$: 0.1(1); 0.5(2); 1(3); 2(4); and 5(5) for $k = 0.1$

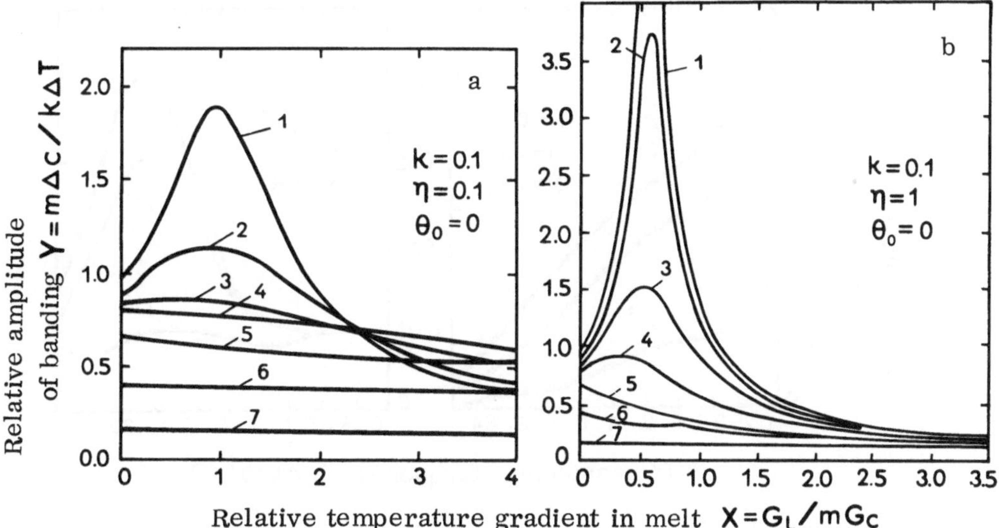


Fig. 1. (a) Relative amplitude of impurity banding in crystal $Y = m\Delta c/k\Delta T$ as a function of relative temperature gradient in melt $X = G_L/mG_c$ for dimensionless frequencies $\gamma = \omega\delta^2/D$ of the temperature fluctuation in the melt as follows: 0.1 (1); 0.2 (2); 0.5 (3); 1 (4); 4 (5); 16 (6); and 64 (7) for $k = 0.1$ and zero latent heat ($\theta_0 = 0$). The mixing parameter is $\eta = V\delta/2D = 0.1$ (b). The same but for a mixing parameter $\eta = V\delta/2D = 1$.

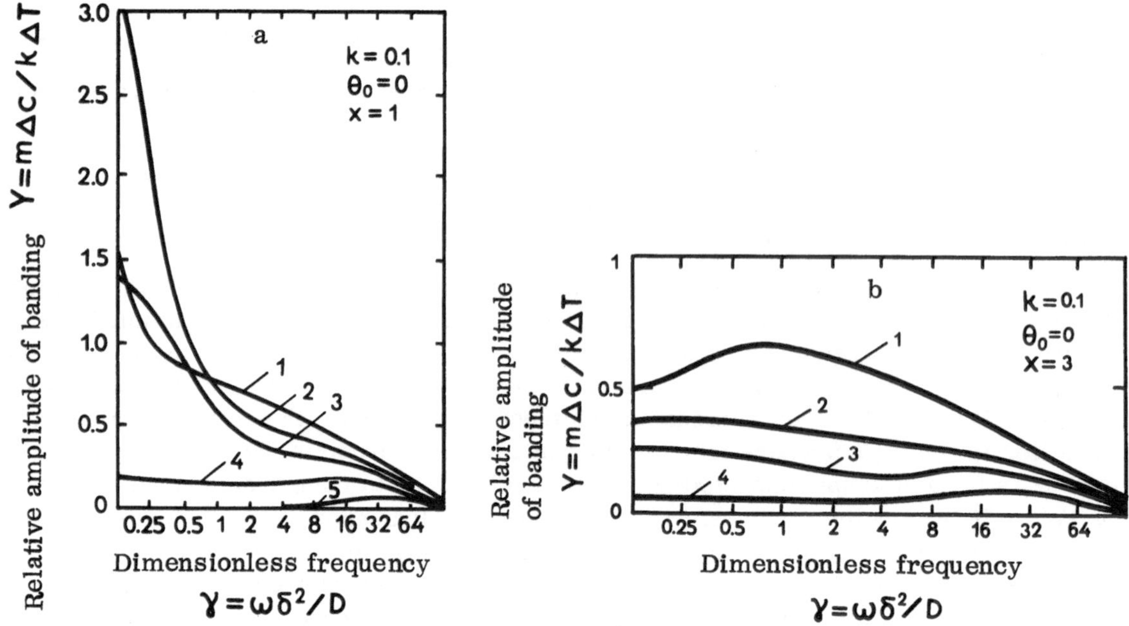


Fig. 2. (a) Relation amplitude of impurity banding $Y = m\Delta c/k\Delta T$ as a function of dimensionless temperature-fluctuation frequency in melt $\gamma = \omega\delta^2/D$ for various values of the mixing parameter $\eta = V\delta/2D$: 0.1 (1); 0.5 (2); 1 (3); 2 (4); and 5 (5) for k = 0.1 and $\theta_0 = 0$; relative temperature gradient in melt $X = G_L/mG_c = 1$ (condition for concentration supercooling). (b) The same for $X = G_L/mG_c = 3$.

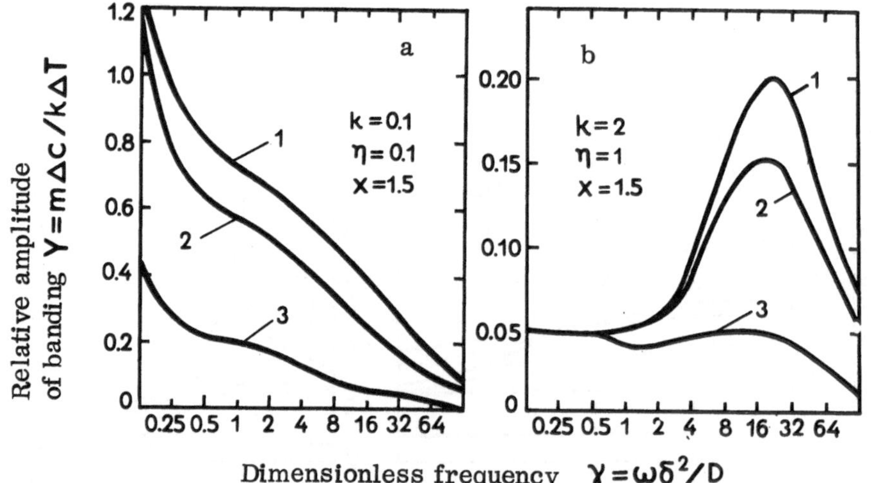


Fig. 3. (a) Amplitude of impurity banding $Y = m\Delta c/k\Delta T$ as a function of dimensionless temperature-variation frequency in the melt $\gamma = \omega\delta^2/D$ for various values of the latent heat parameter $\theta_0 = Q\rho D \cdot [\varkappa_e(-mc_0)]^{-1} = 0$ (1), 1 (2), and 10 (3) for a mixing parameter $\eta = V\delta/2D = 0.1$, k = 0.1, and a relative temperature gradient in the melt $X = G_L/mG_c = 1.5$. (b) The same for $\eta = 1$, k = 2, and X = 1.5.

and $\theta_0 = 0$ for various relative temperature gradients in the melt: Fig 2a corresponds to X=1 and Fig. 2b to X = 3. Parts a and b of Fig. 2 show that the amplitude of the banding has a complicated frequency dependence; as a rule, Y decreases monotonically toward high frequencies for X = 1, while for X > 1, i.e., in the region where there is no concentration supercooling, there is a broad peak at middle frequencies, i.e., the front reacts selectively to temperature fluctuations in this range. A similar effect of frequency has been observed elsewhere [10].

On the whole, mixing affects the banding in two different ways: on the one hand, it displaces the condition for concentration supercooling to higher temperature gradients in the melt, while on the other hand it increases the relative amplitude of the banding somewhat.

The effects of the latent heat parameter $\theta_0 = Q\rho D/\varkappa_e(-mc_0)$ on the amplitude can be seen from Fig. 3 in terms of the frequency response for $\eta = 0.1$, k = 0.1, and X = 1.5 in part a, while Fig. 3b shows results for the case $\eta = 1$, k = 0.1 and X = 1.5; in all instances, the latent heat reduces the relative amplitude of the banding, especially at high frequencies, and the same occurs if the mean impurity concentration in the melt is reduced.

We thus see that banding in crystals can be minimized by avoiding conditions where the response to temperature fluctuations is large; in particular, it is desirable to increase the temperature gradient in the melt in order to minimize the banding.

I am indebted to B. L. Timan for valuable advice during discussions on the results.

## Literature Cited

1. A. I. Landau, Fiz. Met. Metalloved., 6:148 (1958).
2. J. A. M. Dikhoff, Solid-State Electronics, 1:202 (1960).
3. H. Ueda, J. Phys. Soc. Japan, 16:61 (1961).
4. W. Bardsley, J. Boulton, and D. T. J. Hurle, Solid-State Electron., 5:395 (1962).
5. A. Müller and M. Wilhelm, Z. Naturforsch., 19a:254 (1964).
6. Yu. M. Shashkov and N. Ya. Shushlebina, Dokl. Akad. Nauk SSSR, 178:160 (1968).
7. K. Morizane, A. F. Witt, and H. C. Gatos, J. Electrochem. Soc., 113:51 (1966); 113:808 (1966); 114:788 (1967); 15:747 (1968).
8. J. C. Brice and F. A. C. Whiffin, Brit. J. Appl. Phys., 18:581 (1967).
9. B. Cockayne, J. Crystal Growth, 3/4:69 (1968).
10. D. T. J. Hurle, E. Jakeman, and E. R. Pike, J. Crystal Growth, 3/4:633 (1968).
11. Yu. P. Konakov, Fiz. Khim. Obrab. Mat., No. 2, 56 (1970).
12. Yu. M. Shashkov and V. B. Silkin, Dokl. Akad. Nauk SSSR, 187:861 (1969).
13. B. I. Birman, In: Growth and Defects of Metallic Crystals [in Russian], Naukova Dumka, Kiev (1972), p. 177.

# STEADY-STATE DENDRITIC GROWTH

G. E. Nash and M. E. Glicksman

*Transformations and Kinetics Branch, U.S. Naval Research Laboratory*
*Washington, D.C.*

## Introduction

Dendritic growth is observed quite frequently during the solidification of metals. The kinetics of the dendritic growth process is not yet well understood, and even the most rudimentary mathematical theories of this solid−liquid transformation have defied solution.

Most of the theoretical treatments in the literature employ a quasisteady-state model, based on the experimental observation that dendrites grow at a constant axial velocity and the shape of the dendrite in the neighborhood of the tip remains invariant with time. Any steady-state model must permit prediction of both the dendrite shape and growth rate as a function of the supercooling. It is well known that the isothermal model is deficient in this respect, because the growth rate remains attachment and capillarity effects were suggested for inclusion in the theory, in order to furnish a realistic description of the dendritic crystal growth process [3, 4]. A number of attempts were made to solve the boundary value problem describing steady-state nonisothermal dendritic growth [3, 4, 5]; however, no rigorous solution was found and the problem is as yet unsolved.

In this paper a new approach to the problem is presented which allows determination of both the dendrite shape and growth rate in a pure melt as a function of supercooling. The nonlinear steady-state equations are considered as operator equations in an appropriate Banach space, and the shape change induced by the nonisothermality is found by linearizing the equations about the isothermal solution. Of major importance, the conditions necessary for the existence of a solution to the linearized equations provide a theoretical limit on the growth rates. Numerical results are presented for the case where interfacial molecular attachment is very rapid, so that the growth rate is controlled by thermal diffusion and capillarity.

## Theory

### (a) Formulation of the Problem

In this analysis we employ a cylindrical coordinate system $(R, Z, \Phi)$, moving with the dendrite tip, which grows into the supercooled melt in the +Z direction at a constant velocity V. The dendritic interface (a surface of revolution) is described by rotating the curve $Z = Z(R)$ about the Z-axis; the melt is at a uniform temperature $T_0$ far from the dendrite. The problem is to determine the interface shape $Z = Z(R)$ and the range of velocities over which steady-state solutions exist.

If the thermal diffusivities in the solid and liquid are assumed to be equal, then the method of sources [6] may be used to obtain the following nonlinear integrodifferential equation (IDE) for the dendrite shape:

$$\Delta\theta+\lambda\left(\frac{z''}{[1+z'^2]^{3/2}}+\frac{(1/r)z'-D}{[1+z'^2]^{1/2}}\right)=\frac{1}{\pi}\int_0^{\infty}\exp-[z(r)-z(t)]t\,dt\times$$

$$\times\int_0^{\pi}\frac{\exp-\{r^2+t^2+[z(r)-z(t)]^2+2rt\cos\xi\}^{1/2}}{\{r^2+t^2+[z(r)-z(t)]^2+2rt\cos\xi\}^{1/2}}d\xi. \quad (1)$$

Here, r and z are the dimensionless coordinates of points on the interface, $\Delta\theta$ is the dimensionless supercooling and is less than unity, and D and $\lambda$ are dimensionless parameters. Explicit expressions for these quantities are

$$r=\frac{VR}{2\alpha},\quad z=\frac{VZ}{2\alpha},\quad \Delta\theta=\frac{c}{L}(T_m-T_0),\quad \lambda=\frac{V\gamma_{sl}c}{2\alpha\Delta S_f L},\quad D=\frac{2\alpha\Delta S_f}{\mu\gamma_{sl}}, \quad (2)$$

where $T_m$ is the equilibrium freezing temperature on a surface of zero curvature, c is the specific heat of the liquid, $\alpha$ is the thermal diffusivity, $\gamma_{sl}$ is the solid–liquid interfacial free energy, L and $\Delta S_f$ are the latent heat and entropy of fusion, respectively, and $\mu$ is the interfacial molecular attachment coefficient (linear kinetics is assumed to be operative).

Equation (1) is a mathematical statement of the interface temperature condition, i.e., the temperature developed at points on the moving interface from the distribution of latent heat sources must equal the local equilibrium freezing temperature (as influenced by interface curvature) minus a term due to molecular attachment kinetics. This equation, together with the boundary conditions $z'(0) = 0$, from symmetry, and $\lim_{r\to\infty} z(r)$ = solution of IDE with $\lambda = 0$, from thermodynamic considerations, serves to specify the problem.

### (b) Method of Solution

Since the growth rate V appears only in the parameter $\lambda$, it is natural to investigate the behavior of the IDE as a function of $\lambda$. For the special case of $\lambda = 0$, the IDE reduces to an integral equation, the solution of which is given by $z_0 = -r_0^2/2a_0$, where $a_0$ is found from the solution of the transcendental equation $\Delta\theta = a_0\exp(a_0)E_1(a_0)$. This result is equivalent to the celebrated result of Ivantsov [1] for the isothermal problem.

In the more general case $\lambda \neq 0$, it will prove convenient to reformulate the problem slightly by considering the nonisothermal solution to be generated by shifting points $(r_0, z_0)$ on the isothermal interface a distance $\nu(r_0)$ along the normal at $(r_0, z_0)$, as shown in Fig. 1. Then (1) may be regarded as a nonlinear IDE in terms of $v(r_0)$, or, equivalently, as the following operator equation in an appropriate Banach space:

$$\Delta\theta+\lambda M_1[\nu]-M_2[\nu]=0, \quad (3)$$

where $M_1$ and $M_2$ are nonlinear differential and integrodifferential operators, respectively, operating on the normal interface shift, $\nu$.

The formulation of the IDE in a Banach space setting provides a particularly effective means for investigating the character of the solutions as a function of $\lambda$, because the methods of functional analysis become available. The implicit function theorem for operators in a Banach space [7] is our primary tool.

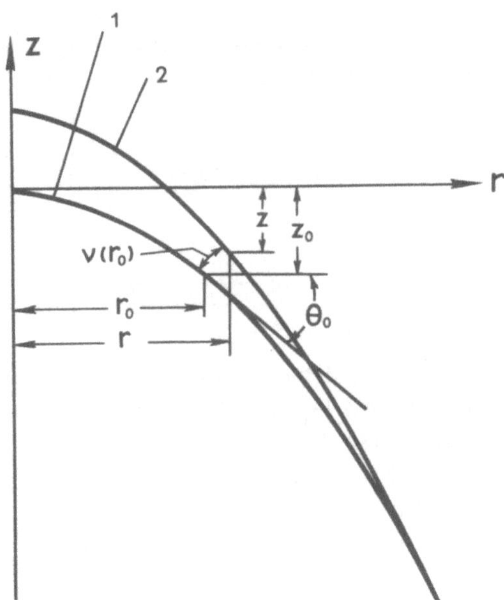


Fig. 1. Generation of the nonisothermal solution by a normal shift of points on the isothermal solution. (1) Isothermal and (2) nonisothermal dendrite surface.

Consider solutions to (3) in the neighborhood of the known solution for $\lambda = 0$. If such solutions exist, then provided that $\lambda$ is sufficiently small, and the operators possess the necessary smoothness properties, a solution can be found with the Newton–Kantorovich technique [8] or with one of its variants. By increasing $\lambda$ and repeating this process, solutions to (3) may be found for a range of $\lambda$ values. An upper bound on the range of permissible $\lambda$ values will be reached when solutions in the neighborhood of the last known solution cannot be found; this corresponds to a condition of local nonexistence. Thus, the conditions necessary for the existence (or nonexistence) of solutions in the neighborhood of a known solution provide the key to the calculation of the maximum growth rate.

These conditions can be obtained with the aid of the implicit function theorem. We state the following result without proof: If for $\lambda = \bar{\lambda}$, (3) has the known solution $\bar{\nu}$, then for any $\lambda$ sufficiently close to $\bar{\lambda}$, a unique solution to (3) exists provided that the linear operator $(L(\bar{\nu}) - \lambda I)$ has a bounded inverse. The operator $L(\bar{\nu}) = [M_1'(\bar{\nu})]^{-1} M_2'(\bar{\nu})$, where $M_1'(\bar{\nu})$ and $M_2'(\bar{\nu})$ are the Frechét derivatives of the operators $M_1$ and $M_2$ evaluated at the known solution $\bar{\nu}$, and I is the identity operator. This result implies that the condition $\lambda = \lambda_{max}$ is reached when the inverse of $[L(\nu_{max}) - \lambda_{max} I]$ becomes unbounded, and provides a rigorous means for calculating the maximum steady-state growth rate.

## Calculation of Nonisothermal Shapes and Steady-State Growth Rates

In principle the procedure outlined above can be used to calculate nonisothermal shapes and maximum growth rates regardless of the size of $\lambda$. However, if $\lambda$ is small, i.e., if the limiting nonisothermal solution differs by very little from the isothermal solution, then this procedure can be simplified. Assuming that $\lambda$ is sufficiently small, (3) can be linearized about the isothermal solution ($\lambda = 0$), and the solution to the linearized equation will provide a good approximation to the actual shape change. Note that this is actually the first step in a Newton–Kantorovich solution to (3), and the effect of subsequent iterations on the shape change for small $\lambda$ is neglected.

The maximum steady-state growth rate can also be estimated from the linearized equation by finding a value of $\lambda = \lambda^*$ such that the linearized equation has no solution. Since it can be shown that the operator $(L(0) - \lambda I)^{-1}$ becomes unbounded when $\lambda = \lambda^*$ [$L(0)$ is defined in the same way as $L(\bar{\nu})$ except that the Frechet derivatives are evaluated at the isothermal solution], it follows that $\lambda^* \to \lambda_{max}$ for $\lambda^*$ sufficiently small.

Upon application of the definition of the Frechét derivative, the linearized equation becomes

$$-\lambda\left(\frac{1+u}{a_0 u^{3/2}}\right)+\lambda\left[\frac{1}{u}(\delta\nu)''+\left(\frac{1}{r_0 u^2}-\frac{D}{u^{1/2}}\right)(\delta\nu)'+\frac{1}{a_0^2}\left(\frac{1}{u}+\frac{1}{u^3}\right)\delta\nu\right]=$$

$$=\frac{-\delta\nu}{u^{1/2}}+\int_0^\infty C^{(1)}(r_0,t)\,\delta\nu(t)dt+\int_0^\infty C^{(2)}(r_0,t)\times$$

$$\times\left\{\left[\frac{1}{u^{1/2}(t)}\left(1+\frac{1}{2a_0}\right)+\frac{1}{a_0 u^{3/2}(t)}\right]\delta\nu(t)+\frac{t}{a_0 u^{1/2}(t)}(\delta\nu(t))'\right\}dt, \qquad (4)$$

where $u = u(r_0) = 1 + (r_0/a_0)^2$, and $a_0$ is obtained from the isothermal solution. The kernels $C^{(1)}(r_0, t)$ and $C^{(2)}(r_0, t)$ are given by relatively complicated expressions which are not reproduced here, although it should be noted that $C^{(2)}(r_0, t)$ has a logarithmic singularity when $r_0 = t$. Equation (4) is a linear IDE in $\delta\nu(r_0)$, which when solved with the boundary conditions $\delta\nu(0) = \delta\nu(\infty) = 0$, yields the shape change required by the nonisothermal interface conditions.

Equation (4) with the above boundary conditions was solved numerically by a finite-difference procedure. In this method, the integrals are approximated by trapezoidal quadrature, with appropriate modifications to allow for the singularity in the kernel $C^{(2)}(r_0, t)$, and the derivatives are approximated by central differences. This numerical procedure results in a matrix equation for the shape change at a discrete set of interface points corresponding to the nodes of the difference scheme. The condition $\lambda = \lambda^*$ is manifested numerically by the occurrence of an eigenvalue in the matrix.

Values of $\lambda^*$ and shape changes have been calculated for dimensionless supercoolings ranging from 0.01 to 0.8. Extensive checks were performed to verify the convergence of the finite difference procedure. A typical result for the shape change and interface temperature is shown in Fig. 2 for $\Delta\theta = 0.8$ and $D = 0$ (infinitely rapid molecular attachment). Calculated values for $\lambda^*$ for $D = 0$ are shown in Fig. 3 as a function of $\Delta\theta$. Shown for comparison are the corresponding values of $\lambda^*$ calculated by the "modified Ivantsov" method [9], and experimental results for ice dendrites [10, 11].

## Summary and Conclusions

1. An analysis is described which permits rigorous calculation of nonisothermal dendrite shapes and maximum steady-state growth rates as a function of supercooling.

2. An approximate procedure, valid when the shape change induced by the nonisothermality is small, is presented along with some numerical results.

3. Values of $\lambda^*$ calculated by this method agree well with published data for ice dendrites at the measured value of $\gamma_{sl} = 30$ ergs/cm$^2$.

4. The maximum growth rates predicted by this method approach those obtained from the "modified Ivantsov" theory at relatively large values of normalized supercooling.

Additional work is in progress that includes a detailed mathematical investigation of the validity of the linearization procedure and additional calculations for $D > 0$ (finite interfacial molecular attachment).

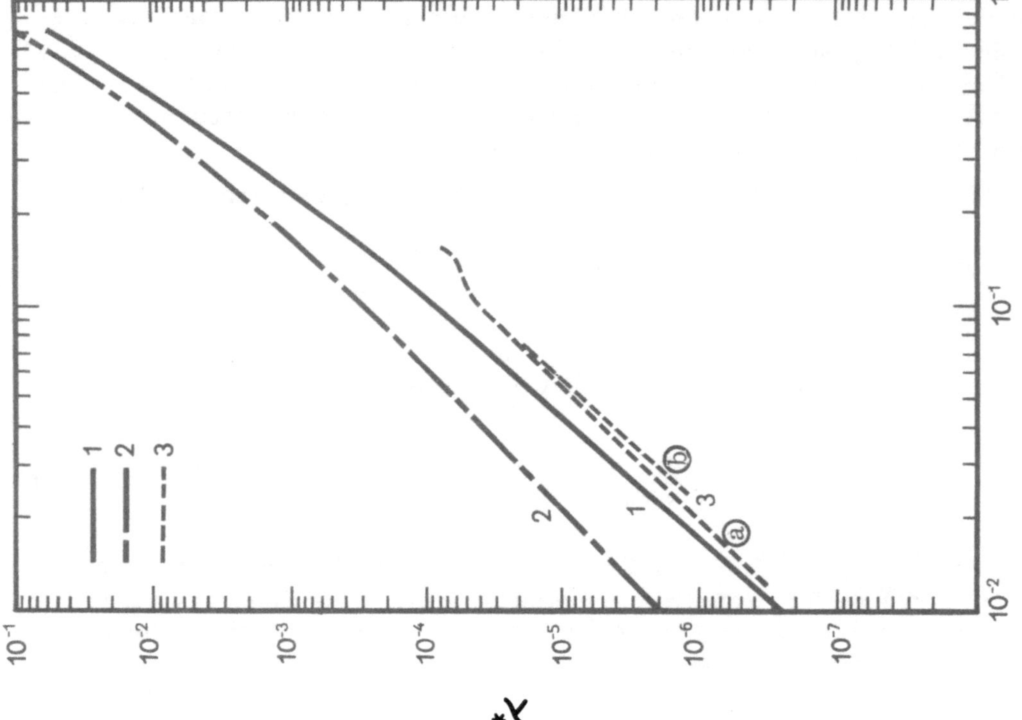


Fig. 2. Normal shift $\nu$ and interface temperature variation $\theta_i - \theta_m$. $\theta_i$ is the dimensionless interface temperature for $\Delta\theta = 0.8$. These results are valid for $\lambda < \lambda^*/10$. (1) Interface temperature; (2) normal shift.

Fig. 3. $\lambda^*$ as a function of $\Delta\theta$ for $D = 0$. (1) This analysis; (2) modified Ivantsov; (3) ice data, $\gamma_{sl} = 30$ ergs/cm$^2$. (a) Pruppacher, (b) Lindenmeyer.

## Literature Cited

1. G. P. Ivantsov, Dokl. Akad. Nauk SSSR, 58:567 (1947).
2. G. Horvay and J. W. Cahn, Acta Met., 9:695 (1961).
3. D. E. Temkin, Dokl. Akad. Nauk SSSR, 132:1307 (1960).
4. G. F. Bolling and W. A. Tiller, J. Appl. Phys., 32:2587 (1961).
5. E. Holtzman, J. Appl. Phys., 41:1460 (1970).
6. G. P. Ivantsov, Growth of Crystal Vol. 1, Consultants Bureau, New York (1959), p. 76.
7. H. Ehrmann, Enseignement Math., 9:129 (1963).
8. L. V. Kantorovich and G. P. Akilov, Functional Analysis in Normed Spaces, Pergamon Press, New York (1964).
9. M. E. Glicksman and R. J. Schaefer, J. Crystal Growth, 1:297 (1967).
10. C. S. Lindenmeyer and B. Chalmers, J. Chem. Phys., 45:2807 (1966).
11. H. R. Pruppacher, J. Colloid Interface Sci., 25:285 (1967).

# DIFFUSION-LIMITED GROWTH OF ZINC SINGLE CRYSTALS

C. N. Nanev and D. G. Ivanov

*Institute of Physical Chemistry, Bulgarian Academy of Sciences, Sofia*

## 1. Introduction

Chernov [1-3] has considered the stability of facetted forms of crystals under diffusion conditions, namely when the stability is determined by the singular anisotropic surface kinetics. He found that the crystallographic anisotropy in the growth rate may be compensated by nonuniformity in the supersaturation only to a certain extent; when the crystal exceeds a certain critical size, it no longer remains isometric. The first sign of loss of stability is the presence of recesses on close-packed faces.

A study has been made [4] of zinc single crystals grown from the vapor in the presence of hydrogen (gas pressure 10-250 mm Hg); it was found that diffusion undoubtedly plays a major part and that there is a correlation between the size of the basal face at which instability sets in and the theoretical critical size deduced by Chernov [2].

The apparatus and method previously described [4] have been used in a further study with the hydrogen replaced by the more inert argon, which is also of lower thermal conductivity. The growth of zinc single crystals in the presence of argon is similar to that in hydrogen.

## 2. Experiment

Figure 1 shows the apparatus; the single crystal C is grown in the heated glass ampoule A; to the left, in the region cooled by the Al rod, polycrystalline zinc is deposited, which then melts, and a single crystal is produced by slow cooling, which subsequently grows from the vapor state with a closely defined temperature difference $\Delta T$ between the crystal and the rest of the ampoule, which contains the vapor source [5, 6]. The equipment can be operated under vacuum (i.e., with the vapor alone) and in the presence of gas.

We used spectrally pure zinc, which was further purified by redistillation three times at a pressure of $10^{-6}$ mm Hg. The argon was also carefully purified, with traces of oxygen, carbon dioxide, and nitrogen removed by means of active nickel, caustic potash, and molten calcium. The gas was dried and passed through a trap cooled to $-80°C$.

## 3. Results and Discussion

The critical size a of an edge of a basal face was measured as follows, this size being the one above which edges and vertices grow preferentially (Fig. 2). The initial specimen was a very small zinc single crystal, whose basal face remained smooth during the growth in argon until the critical size was reached. A recess was formed at once if the initial size was greater than the critical value. The results (Table 1) have been derived by averaging over some dozens

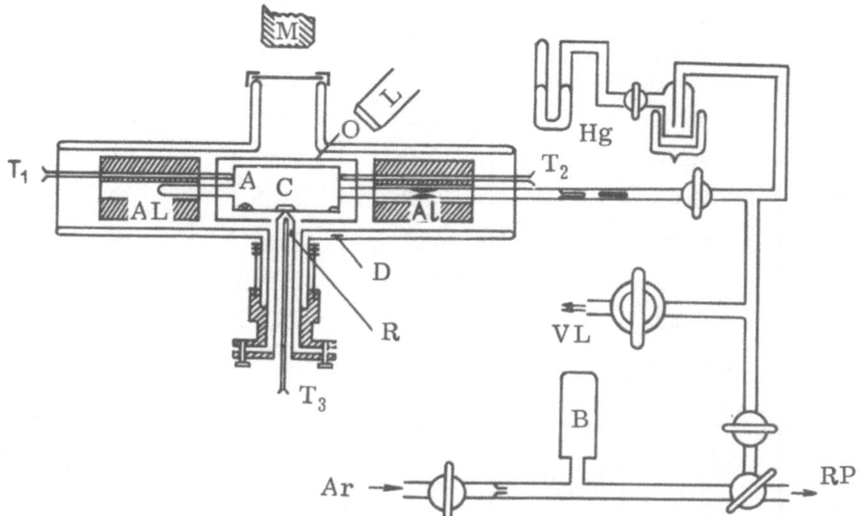


Fig. 1. The apparatus: C, crystal; A, glass ampoule; O, oven; R, cooling rod; D, Dewar vessel; Al, aluminum heater blocks; $T_1$, $T_2$, $T_3$, thermocouples; M, stereoscopic microscope; L, lamp; Ar, inert-gas supply; B, reservoir bulb; VL, vacuum line; RP, rotary pump; Hg, mercury manometer.

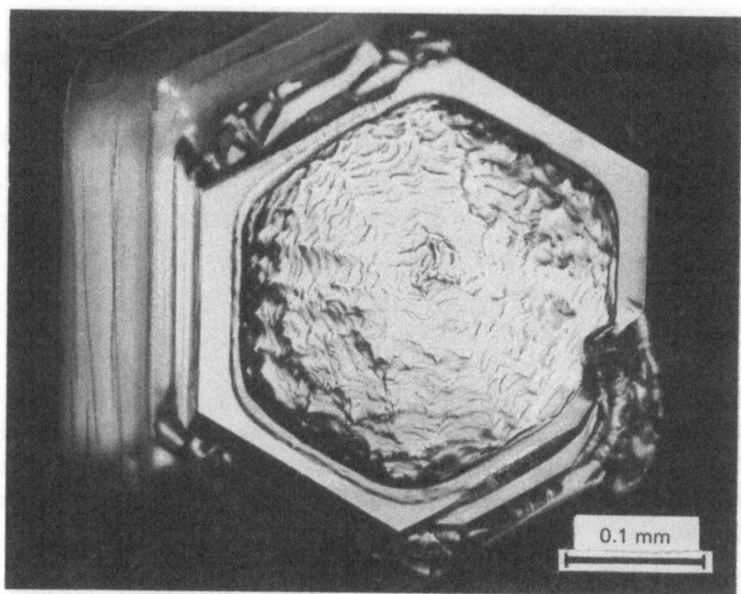


Fig. 2. Recess on a basal face of a zinc single crystal grown in the presence of argon at 250 mm Hg, $\Delta T = 20°C$.

TABLE 1

| Argon pressure, mm Hg | $a$ [$10^{-2}$ cm] | D [$cm^2$/sec] | $V_{0001}$ [$10^{-7}$cm/sec] | $V_{10\bar{1}1}$ [$10^{-7}$cm/sec] | $\beta_{0001}$ [cm/sec] |
|---|---|---|---|---|---|
| Vacuum | — | — | 22±5 | 31.2* | 209 |
| 5 | 3.3±0.5 | 166 | 16±3.5 | 23±5.5 | 152 |
| 10 | 3.0±0.5 | 83 | 10±2 | 16±8.5 | 95 |
| 50 | 2.1±0.5 | 16.5 | 6±1.5 | 7±3 | 57 |
| 100 | 1.8±0.5 | 8.3 | 2.9±0.4 | 4.8±0.2 | 27.5 |
| 150 | 1,5±0.5 | 5.5 | 2.6+0.9 | 4,5±1.5 | 24.7 |
| 250 | <0.8±0.5 | 3.3 | 1.6±0.4 | 3.8±1.5 | 15.2 |

| Argon pressure, mm Hg | $\frac{\beta(\varphi_c)-\beta_{0001}}{\beta(\varphi_c)}$ | $l_{exp}$ [cm] | $l_{theor}$ [cm] | $L_{theor}$ [cm] | $L_{theor}/l_{exp}$ |
|---|---|---|---|---|---|
| 5 | 0.063 | 0.057 | 5.6 | 0.353 | 6.2 |
| 10 | 0.076 | 0.052 | 4.6 | 0.350 | 6.7 |
| 50 | 0.063 | 0.036 | 1.57 | 0.099 | 2,8 |
| 100 | 0.088 | 0.031 | 1.48 | 0.130 | 4.2 |
| 150 | 0.085 | 0.026 | 1.12 | 0.095 | 3.6 |
| 250 | 0.105 | <0.014 | 1.08 | 0.113 | 8.1 |

*One measurement only.

of measurements for each gas pressure. The measurements were made with the crystal at 400°C and the temperature difference $\Delta T = 20°C$.

The critical size falls as the supersaturation increases, i.e., as $\Delta T$ increases (within the range 15 to 45°C), as the theory indicates [3].

Table 1 gives the critical size $l_{theor}$ derived from the theory for growth in the presence of hydrogen:

$$l_{theor} = \frac{D\Omega(c_\infty - c_e)}{\varepsilon V_{0001}}, \tag{1}$$

where D is the diffusion coefficient,* $\Omega$ is the specific volume of a zinc atom in the crystal, and $c_\infty$ and $c_e$ are the true and equilibrium concentrations respectively within the bulk of the medium and at the surface of the crystal, $\varepsilon = 0.2$ is Seeger's dimensionless parameter [7], which is independent of the supersaturation and crystal size, and $V_{0001}$ is the growth rate for a basal face of subcritical size.

The $l_{theor}$ given in Table 1 are larger by factors of 50-100 than the observed values $l_{exp}$ ($l_{exp} = a\sqrt{3}$); similar results were obtained in the presence of hydrogen [4]. Therefore, hydrogen has no particular specific effect, while the increase in critical size with the gas pressure indi-

* Calculated from temperature data and the argon pressure under assumption of a mixture of ideal gases.

cates that diffusion is the sole reason for stability loss. The large difference between $l_{theor}$ and $l_{exp}$ occurs because the value given by (1) corresponds to zero supersaturation at the center of the face [3], although a {0001} face in fact becomes unstable long before such a size is reached, i.e., at higher supersaturations over the center. Therefore, in the next stage we determined the critical size from the more precise parameter

$$L_{theor}=\frac{D\Omega(c_{\infty}-c_{e})}{\varepsilon V_{0001}}\cdot\frac{\beta(\varphi_{c})-\beta_{0001}}{\beta(\varphi_{c})}, \tag{2}$$

where $\beta(\varphi_c)$ and $\beta_{0001}$ are the kinetic coefficients for a vicinal face on the recess and on a close-packed basal face. These are governed by the growth rate of the corresponding face (of subcritical size):

$$V=\Omega\beta(c_{\infty}-c_{e}). \tag{3}$$

We estimated $\beta(\varphi_c)$ by means of interferometric measurements on the inclination of a vicinal face in a recess with respect to the basal face; on average, the results was about 3-4°. On the assumption that such a face consists of steps formed by {0001} and {$10\bar{1}1$} faces (see below), we determined the growth rate geometrically, and the values shown for the {0001} and {$10\bar{1}1$} faces are a means from some dozens of measurements.

The $L_{theor}$ given in Table 1 are of the same order as $l_{exp}$; although $L_{theor}$ is still approximate, it is far more accurate then $l_{theor}$.

Very probably, we observe only the initial stage in the loss of absolute stability in the presence of a gas; therefore, one may sometimes observe nonstationary effects on crystals of critical size: a recess appears at the middle of a basal face, which shrinks, then expands again, and so on. This periodicity persists until the crystal becomes reasonably large and the recess stabilizes.

The exact shape of the recess is also very important. During growth in hydrogen [4], deep recesses arise, which commonly have slightly raised edges. In the presence of hydrogen, these recesses tend to be rather flatter, and it is difficult to establish the shape with an optical microscope. We therefore used a scanning electron microscope to examine the shape of the steps on the basal face around the recess. The edge of the recess was not rounded but dentate,

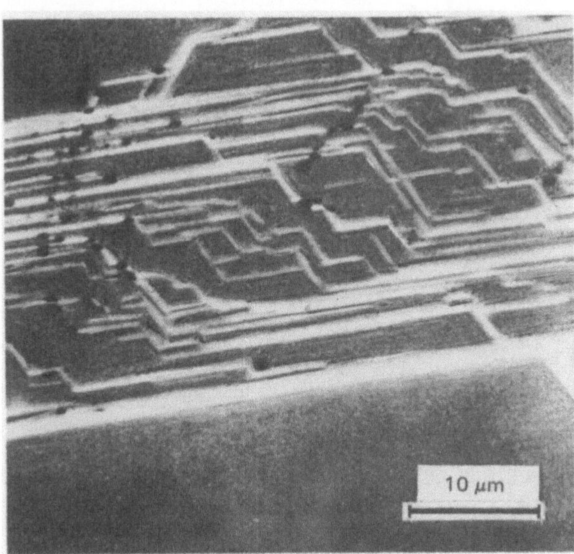


Fig. 3. Steps on the basal face of a zinc crystal forming a recess (the latter lies at the bottom in this scanning electron micrograph). Argon pressure 100 mm Hg, $\Delta T = 20°C$.

and most of the faces forming the fronts of steps were from the first-position zone, probably the one belonging to the $\{10\bar{1}1\}$ pyramid. The shape of the vicinal face in a recess is usually concave, but sometimes convex forms are found (Fig. 3), as is predicted by the steady-state theory. The exact form is probably dependent also on the defect distribution in the crystal. For instance, at low argon pressures (10 mm Hg), a basal face grows only from individual centers at the edges, and these hexagonal stepped growth pyramids appear to arise at defects. High inert-gas pressures (50, 100, or 150 mm Hg) are required to produce recesses in completely closed form.

## 4. Conclusions

If a zinc single crystal grows from the vapor under vacuum, the basal face remains smooth at all supersaturations and crystal sizes [6, 8]; on the other hand, argon (or hydrogen) results in the face being damaged above some critical size because a recess appears. The same effect has been observed for cadmium single crytals growing in inert gases. Loss of shape stability is due to diffusion, while the exact composition of the gas is of only minor importance. Also, our results show that the critical size is dependent on the supersaturation, as the theory would require. This evidence goes with the agreement between theory and experiment for the critical size to confirm Chernov's theory. The only unexplained point is the shape of the recess.

We are indebted to Academician R. A. Kaishev for interest in the work.

## Literature Cited

1. A. A. Chernov, Kristallografiya, 7:895 (1962).
2. A. A. Chernov, Kristallografiya, 8:87 (1963).
3. A. A. Chernov, Kristallografiya, 16:844 (1971).
4. C. Nanev and D. Ivanov, J. Crystal Growth, 3/4:530 (1963); C. Nanev, D. Ivanov, and D. Nenov, Kristall u. Technik, 3:567 (1968).
5. I. N. Stranskii, Z. Phys. Chem., 38:451 (1938); R. Kaishev, L. Keremidchiev, and I. N. Stranskii, Z. Metallkunde, 34:201 (1942).
6. R. Kaishev and C. Nanev, Phys. Status Solidi, 10:779 (1965).
7. A. Seeger, Phil. Mag., 44:348 (1953).
8. C. Nanev, Phys. Status Solidi, 16:777 (1966).

# FACTORS GOVERNING CRYSTAL GROWTH AND DISSOLUTION SHAPES IN MOLTEN METALS


**L. M. Kolganova, A. M. Ovrutskii, and E. V. Finagina**

*Dnepropetrovsk University*


The growth form of a crystal is dependent on the structure of the interface; it may be facetted if the interface is smooth on an atomic scale or rounded if the surface is rough.

It is possible for a rough surface to give way to a smooth one when the solution concentration or temperature alters [1]. It has been found that the shapes of metal crystals alter when the structure of the liquid–crystal interface is changed [2]. Jackson's criterion [3], which characterizes the growth form for crystals growing from the melt, can also be applied to solutions. Facetted growth forms in molten metals are observed when the deposition entropy exceeds 4 cal/mole · deg.

During researches on the kinetics and growth shapes of metal crystals [4], it was found that there are marked differences in growth rate for bismuth in the Bi – Sn and Bi – Pb systems on the one hand and in the Bi – In, Bi – Ga, and Bi – Tl systems on the other; the growth forms of the Bi in all these systems were facetted.

It has also been found that there are threshold supercoolings below which the growth rate is zero for bismuth in Bi – Sn alloy at various concentrations and in Bi + 25 wt.%; in such a system, the surface of a bismuth crystal dissolving in a melt heated slightly above the liquidus is rough on a macroscopic scale.

The kinetic data for bismuth in Bi + 16% In, Bi + 9% Ga, Bi + 18% Tl alloys* indicate that the growth is limited by diffusion even at very low supercoolings.

When a crystal dissolves in one of these systems, the surface is smooth on a macroscopic scale at arbitrarily small melt superheatings; a crystal growing at a low supercooling has more faces than are found in Bi – Sn and Bi – Pb systems. The sizes reached by bismuth crystals before growth-shape distortion in Bi – Sn and Bi – Pb alloys were 2-3 times those found in Bi – In, Bi – Ga and Bi – Tl alloys.

Here we report direct microscopic observations in reflected polarized light [5] on the growth and dissolution forms of bismuth crystals in Bi – Sn, Bi – Pb, Bi – Ga and Bi – In systems in relation to concentration of the second component or traces of a third component, as well as researches on the growth of silver and indium antimonide crystals in Ag – Bi and In – Sb melts.

* All figures are in wt.%.

In some cases we measured the heat of dissolution for the primary bismuth crystals in order to compare the observed growth and dissolution forms with those calculated from the heat and entropy of deposition.

The heat of dissolution was measured by comparing the fall in temperature on dissolving a bismuth crystal with the fall caused by rapid introduction of a known weight of tungsten or nickel initially at room temperature. The thin-walled crucible containing the melt was placed in a tubular oven and was surrounded by asbestos insulation, while the thermo-emf of a thermocouple immersed in the melt was passed to a sensitive potentiometer. The thermal insulation meant that the melt, which was stirred with a thin-walled capillary, received only a little heat from the oven during the dissolution of the bismuth crystals. The melts were heated to 10-15° above the liquidus temperature. Measurements of the latent heat of fusion for pure bismuth, tin, and lead with this equipment gave results in agreement with published values.

The results were as follows: bismuth-rich alloys gave macroscopically smooth surfaces when crystals of the primary phase dissolved (Fig. 1b); distortion occurred in growing primary crystals when a certain critical size was reached, as in scheme I of Fig. 2 (see also Fig. 1a).

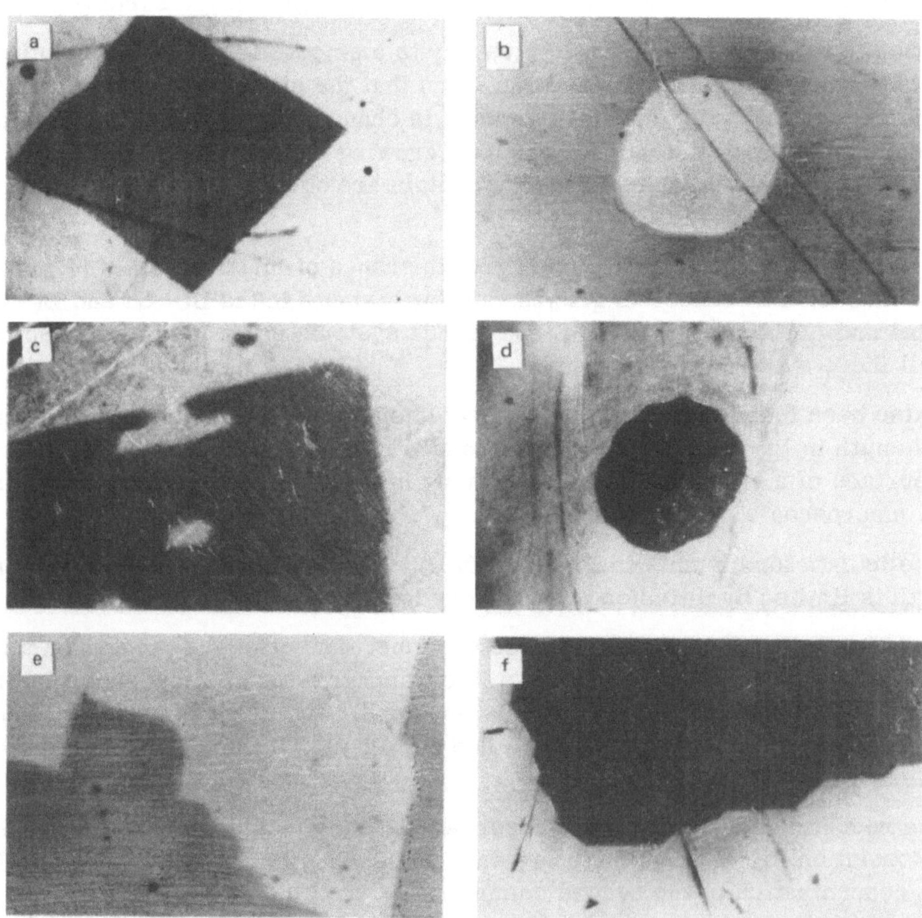


Fig. 1. Growth and dissolution forms of bismuth crystals: (a, b) growth and dissolution in Bi+16% Ni alloy (the bismuth is the light area in part b); (c, d) growth and dissolution in Bi+16% Pb; (e, f) growth and dissolution in Bi+25% Pb (bismuth light in part e).

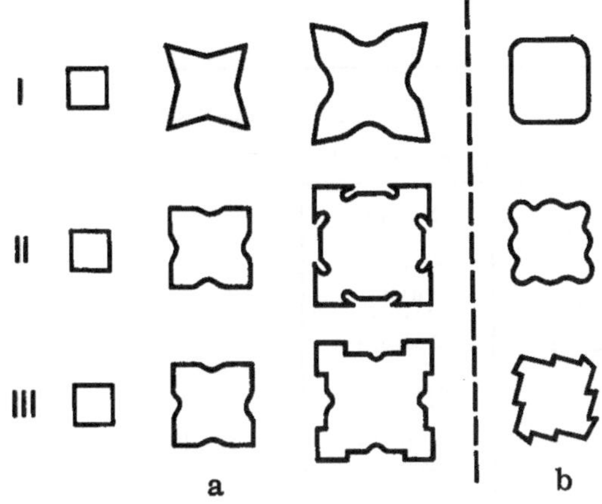


Fig. 2. Growth and dissolution forms of bismuth crystals (schematic): (a) growth; (b) dissolution; (I) ΔS < 5 cal/mole · deg (but more than 4); (II) ΔS ≃ 5 cal/mole · deg; III) ΔS > 5 cal/mole · deg.

The dissolution forms in Bi–Sn and Bi–Pb alloys and the distortion of bismuth crystals during growth varied with the concentration of the second component; in certain narrow ranges (3.5–4% Sn and 16–17% Pb), the growth and dissolution forms were as shown schematically in part II of Fig. 2. Here the characteristic feature is that the solution is trapped by the macroscopic steps during growth (Fig. 1c) and that rippled surfaces arise during dissolution (Fig. 1d). Higher contents of tin or zinc cause the distortion of the bismuth crystal (Fig. 1e) to occur as in part III of Fig. 2, while dissolution gave macroscopic steps (Fig. 1f).

Addition of indium or gallium between 0.5 and 5 wt.% to Bi–Sn (20% Sn) or Bi–Pb (25% Pb) did not alter the dissolution form nore the type of distortion in the growth form, which corresponded to scheme III of Fig. 2. There was also no change in the dissolution form and the distortion of the bismuth crystals (scheme I) in Bi–In and Bi–Ga alloys on introducing up to 1 wt.% of tin or lead.

The addition of 1.5 wt.% tin to Bi + 16% In preduced marked changes in the growth and dissolution forms (from scheme I to Scheme III), and there was an increase in the critical size of the undistorted crystals by a factor 2–3 (Fig. 3).

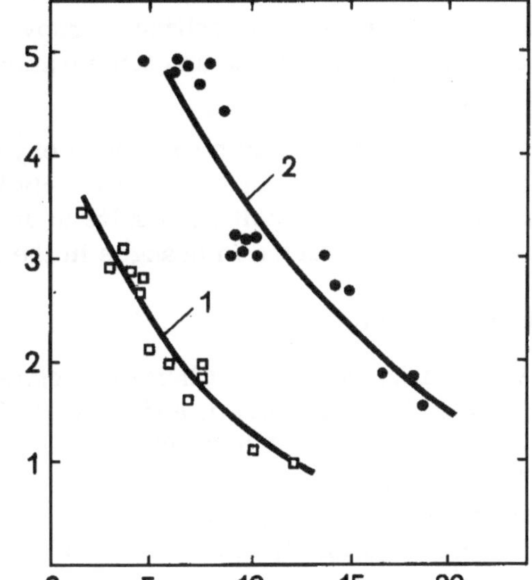


Fig. 3. Critical bismuth crystal size as a function of supercooling: 1) Bi + 16% In; 2) Bi + 15.5 In + 1.5% Sn.

TABLE 1

| Composition, wt.% | $T_{liq}$, °C | $\Delta H_{sol}$, kcal/mole | ΔS, cal/mole·deg | Form of surface on dissolution | Transition range |
|---|---|---|---|---|---|
| Bi (pure) | 271 | 2,6 | 4.8 | Smooth | |
| Bi+3.5%Sn | 257 | — | — | Smooth | |
| Bi+3.75%Sn | 256 | — | — | Wavy | |
| Bi+4%Sn | 255 | 2.65±0.1 | 5.0±0.2 | Stepped | 3.5—4%Sn |
| Bi+20%Sn | 200 | 2.85±0.1 | 6.0±0.2 | Dentate | |
| Bi+16%Pb | 231 | — | — | Smooth | |
| Bi+16.5%Pb | 229 | — | — | Wavy | |
| Bi+17%Pb | 227 | 2.5 ±0.1 | 5.0±0.2 | Stepped | 16—17%Pb |
| Bi+25%Pb | 198 | 2.45±0.1 | 5.2±0.2 | Stepped | |
| Bi+16%In | 192 | 2.25±0.1 | 4.85±0.2 | Smooth | |
| Bi+28%In | 150 | 2.05±0.1 | 4.9 ±0.2 | Smooth | |
| Bi+15.5%In +1.5%Sn | 190 | 2.35±0.1 | 5.1 ±0.2 | Stepped | 1—1.5%Sn |
| Bi+9%Ga | 234 | 2.4±0.15 | 4.7 ±0.3 | Smooth | |
| In+6.5%Sb | 300 | — | — | Smooth | |
| In+2.5%Sb | 215 | — | — | Stepped | |
| Ag+92%Bi | 350 | $L_{mp}$ =2.7 | $L/T_l$=4.0 | Smooth | |
| Ag+96%Bi | 277 | $L_{mp}$ =2.7 | $L/T_l$=4.9 | Smooth | |

Similar effects have been observed for indium antimonide crystals growing and dissolving in the In−Sb system.

No changes in dissolution form or distortion of the facetted forms for silver crystals were found for the Ag−Bi system (here there may be a transition from rounded to facetted growth forms) [2]. It would seem that the conditions required to produce such changes are not attained in an alloy of eutectic composition.

Table 1 gives the results.

These show that the change in dissolution shape and the type of distortion in growing primary crystals can vary sharply over narrow concentration or temperature ranges, which appear to be determined by deposition entropy. The dissolution and growth forms of bismuth crystals alter at concentrations and temperatures for which the deposition entropy is 5 cal/mole·deg.

The results indicate that a flat-faced crystal with a comparatively low deposition entropy has kinetic coefficients that are comparatively large; any increase in the deposition entropy due to change in melt composition results in an increase in the activation energies of surface processes, which is reflected in changes in the growth and dissolution forms.

## Literature Cited

1. V. V. Voronkov and A. A. Chernov, Kristallografiya, 11:662 (1966).
2. I. V. Salli, A. M. Ovrutskii, and L. M. Kolganova, Kristallografiya, 15:848 (1970).
3. C. A Jackson, In: Problems in Crystal Growth [Russian translation], Mir (1968), p. 13.
4. A. M. Ovrutskii and I. V. Salli, In: Growth and Defects of Metallic Crystals [in Russian], Naukova Dumka (1972), p. 95.
5. A. M. Ovrutskii and I. V. Salli, Kristallografiya, 15:533 (1970).

# GROWTH OF NAPHTHALENE AND p-DIBROMOBENZENE CRYSTALS IN THIN FILMS OF MELT


A. M. Ovrutskii and V. V. Podolinskii

*Dnepropetrovsk University*


The growth of crystals of organic substances is of considerable interest, since the molecules are comparatively large, and it is therefore possible to observe growth steps containing small numbers of molecular layers and thus to detect effects associated with small crystal sizes.

The substances used here were monoclinic; naphthalene may be considered as having deformed face-centered close packing [1]. The unit cell has the following dimensions: a = 8.22, b = 5.99, c = 8.64 Å, $\beta$ = 122°38' [2]. The unit cell of p-dibromobenzene is elongated along one axis: a = 15.36, b = 5.75, c = 4.10 Å, $\beta$ = 112°38' [2]. An analysis has been performed [3] for the possible facets of naphthalene. The binding energies of the molecules in the surfaces have been calculated for a variety of planes. It was considered that (001) plates should be bounded by (110) and (201) simple forms, which is in agreement with experiment [1].

We have examined the growth forms of naphthalene and p-dibromobenzene in relation to supercooling, thickness of melt layer, and concentration of the second component (camphor).

## Methods

The crystals were observed under the microscope in reflected polarized light; a molten drop of the substance was placed between a cover glass and a metal mirror (usually chromium-plated) to give a film of thickness between 0.01 and 0.1 mm. An ultrathermostat was used to provide temperature control. There were means of switching the working chamber from one thermostat in another, which was convenient in producing a specified supercooling quickly.

The growth forms in thin films ($10^{-5}$–$10^{-3}$ cm) were examined by using a lens instead of the cover glass, which produced a wedge, with the thickness measurable from Newton's rings.

We used specimens of chemically pure grade, while the naphthalene was additionally purified by sublimation.

## Results and Discussion

If the supercooling is low, small p-dibromobenzene crystals are of rectangular cross section, while the observed sections of small naphthalene crystals were close to regular hexagons. The facetted growth forms became unstable as the size of the primary crystals increased (Fig. 1a), while the maximum size of the undistorted crystals was much larger for naphthalene than it was for p-dibromobenzene (isometric naphthalene crystals of fairly large size are comparatively rare). Figure 2 shows the critical size as a function of supercooling for p-dibromobenzene.

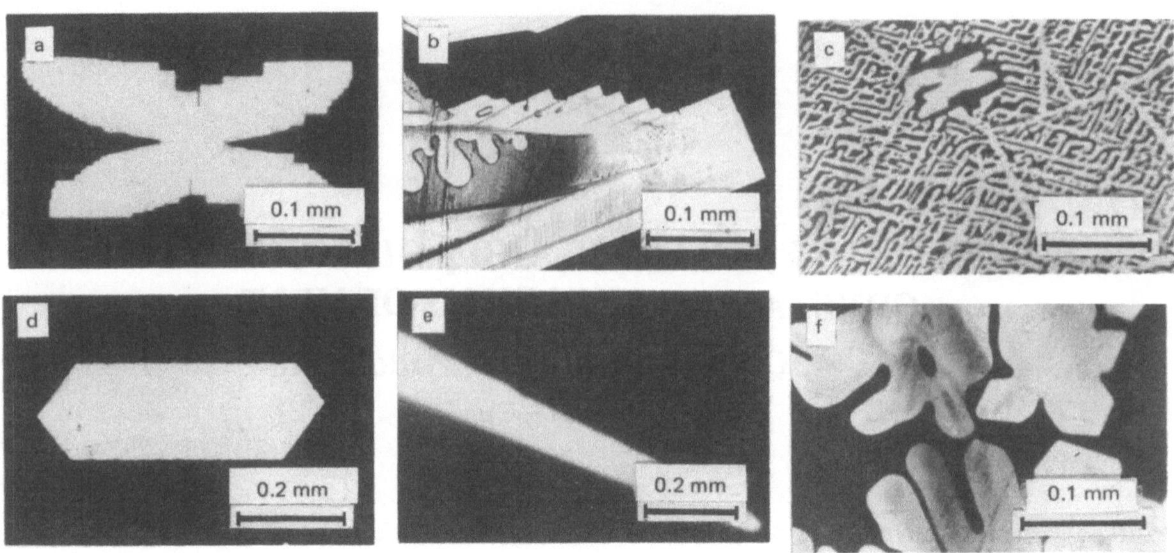


**Fig. 1.** Growth forms of (a–c) p-dibromobenzene and (d–f) naphthalene.

One frequently observes naphthalene and p-dibromobenzene crystals growing as plates, whose widths may vary, but which do not alter during the growth. Thin films give narrower plates. Low degrees of supercooling (1-2°C, the exact value being dependent on the camphor concentration) resulted in dihedral corners at the growing ends of the naphthalene plates, which persisted, so the cross section as a whole remained hexagonal (Fig. 1d). On increasing the supercooling somewhat, one of the planes at the end of the plate was lost, and the vertex advancing into the liquid became rounded (Fig. 1e). Figure 3 shows the supercooling required to produce this effect as a function of the concentration.

These changes in shape occur because the growth is restricted in certain directions; if a hexagonal platy naphthalene crystal is oriented at an angle to the plane of the film, then it can grow only in one direction to produce an elongated plate when two faces have met the cover glass and metal substrate. The edge of the plate in contact with the substrate has a supercooling larger than that for the upper surface in contact with the glass on account of the rapid heat removal through the metal, which leads to the loss of one face and the formation of

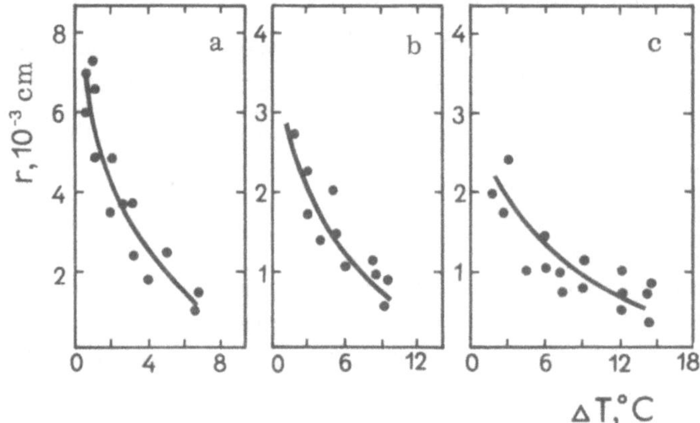


**Fig. 2.** Critical size r (half the length of a short face) as a function of supercooling for melts containing the following mol.% of camphor: (a) 9.3; (b) 29.2; (c) 53.8.

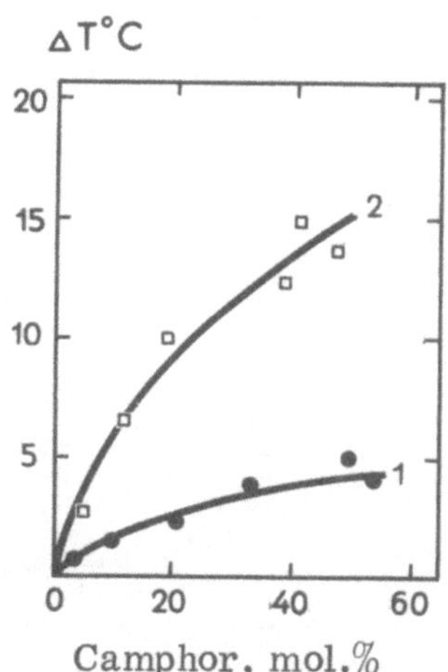


Fig. 3. Supercooling required to change growth form in relation to camphor concentration: (1) naphthalene; (2) p-dibromobenzene.

a vertex extended into the liquid. Sometimes one observes acicular crystals of naphthalene and p-dibromobenzene of width of less than 1 $\mu$m, and it would seem that in that case the plane of the plates is perpendicular to the surfaces bounding the liquid film.

If the supercooling is increased somewhat when a platy crystal is growing, one can observe growth layers extending out from the crystallization front, and it would seem that the end face of the plate becomes thinner. These growth layers frequently have rounded boundaries, and they sometimes develop as rounded dendrites (Fig. 1b).

The following interesting effect was observed during the growth of p-dibromobenzene crystals; if the temperature of the melt is reduced so that the supercooling becomes larger than that indicated by curve 2 of Fig. 3, the crystal becomes distorted and the flat-faced parts at the vertices become smaller (it would seem also that the thickness of the plate decreases), and then there is an explosive increase in the growth rate for one of the vertices, which is such that it becomes impossible to observe the growth form. The rapidly-growing vertex develops as a dendrite and very quickly extends as a network throughout the liquid film. It can then be seen that the resulting crystal has a rounded appearance (Fig. 1c). After a certain interval, faces appear on all parts of the branched crystal, which may be due to increase in thickness due to the growth of new layers, or else to the reduction in the supercooling.

The minimum width of a facetted part at the vertex is about 5 $\mu$m before the shape changes; the thickness would appear to be much as for the growth layers, i.e., not more than 0.1 $\mu$m. The latter figure has been estimated from the difference in contrast for a part of the crystal covered by a single layer. The basis for this was that the change in interference color due to the birefringence from black to white corresponds to a path difference of 2650 Å [4].

If these changes in shape are due to size reduction in the growing faces, it is important to determine the size at which the transition occurs, for which purpose growth in wedges may be employed. When naphthalene grows in this way, it is clear (Fig. 4) that the temperature at which the crystal is in equilibrium with the liquid increases with the film thickness, and there is a transition from the facetted form to the rounded form (Fig. 1f), which occurs with pure naphthalene and with the mixtures. The temperature of the chamber was reduced slowly during the experiment, so the faces at first advanced slowly in the wedge, and the transition point was determined with the crystal in essence in equilibrium with the medium. The variation in equilibrium temperature with thickness (Fig. 4) agrees closely with the values given by Thompson's formula. The film thickness corresponding to the transition point is dependent on the concentration of the second component (we used single crystals of size small by comparison with the liquid volume for this purpose). Figure 5 shows the results.

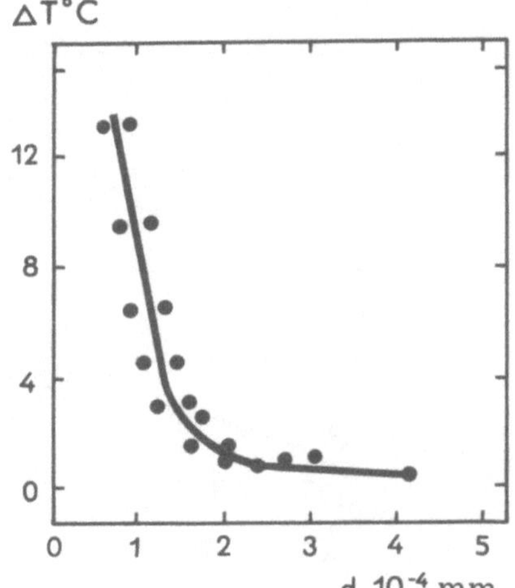


Fig. 4. Temperature for equilibrium between naphthalene crystals and melt in relation to film thickness in a wedge.

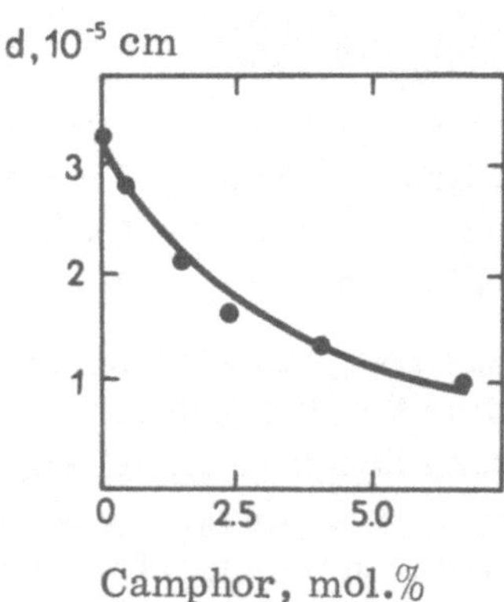


Fig. 5. Wedge thickness at the point of change in shape for naphthalene crystals as a function of camphor concentration in the melt.

It was also of interest to establish whether the substrate affects the point of transition from the facetted form to the rounded one. For this purpose, the glass lens and metal substrate were insulated from the liquid with thin sheets of mica or PTFE. There was no obvious change in the results.

It proved impossible to detect the change from facetted to rounded growth forms for p-dibromobenzene in wedges; the structure became extremely fine-grained in the thinnest part of the wedge and it was impossible to determine the crystal shape.

We thus have established that the growth form changes from facetted to rounded when the size of the growing faces decreases for these two substances; at first sight it may seem surprising that this effect should be observed for crystals with such high entropies of melting ($\Delta S = 12.8$ cal/mole · deg for naphthalene, or 13.5 for p-dibromobenzene). However, it follows from [1, 3] that the binding energy in the surface layer can vary considerably from one face to another because the forces between the long molecules are dependent on the mutual orientation. Therefore, Jackson's roughness parameter [5]

$$\alpha = \xi \frac{L}{kT_e}$$

may be comparatively small for the end faces of plates of these two substances, where L is the latent that heat of transition per molecule, k is Boltzmann's constant, and $\xi$ is the part of the total binding energy arising from interaction with other molecules in the surface layer. Further, the face size affects the growth form because the edges may have fairly broad zones where the structure is rough.

In fact, the edges resemble steps in that they should be subject to diffuseness on account of thermal fluctuations for $T > 0$, and therefore they should be rounded on a microscopic scale [6]. If the face size is reduced until the width of the rounding zone is approached, one might therefore expect a change in growth mechanism.

The width of this edge zone should be comparable with the spread in the width of a step as regards order of magnitude, or rather should be somewhat greater than the latter because several steps should be involved in rounding for any given edge, and the broadening of each step in such a situation would be somewhat less than that for a free step. A step at a crystal-melt boundary may occupy a band of width several times the interatomic distance, so one concludes that the rounding radius for an edge should be at least ten times the interatomic distance. The roughness found on the ends of columnar crystals [7] indicates that the growth rate is linearly dependent on the supercooling when the ends have sizes of $10 \times 10$ interatomic distances, while the rough structure gives way to a smooth one for ends whose sizes exceed $2 \cdot 10^2 \times 2 \cdot 10^2$ interatomic distances. Although the roughness of a rectangular end should set in at sizes smaller than those for a square, the above results do not conflict qualitatively with our observations. In our case, the facets on the ends of the plates were seen at thicknesses representing some hundreds of times the intermolecular distance.

In conclusion the following observation may be made. The main reason for the change in growth form from facetted to rounded is reduction in thickness of the platy crystal; however, if one employs supercooling variation, there are two factors to be considered, namely the reduction in the size and the increase in the supercooling at the crystallization front. The surface structure of a small face may be such that a small increase in the supercooling results in a normal growth mechanism. It would seem [8] that the supercooling at which normal growth sets in is governed by the roughness parameter. The surface structure changes when the face size is reduced, which may result in a fall in this supercooling, at least when the distances between the roughness zones are small. At the same time, size reduction causes the supercooling at the crystallization front to increase because of the change in the balance between the surfaces processes (molecular attachment) and the bulk processes (heat or material transport). If the face sizes are sufficiently small, the surface is rough even under equilibrium conditions. It may be that the threshold supercooling at which the faces on a large crystal start to grow is determined by the width of the roughness zones at the vertices and edges.

## Literature Cited


1. P. Hartman, In: Physics and Chemistry of Solid-State Organic Compounds [Russian translation], Mir, Moscow (1967), p. 342.
2. A. I. Kitaigorodskii, Organic Chemical Crystallography [in Russian], Izd. AN SSSR, Moscow (1955), pp. 355 and 419.
3. P. Hartman and W. G. Perdok, Acta Cryst., 8:525 (1955).
4. G. R. Bargina, E. D. Dukova, I. P. Korshunov, and A. A. Chernov, Kristallografiya, 8:758 (1963).
5. C. A. Jackson, In: Solidification of Liquid Metals [in Russian], Metallurgiya, Moscow (1962), p. 200.
6. A. A. Chernov, Usp. Fiz. Nauk, 73:277 (1961).
7. K. A. Jackson, This volume, p. 115.
8. D. E. Temkin, In: Mechanisms and Kinetics of Crystallization [in Russian], Nauka i Tekhnika, Minsk (1964), p. 86.

# EFFECTS OF A MOVING MAGNETIC FIELD ON THE STABILITY OF A CRYSTALLIZATION FRONT IN A MELT

K. M. Rozin, V. V. Antipov, and N. V. Isakova

*Moscow Institue of Steels and Alloys*

An external moving magnetic field, which may include rotational and translational components, can provide control of many processes in crystal growth; if the moving field interacts with the Foucault currents in a conducting melt, one gets forced circulation in the liquid, which affects the crystallization mechanism substantially. Such a field reduces the curvature of the crystallization front, reduces the thickness of the layer of concentrated impurity solution ahead of the front, brings the impurity partition coefficient closer to the equilibrium value, alters the growth-rate dependence of the partition coefficient, and also reduces the temperature gradient ahead of the crystallization front [1-8]. These effects occur with smooth fronts and can increase the crystallization rate considerably, while improving the purification and the growth conditions generally [3-5, 9].

However, this effect applies mainly to the stable region, and as there is considerable interest in accelerated crystal growth, and also in growth at elevated dope contents, where concentration supercooling can occur, it appeared desirable to present some results on the effects of magnetic fields on the stability of crystallization fronts.

Here we report the growth-front stability for single-crystal zinc growing from the melt containing a certain amount of cadmium. The crystallization in this system at zero field is well understood.

The crystals were grown by Bridgman's method in the vertical system shown in Fig. 1; the porcelain tube 1 contains the ampoule 2, while the heater system 3 provides a vertical temperature gradient of 18 deg/cm. The ampoule 2 is moved by the clock motor 4 at 0.25, 0.35, 0.55, or 0.75 mm/min. Pole 5 of the electromagnet lies close to the solid-liquid interface. The electromagnetic shield and windings given are protected from the heat by the water-cooled thermal screen 6. This electromagnet was supplied with current with a phase shift between the windings, which resulted in a moving magnetic field of $12 \cdot 10^3$ A/m at the crystallization front, which produced vigorous circulation in the melt. The quartz tube had a diameter of 15 mm and a length not less than 120 mm. The cadmium content of the melt varied from 0.01 to 0.40 wt.%. The deviation of the axis of the finished crystal from the growth direction did not exceed 10°.

The transition from the stable state to the unstable one was clearly seen from the cellular structure, which can be seen on decanting the melt or by etching a cross section. We used both methods. The etching agent was a 20% solution of chromic anhydride in water, while the

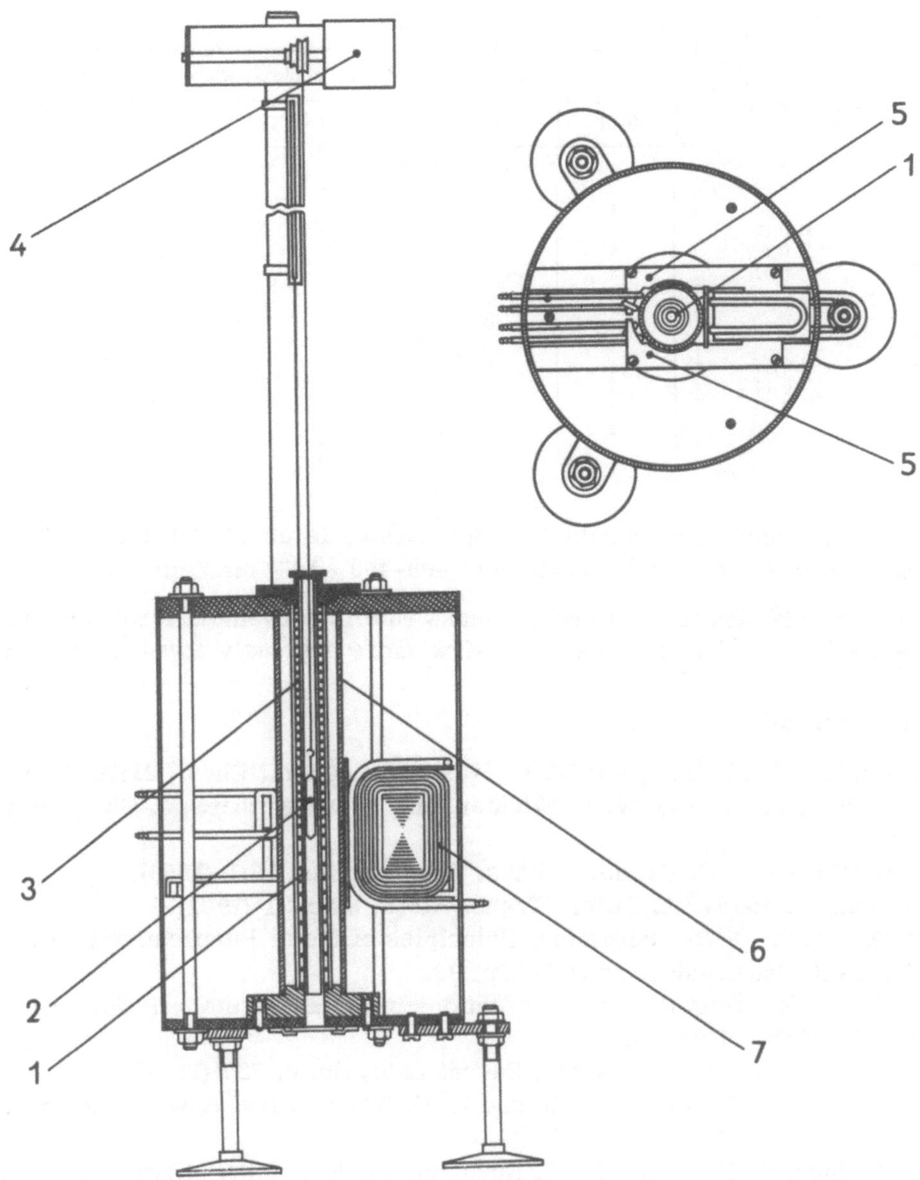


Fig. 1. Apparatus for growing single crystals in an external moving magnetic field: (1) porcelain tube; (2) ampoule containing specimen; (3) heater system; (4) motor; (5) electromagnet pole; (6) thermal screens; (7) electromagnet winding.

etching time was 30-40 sec. The structure was monitored on three specimens for each growth condition, and the results for identical working conditions were in complete agreement.

Table 1 gives the results, where s denotes a smooth front for the corresponding cadmium content and crystallization rate, while c denotes a cellular structure.

It is clear that the cellular structure is displaced toward higher growth rate and impurity contents.

These data show that stability is lost when some critical value is reached for the product of the impurity concentration and the crystallization rate. In the absence of the field, the value

TABLE 1

| Crystallization rate, mm/min | Initial cadmium content, wt.% | | | | | |
|---|---|---|---|---|---|---|
| | 0.01 | 0.05 | 0.10 | 0.20 | 0.30 | 0.40 |
| A. Without field | | | | | | |
| 0.25 | s | s | c | c | c | c |
| 0.35 | s | s | c | c | c | c |
| 0.55 | s | c | c | c | c | c |
| 0.75 | s | c | c | c | c | c |
| B. With moving field | | | | | | |
| 0.25 | s | s | s | c | c | c |
| 0.35 | s | s | s | c | c | c |
| 0.55 | s | s | c | c | c | c |
| 0.75 | s | s | c | c | c | c |

for this quantity, which has been called the mass rate, is about 200-225 · $10^{-4}$ mm/min, while in an alternating field it is roughly doubled at 400-450 · $10^{-4}$ mm/min.

Therefore, this concept of a critical mass rate is convenient for describing crystallization-front stability. Further studies may define more precisely how it should be used.

## Literature Cited


1. J. A. Burton, P. C. Prim, and W. P. Slichter, J. Chem. Phys., 21:1987 (1953).
2. J. A. Burton, E. D. Kolb, W. P. Slichter, and J. D. Struthers, J. Chem. Phys., 21:1990 (1953).
3. I. B. Mullin and K. F. Hulme, J. Phys. Chem. Solids, 17:1 (1960).
4. W. G. Johnson and W. A. Tiller, Trans. AIME, 221:331 (1961).
5. K. M. Rozin, In: Physicochemical Principles of Metal Purification by Crystallization [in Russian], Nauka, Moscow (1970), p. 92.
6. K. M. Rozin, In: Growth and Imperfections in Metal Crystals [in Russian], Naukova Dumka, Kiev (1968), p. 203.
7. G. V. Indenbaum and K. M. Rozin, Zavod. Lab., No. 6, 727 (1967).
8. A. A. Bel'skii, V. N. Vigdorovich, and R. V. Ivanova, Izv. Akad. Nauk SSSR, Metally, 5:85 (1969).
9. V. V. Antipov, K. M. Rozin, M. A. Rusovich, M. P. Shiskol'skaya, and V. T. Shimanyuk, In: Growth and Defects of Metallic Crystals [in Russian], Naukova Dumka, Kiev (1972), p. 301.

PART IV

# IMPURITY TRAPPING

# TRAPPING DURING GROWTH FROM SOLUTION

I. V. Melikhov

*Lomonosov University, Moscow*

The entry of impurities into a crystal growing from a solution has long been the object of discussion [1-3]; however, until recently research on this topic has been concerned with the comparatively narrow range in growth rate V of $10^{-8}$-$10^{-4}$ cm/sec [4], which satisfies electronic and optical requirements, but not many requirements in chemical technology, hydrometallurgy, radiochemistry, and hydrochemistry, where the relevant range is from $10^{-12}$ to $10^{-5}$ cm/sec, and therefore the information has failed to yield a sufficiently general picture of impurity trapping. For this reason, recent years have seen attempts to define in detail the entry mechanisms over wider ranges in crystallization conditions [5], and this has served to elucidate the major features [5, 6]. However, from the practical viewpoint the researches are still in a preliminary state.

## 1. Basic Experimental Data

It is usual [7-20] to determine the following trapping coefficients in examining uptake from solution: the integral coefficient $\bar{K}$ for the steady-state process and the differential coefficient K for the nonstationary one:

$$\bar{K}=\frac{x\rho}{yc_s} \quad \text{and} \quad K=\frac{\rho}{c_s}\frac{dx}{dy}, \tag{1}$$

where x and y are the amounts of impurity and host substance entering the solid, $c_s$ is the mean concentration of the impurity in the solution, and $\rho$ is the crystal density. Also, it is common practice to measure the crystal size and shape [17, 20, 21] and any parameters that are affected by the impurity [18, 19]. Sometimes measurements are made of the local impurity concentration by x-ray spectral or electron-probe means [21], or else by autoradiography [22] and layerwise dissolution or sectioning [23].

The ultimate result has been to show that crystals growing from water, organic solvents, and mixed solutions take up impurities as solid solutions (at the molecular level) or as colloid dispersions [24], with a tendency for the latter to accumulate at block boundaries [22, 25] or else in inclusions of the mother liquor [26]. The main form of uptake arises from nonequilibrium trapping, whose major features can be interpreted by assuming that the bulk of the growing crystal inherits the properties of the surface layer [27-30]. The essence of this phenomenon is as follows. Each fresh part of the solid produced at the interface tends to exist in adsorption equilibrium; such a part is built into the finished crystal, which involves a certain amount of relaxation, which tends to redistribute the impurity between the layer and solution in accordance with the laws for bulk equilibrium. This redistribution can occur if the part remains near the surface for a sufficient period. If the growth is rapid, the time is inadequate

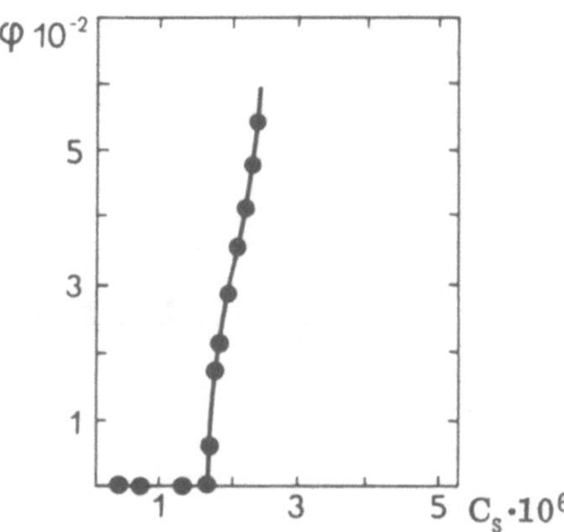


Fig. 1. Adsorption isotherm for the dye Ponso-G for stable potassium sulfate crystals [32], T = 25°C, Re = $10^4$, with $\varphi$ the degree of surface covering and $C_s$ the concentration in the solution.

for relaxation, and so the solid retains all the accumulated material. If the growth is slow, the relaxation goes to completion and the crystal has the equilibrium composition.

The correlation between the surface and bulk properties of a growing crystal is confirmed by the following lines of evidence [17, 28-33]:

(a) When Ce (III) is trapped by $K_2SO_4$ from neutral aqueous solution under conditions of mass crystallization at growth rates between $10^{-5}$-$10^{-7}$ cm/sec, there is an increase in the trapping coefficient as the growth rate becomes lower, which can be interpreted quantitatively if we assume that each layer takes up cerium at a characteristic rate of $\alpha = (2.0 \pm 0.4) \cdot 10^{-4}$ cm/sec, this material being then transferred completely to the bulk of the crystals [32]. If on the other hand a $K_2SO_4$ crystal is placed in a saturated solution containing cerium, one can detect the adsorption, which has a characteristic rate of $\beta = (1.5 \pm 0.2) \cdot 10^{-4}$ cm/sec [31]. The agreement between $\alpha$ and $\beta$ reflects the fact that the crystal traps the impurity adsorbed on the surface during growth in the same way as at equilibrium.

(b) Figure 1 shows the adsorption isotherm for Ponso-G dye taken up by stable potassium sulfate crystals from a saturated aqueous solution, and it is clear that there is a concentration limit below which there is no adsorption. The dye enters the $K_2SO_4$ crystal during growth only when the concentration exceeds some limit (Fig. 2). The lower limits for trapping and adsorption coincide [32], which shows that the growing crystal inherits the adsorbed material. This means that the trapping coefficient is dependent on various kinetic and thermodynamic parameters of the processes in the solution, at the surface, and in the bulk. These parameters can be grouped in such a way as to define four limiting trapping conditions (liquid-phase kinetic, liquid-phase diffusion, kinetic-adsorption, and migration), which differ in the predominant group of

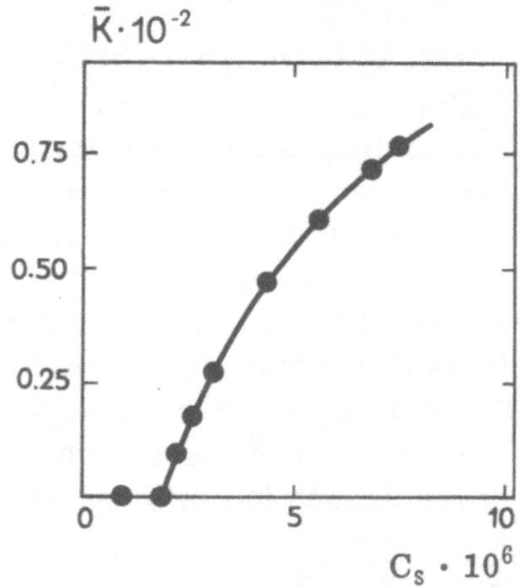


Fig. 2. Trapping of Ponso-G by potassium sulfate crystals in relation to dye concentration in the solution [32], T = 25°C, Re = $10^4$, supersaturation 0.15.

kinetic parameters that affects K. The conditions under which a particular state will predominate cannot as yet be predicted, but each such state has definite features readily established by experiment. For instance, the liquid-phase kinetic state has the trapping coefficient controlled solely by the kinetics of solvation, complexing, hydrolysis, and redox processes in the solution, and then the trapping is dependent on the age of the solution, i.e., on the time elapsed from preparation to start of crystallization [34]. The liquid-phase diffusion state has the trapping controlled by mass transport in the solution, and here K is dependent on the speed of the solution with respect to the crystal. The migration state reflects the change occurring after some alteration that causes instantaneous growth halt, e.g., dilution. These limiting conditions are so different that it is reasonable to consider them independently.

## Kinetic-Adsorption Mode

In this mode, the change in the trapping parameters is due in the main to interaction of the impurity with the surface; this can occur under a wide range of growth conditions, as is clear from data on the mass crystallization of mixed solutions (Table 1). Here we give the macroscopic Reynolds number Re at which the mass transport in the solution does not influence the trapping within the range of normal growth rates V listed. Outside this range in V, particularly on the high-rate side, the trapping gives way to the liquid-phase diffusion mode, while on the small side one gets the migration state. Therefore, the kinetic adsorption mode occurs for Re and V common under laboratory and industrial conditions.

Surface-Process Kinetics and Interface Structure. Slow interaction between an impurity and a solid surface is common during growth from solution, as is clearly seen for highly soluble substances with high kinetic growth coefficients. When such a substance crystallizes, one can adjust V over the range from $10^{-4}$ to $10^{-10}$ cm/sec by altering the supersaturation by 10-30%, i.e., with a liquid phase of essentially constant composition, whereupon the kinetic factors become the dominant ones. In that case, isothermal uptake is described

TABLE 1. Realization Conditions for the Kinetic-Adsorption Mode

| System | Re | V, cm/sec |
|---|---|---|
| $NaCl-Pb(II)-H_2O$ | $10^3$ | $10^{-6}$—$10^{-8}$ |
| $NaCl-Cd(II)-H_2O$ | $10^3$ | $10^{-6}$—$10^{-8}$ |
| $NaCl-Ag(I)-H_2O$ | $10^3$ | $10^{-5}$—$10^{-7}$ |
| $K_2SO_4-Ce(III)-H_2O$ | $10^4$ | $10^{-5}$—$10^{-8}$ |
| $K_2SO_4-La(III)-1.5NHNO_3$ | $10^4$ | $10^{-4}$—$10^{-7}$ |
| $BaSO_4-Sr(II)-H_2O$ | $10^4$ | $10^{-6}$—$10^{-8}$ |

TABLE 2. Characteristics of the Kinetic-Adsorption Mode, T = 25°C [36]

| System | $K_1 \cdot 10^{-2}$ | $K_2 \cdot 10^{-2}$ | $K_3 \cdot 10^{-2}$ | $\beta_1 \cdot$ cm/sec | $\beta_2 \cdot$ cm/sec |
|---|---|---|---|---|---|
| $K_2SO_4-Ce(III)-H_2O$ | 30 | $5 \cdot 10^2$ | $4.5 \pm 0.8$ | $5 \cdot 10^{-11}$ | $5 \cdot 10^{-9}$ |
| $K_2SO_4-Ce(III)-1N\ HNO_3$ | — | $1.7 \pm 0.2$ | $0.45 \pm 0.05$ | — | $(2.2 \pm 0.2) \cdot 10^{-5}$ |
| $K_2SO_4-La(III)-1.5N\ HNO_3$ | — | $0.45 \pm 0.05$ | $1.4 \pm 0.1$ | — | $(1.0 \pm 0.3) \cdot 10^{-4}$ |
| $NaCl-Ag(I)-H_2O$ | $105 \pm 5$ | $1.40 \pm 0.2$ | $0.48 \pm 0.05$ | $\sim 10^{-8}$ | $(1.2 \pm 0.1) \cdot 10^{-5}$ |
| $NaCl-Pb(II)-H_2O$ | — | $7.0 \pm 0.9$ | $2.0 \pm 0.1$ | $\sim 10^{-9}$ | $(4.1 \pm 0.2) \cdot 10^{-7}$ |

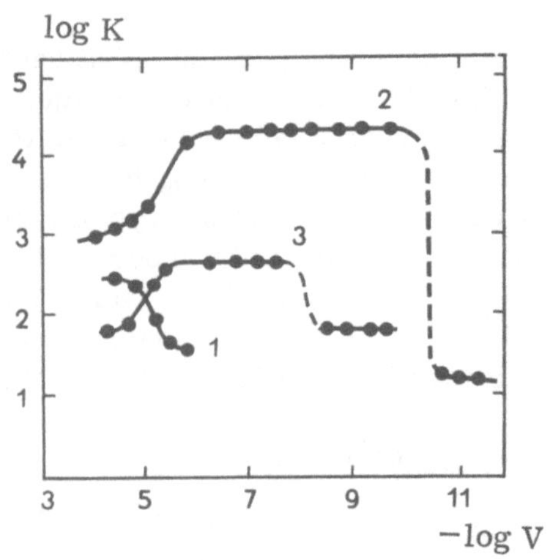


Fig. 3. Trapping coefficient as a function of solute concentration [36] for the following systems: $K_2SO_4$ – La(III) – 1.5 $HNO_3$ (1); $K_2SO_4$ – Ce (III) – $H_2O$ (2); NaCl – AgCl – $H_2O$ (3).

approximately by

$$K = K_1 + \sum_{i=1}^{3} (K_{i+1} - K_i) \cdot e^{-\beta_i/V}, \qquad (2)$$

where $K_i$ and $\beta_i$ are empirical constants (Table 2).

Formula (2) reflects the growth rate ranges in which any change in the conditions does not affect the trapping. These ranges follow one another in such a way that K changes rather sharply from one limiting value $K_i$ to another (Fig. 3). This change in K is due to surface processes of characteristic rate $\beta_i$. Such processes cannot make themselves felt for $V \gg \beta_i$, but but they are completely dominant for $V \ll \beta_i$.

If the change in supersaturation is considerable, one finds that $K_i$ and $\beta_i$ are dependent on the concentration C of the host substance; the form of $\beta_i(C)$ has not been examined, but $K_i(C)$ is represented by

$$K_i = \lambda_i \left( \frac{\rho}{C - b_1 L + b_2} \right)^n, \qquad (3)$$

where $\lambda_i$ is a trapping parameter called the cocrystallization coefficient [35], while n, $\beta_1$, and $\beta_2$ are empirical coefficients (Table 3), and L is the solubility of the substance.

Equations (2) and (3) reflect the elementary interaction between the impurity and the interface; (2) indicates uniformity in the centers responsible for each of the surface processes. If this were not so, the measurements would not be described by a sum of exponentials. Formula (3) characterizes the disposition of these centers at the interface. For instance, with

TABLE 3. Coefficients in (3) for 20-25°C

| System | $\lambda_1$ | $b_1$ | $b_2$ | n | Ref. |
|---|---|---|---|---|---|
| $BaCrO_4$—Ra (II) — $H_2O$ | 55.0±0.5 | 0 | 0 | 1 | 7 |
| KCl—Rb (I) — acetone | 0.50±0.03 | 0 | 0 | 1 | 8 |
| $U(C_2O_4)_2 \cdot 6H_2O$—Ce (III) — $H_2C_2O_4$—$H_2O$ | 2.5±0.2 | 0 | 0 | 1 | 11 |
| AgCl—Tl (I) — $NH_4OH$ | $(1.0\pm0.5)\cdot 10^{-2}$ | — | — | 0 | 14 |
| $U(C_2O_4) \cdot 6H_2O$—Ca (II) — $H_2O$—$H_2C_2O_4$ | 3.1±0.5 | — | — | 0 | 13 |
| $BaSO_4$—Sr (II) — $H_2O$ | $(6.5\pm0.5)\cdot 10^3$ | 1 | >0 | 1 | — |

$n = 1$ and $b_{1,2} = 0$, we get that (3) is analogous to the isotherm for an exchange between the host substance and the impurity. Such exchange is possible on areas fully filled by the host substance. Therefore, for $n = 1$ and $b_{1,2} = 0$ we have that the centers occur in a close-packed surface homogeneous layer of solid phase (presumably monomolecular) or else in the first underlying layer. If $n = 0$, the process is dominated by the properties of centers free from the main substance, which therefore do not respond to the presence of the latter in the solution. Such centers may be localized near the surface, particularly at points occupied mainly by the solvent. Finally, for $n = 1 = b_1$ and $b_2 > 0$, we have that (3) is identical with the simplified formula due to Chernov and Voronkov [37] for trapping of an impurity by centers on end faces formed by close-packed surface monolayers in the growing crystal.

These results can be explained if we employ the sorption model for the interface [29]; the latter is taken as being smooth at the molecular level and including a layer of adsorbed solution, within which one can identify a monolayer (Volmer layer) in direct contact with the close-packed surface monolayer of the crystal (Paneth layer) and with the ends of the steps (Fig. 4). One of the $K_i$ characterizes the equilibrium incorporation of the impurity into the Volmer, Paneth, and first covered layers (the last is denoted by $\Gamma$). The rates of equilibration between each of these layers are determined by the $\beta_i$. If the growth is rapid, each of the monolayers laid down at the interface takes up all the impurity accumulated by the Volmer layer, and this impurity does not have time to become redistributed between the phases and thus enters the bulk of the crystal completely. This state is defined by (3) with $n = 0$. At lower growth rates, i.e., comparable with the rate of accumulation of impurity at the ends of steps, the trapping coefficient varies as the supersaturation alters in accordance with the Chernov-Voronkov formula. Any fall in V results in an approach to equilibrium with the solution for the Paneth layer and then for layer $\Gamma$, with subsequent complete retention of the trapped impurity. Here $K_i$ begins to reflect the exchange between the impurity and the host substance.

This model accentuates the effects of surface regions of extent about ~10 Å, within which the properties of the crystal vary sharply along the normal to the surface and therefore cannot be described within the continuum approximation. The model does not rule out the possibility of long-range adsorption forces, which can influence $\lambda_i$ and $\beta_i$.

Effects of Growth Conditions on $K_i$ and $\beta_i$. These parameters are dependent on temperature, solution composition, and also on any fields and mechanical loads [5, 10, 38]. Most research has been done on the effects of complexing agents and impurities. Such substances have effects that can be explained in part quantitatively within the framework of the quasichemical approach, in which the solution is considered as containing a set of forms of the impurity differing in immediate environment and each capable of interacting in accor-

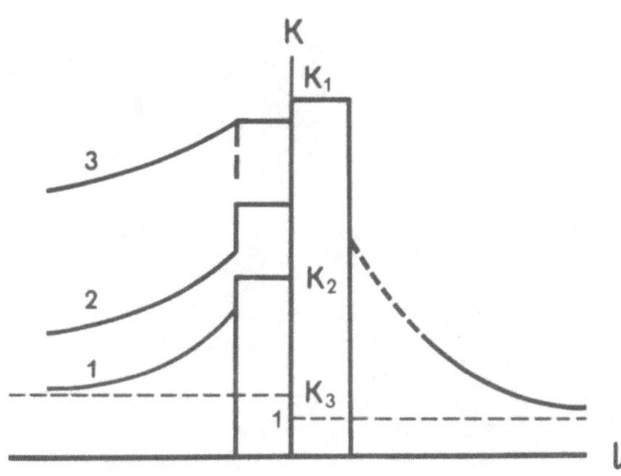


Fig. 4. Scheme for boundary-zone structure; K is the local partition coefficient, $l$ is the distance from the interface, and $K_1$, $K_2$, and $K_3$ are the coefficients for the adsorption and for the incorporation into the Paneth layer and into layer $\Gamma$. The serial numbers on the curves represent successively increasing growth rates.

dance with the laws of chemical kinetics. For instance, it has been found [35] that strontium is taken up by barium nitrate crystals from solutions of constant ionic strength in the presence of sodium ethylenediaminetetraacetate in accordance with

$$K_i = K_{i0}\left[\frac{1 + C_d/(c_s - C_d)}{1 + \sigma C_d/(c_s - C_d)}\right], \tag{4}$$

where $K_{i0}$ and $\sigma$ are empirical constants, while $C_d$ is the concentration of the sodium salt. Formula (4) coincides with the solution to the quasichemical problem for uptake of an uncomplexed form of strontium from a mixture with barium ethylenediaminetetraacetate. This result goes with many other similar cases [39] to show that ligands of size considerably exceeding the size of the host molecule are taken up only slightly by crystals. As a result, the $K_i$ are affected if the complexes are highly stable and if the complexing agent is present in amounts comparable with the amount of impurity.

In research on the concentration dependence of the $K_i$ it has been found that trapping can be accentuated or reduced [5, 22] as the impurity concentration increases. The first case is usually ascribed to the formation of colloidal solid solutions, while the other is ascribed to internal adsorption or the formation of restricted solid solutions [5], with intermediate cases representing complete miscibility. Little is known about the features of these forms of trapping from solution.

## Migration State

In this state, the surface is almost in equilibrium with the solution at any time, while the growth rate V is comparable with the rate of diffusion of the impurity from the bulk of the solid phase into the solution, the condition thus being

$$V \ll D/a, \tag{5}$$

where D is the impurity diffusion coefficient and a is the thickness of the monolayer. Here the migration is responsible for eliminating the effects of inherited surface properties, and also of secondary processes that result in impurity segregation at defects, which do not make themselves felt during growth. The following examples confirm this.

(a) Growing (100) faces of barium sulfate crystals in aqueous solution show appreciable sorption of strontium, with the trapping factor dependent on the growth rate (Fig. 5). Electron microscopy indicates that the diagonals of each face generate constantly-acting sources of

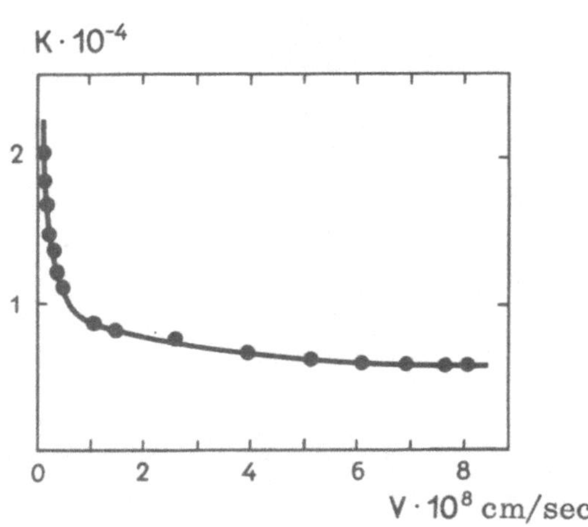


Fig. 5. Trapping of strontium by (100) faces of barium sulfate crystals. T = 25°C, Re = $2 \cdot 10^4$. $5 \cdot 10^{-3}$ M solution of potassium chloride.

parallel and almost equidistant steps of height $h \simeq 5 \cdot 10^{-6}$ cm, which are separated by terraces of width $\gamma = 5 \cdot 10^{-5}$ cm; the ends of these steps trap the strontium under almost equilibrium conditions, since the sorption rate is almost independent of the growth rate over the range $V$-$5 \cdot 10^{-3}$-$10^{-7}$ cm/sec, and here the trapping factor attains one of the limiting values $K_i = 6.5 \cdot 10^3$.

At lower growth rates, the crystals tend to take up more strontium, and this continues when the crystallization has been halted by diluting the solution with water. At short times after diluter, the sorption is described by the diffusion equation for a semiinfinite homogeneous space with the diffusion coefficient $D = 2 \cdot 10^{-14}$ cm$^2$/sec; at larger times, there is complete homogenization by diffusion within the crystals, and the equilibrium partition coefficient $K_e = 1.0 \cdot 10^5$ applies. Therefore, all the conditions for the migration state occur in the sorption of strontium by growing $BaSO_4$ crystals.

(b) Crystals of magnesium hydroxide deposited from highly supersaturated solution trap cobalt hydroxide mainly in the late state of crystallization, when the growth is almost completed. At this stage, the crystals have a size of about $10^{-6}$ cm and are covered by layered faces, which present internal surfaces capable of taking up the cobalt. The cobalt ions diffuse rapidly into the crystals ($D > 10^{-10}$ cm$^2$/sec), and they slowly form internal adsorption inclusions as chains of molecules extending along the common edges of intergrown crystals from the periphery towards the center. At higher cobalt concentrations, these linear inclusions expand into planar ones and tend to link the contacting faces of the crystals. This information has been derived from electron micrographs, x-ray studies, and radiometric examination of cobalt trapping by magnesium hydroxide [23].

These examples show that the migration state is common in growth from solution, which is due to the rapid impurity diffusion in freshly-formed crystals, although diffusion is also rapid in stable solid organic substances and hydrated crystals (Table 4).

Quantitative studies have been made on this state only in certain rare instances; however, there are reasons for believing that the trapping picture is fairly complicated on account of the effects on the impurity from minor details of the relief of the interface and from the fall in mobility during aging of the solid. This is confirmed by the above studies on the uptake of strontium by $BaSO_4$. The uptake factor as a function of growth rate for (100) faces in the range $10^{-9} < V < 10^{-8}$ cm/sec (Fig. 5) is described by

$$K_i = (6.5 \pm 0.5) \cdot 10^3 + (4.0 \pm 0.1) \cdot 10^{-5}/V. \tag{6}$$

The first term in (6) reflects the uptake by the ends of steps, while the second represents the additional uptake due to impurity migrating from the solution via the interface, particularly via terraces between steps. The first has been examined by Chernov [45] and occurs when the tangential speed of a growing layer is much larger than the normal growth rate, and also larger

TABLE 4. Room-Temperature Bulk-Diffusion Coefficients for Impurities in Crystals

| Solid phase | D, cm$^2$/sec | Ref. |
|---|---|---|
| Perfect ionic crystals | $10^{-21}$–$10^{-24}$ | [40] |
| Freshly-prepared ionic crystals | $10^{-14}$–$10^{-17}$ | [33] |
| Crystals of organic substances | $10^{-10}$–$10^{-16}$ | [41, 42] |
| Hydrated crystals | $10^{-9}$–$10^{-18}$ | [43, 44] |

than the impurity migration rate. Then (6) follows from [45] for $V \gg D/\gamma$ and $h \gg a$. The second case is expected when the migration and tangential-growth rates are comparable. These conditions occur in the sorption of strontium where $V\gamma/h \ll D/a$, so one assumes that each part of the enriched layer takes up an additional amount of impurity, which is present (in the main) very close to the ends of the macroscopic steps.

The interface can also alter the properties of the solid phase, and this itself can influence the impurity sorption; if one assumes that this effect occurs over a fairly extended region, we can apply the continuum approximation and neglect the periodicity in the motion of step ends to get for any part of the deposited layer far from an end that

$$\left.\begin{aligned}&\frac{\gamma V}{h}\frac{\partial c_c}{\partial l}-\frac{\partial^2(\alpha c_c D)}{\partial l^2}=0 \quad \text{for} \quad l>0,\\ &\alpha=\alpha_0,\ c_c=K_i c_s,\ \text{and}\ \frac{d(\alpha c_c D)}{dl}=\frac{V\gamma}{h}(K_s c_s-c_{T\infty}) \quad \text{for} \quad l=0,\\ &\alpha\to 1,\quad c_c\to c_{T\infty} \quad \text{for} \quad l\to\infty,\end{aligned}\right\} \tag{7}$$

where $l$ is the distance from the interface reckoned into the crystal along the normal to the end plane of a macroscopic step, $c_c$ is the concentration in the bulk of the crystal at a distance $l$, and $\alpha$ is the thermodynamic activity. The solution to (7) can be used with $D(l)$ and $\alpha(l)$ in conjunction with measurement to determine the state of the impurity as a function of $l$. In particular, the solution to (7) for d = constant and linear variation in $\alpha$ over a layer of thickness $l_0$ takes the form

$$K_i=\frac{c_{T\infty}}{c_s}=K_e\left(\frac{A+1}{A\alpha_0+\alpha_0^{-A}}\right), \tag{8}$$

where

$$A=\frac{l_0 V\gamma}{hD(\alpha_0-1)}.$$

If $\alpha_0 A \gg \alpha_0^{-A}$, (8) coincides with (6) if we assume that $\alpha = 16$, $l_0 = 5 \cdot 10^{-6}$ cm and that $D = 2 \cdot 10^{-14}$, $h = 5 \cdot 10^{-6}$ cm, and $\gamma = 5 \cdot 10^{-5}$ cm, the latter values being given by experiment. The result for $l_0$ is unreasonably large, so one has to assume that the cause of variation in $K_i$ with growth rate is localized directly at the step surface. Then if the end faces grow by layers, and the layer steps take up impurity with partition coefficient $K_1 = 6.5 \cdot 10^3$, while the additional enrichment is in accordance with Chernov's model, as modified to account for the migration state, we get

$$\left.\begin{aligned}&D\frac{\partial^2 c_c}{\partial l^2}=\frac{\partial c}{\partial t} \quad \text{for} \quad 0<l<h_c \quad \text{and} \quad 0<t<hh_c^2/V\gamma\gamma_c;\\ &c_c(l,0)=K_1 c_s,\quad c_c(0,t)=K_e c_s,\quad c_c(h,t)=\bar{c}_c=\text{const};\end{aligned}\right\} \tag{9}$$

and therefore for $V\gamma\gamma_c > \pi^2 hD$ we have that

$$K_i=K_1+(K_e-K_1)\frac{Dh}{\gamma\cdot\gamma_c V}. \tag{10}$$

Here $h_c$ and $\gamma_c$ are the height of a step and the distance between steps on the end faces, while $\bar{c}_c$ is the mean concentration within a layer. Equation (10) coincides with (6) for $\gamma_c = 5 \cdot 10^{-6}$ cm, and this result is quite acceptable for macroscopic steps, so the experimental data conflict with the view that there is a considerable change in the state of the strontium in the surface region, although the evidence does confirm that there is additional enrichment of each layer after the growth via end faces.

The diffusion coefficient for strontium does not vary with time within this range of V and is only slightly dependent on the latter; however, if V is varied widely, there is a change in the mobility of the impurity, as well as a change in the diffusion coefficient for the host substance. This change has been observed as follows [33]. At a certain point in the trapping, one adds a complexing agent that increases the solubility of barium sulfate or else simply extra water. The amount of added water or complexing agent is chosen on the basis of the crystallization kinetics to leave the solution saturated after dilution, which means that the growth rate rapidly falls. Tests have shown that minor deviations of the concentration from equilibrium do not influence the result, since at such deviations the rates of growth or dissolution of $BaSO_4$ are negligibly small. Then, when the growth has ceased, microscopic amounts of strontium nitrate, potassium sulfate, and barium nitrate are added, which are labeled respectively with the radioisotopes $^{89}Sr$, $^{35}S$, and $^{133}Ba$. Then the fluxes of each of these tracers into the solid are determined for a period sufficient for the $BaSO_4$ to take up some dozens of surface monolayers. It has been found that the tracers obey the diffusion equation for a homogeneous medium, with the diffusion coefficient dependent on the concentration of the host substance in the solution before dilution (Fig. 6). The time stability of the diffusion coefficient indicates that the results are governed by the state of the freshly-formed parts of the solid before dilution, and this can be described by saying that the crystal inherits the vacancies in the quasiequilibrium surface layers, and the concentration of these vacancies is determined by the initial disorder in the freshly-formed parts of the crystal and the kinetics of vacancy healing by host substance entering from the solution. The initial state of disorder becomes more pronounced as the concentration of the host substance decreases, while the scope for retaining defects diminishes. As a result, D(C) is not a monotone function. Over a certain period, the vacancies inherited by the crystal combine into dipolons, which migrate to the interface, where they are filled by host substance. As a result, the flux of material to the ends of the steps is accompanied by a flux through the terraces, which does not affect the volume of the crystal but does alter the density. This effect is detectable by measuring the local self-diffusion coefficient, the electrical conductivity, and the dielectric loss for freshly-prepared crystals [33].

This elimination of the excess vacancies reduces the concentration of these to the level determined by the mechanical factors; if the latter have a considerable effect, e.g., when the crystal moves freely with the solution, with the result that it collides with other crystals and with the walls of the vessel, the level of the residual defects may remain extremely high. For

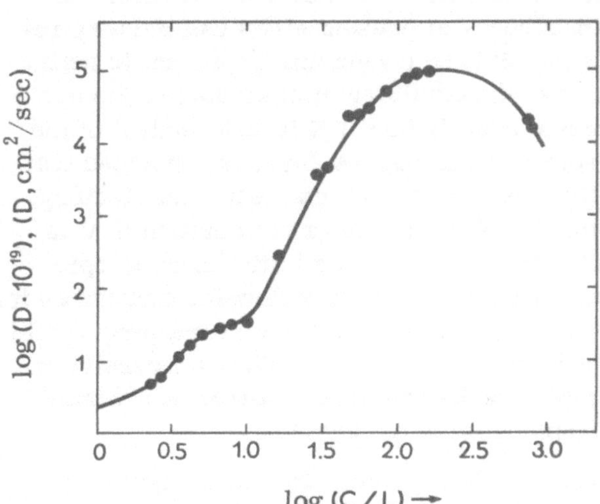


Fig. 6. Self-diffusion coefficients for barium and sulfur in freshly-prepared barium sulfate crystals at 25°C, crystals prepared by mixing $10^{-2}$ M solutions of $K_2SO_4$ and $BaCl_2$. The values for D were determined during the growth of crystals formed immediately on mixing the reagents. The abscissa is the concentration C of the solution before dilution, while L is the solubility of the substance.

instance, crystal collision transfers the NaCl−Ag (I)−$H_2O$ system to the migration state for $V < 10^{-4}$ cm/sec [46]. The behavior of the vacancies at the surface clearly determines the response of the migration state in many instances of trapping, since the temperature range 0-100°C commonly employed in sorption studies is the one in which the vacancy mechanism for impurity diffusion is the main one. The effects of nonequilibrium vacancies and interface fine structure result in complexity in the migration state.

## Liquid-Phase Diffusion and Liquid-Phase Kinetic States

Impurity diffusion in a solution alters the trapping coefficient at growth rates of $10^{-4}$-$10^{-5}$ cm/sec with natural or forced convection [5, 32, 47]. At such rates, the available evidence can often be described formally by means of Chernov's model, which assumes that there is an immobile layer of solution near the crystal (thickness $\delta$ of $10^{-3}$-$10^{-5}$ cm), whose properties are identical with those of the free liquid.

For instance, when cerium is taken up by a $K_2SO_4$ crystal at V of $10^{-4}$-$10^{-5}$ cm/sec, the effective trapping coefficient K is [21] described by

$$K = \frac{K_e}{K_e + (1 - K_e)\exp(-V\delta/D_s)} \tag{11}$$

with $\delta \simeq 5 \cdot 10^{-4}$ cm; (11) coincides with the solution to Fick's equation with a mobile boundary bearing a Nernst layer [48] subject to the diffusion coefficient $D_s$ for the solution.

Very little is known about the liquid-phase kinetic state, although slow hydrolytic, redox, solvation, and complexing processes are common in solutions, which indicates that the state should be important.

## 2. Measurement Purposes

The above evidence indicates that we now have a sufficient data on impurity uptake from solutions to define the main factors that govern trapping for individual model systems; also, a theory has been devised that can explain observations and stimulate fresh studies, and therefore provide a basis for more refined techniques.

In general, we require very detailed studies of a variety of model systems in order to define the relationship between trapping and the properties of pure solids, as well as the effects of detailed growth conditions. Such systems should be chosen on the basis of the theory of equilibrium solid-solution formation. However, that theory at present gives satisfactory results only for alkali halide and molecular crystals [49, 50], so it remains important to define empirical laws for cases where there is no doubt as to the equilibrium character of the process. It is usually assumed that the results represent equilibrium if K is independent of the growth rate, which is usually adjusted over a narrow range in supersaturation, provided that the crystal shows no growth pyramids [51]. The first criterion is by no means satisfactory, since (2) indicates that constancy in K can occur for $K = K_1$ if the range of variation in V is insufficiently wide. The second criterion is inapplicable if differences in the macroscopic orientation of the crystal faces do not result in substantially different molecular structures for the surface layers (Paneth layers). Therefore, measurements on equilibrium trapping should include definite demonstration of the equilibrium condition, e.g., by employing successive cycles or different solvents [52]. Unfortunately, such checks are rarely performed, which makes it difficult to discuss data on equilibrium trapping for most systems.

Any research on nonequilibrium trapping should include examination of the relief of the interface during the growth, as well as measurement of diffusion rates and the forms taken by impurities in the phases. The purpose of such studies is to define conditions under which one attains the limiting modes, which can then be utilized to define the parameters of the elementary processes. Each limiting mode has its own specific features.

(a) In the kinetic-adsorption state, it is important to accumulate data on K(V) over a wide range in growth rates; at $V > 10^{-3}$ cm/sec, it is possible to use single crystals for the purpose [17], while at $V < 10^{-3}$ cm/sec one can employ a set of evenly facetted crystals. The extensive total surface in the latter case provides for an appreciable increase in the total crystal mass over a reasonable time even at $V \sim 10^{-11}$ cm/sec [32]. Unambiguous interpretation of K(V) is best based on demonstration of nonuniform impurity distributions over the various faces. For this purpose one can use autoradiography or electron microscopy, since in principle it is then possible to detect nonuniformities of scale $10^{-4}$-$10^{-6}$ cm/sec without necessarily transferring the crystals to vacuum [53], while isotope exchange allows one to define the temperature dependence of the impurity distribution at the surface in relation to the probability of migration in solution [54]. The quasichemical approach [55] appears promising in the description of the kinetic-adsorption and liquid-phase kinetic modes. In that approach, the interaction with the surface is considered as a chemical reaction, and the results are presented in terms of rate constants and the equilibrium local composition or the impurity distribution function in terms of activation energy or heat of reaction. The quasichemical approach treats (2) and (3) as due to i successive first-order reactions. The approach also allows one to approximate the behavior of molecules at the interface. In particular, it indicates molecular smoothness for the crystals in the systems listed in Table 2.

(b) In research on the liquid-phase diffusion state, it is important to define the hydrodynamic circumstances and the impurity distribution near the crystal surface, since there is evidence that the liquid near the surface has special properties in a film of thickness comparable with $\delta$. For example, it has been found that water passing through concentrated crystal suspensions shows deviations from D'Arcy's law [56]. Also, the shapes taken by thin films of liquid on crystals in tangential air flows would indicate anomalous elasticity in these films up to thicknesses of $10^{-5}$ cm [57]. Further, self-diffusion parameters have been measured for ions passing through porous membranes, and it has been found that the mobility alters near the surface [58]. No studies have so far been made on the effects of these phenomena on mass transfer at crystal surfaces.

(c) In the migration state, one needs evidence on the concentration distribution and diffusion coefficients for surfaces under a variety of conditions; there are theoretical reasons for considering that double electrical layers are important here, and this is supported by the scanty experimental data [33, 38, 55]. It is clear that such evidence should be extended by examining concentrated crystal suspensions, where the mass of the host material in the surface layer is reasonably large.

Another important aspect of limiting conditions is the effect of any given factor on the results of uptake, where a detailed study of the mechanism is required. Studies of this type have been begun by Zhmurova and Khaimov-Mal'kov [17], who examined how supersaturation affects the trapping coefficient (either directly or via the growth rate). For this purpose, the trapping was performed in the kinetic-adsorption mode, while the growth rate was varied by adding the adsorbed substance at a constant supersaturation. It was found that the additive did not affect K, which means either that there is no correlation between K and V or that V and $\beta_i$ vary in proportion. The second is the less probable. It was therefore concluded that the supersaturation affects the trapping directly, so the elementary acts of uptake could be judged from the form of K(C). Similar techniques should provide more reliable evidence on trapping conditions and should accelerate the discovery of means of controlling crystal composition, particularly for growth from solution.

## Literature Cited


1. A. V. Shubnikov, How Crystals Grow [in Russian] (1935).
2. H. Buckley, Crystal Growth [Russian translation], IL (1954).
3. G. G. Lemmlein, Sector Structure in Crystals [in Russian], Izd. AN SSSR (1949).
4. N. N. Sheftal', In: Growth of Crystals, Vol. 1, Consultants Bureau, New York (1958), p. 5.
5. G. Walton, Formation and Properties of Precipitates, Pergamon Press (1965), p. 79.
6. A. A. Chernov, Usp. Fiz. Nauk, 73:317 (1961).
7. Z. Walter and H. Schmidt, J. Amer. Chem. Soc., 50:3266 (1928).
8. N. B. Mikheev and G. L. Shmanenkova, Zh. Neorg. Khim., 10:244 (1965).
9. V. I. Grebenshikova and V. N. Bobrova, Radiokhimiya, 3:377 (1961).
10. L. Gordon and K. Rowley, Anal. Chem., 29:35 (1957).
11. D. N. Bykhovskii and A. A. Grinberg, Radiokhimiya, 2:164 (1960).
12. D. N. Bykhovskii, Radiokhimiya, 3:535 (1961).
13. D. N. Bykhovskii, Dokl. Akad. Nauk SSSR, 145:845 (1962).
14. L. Gordon, J. J. Peterson, and B. Burtt, Anal. Chem., 27:1770 (1955).
15. J. A. Hermann and J. E. Suttle, Treatise on Analytical Chemistry, Parts 1 and 3, Interscience, New York (1961), p. 32.
16. D. I. Klein and B. Fontal, Talanta, 12:35 (1955).
17. E. I. Zhmurova and V. Ya. Khaimov-Mal'kov, Kristallografiya, 15:136 (1970).
18. G. A. Andreev, Kristallografiya, 12:104 (1967).
19. C. Fujiwara and K. Nagashima, Japan Analyst, 15:980 (1966).
20. G. D. Botsaris, E. A. Mason, and R. C. Reid, Amer. Chem. Eng. J., 13:764 (1967).
21. I. V. Melikhov and S. G. Babayan, Radiokhimiya, 6:153 (1964).
22. O. Hahn, Applied Radiochemistry [Russian translation], LM (1947).
23. I. V. Melikhov and M. Ya. Belousova, Radiokhimiya, 13:802 (1971).
24. V. G. Khlopin, Selected Works [in Russian], Vol. 1, Izd. AN SSSR, Moscow (1957), p. 173.
25. D. Balarev, Freiberg Forschung, A 123:333 (1959).
26. A. V. Belyustin and S. S. Fridman, Kristallografiya, 13:363 (1968).
27. R. N. Hall, J. Phys. Chem., 57:836 (1953).
28. G. Bliznakov, A. Phys. Chem., 299:372 (1958).
29. A. A. Chernov, Dokl. Akad. Nauk SSSR, No. 5, 470 (1960).
30. I. V. Melikhov, Radiokhimiya, 3:513 (1961).
31. I. V. Melikhov and E. K. Kirkova, Radiokhimiya, 6:5 (1964).
32. I. V. Melikhov, E. K. Karkova, and S. G. Babayan, Radioelement Coprecipitation and Adsorption [in Russian], Nauka, Moscow (1965), p. 34.
33. I. V. Melikhov, Zh. Bukovich, and B. D. Nebylinyn, Dokl. Akad. Nauk SSSR, 199:1350 (1970).
34. I. V. Melikhov, M. Ya. Belousova, and V. M. Peshkova, Zh. Anal. Khim., 25:1144 (1970).
35. N. B. Mikheev and L. M. Mikheeva, Zh. Neorg. Khim., 7:3671 (1962).
36. I. V. Melikhov, Radiokhimiya, 10:513 (1968).
37. V. V. Voronkov and A. A. Chernov, Kristallografiya, 12:222 (1967).
38. A. A. Chernov and A. M. Mel'nikova, Kristallografiya, 16:488 (1971).
39. G. I. Gorshtein, Radiokhimiya, 1:497 (1959).
40. B. B. Lidiard, Sci. Progr., 56:103 (1958).
41. J. N. Sherwood and D. J. White, Phil. Mag., 16:957 (1967).
42. E. Hoinkis and H. W. Levi, Naturwissenschaften 53:500 (1966).
43. R. Calvet and H. Chaussidon, Bull. Groupe Franç. Argiles, 19:91 (1957).
44. R. M. Barrer, R. F. Bartholomew, and L. V. C. Rees, J. Phys. Chem. Solids, 24:51 (1963).
45. A. A. Chernov, In: Growth of Crystals, Vol. 3, Consultants Bureau, New York (1962), p. 35.
46. I. V. Melikhov and G. Zwald, Radiokhimiya, 10:120 (1968).
47. N. Richt, R. Sizmahn, and P. Hidalgos, Z. Phys. Chem., 25:351 (1960).

48. J. A. Burton, P. C. Prim, and W. P. Slichter, J. Chem. Phys., 21:1987 (1953).
49. A. I. Kitaigorodskii, Zh. Strukt. Khim., 1:324 (1960).
50. P. Brauer, Z. Electrochem., 57:749 (1953).
51. F. A. Trambore and P. E. Freeland, J. Electrochem. Soc., 103:458 (1961).
52. I. V. Melikhov, Proceedings of the All-Union Conference on Methods of Producing Especial-Purity Substances [in Russian], Izd. NIITEKhIM, Moscow (1967), p. 96.
53. E. P. Senchenko, S. E. Bokshtein, and L. M. Moroz, Zav. Lab., 35:57 (1969).
54. R. P. Eischens, J. Amer. Chem. Soc., No. 12, 6167 (1952).
55. F. Kröger, Chemistry of Imperfect Crystals [Russian translation], Mir, Moscow (1969).
56. P. I. Andrianov, Bound Water in Soils [in Russian], Izd. AN SSSR, Moscow (1947).
57. V. V. Karasev and B. V. Deryagin, Dokl. Akad. Nauk SSSR, 101:289 (1955).
58. O. Grigor'ev, Electrokinetic Parameters of Capillary Systems [in Russian], Izd. AN SSSR, Moscow (1956).

# TRAPPING DURING GROWTH FROM A MELT


**J. Barthel**

*Central Solid State Physics and Materials Science Research Institute, German Academy of Sciences, Dresden*


## 1. Introduction

The composition and homogeneity or otherwise of a material can have substantial effects on applications; important effects arise from the distributions of impurities or doping elements produced by crystal growth from melts. Applications of this type have given rise to various trends in research, which are based on various methods of crystallizing substances and systems. However, they all involve a unified viewpoint on the basic processes in melt–crystal transition. The first part of this paper presents experimental data on the vertical pulling of single-phase crystals, while the second part deals with current views on basic crystal growth processes.

## 2. Experimental Data on Impurity Element Distributions

### 2.1. Cellular Structure

A cellular structure can arise from a nonuniform impurity distribution in the form of chains extending along the growth direction [1, 2]. These chains are seen in cross section as elongated or hexagonal networks composed of lines with elevated or reduced impurity concentrations (Fig. 1), the lines being enriched when the partition coefficient k at the boundary is $<1$ or depleted when $k > 1$. In extreme instances there are no closed lines but merely spots of altered concentration (Fig. 2), which appear to be located on more or less well-defined lines, which may sometimes be inappreciable [3]. It is clear that the cells do not occur throughout the cross section, and the boundaries of the grains and subgrains revealed by autoradiography here represent closed curves, which are clearly visible even in parts of the crystal where no cellular structure is visible (Fig. 2).

The cellular structure arises from supercooling itself dependent on the concentration, which results in instability in the growth front; theoretical studies have been made [4, 5] on the conditions for onset of instability for a rough growth surface. In practice, the following condition for a cellular structure is found to be sufficient, which is derived from simple arguments:

$$G/v < \frac{mc_0(1-k)}{Dk},$$

where G is the temperature gradient in the melt, v is the crystallization rate, m is the slope of the liquidus line, $c_0$ is the concentration in the melt, and D is the diffusion coefficient in the melt.

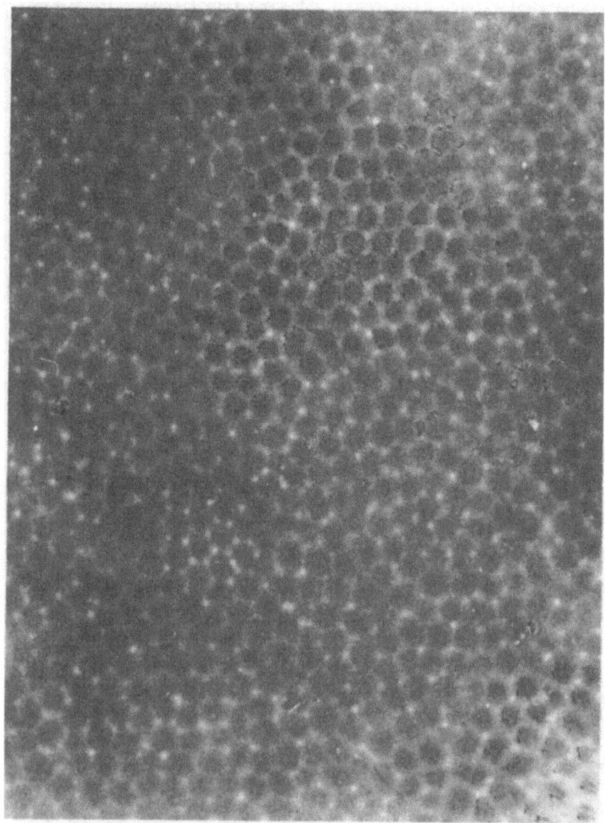

Fig. 1. Cellular structure: autoradiograph of a cross section of an aluminum crystal containing iron [2].

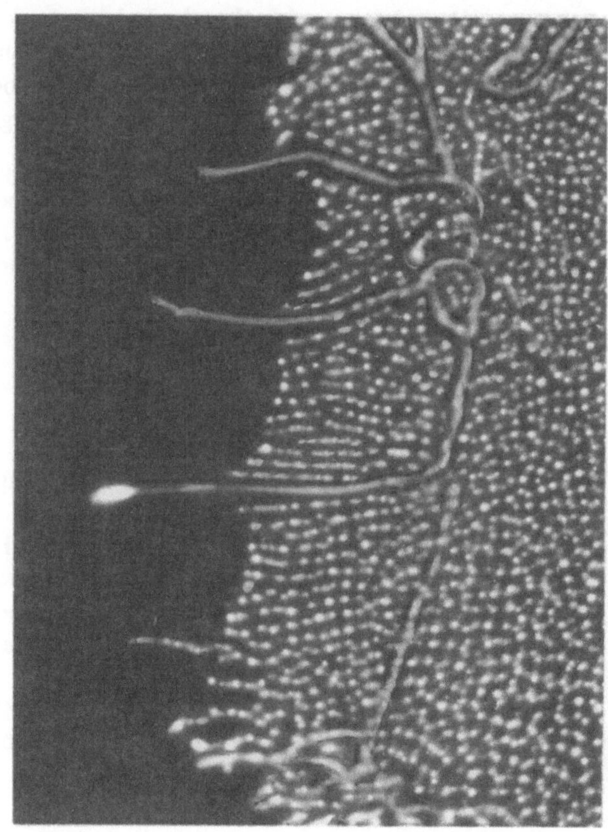

Fig. 2. Cellular structure with localized (point) enrichment in the impurity element and transition to stable growth. The grain boundaries are indicated by strings of regions enriched in the impurities. Cross section of an aluminum crystal containing iron.

A logical theory of the stability requires more detailed analysis of the problem, in particular the boundary energy; the major importance of this energy becomes clear on considering instability at grain boundaries (Fig. 2), which occurs when the phase boundary is otherwise stable. The theory also requires a further refinement, namely extension to crystallographically anisotropic forms of growth front.

## 2.2. Growth Channels of the First Kind

Macroscopically, such channels seem as regions of altered impurity concentration, and the effective partition coefficient may exceed 1 (Fig. 3) [6]. The growth front may be examined by detaching the crystal from the melt and by examining the growth bands, which indicates that the channels are related to the formation of low-index planes at the growth front. Therefore, the impurity channels are due to the scope for facetted growth. The latter in turn is determined by the type of crystal and the orientation. The size and disposition of the impurity core are substantially dependent on the curvature of the growth front and the supercooling at the faces [7, 8]. Usually, the faces lie at the middle of the crystal, although additional faces do occur at the edge of the growth front during the production of dislocation-free semiconductor crystals [9]. The results from various lines of research indicate clearly that the growth kinetics must be the controlling factor responsible for the smooth or rough surfaces.

## 2.3. Rotational Nonuniformity

This type of nonuniformity has elevated levels of the impurity along a single-start or multiple-start spiral surface in a crystal that is rotated during growth; longitudinal sections show this nonuniformity as a system of bands originating at the phase boundary and having elevated or reduced impurity concentrations (Fig. 4), whose periodicity is determined by the ratio of the pulling rate to the rotational speed [10]. A cross section in general shows spirals, but more complicated forms for the phase boundary [11] can result in other figures (Fig. 5). The bands containing elevated impurity levels show not only ordinary enrichment but also special deposits (Fig. 5). An impurity channel is seen because the growth bands, which reflect the shape of the crystallization front, produce straight lines in the section [12].

The changes in growth rate occurring on account of external perturbations indicate that the banding is due to the periodic variation in the crystallization rate arising from rotation in an unsymmetrical temperature distribution [13]. So far it has proved impossible to suppress such rotational nonuniformity, although it has been found that the growth rate at the center fluctuates much less than that at the periphery.

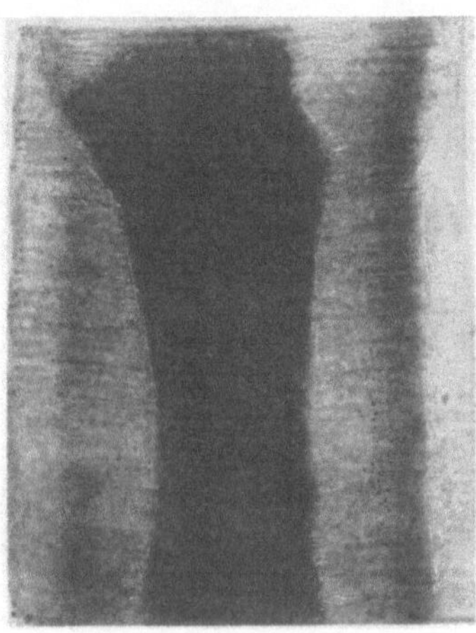

Fig. 3. Growth channel of the first kind in the middle of a crystal and growth channel of the second kind at the periphery; autoradiograph of a germanium crystal doped with antimony, longitudinal section [6].

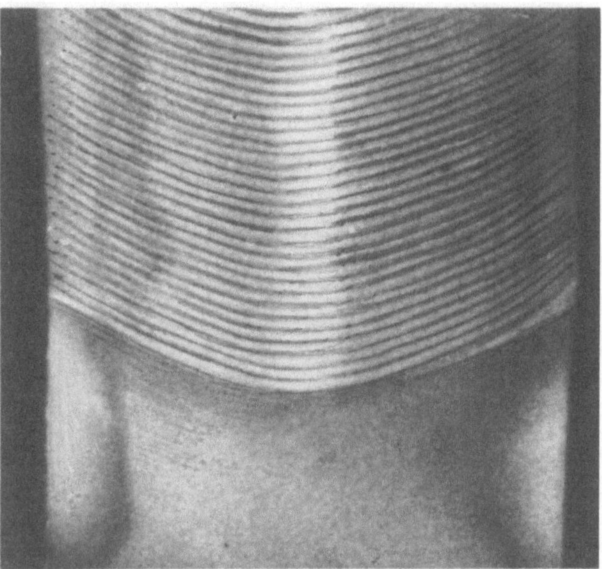

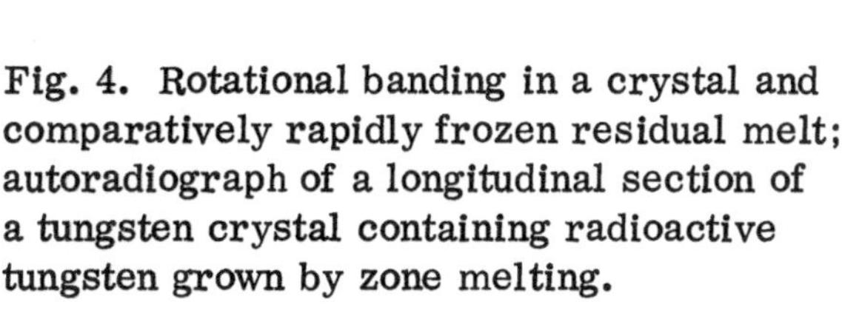

Fig. 4. Rotational banding in a crystal and comparatively rapidly frozen residual melt; autoradiograph of a longitudinal section of a tungsten crystal containing radioactive tungsten grown by zone melting.

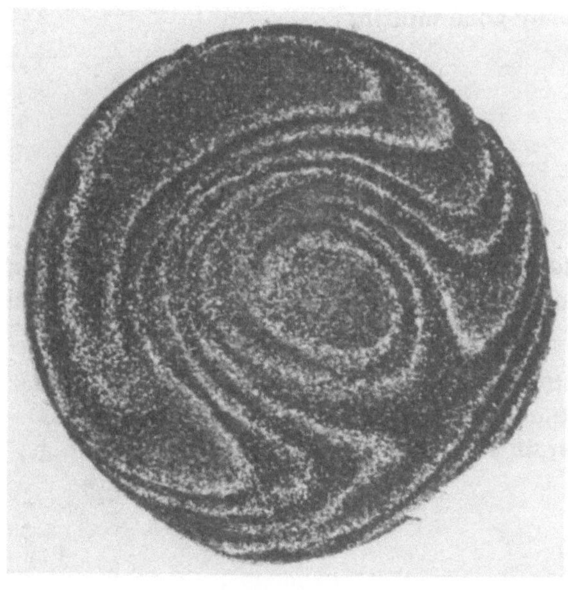

Fig. 5. Rotational banding, cross section of a crystal grown by zone melting and containing oxygen, with deposited oxide.

## 2.4. Convective Nonuniformity

The bands of elevated and reduced impurity content described above are simulated by similar bands frequently observed in crystals grown without rotation; in that case, the bands also reproduce approximately the shape of the crystallization front at successive instants [14]. The distances between the bands in general are variable (Fig. 6). Also, cross sections show that there may be several systems that lie in different parts of the cross section and which may overlap.

Measurements on the growth rate and temperature distribution show that these effects are due to fluctuations in the instantaneous crystallization rate, which themselves are due to temperature fluctuations; it has also been found that partial melting can occur. Studies on the convection have indicated that the temperature fluctuations are related to turbulence [15]. The critical turbulence parameter is the Rayleigh number

$$R_a = \frac{l^3 \beta g \Delta T}{\alpha \nu},$$

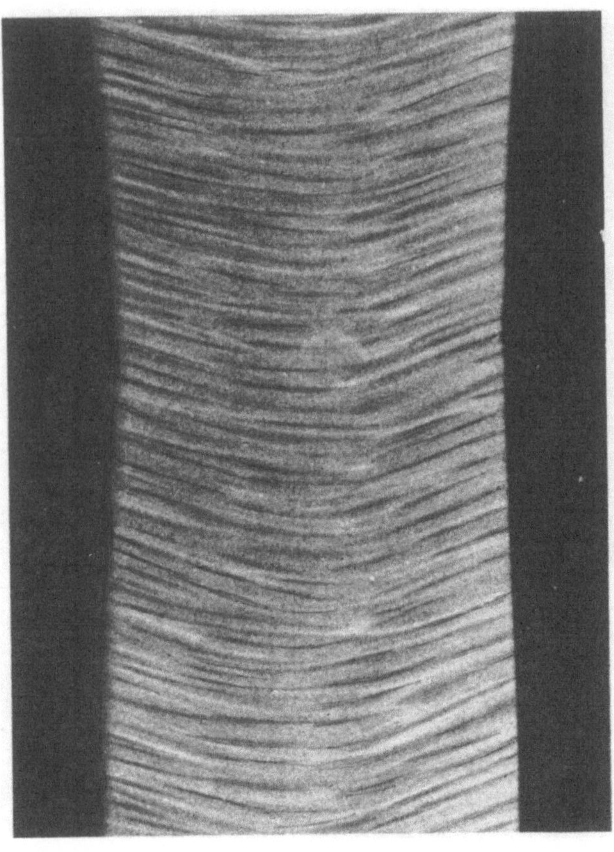

Fig. 6. Convective banding, autoradiograph of a longitudinal section of a molybdenum crystal containing tungsten grown by zone melting.

where $l$ is the characteristic length, $\beta$ is the critical expansion coefficient, g is the acceleration due to gravity, $\alpha$ is the thermal conductivity of the melt, $\Delta T$ is the temperature gradient in the melt, and $\nu$ is kinematic viscosity.

If the Rayleigh number falls below some critical value, it may be possible to suppress the convective nonuniformity, e.g., by reducing the characteristic length of the melt or the temperature gradient, i.e., the density differences, or else by increasing the viscosity, e.g., by means of a magnetic field [14, 15].

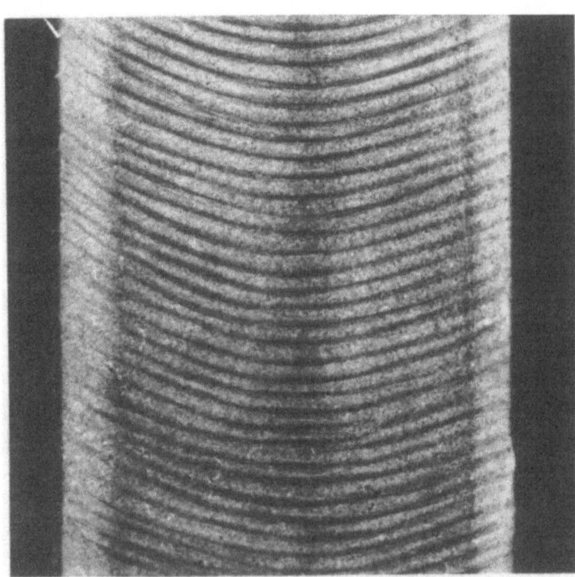

Fig. 7. Convective banding with longitudinal convective nonuniformity; autoradiograph of a longitudinal section of a molybdenum crystal containing tungsten grown by zone melting.

Convective banding with strict periodicity has been observed (Fig. 7) in the transitional region [16]; under other conditions, it has proved possible to detect sinusoidal temperature variations [17].

## 2.5. Growth Channels of the Second Kind

Baralis [6] used this term to denote concentration variations in the cross section not due to faces at the growth front. Similar effects can be seen in longitudinal sections (Figs. 3, 4, 6, and 7). Figure 8 shows the effect particularly well. Such nonuniformity also may have no correlation with the banding (Fig. 9) and is an element in the structure that persists along the crystal [6].

The causes must be discussed in terms of the spatial distribution of the impurity, particularly the effects of the mean concentration on the partition coefficient.

Baralis [6] examined a system with k less than 1 and observed a rise in the mean concentration following the deepest depressions on the crystallization front for crystals with convective nonuniformity. If the rotational speed is comparatively high, it is possible to produce a broad band with an elevated impurity content (Fig. 3).

This nonuniformity has been ascribed to nonuniform mixing of the melt over the cross section, as has previously been supposed [16] in a discussion of the causes of regions of low concentration in systems having k greater than 1 (Figs. 7 and 9). However, Figs. 4 and 8 show that regions of elevated impurity content can occur for $k > 1$ in relation to recesses on the growth front when there is rotational and convective banding. Correspondingly, Baralis observed regions of reduced concentration for rotated crystals when k was less than 1. However, this conflict was overlooked in [6]. The mechanism must be different from the one appearing in the above explanation, and it would seem that if one quenches the melt at the end of the pulling, the concentration distribution in the product is uniform in all the above cases. Therefore,

Fig. 8. Convective banding with central banded nonuniformity, autoradiograph.

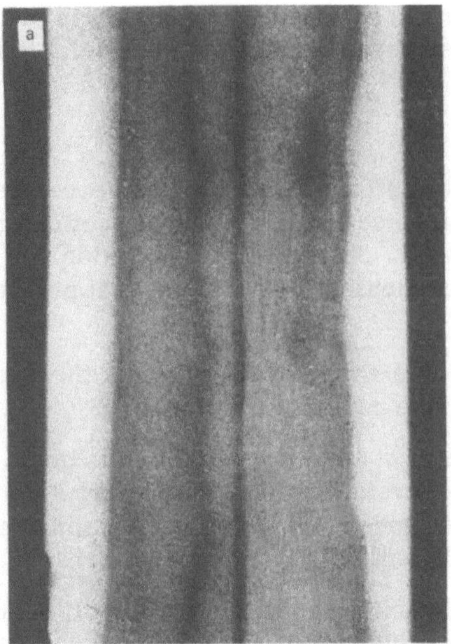

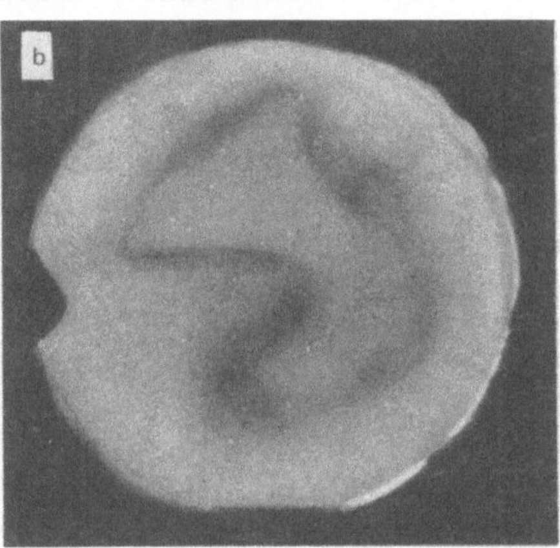

Fig. 9. Convective nonuniformity; autoradiographs of (a) longitudinal section and (b) cross section. Molybdenum crystal containing tungsten grown by zone melting.

the differences in concentration within the crystal cannot be explained in terms of ones in the melt.

There are clearly differences in the causes of the various types of nonuniformity, so it is best not to call all these forms channels of the second kind. For reasons discussed below, nonuniformities previously called convective are best called vertical banding.

## 3. Basic Impurity Trapping Processes

### 3.1. Mechanisms

The entire process of crystal growth and impurity trapping involves three groups of phenomena, which include reactions at the grain boundary and transport processes in both phases. The most important are the processes at the boundary and the transport in the melt. In what follows we survey experimental and theoretical results on these topics.

### 3.2. Phase-Boundary Partition Coefficients

Under equilibrium conditions, the laws of thermodynamics apply to impurity trapping, and the distribution between the crystal and the melt can be derived from the phase diagram by means of the equilibrium partition coefficient $k_0 = c_S/c_L$, where $c_S$ and $c_L$ are the concentrations of the impurity in the solid and liquid phases. This partition coefficient is dependent only on the composition and on the pressure. The value is best determined from thermodynamic parameters if the impurity concentration is very low. In the case of a binary system not yet examined by experiment, the value can be estimated from the positions of the elements in the periodic system [18].

Chernov's estimates [19] show that deviations from equilibrium distribution occur at realistic growth rates on account of the kinetic processes, so the kinetic partition coefficient $k_i$ should be dependent on the kinetics of the surface processes and thus ultimately on the orientation of the surface, the supercooling, and the growth rate.

The transport processes in the melt make it difficult to measure $k_i$; however, some evidence on the deviation of $k_i$ from $k_0$ can be obtained if the crystal orientation alone is varied while the other factors remain constant. Evidence on this comes from a variety of sources (see [20] for a bibliography), but from the current viewpoint one cannot take these results as representing proof, since the interpretations neglect the occurrence of growth channels of the first kind. In particular, Spittle et al. [20] observed that the effective partition coefficient was dependent on the orientation at medium growth rates in systems where the growth front consisted entirely of an atomically rough surface, where growth channels of the first kind thus could not occur (the face effect). Kramnacker and Lange [21] attempted to measure $k_i$ directly by quenching the unused melt and determining the impurity concentration in the boundary zones; they found that the coefficient was very much dependent on the growth rate for rough surfaces. Brice's analysis [22] of the face effect showed very good agreement with simple theoretical models, but a more detailed discussion of the method [23] throws doubt on it and thus on the results. The Burton–Prim–Slichter theory [4] indicates similar marked dependence of the partition coefficient on the growth rate.

Growth channels of the first kind show unambiguously that the kinetic partition coefficient is dependent on the growth kinetics, namely on whether the growth occurs on a rough surface or an atomically smooth one. Here the supercooling dependence of $k_i$ was not clearly established for a smooth front. Certain measurements [24] indicate that the impurity concentration varies at the periphery of the impure core.

Zhmurova and Khaimov-Mal'kov [25] found dependence on the supersaturation at high values.

Theoretically speaking, the problem has not been solved for atomically smooth faces or for rough ones; Chernov and Voronkov [25] performed model calculations on the incorporation of impurity atoms at isolated lattice nodes in a semiconductor and at groups of nodes for particular stages in the process, but their results were of extremely preliminary type. Studies have also been made by the methods of irreversible thermodynamics without consideration of the detailed kinetics at rough growth fronts. Jindal and Tiller [26] in particular discussed the case of high supersaturations and pointed out various errors in earlier studies. If the supersaturation is low, which corresponds to real growth conditions, it has been shown [27] that the deviations of the kinetic coefficient from $k_0$ should be very small, and then in general one would expect $k_i \approx k_0$ at the growth rates occurring in practice.

### 3.3. Effective Partition Coefficient

Energy and mass transport occur during crystallization, and the effects of transport processes on impurity uptake are known for very simple models [28]. If pure diffusion applies in the steady state, the impurity distribution ahead of the crystallization front becomes such that the effective partition coefficient is 1. Then $k_{eff} = c_S/c_{L_\infty}$, where $c_{L_\infty}$ is the concentration in the melt at a large distance from the interface. If there is convective mixing ahead of the phase boundary, the internal friction in the liquid retards the convection, so the transport occurs mainly by diffusion. If one assumes that the concentration is kept equal throughout the main body of the melt by convection, whereas near the boundary there is a layer of thickness $\delta$ where the transport occurs only by diffusion, then the uptake can be described by means of an effective partition coefficient

$$k_{eff} = c_S/c_{L\infty} = k_i\{k_i + [1 - k_i] \exp(-R\delta/D)\}^{-1},$$

where D is the diffusion coefficient in the melt. The value of $k_{eff}$ is 1 for high values of $R\delta/k_{eff}$, whereas the value tends to $k_i$ for $R\delta \to 0$; in the case $k_i < 0.5$, the derivative of the effective coefficient with respect to $R\delta$ at the critical value of the latter is

$$(R\delta_{int}) = -D \cdot \ln \frac{k_i}{k_i - 1}$$

and the maximum value is $\delta k_{eff}/\delta(R\delta) = 1/4D$; in the other case of $k_0 \geq 0.5$, the maximum value of $k_{eff} = f(R\delta)$ occurs for $R\delta = 0$.

Evidence on $\delta$ can be obtained only by examining the convective mixing in the melt.

### 3.4. Laminar Convection in the Melt

The convection produced by external forces and density differences is very important in many applied instances, although comparatively little is known about the details, mainly on account of theoretical and experimental difficulties.

We have long understood the convection conditions arising in a simple model case, namely when a plate rotates in an unbounded liquid; a similar geometry applies to vertical pulling of a rotating crystal. Theoretical analysis indicates that there is a boundary layer throughout the cross section, within which the speed of the liquid increases from zero to some limit. The thickness of this layer is $\delta \sim (\nu/\omega)^{1/2}$, where $\nu$ is the kinematic viscosity and $\omega$ is the speed of rotation. Then the diffusion boundary layer $\delta_c$ is defined as the layer within which the concentration remains constant, and this is defined by

$$\delta_c = 1.6 \cdot \nu^{1/6} \cdot D^{1/3}/\omega^{1/2} \sim \delta/Sc^{1/3},$$

if the Schmidt criterion $Sc > 1$; more exact theories also give relationships for arbitrary Sc [29]. The result for $\delta_c$ can be used in calculating $k_{eff}$.

There has as yet been no theoretical analysis of conditions for pure or mixed thermal convection in vertical crystal growth, although corresponding researches have been performed for horizontal growth [30], but these cannot be discussed here.

Qualitative concepts on the convective fluxes are based on model experiments with transparent liquids; Carruthers' data [31] indicate that the thickness of the boundary layer varies over the cross section, and the thickness of the diffusion layer is itself determined by the diffusion perpendicular and parallel to the phase boundary. As yet, no calculations have been performed, but there are reasons for believing that the diffusion layer is of maximal thickness where the flow lines pass far into the melt, and one can explain the origin of the convective nonuniformity on this basis. The areas in the crystal mostly conform with the flux distribution described by Carruthers. However, in certain instances (Fig. 9) there is disagreement, and more detailed research on such convective nonuniformity provide new evidence on the flux distribution.

## 3.5. Turbulence

The theory of turbulence in a melt is substantially more complicated than the theory of simple convection; there are many papers on the topic, including some recent ones [30, 32]. The main parameter governing the turbulence onset is the Rayleigh number, which has been discussed also for horizontal growth [30]. The fluctuations in temperature arising from the turbulence are given by the following formula subject to certain assumptions [30]:

$$\bar{T}=\frac{2}{3\pi}(\Delta T)^2\frac{\ln \mathrm{Nu}}{\mathrm{Nu}},$$

where $\bar{T}$ is the rms amplitude of the temperature fluctuations, $\Delta T$ is the vertical temperature difference, and Nu is the Nusselt number.

It is also clear that the turbulence appears particularly as eddies superimposed on the ordinary convection, which is not thereby substantially distorted. This explains the strict periodicity in the convective banding even within the transition region (Fig. 7), as well as the conservation of the convective nonuniformity, which is independent of the banding. This nonuniformity vanishes if the turbulence is strong.

## 3.6. Concentration Distribution and Temperature Fluctuations

The nonstationary approximation must be used in any theoretical examination of turbulent convection and the related temperature fluctuations, particularly as regards variations in growth rate and effective partition coefficient; Hurle, Jakeman, and Pike [33] used a simple model in which there is an immobile diffusion layer of thickness $\delta$ ahead of the front, while the frequency of the temperature fluctuations is constant, and any variation in the flux distribution in the melt is neglected. Numerical results were obtained for metal and semiconductor crystals under various growth conditions for various partition coefficients. The amplitudes of the fluctuations in growth rate and concentration are proportional to the amplitude of the temperature fluctuations. The dependence of the growth-rate change on the temperature-fluctuation frequency has a broad peak, which is usually more prominent for a metal than for a semiconductor. The amplitude of the concentration fluctuations decreases towards high frequencies for semiconductors. In the case of a metal, the frequency dependence may show a peak, which can be particularly pronounced if k is larger than 1.

The major significance of this study is that it allows one to estimate the effects of various parameters of the material and technological factors on the concentration nonuniformity associated with the growth.

These nonuniformities should also affect the mean impurity content; the mean effective partition coefficient is defined by

$$\bar{k}_{eff}=c_S/c_{L\infty}, \qquad \bar{c}=\frac{f}{v}\int_0^{v/f} c_S(x)dx,$$

where f is the frequency of the concentration fluctuations, v is crystal pulling rate, and $c_S(X)$ is the concentration distribution along the crystal [16, 33]. Barthel and Eichler [16] have shown that this mean effective coefficient jumps to 1 for $k_i > 1$ when the convective banding appears. Hurle [33] has shown that such a change is of major importance and is maximal for metallic systems having $k > 1$.

The amplitude and frequency of the temperature fluctuations vary over the cross section of the crystal; the most marked changes in the growth-band systems are seen in cross sections

[14]. The effective partition coefficient is dependent on these various quantities, which explains the nonuniformities seen in the central banded sections described in Section 2.5. A more detailed comparison of the measurements for the various $k_i$ with the theory indicates general agreement, which eliminates the conflict discussed in relation to channels of the second kind.

### 3.7. Crystallization Rate and Rotational Nonuniformity

The first equations for this type of banding were derived from a simplified phenomenological calculation of the instantaneous crystallization rate for growth in an unsymmetrical temperature distribution; in that case, the correlation between the external and internal temperature distributions was neglected, as was the convection, but the face effect was incorporated, and this provided an explanation of the changes in growth rate with time and position [8].

The results show that even very slight asymmetry can produce considerable variations in growth rate, which increase with the distance from the axis of rotation and with the curvature of the phase boundary. This explains the commonly-observed variation in mean concentration over the cross section. Also, the analysis defines the conditions for partial melting of the growing crystal during rotation. The calculations also provide a ready explanation for the slight variations in growth rate occurring within the faces near the central part of the growth front.

## Conclusions

The above results show that we now possess all the major concepts that define impurity trapping during crystallization, as well as the factors responsible for concentration nonuniformity and the parameters governing all such phenomena. Theoretical description of such phenomena involves major mathematical difficulties, and here only the first steps have been taken, although the principles are clear.

## Literature Cited


1. F. Weinberg, Crystal Growth, H. S. Petser, ed., Pergamon Press, Oxford (1957), p. 639.
2. J. Barthel, Kristallisation, VEB Deutscher Verlag für Grundstoffindustrie, Leipzig (1969), p. 164.
3. J. Barthel, J. Kunze, and R. Scharfenberg, Phys. Status Solidi, 6:529 (1964).
4. R. L. Parker, Solid State Physics, Academic Press, New York and London, Vol. 25, p. 152.
5. R. R. Sekerka, Crystal Growth, North Holland, Amsterdam (1968), p. 71.
6. G. Baralis and M. C. Perosino, Crystal Growth, North Holland, Amsterdam (1968), p. 651.
7. J. C. Brice, J. Crystal Growth, 6:205 (1970).
8. J. Barthel and M. Jurisch, J. Crystal Growth, 11:293 (1971).
9. T. F. Cliszek, J. Crystal Growth, 10:263 (1971).
10. A. F. Witt and H. C. Gatos, J. Electrochem. Soc., 113:808 (1956).
11. W. Neumann, Diploma Work, Mining Academy Freiberg (1969).
12. J. A. M. Dikhoff, Solid-State Electronics, 1:202 (1960).
13. K. Morizane, A. Witt, and H. C. Gatos, J. Electrochem. Soc., 114:738 (1967).
14. J. Barthel and R. Scharfenberg, Crystal Growth, H. S. Peiser, ed., Pergamon Press, Oxford (1957), p. 133.
15. H. P. Utech, W. S. Power, and J. C. Early, Crystal Growth, H. S. Peiser, ed., Pergamon Press, Oxford (1957), p. 201.
16. J. Barthel and K. Eichler, Kristall u. Technik, 2:205 (1967).
17. D. T. J. Hurle, Phil. Mag., 13:305 (1966).
18. L. Kuchar, Reinstoffprobleme, Vol. 1, E. Rexer, ed., Akademieverlag, Berlin (1966), p. 523.

19. A. A. Chernov and B. Ya. Lyubov, In: Growth of Crystals, Vol. 5a, Consultants Bureau, New York (1968), p. 7.
20. J. A. Spittle, M. D. Hunt, and R. W. Smith, Crystal Growth, North Holland, Amsterdam (1968), p. 617.
21. M. Kramnacker and W. Lange, Kristall u. Technik, 4:207 (1969).
22. J. C. Brice, J. Crystal Growth, 10:205 (1971).
23. R. M. Sharp and A. Hellawell, J. Crystal Growth, 5:155 (1969).
24. A. F. Witt and H. C. Gatos, J. Electrochem. Soc., 115:70 (1968).
25. A. A. Chernov, Reinststoff in Wissenschaft und Technik, M. Balarin, ed., Akademieverlag, Berlin (1972).
26. B. K. Jindal and W. A. Tiller, J. Chem. Phys. 49:4632 (1968).
27. G. Baralis, Crystal Growth, North Holland, Amsterdam (1968), p. 627.
28. W. A. Tiller, The Art and Science of Growing Crystals, J. J. Gilman, ed., John Wiley, New York (1963), p. 304.
29. A. S. Emanuel and D. R. Olander, Int. J. Heat Mass Transfer, 7:539 (1964).
30. J. R. Carruthers, J. Crystal Growth, 2:1 (1968).
31. J. R. Carruthers, J. Electrochem. Soc., 114:959 (1967).
32. E. Jakeman, Phys. Fluids, 11:10 (1968).
33. D. T. J. Hurle, E. Jakeman, and E. R. Pike, Crystal Growth, North Holland, Amsterdam (1968), p. 633.

# ASPECTS OF HIGH-SPEED SOLID-SOLUTION CRYSTAL GROWTH

D. E. Temkin

*Baikov Central Ferrous Metallurgy Research Institute, Moscow*

In this study, particular attention has been given to transitions in binary condensed phases, especially features involved in component redistribution. From this viewpoint, high growth rates are ones that exceed the characteristic atomic diffusion rates. Under such conditions, a detailed description of the microscopic processes at the phase boundaries is required in any analysis. Such a description can be given in terms of the frequencies of attachment to various configurations at the phase boundaries, and correspondingly the detachment frequencies and the frequencies of diffusion jumps. However, enormous difficulties arise in any detailed description on account of the large number of possible configurations. This means that highly simplified models have to be employed in order to define even qualitative features. Here we discuss results obtained from two such models (Fig. 1): a one-dimensional chain and a phase boundary with one kink on a [100] step (simple cubic lattice, (100) face). The latter model is applicable to the growth of an atomically smooth face with a low step density and with kinks on the steps. Even for such simple models, only recently have reasonably accurate results been obtained for various kinetic parameters [1-7].

We now discuss these models briefly. Consider a binary system containing components A and B constituting the phases $\alpha$ and $\beta$; there are two semiinfinite chains, one of which corresponds to the $\alpha$ phase and the other to the $\beta$ phase. In each of these, the A and B atoms are randomly distributed over the nodes, while the concentration of B atoms is C. At the start, the $\alpha$ chain is linked to the $\beta$ chain, the result being a chain infinite in both directions (Fig. 1a), which contains the $\alpha$ and $\beta$ phases together with an interphase boundary (the $\alpha$ phase lies to the left of the boundary in Fig. 1). The transition from one phase to another at the boundary is accompanied by displacement of the boundary to the right or left on each atomic transition, the distance being the same for both phases. The frequencies of transition of A and B atoms from the $\beta$ phase to the $\alpha$ phase are respectively $\nu_{+A}$ and $\nu_{+B}$, while those for the reverse transitions are correspondingly $\nu_{-A}$ and $\nu_{-B}$ (Fig. 1c). The atomic diffusion is determined by the frequency $\nu_{\beta}$ of particle exchange between the closest A and B atoms in the $\beta$ phase and similarly by $\nu_{\alpha}$ for the $\alpha$ phase (it is assumed that there is no exchange via the boundary). In the case of the single kink (Fig. 1b), the phase transition involves one-dimensional motion of a kink along a step, but a difference from the chain case is that there is three-dimensional atomic diffusion.

As the transition frequency is monotonically dependent on the temperature T, these models correspond to the equilibrium diagram of Fig. 2a [8]; the concentrations $C_{\alpha}$ and $C_{\beta}$ are expressed as follows in terms of the transition frequencies, as is the concentration $C_0$ at which

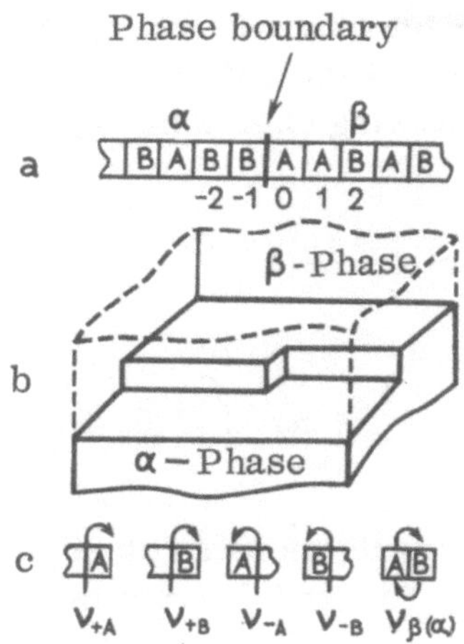


**Fig. 1.** Models: (a) One-dimensional chain; (b) isolated kink; (c) various processes, with frequencies.

the free energies of the $\alpha$ and $\beta$ phases per particle are equal:

$$C_\alpha = \frac{\nu_{+B}(\nu_{+A}-\nu_{-A})}{\nu_{+A}\nu_{-B}-\nu_{-A}\nu_{+B}},\qquad C_\beta = \frac{\nu_{-B}(\nu_{+A}-\nu_{-A})}{\nu_{+A}\nu_{-B}-\nu_{-A}\nu_{+B}},\qquad C_0 = \ln\frac{\nu_{-A}}{\nu_{+A}}\Big/\ln\frac{\nu_{-A}\nu_{+B}}{\nu_{+A}\nu_{-B}}. \tag{1}$$

Then $\nu_{+A} > \nu_{-A}$, $\nu_{+B} < \nu_{-B}$ for $T_A > T > T_B$.

The equilibrium partition coefficient is $K_e = C_\alpha/C_\beta = \nu_{+B}\,\nu_{-B}$; comparison of expressions (1) for $C_\alpha$ and $C_\beta$ with the corresponding thermodynamic relationships gives the following relation between the frequencies and the chemical potentials [8]:

$$\nu_{+A}/\nu_{-A} = \exp[(\mu_{A,\beta}-\mu_{A,\alpha})/kT],\qquad \nu_{+B}/\nu_{-B} = \exp[(\mu_{B,\beta}-\mu_{B,\alpha})/kT], \tag{2}$$

where $\mu_{i,j}$ are the chemical potentials of the j phases ($j = \alpha, \beta$), each phase consisting of component i (i = A, B).

## 1. Diffusionless Transition Kinetics

If one assumes that there is no atomic diffusion in the model system, but that the phase boundary shifts, then the transition is diffusionless in the exact sense; this situation is very close to the actual one in certain solid-state transitions [9].

We now consider a $\beta \rightleftarrows \alpha$ transition in a one-dimensional chain for $\nu_\alpha = \nu_\beta = 0$; at the start, t = 0, the $\alpha$ and $\beta$ chains, which have concentration C, are connected, and the phase boundary between them performs a type of random walk, during which it drifts to the right or left. In the first case, there is a diffusionless $\beta \to \alpha$ transition, while in the second there is a corresponding $\alpha \to \beta$ transition. The drift rate is v(t) (t is time), and we have to examine the distribution of the components with respect to the moving boundary. The latter can be described in terms of an infinite sequence of distribution functions W(n; t), W(n, k; t) etc., which have the following meaning: W(n; t) is the probability of finding a B atom at node n of a chain reckoned from the phase boundary at time t, while W(n, k; t) is the probability of finding B atoms at nodes n and k simultaneously, and so on. In what follows, it is assumed that the $\beta$ phase corresponds to n = 0, 1, 2, ..., and the $\alpha$ phase to n = −1, −2; ...; the rate v(t) (in reciprocal-time units) is the resultant of the atomic fluxes from the $\beta$ phase to the $\alpha$ phase and is expressed in terms of W(0; t) and W(−1; t):

$$v(t) = [1-W(0;t)]\nu_{+A} + W(0;t)\nu_{+B} - [1-W(-1;t)]\nu_{-A} - W(-1;t)\nu_{-B}. \tag{3}$$

In this system, there is a period of nonstationary phase-boundary motion when the chains have been linked, while a steady-state transition rate is set up for $t \to \infty$ [8]. Analysis of this situation [10, 11] via the solution of [7] gives the following expressions for v and W(n) for the $\beta \to \alpha$ transition in the one-phase $\alpha$ region:

$$v = \frac{\nu_{+A}\nu_{+B}(1-\eta_1)}{\nu_{+B}+C(\nu_{+A}-\nu_{+B})}, \quad \text{where } \eta_1 = C\frac{\nu_{-B}}{\nu_{+B}} + (1-C)\frac{\nu_{-A}}{\nu_{+A}} \quad (\ll 1); \tag{4}$$

$$W(n) = \begin{cases} C & \text{for } n \leqslant -1 \text{ (in the } \alpha \text{ phase)}, \\ C + C(1-C)\left[\dfrac{\nu_{+A}-\nu_{-A}-\nu_{+B}+\nu_{-B}}{\nu_{+B}+C(\nu_{+A}-\nu_{+B})}\right]\eta_1^n & \text{for } n \geqslant 0 \text{ (in the } \beta \text{ phase)}, \end{cases} \tag{5}$$

The steady-state $\alpha \to \beta$ transition can occur for concentrations and temperatures corresponding to the $\alpha$-phase region in the phase diagram (Fig. 2a), and the rate is

$$v = -\frac{\nu_{-A}\nu_{-B}(1-\eta_2)}{\nu_{-B}+C(\nu_{-A}-\nu_{-B})}, \quad \text{where } \eta_2 = C\frac{\nu_{+B}}{\nu_{-B}} + (1-C)\frac{\nu_{+A}}{\nu_{-A}} \quad (\ll 1). \tag{6}$$

In the two-phase region (Fig. 2a) ($C_\alpha < C < C_\beta$), there is no steady-state transition having $v \neq 0$, and the solid line in Fig. 2b shows schematically the C dependence of v. We have $\eta_1 = 1$ and $\eta_2 = 1$, with $v = 0$, at the boundaries of the one-phase regions with the two-phase one, i.e., for $C' = C_\alpha$ and $C = C_\beta$. Further, (4) and (6) are confirmed by computer simulation of a moving phase boundary in a one-dimensional chain [8].

The diffusion coefficient for the phase boundary has been calculated [12] for low C.

A study has been made [8] of nonstationary diffusionless transition subject to an additional assumption that there is no correlation [$W(n, k; t) = W(n; t)W(k; t)$, etc.], which gave the following results. If C corresponds to the one-phase regions ($C < C_\alpha$ or $C > C_\beta$), then for $t \to \infty$ a steady state having $v \neq 0$ is set up, as (4) and (6) show. In the two-phase region, ($C_\alpha < C < C_\beta$), one gets the diffusion mode of growth for t large, with the concentrations at the phase boundary close to the equilibrium values, while the growth rate tends to zero in accordance with the law $v(t) \simeq zt^{-1/2}$, where z is the root of a certain transcendental equation; z decreases monotonically from $\infty$ to $-\infty$ as C goes from $C_\alpha$ to $C_\beta$ and becomes zero at $C = (C_\alpha + C_\beta)/2$.

This pattern of diffusionless transition is similar to the one characteristic of a transition accompanied by redistribution of components by diffusion [13]:

1. The steady state is set up under conditions corresponding to the one-phase regions in the phase diagram; the distribution of the concentrations with respect to the phase boundary is nonuniform and is analogous to that shown in Fig. 3a.

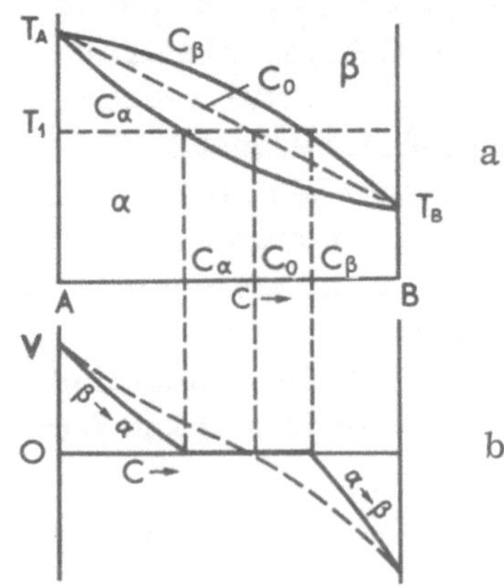


Fig. 2. (a) Phase-equilibrium diagram; (b) schematic dependence of the growth rate on concentration for a disordered distribution (solid line) and an ordered distribution (broken line).

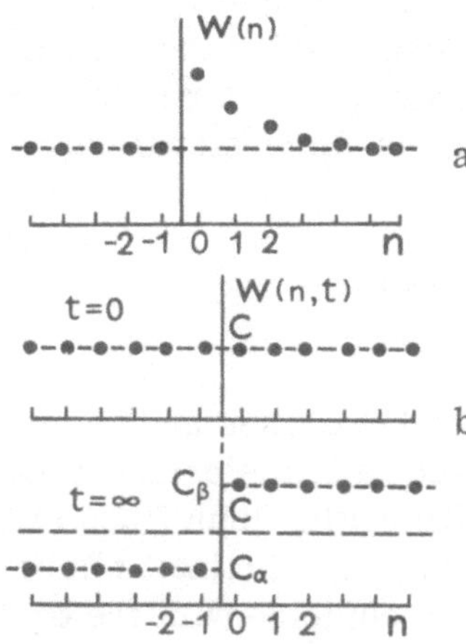


**Fig. 3.** Distribution of component B with respect to the phase boundary in a steady-state $\beta \to \alpha$ transition: (a) in the $\alpha$ region; (b) in the two-phase $\alpha + \beta$ region.

2. The rate v(t) in a two-phase region for $t \to \infty$ decreases to zero as $t^{-1/2}$, while the system tends to a state of two phases of equilibrium compositions $C_\alpha$ and $C_\beta$ around the phase boundary (Fig. 3b).

These effects arise from fluctuations and kinetic interaction of such fluctuations with the phase boundary; for instance, consider the $\beta \to \alpha$ transition, with the B atoms less ready than the A atoms to pass from the $\beta$ phase to the $\alpha$ phase, i.e., $(\nu_{+B} - \nu_{-B}) < (\nu_{+A} - \nu_{-A})$, which means that the phase boundary passes with difficulty through the parts of the $\beta$ phase containing elevated amounts of B, and therefore there is an increased probability of observing such parts ahead of the phase boundary. The contributions from the various types of fluctuation to the mean boundary speed are determined by the probabilities of such fluctuations and the time needed to overcome the boundary. Although the probability of any deviation from the mean value C is inversely related to the deviation, the time needed to pass through such a region may change more rapidly than does the probability. The possibility of such a situation is critically dependent on C, as is clear from the following expressions for the time $t_p$ for the boundary to pass from left to right along a part of the chain of length $l$:

$$t_p = \frac{l[C\nu_{+A} + (1-C)\nu_{+B}]\ [1 + \eta_1(\eta_1^l - 1)/l(1-\eta_1)]}{\nu_{+A}\nu_{+B}(1-\eta_1)}.$$

This $t_p$ is defined from the instant when the boundary first reaches the right-hand end of this chain of length $l$ and is averaged over all possible structures and concentrations in such a chain; the $\eta_i$ are as in (4). Further $\eta_1 < 1$ in the $\alpha$-phase region, and $t_p$ is proportional to $l$ for $l$ large, which results in a mean speed $v = l/t_p$ as defined by (4). In the two-phase region, we have $\eta_1 > 1$, and $t_p$ increases as $\eta_1^l$ for $l$ large, while the mean speed tends to zero as $l\eta_1^{-l}$.

There are no fluctuations in a completely ordered system, and the mean boundary speed does not become zero throughout the two-phase region, but only on the $C - C_0$ line, as is shown schematically in Fig. 2b. This is confirmed by analysis and by computer simulation [8].

These results conflict with the usual thermodynamic viewpoint on diffusionless transition [15], which is envisaged as similar to that in a one-component system, while the $C_0(T)$ line plays the same part as the equilibrium point in a pure substance.

These features of diffusionless transition have been examined on the one-dimensional model, but qualitative arguments [16] also indicate that these features persist also in system of more dimensions; e.g., when a step moves along a phase boundary (two-dimensional systems and for a phase interface (three-dimensional system). In both of these cases, the speed of the boundary (step or interface) should tend to zero during the transition in the two-phase region because the boundary encounters fluctuations steadily more difficult to overcome. The kinetics of diffusionless motion of a rectilinear step may be examined if the hanging configurations are neglected, and this shows that the steady-state speed of a step tends to zero at the boundaries of the one-phase and two-phase regions, as in the one-dimensional case. This topic will be considered in more detail in a forthcoming paper.

## 2. Effects of Deviation from Ideal Behavior on Diffusionless Transition

So far we have envisaged a model in which the $\alpha$ and $\beta$ phases are ideal solutions; in that case, the kinetic lines separating the regions of steady-state diffusionless transition from the region where there is no such transition coincide with the equilibrium lines in the phase diagram. If the solutions are not ideal [17], the $C_{\alpha,k}$ and $C_{\beta,k}$ kinetic lines, where the steady-state rate of pure diffusionless growth is zero, as a rule do not coincide with the equilibrium $C_\alpha$ and $C_\beta$ lines (Fig. 4a). Coincidence occurs only when the mixing energies of the two phases are equal. The kinetic lines may lie in the one-phase regions or in the two-phase one, and in particular may intersect the equilibrium lines.

## 3. Effects of Diffusion on Transition Kinetics

We have already seen that diffusionless transition (for $\nu_\alpha = \nu_\beta = 0$) is qualitatively similar to a transition with $\nu_\alpha$ and/or $\nu_\beta$ different from zero. In our one-dimensional model, v is independent of $\nu_\alpha$ and $\nu_\beta$ and is described by (4) and (6). On the other hand, the component distribution is dependent on these frequencies, and, for example, for the $\beta \to \alpha$ transition we have instead of (5) that

$$W(n)=\begin{cases} C \quad \text{for} \quad n\leqslant -1, \\ C+C\left[\dfrac{\nu_{+A}-\nu_{-A}-\nu_{+B}+\nu_{-B}}{\nu_{+B}}\right]\left(\dfrac{\nu_{-A}+\nu_\beta}{\nu_{+A}+\nu_\beta}\right)^n \quad \text{for} \quad n\geqslant 0, \end{cases} \tag{7}$$

which applies up to terms in $C^2$ for C small [10]. In the two-phase region, the nonstationary rate and the component distribution W(n, t) will also be dependent on $\nu_\alpha$ and $\nu_\beta$.

If the solutions are not ideal [17], the effects of diffusion are extremely specific (Fig. 4b). If there is no diffusion at all, the rates of the $\beta \to \alpha$ and $\alpha \to \beta$ steady-state transitions become zero respectively on the $C_{\alpha,k}$ and $C_{\beta,k}$ kinetic lines. If there is component diffusion (even if

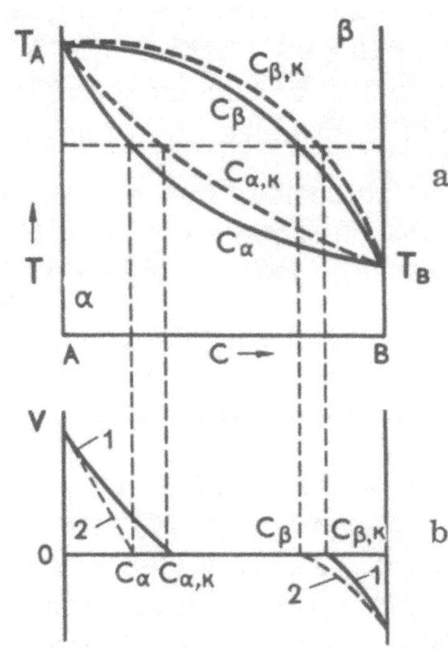


Fig. 4. (a) Phase-equilibrium diagram (nonideal solutions) and kinetic lines for diffusionless transition at zero rate; (b) steady-state phase-boundary speed as a function of concentration in the absence of diffusion (curve 1) and for a finite diffusion rate (curve 2).

**slow), then the steady-state rate tends to zero as the $C_\alpha$ and $C_\beta$ lines are approached in the equilibrium diagram.**

**Consideration of the diffusion results in qualitative changes in the kinetics if the dimensions involved in the diffusion exceed those involved in the phase boundary; for instance, one-dimensional motion of a kink or two-dimensional motion of a step accompanied by three-dimensional diffusion results in the atoms of the component that hinders the boundary motion (kink or step) being displaced from the line of motion of the boundary by lateral diffusion, and therefore the steady-state transition rate differs from zero even in the two-phase region and is determined by the lateral diffusion rate [18].**

## 4. Component Redistribution during Transition and Impurity Trapping

**We now determine the partition coefficient K and the ratio of the concentration of the B atoms in the $\alpha$ phase to the same in the $\beta$ phase at the interface; in our models we have $K = W(-1; t)/W(0; t)$, and (5) shows that $K \neq 1$ for the pure diffusionless case, the value being**

$$K = K_e \Big/ \left[1 + \frac{v}{\nu_{-B}}\right]. \tag{8}$$

**In the two-phase region, K tends to the equilibrium value for $t \to \infty$, that value being $K_e = C_\alpha/C_\beta = \nu_{+B}/\nu_{-B}$, and (8) coincides with an equation derived previously [19].**

**This difference of K from 1 is responsible for many of the difficulties in discussing diffusionless transition. It is quite clear that all the atoms remain in place in such a transition, so we may introduce a different coefficient K', which is defined as the ratio of the concentration immediately after the transition to that directly before the transition at the same point in the system, in which case K' = 1 in the diffusionless case. K' may be called the trapping coefficient. It is difficult to give clear-cut definitions of the before and after concepts for the transition, and correspondingly of K'; this is unimportant if $\nu_\alpha$ and $\nu_\beta = 0$ but is important if $\nu_\alpha$ and $\nu_\beta \neq 0$; however, it is clear that $K \simeq K' \simeq K_e$ if the speed of the boundary is low, and thus that the difference between these coefficients is small. If the speed is high, K and K' become distinctly different even for diffusionless growth, which is a point that is usually neglected, since it is commonly assumed that K = 1 for diffusionless transition [20, 21].**

**The difference between K and K' corresponds to a difference between the W(n) for a coordinate system moving with the phase boundary and the distribution V(n) in an immobile system. The steady-state W(n) for the $\beta \to \alpha$ transition in the chain model is described by (5) or (7), while V(n) is the probability of observing a B atom at a given node if the phase boundary lies n nodes to the left of the relevant one (n = 0 when the boundary is directly to the left of the given node). Also, V(n) describes the concentration change at a given point when the phase boundary passes that point, and for a chain with diffusion in the $\beta$ phase alone ($\nu_\alpha = 0, \nu_\beta \neq 0$, but small) we have that V(n) takes the following form [23] up to terms in $C^2$ and $\nu_\beta^2$:**

$$V(n) = \begin{cases} C[1 + \nu_\beta(\nu_{-B} - \nu_{-A})/\nu_{+B}(\nu_{+A} - \nu_{-A})] & \text{for} \quad n = -1, \\ C[1 + \nu_\beta(\nu_{+A} - \nu_{+B})/\nu_{+B}(\nu_{+A} - \nu_{-A})] & \text{for} \quad n = 0, \\ C \ \text{otherwise}. \end{cases} \tag{9}$$

**The moving boundary transports some quantity M of component B: $M = \sum_{n=-\infty}^{\infty} [V(n) - C] \simeq C\nu_\beta(\nu_{+A} - \nu_{-A} - \nu_{+B} + \nu_{-B})/\nu_{+B}(\nu_{+A} - \nu_{-A})$; we have V(n) = C for all n if $\nu_\beta = 0$, and then M = 0.**

In the case of a $\beta \to \alpha$ transition in a semiinfinite chain, we have an initial nonstationary part for the component distribution [24], where

$$C(n) \simeq C\left[1 - \frac{\nu_\beta(\nu_{+A} - \nu_{-A} - \nu_{+B} + \nu_{-B})}{\nu_{+A} - \nu_{+B}}\left(\frac{\nu_{-A}}{\nu_{+A}}\right)^n\right] \quad \text{for } n \geqslant 0. \tag{10}$$

Here C(n) is the concentration of B in the $\alpha$ phase at a distance n from the end of the chain. It is readily verified that M takes the value $\sum_{n=0}^{\infty}[C - C(n)]$, i.e., is the amount of component B lacking in the initial part.

A study has been made [25] of impurity trapping in the motion of an isolated kink with allowance for diffusion in the initial $\beta$ phase; the effective trapping coefficient $K_3$ is defined as the ratio of the concentration of B in the $\alpha$ phase far from the kink to the concentration $\beta$ in the initial phase. If $\beta$ is small, we have up to terms of order $\nu_\beta^2$ that

$$K_3 \simeq 1 - \frac{2\nu_\beta \nu_{+A}}{(\nu_{+A} - \nu_{-A})^2} \frac{(1-K)}{K}, \tag{11}$$

where $K = \nu_{+B}/(\nu_{+A} - \nu_{-A} + \nu_{-B})$ is the partition coefficient for $C \to 0$ as (8) shows; we have the pure diffusionless process for $\nu_\beta = 0$, and then $K_3 = 1$.

## 5. The Kinetic Condition for Diffusionless Growth

By diffusionless transition we here envisage a process in which the atoms do not exchange places by diffusion; on this basis we can formulate the condition for diffusionless growth.

During the phase transition, the phase boundary on average advances at a certain speed, but fluctuations may cause it to retreat; in the case of one-dimensional motion along a chain (or a kink on a step), the entire boundary retreat, whereas a step or surface may retreat only in certain sections. The corresponding characteristic time is $\tau_p$, which is related to the characteristic length $l$ by $\tau_p \sim l/v$, where v is the speed of the boundary. In fact, the start of the conversion at a given point (node) occurs when the boundary reaches the corresponding node, namely it is in the initial phase $\beta$ at the interface. The end of the transition occurs when the boundary converts the node from the $\beta$ phase to the $\alpha$ one for the last time. If the boundary is not to return any more to the position corresponding to the onset of the transition, it must be displaced in the appropriate sense by a distance of the order of $l$. During $\tau_p$, the node and the corresponding B atom passes from the $\beta$ phase to the $\alpha$ one and back from $\alpha$ to $\beta$. Part of the time the atom is thus in the $\beta$ phase and part in the $\alpha$ phase, and the probabilities of diffusion jumps may differ for these two states, which must be incorporated in determining the characteristic diffusion-jump time $\tau_T$. The order of magnitude is $\tau \sim a^2/D$, where a is the interatomic distance and D is the average diffusion coefficient in the boundary region. In general $\tau_p/\tau_D \ll 1$ for diffusionless growth.

The characteristic length $l$ can be estimated from a condition that implies that the increase in the free energy due to deviation of the boundary is approximately equal to kT; this condition gives $l \equiv l_1 \sim a kT/\Delta\mu_A$ for a one-component chain, where $\Delta\mu_A = \mu_{A,\beta} - \mu_{A,\alpha}$ is the free-energy change for a $\beta \to \alpha$ transition (per particle). From (2) we have for $\Delta\mu_A/kT \ll 1$ that $l_1 \sim a\nu_{-A}/(\nu_{+A} - \nu_{-A})$, and a direct calculation of the mean deviation length gives the same result, but for arbitrary $\nu_{-A}$ less than $\nu_{+A}$.

The condition for diffusionless growth has previously been derived [24] for a chain and for a kink by introducing a probability Q that the B atom at the given node at the start is there also at the end of the transition, provided also that the transition time is such that the atom performs not even one diffusion jump. The low-concentration case was envisaged ($C \ll 1/l_1$), with diffusion only in the $\beta$ phase. The result for diffusionless growth is $(1 - Q) \ll 1$, which gives

$$\frac{\nu_\beta}{K}\frac{(\nu_{+A}+\nu_{-A})}{(\nu_{+A}-\nu_{-A})^2}\ll 1, \tag{12}$$

where $K = \nu_{+B}/(\nu_{+A} - \nu_{-A} + \nu_{-B})$ is the partition coefficient; a condition similar to (12) was derived similarly for a kink. The 1/K factor in (12) reflects the effect of the impurity atom on $\tau_p/\tau_D$, since the value is different from that for a one-component chain.

When a step moves while having a high kink density, the estimate for $l$ is [26] $l \equiv l_2 \sim (kT)^{2/3}(q\alpha_c\Delta\mu_A)^{-1/3}$, where $\alpha_c$ is specific free energy of the corner in a step, while q is the surface atomic density. As $\tau_p \sim l_2/v$ (v is step speed), we have $\tau_D \sim a^2/D$, and we introduce the 1/K factor by analogy with (12) to get the condition for diffusionless step motion as

$$l_2 D/Ka^2 v \ll 1. \tag{13}$$

When a phase interface moves by the normal mechanism, the deviations of the parts due to fluctuation may be characterized via the length $l \equiv l_3 \sim (kT/\alpha_s)^{1/2}$, where $\alpha_s$ is the specific surface free energy, and then the condition for diffusionless growth takes the form

$$(kT/\alpha_s)^{1/2}\, D/Ka^2 v \ll 1. \tag{14}$$

## 6. Nonequilibrium States

The diffusionless process is fast and does not allow atoms to be exchanged by diffusion, so the product is a new phase having a nonequilibrium composition, and composition equilibration thus requires such exchange. The properties of the unstable phase reflect the state of the initial phase and the transition mechanisms [14]. Examples of such features are the short-range order and the long-range order.

A one-dimensional model has been employed [10, 11] to examine the conversion of a $\beta$ phase free from short-range order to an $\alpha$ phase whose short-range order at equilibrium is represented by the parameter $\sigma_e$. As by definition there is no diffusion in the $\alpha$ phase, the short-range order can be set up only by particle selection at the phase interface, which occurs as follows.

We assume that the short-range order in the $\alpha$ phase is such that an A atom is surrounded mainly by B atoms. If an A atom passing from $\beta$ to $\alpha$ is followed by A (an irregular particle), then the latter will have a probability of returning to $\beta$ higher than for a regular B particle, so replacement by A will follow the regular transfer in the $\beta$ phase. If the diffusion in the $\beta$ phase is infinitely rapid ($\nu_\beta = \infty$), then short-range order $\sigma_\infty$ is formed in the $\alpha$ phase growing at a finite rate, this order differing from $\sigma_e$. The $\sigma$ parameter increases monotonically with $\nu_\beta$, and it is small by comparison with $\sigma_\infty$ under conditions close to those of (12), i.e., for diffusionless growth.

A qualitatively different situation occurs in the growth of an ordering crystal, which was first examined by Chernov [1, 27, 28] for growth of a crystal from a phase that provides for rapid particle sorting. When a crystal grows below the Curie point $T_c$ for the ordering, the long-range order parameter falls as the supersaturation increases, and thus it becomes zero at some finite supersaturation. In the present instance, this kinetic phase transition resembles a second-order thermodynamic transition. This case has been examined [29] for a model with one kink on the basis of diffusion in the initial disordered phase. It is clear that, if the diffusion is slow, the kinetic phase-transition point (the transition from growth of an ordered crystal to growth of a disordered one) corresponds to the conditions considered above for diffusionless growth, i.e., conditions when the left side of (12) is of the order of unity.

## 7. Some Experimental Results

1. There is no doubt that diffusionless transition is possible in the solid state; the most direct evidence for this would be direct observation of the short-range and long-range order in the new phase, particularly if these corresponded to those of the initial phase. This has been observed for the martensite transition in Cu–Al [9, 30]: the initial ordered $\beta_1$ phase was close in composition to $Cu_3Al$ and was converted to a $\gamma'$ phase whose ordered structure corresponded to the $\beta$ phase, while the transition from $\beta_1$ to $\gamma'$ did not involve the atoms changing places.

The analysis indicated that there is a region in the (T, C) plane, which may be called the kinetic two-phase region,* in which the growth rate tends gradually to zero, which indicates that there is kinetic difficulty in complete conversion of the initial phase, which is indicated by several lines of experimental evidence. The isothermal $\gamma \rightarrow \alpha$ transition (not of martensite type) in Fe–Cr–Ni alloys [31] is such that there are temperatures below which the transition goes to completion, but above which it does not. The evidence for the absence of appreciable component redistribution is indirect in that there is virtually no change in the temperature of the $\gamma \rightarrow \alpha$ martensite transition following quenching after isothermal $\gamma \rightarrow \alpha$ transition. Alloys containing Fe with 5-20% Ni show an isothermal $\gamma \rightarrow \alpha$ transition that does not go to completion [32], although the degree of conversion increases at low temperatures.

There is also evidence for a qualitative difference between the modes of transition in the one-phase and two-phase regions, as is clear from the bulk $\beta \rightarrow \alpha$ transition in Cu alloys containing 37-39 at.% Zn [33]. The specimens were quenched from the $\beta$ region and heated rapidly to the required temperature in the $\alpha$ or $\alpha + \beta$ region. The $\beta \rightarrow \alpha$ transition was not observed† in the two-phase region. The measured growth rate of the $\alpha$ phase was of the order of 1 cm/sec (this ensures that no diffusion occurred), but the exact speed was dependent on the temperature and it did not become zero at the boundary of the one-phase region. The latter may be due to the nonideal behavior of the alloys (see curve 1 in Fig. 4b).

2. At present there is no direct evidence for diffusionless crystallization in melts; also, it is even an open question whether such a process is possible, since (14) indicates that $K \sim 1$, $(kT/\alpha_s)^{1/2} \sim a$, if we put K as about 1, and diffusionless crystallization requires a growth rate $v \gg D/a$ (where D can be taken as equal to the diffusion coefficient in the melt). On the other hand, existing views [35] indicate that the maximum growth rate may be of the order of D/a. However, it is possible that the transition at the interface involves the atoms moving smaller distances and with higher frequencies, in which case the estimate for the maximum rate would be too low.

High crystallization rates were produced at high initial supercoolings, or else by rapid cooling (quenching) of small droplets and thin films. The latter approach has yielded many interesting results (see [35, 36, 38] for reviews).

Here we may note only [37], in which Zn–Cd alloys could be made to crystallize following quenching without composition change up to 5 wt.% Cd. The solubility of Cd in Zn is of retrograde type, the maximum value being 2.6 wt.%, so any alloy containing more than 2.6 wt.% Cd must solidify within the equilibrium or metastable two-phase region. However, it proved im-

* At low C, this region coincides with the equilibrium two-phase region, which applies also for ideal solutions or for solution with identical mixing energies.

* The massive $\alpha$ phase was produced [34] in the two-phase region, but only close to the boundary with the single-phase region.

possible [36] to quench alloys containing higher Cd contents and obtain a continuous series of solid solutions. It may be that this is due to the difficulty in obtaining complete conversion within the kinetic two-phase region.

## Literature Cited


1. A. A. Chernov, Usp. Fiz. Nauk, 100:277 (1970).
2. A. A. Chernov, Adsorption et Croissance Cristalline, Colloq. Intern. CNRS, No. 152, Paris (1965), p. 283.
3. A. A. Chernov, Suppl. J. Phys. Chem. Solids (1967), p. 25.
4. A. A. Chernov, Dokl. Akad. Nauk SSSR, 170:580 (1967).
5. J. J. Lauritzen Jr., E. Passaglia, and E. A. DiMarzio, J. Res. NBS (USA), 71A:245 (1967).
6. J. J. Lauritzen Jr., E. A. Di Marzio, and E. Passaglia, J. Chem. Phys., 45:4444 (1966).
7. A. A. Chernov, Biofizika, 12:297 (1967).
8. D. E. Temkin, Kristallografiya, 17:77 (1972).
9. G. V. Kurdyumov, In: Problems of Metallography and Metal Physics [in Russian], Metallurgizdat, Moscow (1949), p. 132.
10. D. E. Temkin, J. Crystal Growth, 5:193 (1969).
11. D. E. Temkin, Kristallografiya, 14:423 (1969).
12. D. E. Temkin, Fiz. Tverd. Tela, 13:3381 (1971).
13. G. I. Ivantsov, Dokl. Akad. Nauk SSSR, 81:172 (1951).
14. D. E. Temkin, In: Problems of Metallography and Metal Physics [in Russian], Metallurgiya, Moscow (1972), p. 35.
15. A. A. Popov, In: Problems in Metallography and Heat Treatment [in Russian], Mashgiz (1956), p. 5.
16. D. E. Temkin, In: Mechanism and Kinetics of Crystallization [in Russian], Minsk (1972).
17. D. E. Temkin, Kristallografiya, 18:906 (1973).
18. D. E. Temkin, Kristallografiya, 18:675 (1973).
19. C. D. Thurmond, In: Semiconductors, N. B. Hanney, ed., Reinhold, New York (1960), p. 160.
20. V. T. Borisov, Dokl. Akad. Nauk SSSR, 150:294 (1963).
21. I. L. Aptekar' and D. S. Kamentskaya, Fiz. Met. Metalloved., 14:358 (1962).
22. B. K. Jindal and W. A. Tiller, J. Chem. Phys., 49:4632 (1968).
23. D. E. Temkin, Kristallografiya, 15:428 (1970).
24. D. E. Temkin, Kristallografiya, 15:421 (1970).
25. D. E. Temkin, Kristallografiya, 17:461 (1972).
26. V. V. Voronkov, This volume, p. 364.
27. A. A. Chernov, Zh. Eksp. Teor. Fiz., 53:2090 (1967).
28. A. A. Chernov and J. Lewis, J. Phys. Chem. Solids, 28:2185 (1967).
29. D. E. Temkin, Kristallografiya, 15:884 (1970).
30. G. Kurdyumov, V. Miretskii, and T. Steletskaya, Zh. Tekh. Fiz., 8:1959 (1938).
31. L. I. Kogan and R. I. Zitin, Fiz. Met. Metalloved., 31:379 (1971).
32. E. I. Estrin and V. I. Soshnikov, Dokl. Akad. Nauk SSSR, 210:826 (1973).
33. D. A. Karlyn, J. W. Cahn, and M. Cohen, Trans. Met. Soc. AIME, 245:197 (1969).
34. T. B. Massalski, A. J. Perkins, and J. Jaklovsky, Metal Trans., 3:677 (1972).
35. I. S. Miroshnichenko, In: Growth and Defects of Metal Crystals [in Russian], Naukova Dumka, Kiev (1972), p. 385.
36. T. R. Anantharaman and C. Suryanarayana, J. Mater. Sci., 6:111 (1971).
37. J. C. Baker and J. W. Cahn, Acta Met., 17:575 (1969).
38. I. S. Miroshnichenko, this volume, p. 344.

# FORMATION OF METASTABLE PHASES IN A RAPIDLY COOLED MELT

**I. S. Miroshnichenko**

*Dnepropetrovsk University*

It has long been known [1, 2] that metastable phases may be formed when a melt is cooled rapidly, but systematic studies on the effects of rapid cooling began only in the 1950s.

The first major research on this was performed by Falkenhagen and Hofmann [3]; molten metal was sucked into a thin metal mold, where it crystallized rapidly. The measured cooling rate was $5 \cdot 10^4$ deg/sec. The elements in the first transition period (Ti, V, Cr, Mn, Fe) were examined for solubility in aluminum, and it was found that the limiting solubility considerably exceeded the equilibrium value indicated by the phase diagram.

In 1959, an equipment was described [4] that provided high cooling rates in conjunction with better structural uniformity in the specimen. Small amounts of molten metal were converted to film form and cooled between copper plates. This technique gave aluminum containing up to 10% manganese or nickel containing up to 1.8 wt.% carbon. A method was also described for estimating the cooling rate and the crystal growth rate.

The results from that period [5-8] showed that the method can expand the solid-solution region very frequently in a system of eutectic or peritectic type. Highly supersaturated solid solutions are most readily obtained in a system where there are intermediate phases formed by peritectic reaction because the crystallization of the incongruent compound can be suppressed quite readily at high cooling rates. The liquid ahead of the growing crystals becomes enriched in the second component far in excess of the eutectic composition. Corresponding enrichment is then observed in the solid solution [9]. The compositions of the phases (solid and liquid) in that case are described by continuing the liquidus and solidus lines into the metastable region (Fig. 1).

The $Fe-Fe_3C$ system is of particular interest, since the production of supersaturated austenite instead of ledeburite would give entirely new properties to iron–carbon alloys. However, it has not proved successful to quench more carbon into austenite than the equilibrium diagram would allow [5, 10]. This has led attention to be directed [4, 8] to analogous systems, namely Ni–C and Co–C. In these, the Ni and high-temperature modification of Co resemble $\gamma$-Fe having face-centered cubic structures and similar lattice parameters, but austenite will dissolve up to 2% carbon, whereas nickel and cobalt will dissolve only 0.5 and 1.0% respectively [11].

High cooling rates (in excess of $10^5$ deg/sec) extended the solid-solution ranges for Ni–C and Co–C alloys, so the carbon contents could be raised to 2%; the metastable solid solutions were accompanied by metastable carbides. The presence of the latter did not prevent the

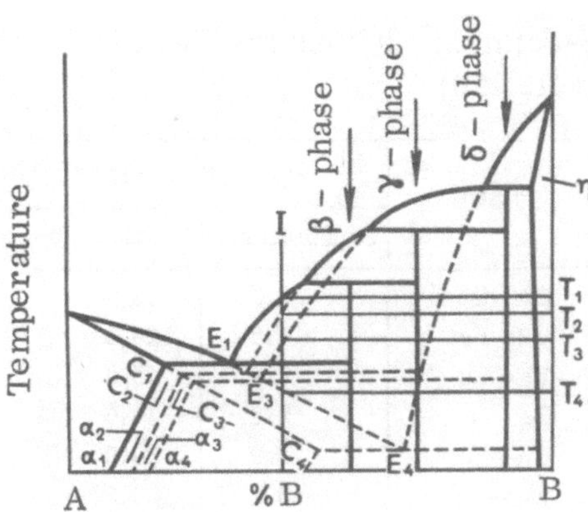


Fig. 1. Phase diagram with metastable-equilibrium lines.

nickel from retaining up to 1.85% carbon, or the cobalt up to 1.65%. If on the other hand graphite was deposited from the liquid (the stable second phase), then the carbon content of the solid solution did not exceed the solubility indicated by the equilibrium between metal and graphite [5, 8]. A similar picture has been observed for Al−Mn and Al−Cr alloys.

It was therefore already clear that extension of the solid-solution region can be obtained not by suppressing the diffusion [12] but by crystallization in accordance with a metastable phase diagram; here the deviations from equilibrium (stable or metastable) may be minor, and the phase compositions may be described with a certain approximation by extending the equilibrium diagram into the metastable region.

In 1960, Duwez published his first paper [13], which provided a powerful impetus to this line of research in the United States. Somewhat later, interest in the subject was aroused in various countries: Britain, France, Italy, the Federal German Republic, Yugoslavia, Japan, India, and Finland.

Professor Duwez and his colleagues improved the available methods, particularly spreading a drop of liquid on a copper substrate by firing at an angle. The droplet spread out over the surface and rapidly cooled. Very thin films could be produced, which provided cooling rates up to $10^8$ deg/sec.

Another impetus to such research was provided [14] by the desire to obtain continuous solid solutions in the Ag−Cu system; Hume-Rothery's criterion indicates that silver and copper should show unrestricted mutual solubility, although in fact the liquid crystallizes with eutectic decomposition. Extremely high cooling rates suppress the eutectic reaction and provide solid solutions throughout the concentration range. More recently, continuous solid solutions have been obtained in other systems of eutectic or peritectic type: Ag−Pt [15], Cu−Rh, Ni−Rh [16], Ge−GaSb [17], and Er−Zr [18], and particular interest attaches to the last, since the continuous series was produced between the low-temperature forms of erbium and zirconium.

However, there are systems such as Cd and Zn that also satisfy Hume-Rothery's rules but which have not yet produced continuous solid solutions. At the present time, we know of over 80 systems in which the continuous primary solid-solution region has been expanded to certain extents; some of these are listed in Table 1.

In nearly all cases where extended regions have been obtained (84%), the metastable supersaturation does not exceed the eutectic composition, because such a composition would make it very difficult to suppress the second stable phase. The existing exceptions (Al−Mn, Al−Fe, Au−Co, etc.) do not conflict with this interpretation. Passage through the eutectic point only hinders the extension of the primary solid-solution region, so in principle it can

TABLE 1. Extension of the Solid-Solution Regions for Certain Systems [19, 20]

| System | Solubility, at.% | | System | Solubility, at.% | |
|---|---|---|---|---|---|
| | equilibrium | attained | | equilibrium | attained |
| Ag—Ge | 9.6 | 13.0 | Cu—Co | 5.5 | 15.5 |
| Al—Cr | 0.45 | 3.0 | Cu—Fe | 4.5 | 20.0 |
| Al—Cu | 2.5 | 18.0 | Fe—Cu | 7.2 | 15.0 |
| Al—Fe | 0.026 | 4.4 | Fe—Ti | 9.8 | 16.0 |
| Al—Mg | 18.9 | 36.8 | Cd—La | 25.0 | 30.0 |
| Al—Mn | 0.9 | 7.7 | Mg—Zr | 0.2 | 0.32 |
| Al—Ni | 0.023 | 7.7 | Ni—C | 2.6 | 8.2 |
| Al—Si | 1.59 | 11.0 | Ni—W | 16.0 | 32.0 |
| Al—V | 0.18 | 0.65 | Pb—Na | 12.0 | 23.8 |
| Au—Co | 23.5 | 42.0 | Pb—Sb | 5.9 | 17.0 |
| Co—C | 3.8 | 7.8 | Sn—Au | 0.2 | 4.0 |
| Co—Cu | 12.0 | 25.0 | Sn—Sb | 10.2 | 15.6 |
| Co—Ga | 11.0 | 18.2 | Sn—Zn | 2.0 | 3.0 |
| Co—W | 14.5 | 17.0 | V—Ni | 8.0 | 18.0 |

occur, particularly if the extrapolated liquidus lines are nearly level and do not diverge far from the eutectic line.

Highly supersaturated solid solutions are produced abruptly rather than gradually, namely when a certain critical cooling rate is attained [9, 21]; this critical rate is dependent on the type of alloy and the composition. For instance, one can obtain an appreciable extension in the Al−Mn system at comparatively low cooling rates (about $10^2$ deg/sec), whereas in many other systems (Ag−Cu, Al−Mg, Ag−Pt, Al−Si) one obtains highly supersaturated solid solutions only at extremely high cooling rates (over $10^7$ deg/sec). If the cooling rate falls below the critical value, the content of the second component in the solid solution falls considerably, usually to values below those permitted by the equilibrium phase diagram.

Very often, quenching from the liquid state is envisaged solely as a means of freezing the high-temperature state, on the assumption that the metastable structures produced by rapid cooling provide information on the short-range order in the melt above the liquidus line [22]. However, it would appear more correct to consider quenching from the liquid state primarily as a crystallization, with all the consequences of that.

For instance, it is familiar that films of metastable solid solutions whose compositions are taken as the same as those of the initial liquid may be fact be not entirely homogeneous; the grain boundaries differ in composition from the grain cores. Dendritic liquation can be seen within the grains [23, 24]. Microscopic inhomogeneity can be seen in alloys having phase diagrams of cigar type (Ge−Si and Be−Sb) even when the cooling rate before the start of crystallization is as high as $10^7$ deg/sec [25].

Unfortunately, detailed structural studies on alloys produced at extremely high cooling rates (over $10^7$ deg/sec) are very difficult; the scanty available data from transmission electron microscopy [26, 27] provide no firm basis for any final conclusions. The depth of the two-phase zone becomes vanishingly small at very high growth rates, and the highly dentate crystallization front becomes nearly as effective as a planar one as regards impurity rejection. The thickness of the enriched zone becomes so small as to make it extremely difficult to detect at the grain boundaries.

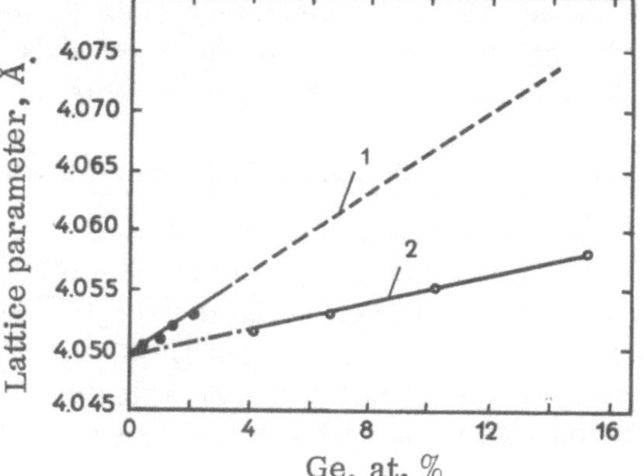


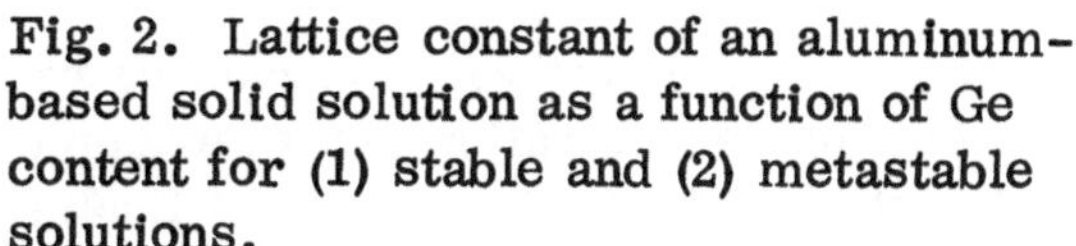
Fig. 2. Lattice constant of an aluminum-based solid solution as a function of Ge content for (1) stable and (2) metastable solutions.

For this reason, it is as yet uncertain which partition coefficient is responsible for establishing the composition of the solid solution, which approaches that of the initial liquid, when the cooling rate is extremely high (the equilibrium-kinetic value* or else the effective value, or the two together). The isolated fragments of experimental evidence (the sudden production of metastable solutions, the exceptionally sharp x-ray lines [21], and the extension of the solid-solution region for alloys having retrograde solidus lines [28]) do not conflict with the scope for diffusionless crystallization [29-31] or for kinetic phase transition [32].

Attempts have been made [33, 34] to derive empirical criteria for the formation of metastable solid solutions; if the atoms differ considerably in structure, one expects that the solid-solution region will be very narrow. Dissolution of one component in the other would raise the free energy so greatly that the formation of some new intermediate phase would be more favorable, or else transition of the liquid to a vitreous state. A notable point in this respect is that almost all alloys that have been made in amorphous form show no extension of the solid-solution region.

Recent studies indicate that there is a further mechanism that can produce metastable solid solutions.

If the thermodynamic driving force is high, the electron energies of solid solutions may be substantially altered, and therefore there may be a jump in the position of the free-energy curve for the solid solution, and hence in the solubility.

Figure 2 shows the lattice parameter of aluminum as a function of germanium content [35] for equilibrium alloys (curve 1) and metastable ones (curve 2). The curves clearly differ considerably in scope, which indicates differences in the electronic state of the germanium atoms in the two types of solution.

The effective volume of the dissolved element and the transition entropy indicate [35] that the electronic state of germanium in metastable solid solution corresponds to the cubic modification of Al.

Similar arguments have been used to conclude [36] that the state of gallium in a metastable aluminum-based solid solution corresponds to face-centered cubic or face-centered tetragonal packing.

* By equilibrium-kinetic partition coefficient is meant the ratio $C_S/C_L$ at the moving crystallization front.

These changes in electronic state for the germanium and gallium atoms result in corresponding polymorphic transitions in response to pressure [37]. One therefore assumes that similar pressures arise when metastable solid solutions are formed [38].

## Crystallization of Metastable Intermediate Phases

There are systems, mainly organic ones, in which even comparatively low cooling rates result in phases that are stable in some other temperature or concentration range; examples are the formation of $Fe_2P$ instead of $Fe_3P$ in the Fe−P system [39], $Zn_3Sb_2$ instead of ZnSb in the Zn−Sb system [11], and $2SbCl_3 \cdot nC_6H_4(CH_3)_2$ instead of $SbCl_3 \cdot nC_6H_4(CH_3)_2$ in the system formed by p-xylene with antimony trichloride [40]. The probability of metastable phase crystallization increases with the cooling rate and with the supercooling.

One therefore naturally supposes that advances in rapid-cooling methods will increase considerably the scope for making new metastable phases.

The first metastable phases made by quenching from the liquid state were the carbides $Ni_3C$ and $Co_3C$ [8], together witht he hexagonal close-packed phase in the Ag−Ge system [33]. At present we know of about 100 new metastable phases [19]. These can vary considerably in structure, namely from the simple cubic type of $\alpha$-Po to the very complex types containing many atoms in the unit cell.

It is obvious that metastable intermediate phases can crystallize only if the melt is appropriately supercooled; for instance, in the system shown in Fig. 1, the $\beta$, $\gamma$, $\delta$, and $\alpha$ phases crystallize in alloy 1 only when the melt is cooled below the temperatures $T_1$, $T_2$, $T_3$, and $T_4$ respectively. Therefore, an alloy supercooled below $T_4$ will be supersaturated in several phases simultaneously ($\beta$, $\gamma$, $\delta$, and $\infty$). The one that crystallizes is determined by the kinetics (the nucleation rate and the growth rate) [41].

Metastable intermediate phases have also been observed in alloys having phase diagrams of simple eutectic type (Au−Tl, Ga−Sn, Pb−Sb, Cd−Bi, Al−Sn, Zn−Ga, Cd−In, Ag−Si, Ag−Ge, Au−Ge, Ga−Al, Al−Ge, Sn−Pb, Bi−Sn, Sn−Zn, Au−Si [19]); these usually crystallize in the composition range around the eutectic point. The composition of the intermediate phase usually corresponds to that of the initial liquid.

One naturally enquires whether these phases are metastable in general or whether they may be stable, e.g., at low temperatures, and be formed by some peritectoid reaction. The latter is readily suppressed at sufficiently low temperatures.

Consider alloy $C_1$ that corresponds in composition to the intermediate phase $\beta$ (Fig. 3); the free-energy curves show that the $\beta$ phase starts to be formed by peritectoid reaction at temperature $T_3$, which is very low, while the temperature for equilibrium with the melt, $T_2$, may be much higher, and therefore the supercooling $\Delta T_1$ required to nucleate the $\beta$ phase directly in the melt is attainable at sufficiently high cooling rates [42]. Although the free energy of the $\alpha + \gamma$ eutectic mixture is less in the temperature range $T_2$-$T_3$, the formation and growth of the $\beta$ phase may be favored by kinetic features. Therefore, it is possible that some of the metastable phases are actually stable at very low temperatures. High supercoolings provide the thermodynamic conditions for direct crystallization from the melt at temperatures above those of the peritectoid reactions.

The mixing energy in the liquid state is often positive in these systems, but the value has a positive temperature coefficient, so zero is reached at low temperatures (such as may be attained by supercooling), which favors nucleation of intermediate phases.

There is an interesting relationship between the structure of the alloy and the crystallization of such metastable phases; the Cd−In, Cd−Sn, In−Sb, Al−Ge, Al−Sn, Bi−Pb, In−Sn,

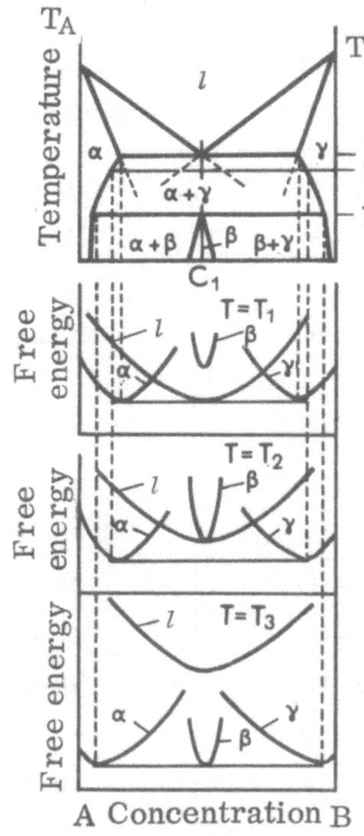


Fig. 3. Phase diagram containing an intermediate compound; relative disposition of the free-energy curves at temperatures $T_1$, $T_2$, and $T_3$.

Pb−Sb, Sb−Zn, and Zn−Sn systems have been examined for the excess partial entropy of the liquid as a function of the content of the second component [43], and it has been found that the curves deviate from monotonic at the points corresponding to formation of metastable phases, which indicates that the formation of stable and metastable phases is determined by the atomic interactions.

However, the above relationship is not obligatory for all instances of metastable phases; the structure of the liquid is important to nucleation, especially in the supercooled state. In some instances, at least, the structure of the supercooled liquid is different from the structure above the melting point. This occurs, for example, with Cu−Fe and Cu−Co alloys, which show unmixing on supercooling [44], which is due to change in the mixing energy, as mentioned above.

Recently, analogies have been drawn between the production of metastable phases at high cooling rates and at high pressures; for instance, one can obtain a hexagonal close-packed phase in the Fe−C system similar to that given by iron at high pressures [45]. In nine instances out of 11, simple cubic phases with the structure of $\alpha$-Po were produced by bismuth and antimony [46, 47], which can take the structure of $\alpha$-Po at high pressures. Metastable nickel carbides can also be produced at elevated pressures [48].

All of this evidence indicates that metastable phases can be produced at high cooling rates, namely ones that do not occur in the equilibrium phase diagram. However, they do exist in the metastable phase diagram, and supercooling or pressure makes it thermodynamically possible for such phases to crystallize.

## Production of Amorphous Alloys

We now know of over 30 binary and ternary systems in which rapid melt quenching can produce a solid amorphous state (Table 2).

Of all the alloys so far examined, only Te−Ga, Te−Ge, Te−In, Te−Cu, and Te−Cu−Au are nonmetallic; in all other instances the base is a typical metal.

TABLE 2. Amorphous Solid Phases Obtained by Quenching from the Liquid State [19]

| System | Composition range, at.% | System | Composition range, at.% |
|---|---|---|---|
| Ag—Si | 17—30 | Pd—Ge | 18—20 |
| Au—Ge | 27 | Pd—Si | 15—23 |
| Au—Si | 15—40 | Pt—Ge | 17—30 |
| Au—Si—Ge | 13.7Ge 9.4Si | Pt—Sb | 33—37 |
| Au—Sn | 29—31 | Pt—Si | 25 |
| Cu—Ti | 30—35 | Rh—Si | 22 |
| Fe—P—C | 13P; 7C | Te—Ag | 33—40 |
| Fe—Pd—P | 10.5 Pd; 20 P | Te—Cu—Au | 25 Cu, 5 Au |
| Mn—P—C | 15 P; 10 C | Te—Ga | 10—30 |
| Nb—Ni | 33—78 | Te—Ge | 10—25 |
| Nb—Ni—Al | 39 N, 13 Al | Te—In | 10—30 |
| Ni—Pd—P | 10—70 Ni, 20 P | Zr—Co | 28 |
| Ni—Pt—P | 40—80 Pt, 25 P | Zr—Cu | 40—75 |
| Ni—Ta | 35—45 | Zr—Ni | 20—40 |
| Pb—Au | 25 | Zr—Pd | 20—35 |
| Pb—Sb | 48 | B | |

In order to produce an amorphous alloy, it is usually necessary to employ extremely high cooling rates; however, Pd−Au−Si, Pd−Ag−Si and Pd−Cu−Si alloys can [49] be obtained in the vitreous state at comparatively low cooling rates ($10^2$-$10^3$ deg/sec). In that case, the vitreous state occurs primarily in the alloy of lowest melting point.

The amorphous state is best demonstrated by combining x-ray methods with electron diffraction and electron microscopy, as well as thermal analysis [50]; the specific heat $C_p$ of the amorphous alloy is measured as a function of temperature, and there is a sharp rise in $C_p$ at the glassification temperature $T_g$. The specific heat then falls, tending toward the extrapolated curve for $C_p$ as a function of T for the liquid.

The heat absorbed by amorphous Au−Ge−Si alloys indicates [51] that the devitrification temperature is reached first, and then, at some higher point, crystallization in the metastable liquid.

The atomic structures of some amorphous solid alloys are similar to those of the corresponding melts [52], which opens a new means of research on liquid structure.

## Morphological Changes

The grain size generally decreases as the cooling rate rises; in addition, the dendrites acquire a finer internal structure. The distances between the axes of the dendrites decrease, while the branches become thinner (Fig. 4a).

The following relationship has been derived [53] between the cooling rate D and the distance d between the twofold axes in the dendrites:

$$d = A \cdot V^{-0.45},$$

where A is dependent on the type of alloy.

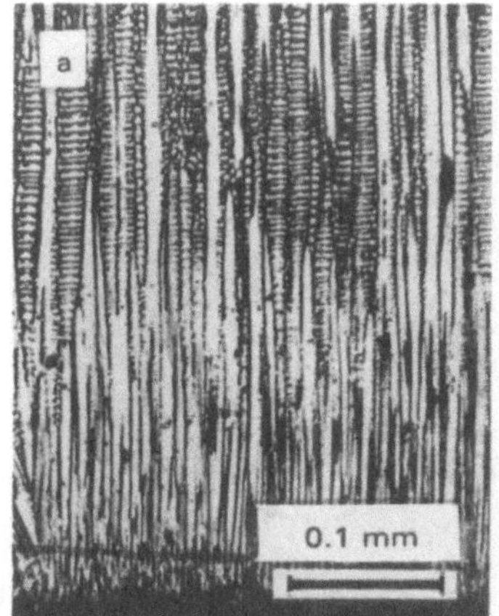

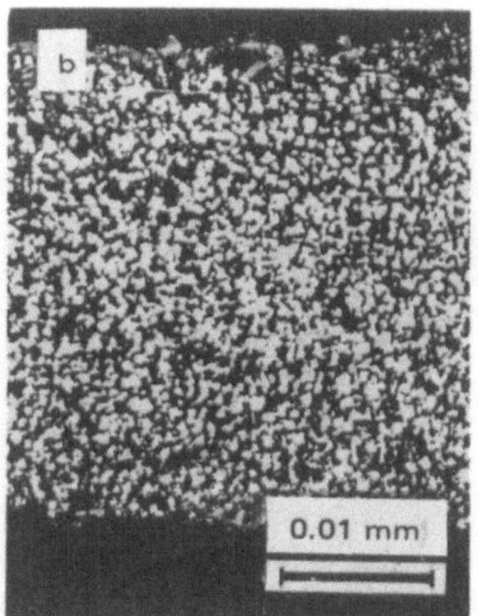

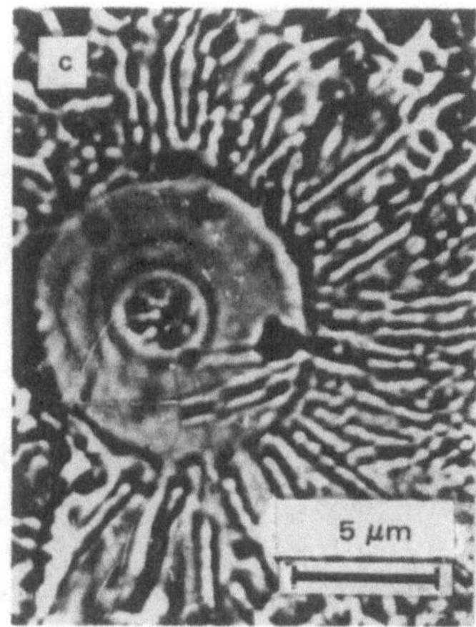


Fig. 4. Microstructures of rapidly-cooled alloys: (a) Kh23N18, cooling rate $10^4$ deg/sec, end of plate $\delta$ = 0.8 mm; (b) Bi+25% Sb, cooling rate $10^7$ deg/sec, end of film $\delta$ = 0.03 mm; (c) Fe + 10% Si, surface of film.

The branching in the dendrites begins to diminish at cooling rates of about $10^4$ deg/sec; in particular, the higher-order branches vanish, and above $10^6$ deg/sec one gets column crystals growing from the cooled surface essentially without branching. The origin of the column zone lies on a surface dendrite. The latter may grow as a spherulite. If the supercooling is sufficient to produce nuclei in the bulk of the melt, one can get spherulites also within the film of material.

At rates above $10^7$ deg/sec, the dendritic growth degenerates completely, and numerous small and more or less equiaxial crystals are formed throughout the film (Fig. 4b).

The surfaces of thin films of certain alloys such as Fe–Si may show large areas without any signs of dendritic structure (Fig. 4c), which lie at the points of best contact with the metal substrate.

One assumes that the dendrite-free parts or grains are formed when the planar front stabilizes, which is possible, in particular, if the growth rate exceeds a value derived from the Mullins–Sekerka relation [54]:

$$R_{cr} = \frac{(k-1)mC_\infty D \cdot L}{k^2 \cdot T_{sn} \cdot \gamma}\,.$$

The value calculated for this critical rate for Fe + 10% Si is 30 cm/sec. Such rates can be attained at the surfaces of thin films.

Therefore, elevated cooling rates at first produce deviations from planar crystallization and dendritic growth forms; but as the rate increases, the dendritic growth degenerates and planar or quasiplanar crystallization again predominates provided that the growth rate attains some critical value (Fig. 5).

Similar morphological changes occur in the crystallization of eutectics; as the cooling rate increases, the internal structure of the eutectic colonies becomes more perfect, and the distances λ between the eutectic plates diminish, the value being given by the following formula [55]:

$$\lambda = A \cdot R^{-n},$$

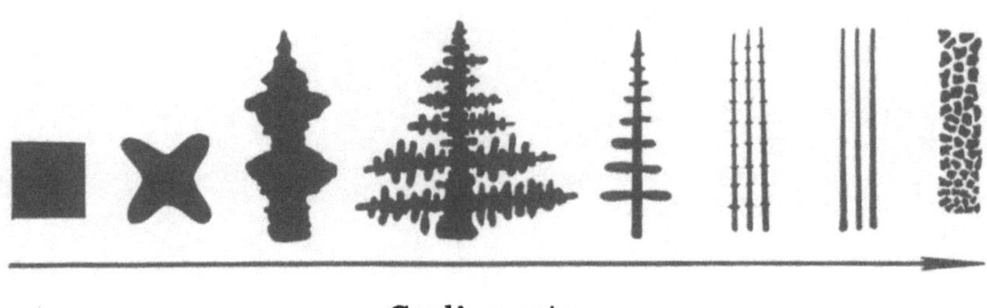


Fig. 5. Effects of cooling rate on growth of dendritic structure.

where R is the growth rate, while A and n are constants taking the values $1.04 \cdot 10^{-5}$ $cm^{3/2}/sec^{1/2}$ and 0.5 respectively.

If the cooling rate is large enough, one can get compatible growth of the eutectic component even in alloys of noneutectic composition [56]; one then gets structures that may be called quasieutectic. If the cooling rate is increased further, the joint growth of the phases degenerates. Instead, a fine-grained two-phase mixture is formed without any signs of eutectic colonies [57].

## Conclusions

Highly supersaturated solid solutions, high defect densities, and fine-grained structure results in considerably increased strength in rapidly cooled alloys.

The metastable phases produced by quenching from the liquid state may have new and valuable properties; for instance, the tetragonal phase in the Au–Ge system [58] is a superconductor, while neither of the components is. Also, the transition temperature is comparatively high.

Further, semiconductor or ferromagnetic behavior has been observed in some amorphous alloys [59].

The production of numerous metastable phases by rapid cooling has provided much new knowledge on the physics of the condensed state and provides means of producing alloys with completely new properties.

## Literature Cited

1. E. H. Dix and W. D. Keith, Trans. AIME (1927).
2. A. A. Bochvar, Metallography [in Russian], Metallurgizdat, Moscow (1945).
3. G. Falkenhagen and W. Hofmann, Z. Metallkunde, 43:69 (1952).
4. I. S. Miroshnichenko and I. V. Salli, Zavod. Lab., 11:1398 (1959).
5. I. S. Miroshnichenko and N. V. Salli, Izv. VUZov, Chernaya Metallurgiya, 8:101 (1960).
6. I. V. Salli and I. S. Miroshnichenko, Dokl. Akad. Nauk SSSR, 132:1364 (1960).
7. N. I. Varach and K. E. Kolesnichenko, Izv. VUZov, 4:131 (1960).
8. I. S. Miroshnichenko, Izv. VUZov, Tsvetnaya Metallurgiya, 1:128 (1964).
9. I. S. Miroshnichenko, In: Mechanism and Kinetics of Crystallization [in Russian], Nauka i Tekhnika, Minsk (1964), p. 138.
10. G. Falkenhagan and W. Hofmann, Archiv Eisenhüttenwesen, 1:73 (1952).
11. M. Hanson and K. Anderko, Structures of Binary Alloys [Russian translation], Metallurgizdat, Moscow (1962).
12. M. I. Novikov, V. M. Glazov, and V. S. Zolotarevskii, In: Structure and Properties of Nonferrous Metals and Alloys [in Russian], Vol. 3, Izd. AN SSSR (1961).

13. P. Duwez, R. H. Willens, and W. Klement, J. Appl. Phys., 31:1136 (1960).
14. P. Duwez, Progr. Solid State Chem., 3:337 (1957).
15. W. Klement and H. L. Lao, Trans. Met. Soc. AIME, 227:1253 (1963).
16. H. L. Luo and P. Duwez, J. Less-Com. Met., 6:248 (1964).
17. P. Duwez, R. H. Willens, and W. Klement, J. Appl. Phys., 31:1500 (1960).
18. R. Wang, Appl. Phys. Letters, 17:460 (1970).
19. T. R. Anantharaman and C. Suryamarayana, J. Mater. Sci., 6:1111 (1971).
20. A. A. Yakunin and N. I. Varich, In: Solid-State Physics [in Russian], Izd. DGU, Dnepropetrovsk (1968).
21. I. S. Miroshnichenko, Dokl. Akad. Nauk SSSR, 164:137 (1965).
22. P. Esslinger, Z. Metallkunde, 57:12 (1966).
23. V. I. Dobatkhin, V. I. Elatin, and V. M. Fedorov, Izv. Akad. Nauk SSSR, Metally, 5:164 (1969).
24. I. S. Miroshnichenko, In: Growth and Defects of Metal Crystals [in Russian], Naukova Dumka (1972), p. 385.
25. I. S. Miroshnichenko and G. A. Sergeev, In: Mechanism and Kinetics of Crystallization [in Russian], Nauka i Tekhnika (1969), p. 40.
26. K. Löhberg and H. Müller, Z. Metallkunde, 60:231 (1969).
27. P. Ramachandrarao, M. Laridjani, and J. W. Cahn, Z. Metallkunde, 63:43 (1972).
28. J. C. Baker and J. W. Cahn, Acta Met., 17:575 (1969).
29. I. L. Aptekar' and D. S. Kamenetskaya, In: Problems in Metallography and Metal Physics [in Russian], Metallurgiya, Moscow (1964), p. 205.
30. V. T. Borisov, Dokl. Akad. Nauk SSSR, 150:2 (1963).
31. D. E. Temkin, this volume p. 334.
32. A. A. Chernov, Uspekhi Fiz. Nauk, 100:329 (1970).
33. W. Klement, J. Inst. Metals, 90:27 (1961).
34. I. I. Varich and I. A. Kravtsov, In: Phase Diagrams for Metallic Systems [in Russian], Nauka (1971), p. 213.
35. B. Predel and G. Schluckebier, Z. Metallkunde, 63:198 (1972).
36. B. C. Giessen, V. Wolf, and N. J. Grant, J. Appl. Cryst., 1:30 (1968).
37. V. K. Grigorovich, Mendeleev's Periodic Law and the Electronic Structures of Metals [in Russian], Nauka (1966), p. 258.
38. V. K. Grigorovich, Electronic Structures and Thermodynamics of Iron Alloys [in Russian], Nauka (1970), p. 42.
39. N. S. Konstantinov, Zh. Russ. Fiz.-Khim. Obshch., Chast' Khim., 41:1220 (1909).
40. B. N. Menshutkin, The Effects of Substituents on Some Reactions of Benzene [in Russian], St. Petersburg (1912).
41. N. N. Sirota, Izv. SFKhA, 20:326 (1959).
42. I. S. Miroshnichenko and G. P. Brekharya, Fiz. Met. Metalloved., 29:664 (1970).
43. P. Ramachandrarao, C. Suryanarayana, and T. R. Anantharaman, Metal. Trans., 2:617 (1971).
44. Y. Nakagawa, Acta Met. 6:704 (1958).
45. R. C. Rahl and M. Cohen, Acta Met., 15:159 (1967).
46. B. C. Giessen, V. Wolff, and N. J. Grant, Trans. Met. Soc. AIME, 242:597 (1968).
47. J. D. Speight, Metal Trans., 3(1):1011 (1972).
48. E. G. Ponyatovskii, M. L. Aptekar', and T. P. Ershova, Dokl. Akad. Nauk SSSR, 171:919 (1966).
49. H. S. Chen and D. Turnbull, Acta Met., 17:1021 (1969).
50. H. S. Chen and D. Turnbull, Acta Met., 48:2560 (1968).
51. H. S. Chen and D. Turnbull, Acta Met., 18:251 (1970).
52. P. Duwez, In: Phase Stability in Metals and Alloys [Russian translation], Mir, Moscow (1970).

53. I. S. Miroshnichenko, A. K. Petrov, V. A. Golovko, G. P. Brekharya, and V. I. Novikov, Zavod. Lab., 38(12):1479 (1972).
54. W. Mullins and P. Sekerka, In: Problems in Crystal Growth [Russian translation], Mir (1968), p. 107.
55. M. H. Burden and H. Jones, J. Inst. Metals, 98:249 (1970).
56. I. S. Miroshnichenko, Izv. Akad. Nauk SSSR, Metally, 5:188 (1968).
57. I. S. Miroshnichenko and A. Ya. Andreeva, In: Metal Physics [in Russian], Vol. 32, Naukova Dumka, Kiev (1970), p. 169.
58. H. L. Luo, M. F. Merriam, and D. C. Hamilton, Science, 145:581 (1964).
59. C. C. Tsuei, Phys. Rev., 170:775 (1968).

# TRENDS IN THE FORMATION OF SUPERSATURATED SOLID SOLUTIONS AT HIGH CRYSTAL GROWTH RATES

**A. I. Dukhin and G. I. Miroshnichenko**

*Bardin Central Ferrous Metallurgy Research Institute, Moscow*

In recent years, there has been increased interest in the crystallization of metals at high growth rates on account of the need to produce materials with properties substantially different from those of equilibrium alloys. Extensive studies have been made on the crystallization and structure of alloys cooled at very high rates (of the order of $10^7$ deg/sec). Although about 100 such systems have been examined, little is known about the crystallization conditions on quenching from the liquid state, and the same applies for the mechanisms of the corresponding phenomena. The results reported here serve to elucidate some characteristic features of alloy crystal'ization under nonequilibrium conditions.

For this purpose we have devised a method and equipment [1] for measuring the temperature of a droplet of melt from point to point as the droplet spreads as a film. The spreading is provided by two copper rods brought together under vacuum. We examined the tin–antimony system. The solubility of antimony in solid tin at the peritectic temperature is 13.0 at.%. The lattice parameter a of the solid solution increases with the antimony content ($da/dc = 0.00175$ per at.%), whereas the c parameter remains almost unchanged. Figure 1 shows a as a function of antimony content for the initial alloys. The largest value occurs for alloys containing 22 at.% antimony. Over the range 0-22 at.%, all the antimony enters the solid solution. Higher contents cause the lattice parameter to fall down to or below the value corresponding to the solubility.

Transmission electron micrographs show that the films have an extremely fine-grained structure and contain granules of the $\beta'$ phase of size 300-5000 Å. Figure 2 shows a clear area in the supersaturated solid solution containing patches of $\beta'$ phase. This pattern corresponds to the onset of decomposition in the supersaturated solution. The inclusions are largely regularly oriented with respect to the matrix. The microdiffraction pattern indicates that the $\beta'$ phase has a layered structure, which may be due to the decomposition arising from supersaturation during the crystallization.

The temperature variation during the cooling was determined by measuring the thermoelectric emf arising at the planes of contact between the liquid metal and the cooling rod or other electrode [2]. The thermo-emf was recorded by an oscilloscope. We examined alloys containing 20 at.% antimony, tin, and aluminum, as well as an alloy of tellurium containing 15 at.% germanium. The conditions also enabled us to measure the temperature on the uncooled lower side of the drop, which in ordinary quenching experiments corresponds to the central plane (center) of the drop during cooling from both sides, At the same time we measured the temperature of the upper cooled side. Table 1 presents the results for tin, tin-antimony alloy, and aluminum.

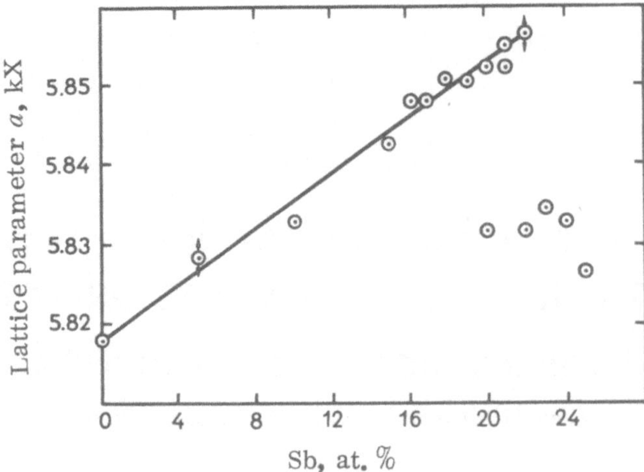


**Fig. 1.** Composition dependence of the lattice parameter for equilibrium and supersaturated solid solutions.

On the whole, the experiment showed that there was considerable supercooling (200-300°C) on the cooled side, which increased with the cooling rate. The uncooled side solidified at temperatures close to the liquidus point.

As there was no supercooling within the quenched film, one supposes that the supersaturated solution arises not on account of the high initial supercooling, which can initiate the production of nuclei of nonequilibrium composition, but of the high rate of crystal growth. In that case, the supersaturation should occur on sequential crystallization if the rate is high enough. A study has been made [3] of the supersaturation in tin–antimony alloys in relation to the speed of the solidification front.

Figure 3 shows the results; the lower broken line indicates the parameter corresponding to the equilibrium solution of antimony in tin (10.3 at.%), while the upper broken line represents the initial antimony content in the alloy (16 at.%).

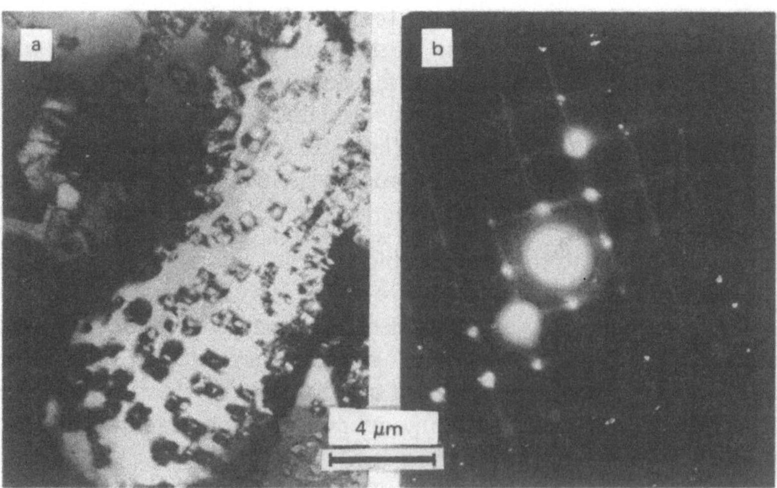


Fig. 2. Electron micrographs of Sn–20 at.% Sb alloy films: (a) ×29,000; (c) electron microdiffraction pattern.

TABLE 1

| Cooled side of droplet | | | Uncooled side | |
|---|---|---|---|---|
| Cooling rate in $10^5$ deg/sec | Crystallization temperature, °C | Super-cooling, °C | Cooling rate in $10^5$ deg/sec | Crystallization temperature, °C |
| | | Sn | | |
| 0.9 | 125 | 110 | 0.8 | 240 |
| 0.6 | 120 | 115 | 1.0 | 230 |
| 0.7 | 100 | 130 | 0.8 | 230 |
| 0.65 | 90 | 145 | 1.0 | 235 |
| | | Sn + 20 at. % Sb | | |
| 0.3 | 275 | 40 | 0.2 | 320 |
| 0.9 | 130 | 185 | 0.2 | 320 |
| 1.0 | 140 | 175 | 0.2 | 315 |
| 1.1 | 235 | 80 | 0.4 | 300 |
| 1.25 | 135 | 180 | 0.17 | 330 |
| 1.3 | 115 | 200 | 0.23 | 300 |
| 1.35 | 235 | 80 | 0.1 | 300 |
| 1.3 | 160 | 155 | 0.6 | 325 |
| 1.4 | 110 | 206 | 1.0 | 315 |
| 1.7 | 120 | 195 | 1.1 | 330 |
| 3.3 | 115 | 200 | 1.24 | 315 |
| | | Al | | |
| 5.0 | 600 | 60 | 6.0 | 660 |
| 10.0 | 360 | 300 | 6.0 | 660 |

The experiment showed that the antimony concentration in the alloy did not attain the equilibrium value for front speeds between 0 and 10 cm/sec; at higher speeds, the lattice parameter increased, and at 10-60 cm/sec it attained the values corresponding to the initial antimony content in the liquid.

These results were obtained for crystals growing in a superheated melt. A study has been made [3] of sequential crystallization in previously supercooled films of gallium–aluminum alloy in order to determine the scope for formation of supersaturated solid solutions at high crystal growth rates. It was found that a supercooling of 30° resulted in a supersaturated solid solution in an alloy crystallizing at a high growth rate. If the alloy was then kept for 12 hr

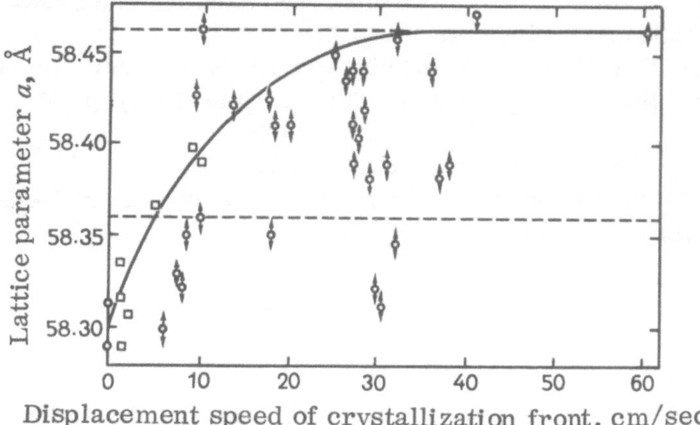


Fig. 3. Lattice parameter as a function of speed of front in quenched liquids.

at room temperature, the solid solution decomposed, and the diffraction pattern corresponded to pure gallium over the angular range $2\theta$ of 78-80°.

We made a cinematographic study of the crystallization in thin films of optically transparent substances (p-chloronitrobenzene and p-bromonitrobenzene) in order to examine the details of supersaturated solid-solution production. The temperatures were measured at the same time. Polarized light enabled us to perform a qualitative phase analysis. The phase diagram for this system is of peritectic type and contains two solid solutions, $\alpha$ and $\beta$.

We found that growth rates of the $\alpha$ phase up to 1 mm/sec resulted in crystallization under virtually equilibrium conditions, with the $\alpha$ and $\beta$ phases stable during subsequent prolonged isothermal annealing. The degree of supersaturation in the $\alpha$ crystals increased with the growth rate, while the delay before decomposition decreased. Rates of 10-15 mm/sec resulted in onset of decomposition within 0.3 sec after the end of crystallization.

Crystals of the $\beta$ phase grew stably up to growth rates of about 16 mm/sec; the region of equilibrium existence of the $\beta$ phase was much wider than that for the $\alpha$ phase.

The degree of supersaturation in the $\beta$ crystals also increased with the rate above 16 mm/sec.

Kinetic-crystallization diagrams serve to explain the supersaturated solid solutions formed with these two organic compounds; these diagrams represent solutions to the kinetic equations [4-6] for crystallization at finite rates. In such cases, the temperature at the crystallization front differs from the equilibrium liquidus temperature, and the liquidus and solidus kinetic curves are displaced with respect to the equilibrium values. This displacement increases with the crystal growth rate. Borisove et al. have made computer calculations on these kinetic diagrams and found that the shapes of the solidus and liquidus curves also vary with the speed of the phase interface.

The experimental results indicate that the range of existence of the solid solution expands when the crystals grow at finite rates, which is explained qualitatively by the kinetic diagrams.

We are indebted to V. T. Borisov and Yu. E. Matveev for assistance.

## Literature Cited

1. V. T. Borisov and A. I. Dukhin, In: Growth and Defects of Metallic Crystals [in Russian], Naukova Dumka, Kiev (1972), p. 408.
2. V. T. Borisov and A. I. Dukhin, In: Mechanism and Kinetics of Crystallization [in Russian], Nauka i Tekhnika, Minsk (1969), p. 176.
3. V. T. Borisov, A. V. Gavrilova, A. I. Dukhin, and Yu. S. Matveev, In: Growth and Defects of Metallic Crystals [in Russian], Naukova Dumka, Kiev (1972), p. 414.
4. V. T. Borisov, Dokl. Akad. Nauk SSSR, 142:69 (1962).
5. V. T. Borisov, Dokl. Akad. Nauk SSSR, 150(2):294 (1963).
6. V. T. Borisov, A. I. Dukhin, and Yu. E. Matveev, In: Problems of Metallography and Metal Physics [in Russian], Vol. 8 (1954), p. 296.

# THE DISTRIBUTION CONSTANT IN HYDROTHERMAL QUARTZ GROWTH

R. A. Laudise, E. D. Kolb, N. C. Lias, and E. E. Grudenski

*Bell Laboratories, Murray Hill, New Jersey*
*Merrimack Valley Works, North Andover, Massachusetts*

## Introduction

The concepts of the distribution constant and effective distribution constant have played important roles in our understanding of impurity segregation during melt growth, but a detailed description of impurity segregation under other conditions has never been given, either for monocomponent growth from the gas phase or for any polycomponent growth. Recently, we have applied the concept of effective distribution constant with appropriate modifications to hydrothermal solution growth [1]. In particular, we have developed relationships for the dependence of the partition constant for $H^+$ in hydrothermal quartz grown from NaOH on growth rate [1] and for the dependence of the concentration of $Fe^{2+}$ on $H^+$ [2]. Since $Fe^{2+}$ causes loss at the important 1.06 $\mu$m (laser) wavelength, these studies have enabled us to prepare high optical transmission quartz [2]. Since $H^+$ is responsible for acoustic loss, we have been able to use our findings to deduce new conditions for high acoustic Q growth at high growth rates [1].

In this paper we briefly review our past studies and show that the effective partition constant for $H^+$ when growth is from RbOH and the effective distribution constant for $Ge^{4+}$ depend in a similar manner on growth rate. In addition, we examine the temperature dependence of the partition constant for $H^+$ in RbOH growth.

## Experimental

Growth in NaOH was generally in large-size commercial autoclaves [1, 2]. The conditions in NaOH were: temperature of crystallization, 345-375°C; fill, 77-90%, $\Delta T$ (temperature difference between dissolving and growth zones), 10-50°; pressure, 10,000-45,000 psi (~670-3000 bar); and solvent, 1.0 M NaOH + 0.025 M $Li_2CO_3$ + 0.1 M $NaNO_2$. In some experiments the NaOH concentration was reduced. For germanium doping and some other experiments Ag cans [2] were used. The RbOH grown quartz was prepared by Kopp and Statts [3], and our data are taken from their paper, which should be consulated for conditions of growth. $[H^+]_s$ was estimated from the absorption coefficient at 3500 cm$^{-1}$ (2.86 $\mu$m) assuming an extinction coefficient of 77.5 liters/mole-cm for $[H^+]_s$ [4].* $[Fe^{2+}]_s$ was estimated from the absorption coefficient at 1.06 $\mu$m assuming an extinction coefficient of 2 liters/mole-cm [5]. Ge was determined by spectrochemical analysis.

*The subscripts s and l refer to the solid and liquid phases.

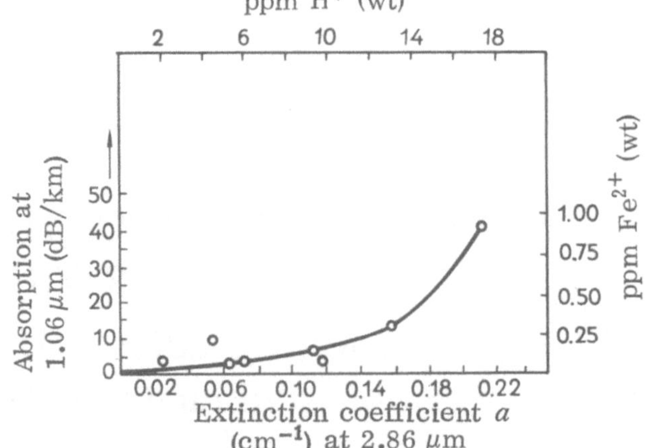


**Fig. 1.** Dependence of $[Fe^{2+}]_s$ on $[H^+]_s$ for NaOH-grown quartz.

## Results and Discussion

Figure 1 shows the dependence of $[Fe^{2+}]_s$ on $[H^+]_s$ obtained from the analysis of a variety of basal plane (0001) and minor rhombohedral $(10\bar{1}1)$ NaOH-grown specimens. The square power dependence can be explained from a consideration of the relevant equilibria and partition constants:

$$[Fe^{2+}]_l + [2H^+]_l \rightleftarrows [Fe^{2+} \cdot 2H^+]_s, \tag{1}$$

$$[Fe^{3+}]_l + [H^+]_l \rightleftarrows [Fe^{3+} \cdot H^+]_s, \tag{2}$$

$$K_1 = \frac{[Fe^{2+} \cdot 2H^+]_s}{[Fe^{2+}]_l\ [H^+]_l^2}, \tag{3}$$

$$K_2 = \frac{[Fe^{3+} \cdot H^+]_s}{[Fe^{3+}]_l\ [H^+]_l}, \tag{4}$$

where $K_1$ and $K_2$ are the equilibrium constants for the respective reactions. $K_1$ and $K_2$ are also the partition constants for $H^+$ which enters quartz interstitially via coupled substitution to $M^{2+}$ and $M^{3+}$ ions,* and are analogous to $k_0$, the equilibrium distribution constant in melt growth. Eliminating $[H^+]_l$, rearranging terms, combining constants, and assuming that $[Fe^{3+}]_l$ and $[Fe^{2+}]_l$ do not change much during the preparation of our samples (a valid assumption when relatively insoluble acmite coats the vessel walls),

$$[Fe^{3+} \cdot H^+]_s^2 = K_5[Fe^{2+} \cdot 2H^+]_s. \tag{5}$$

Since

$$[H^+]_s \gg [Fe^{2+}]_s, \quad [H^+]_s \simeq [Fe^{3+} \cdot H^+]_s, \tag{5a}$$

so that as in Fig. 1

$$[H^+]_s^2 = K_5[Fe^{2+}]_s. \tag{6}$$

* The equilibria and the conclusions drawn from them can be shown to be valid when $Fe^{2+}$ is replaced by $M^{2+}$ and $Fe^{3+}$ by $M^{3+}$, where $M^{2+}$ represents all the +2 ions which absorb at 1.06 μm (e.g., $Fe^{2+}$ + $Cu^{2+}$) and $M^{3+}$ represents all the ions which do not absorb at that wavelength (e.g., $Fe^{3+}$ + $Al^{3+}$).

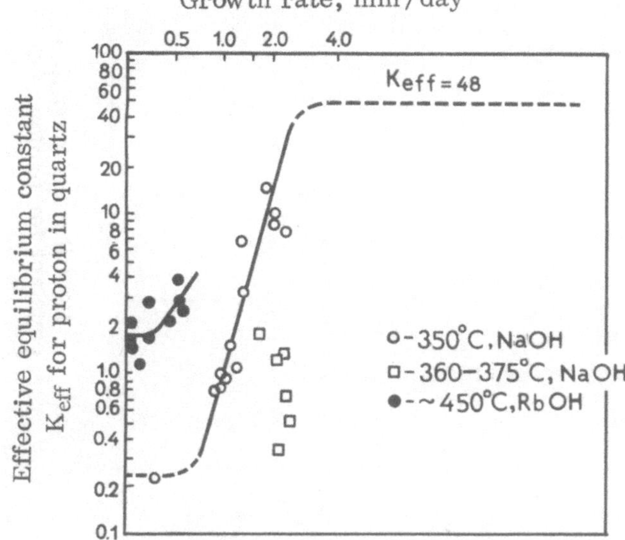


Fig. 2. Logarithm of the effective equilibrium constant for the segregation of $H^+$ vs. the logarithm of the growth rate for NaOH- and RbOH-grown quartz.

The implications for low acoustic and optic loss quartz have been discussed elsewhere [1, 2]. It should be remembered that most of the proton is associated with $[Fe^{3+}]_s$, so that for acoustic loss equation (2) is overriding.

When the rate is such that diffusion is important, by analogy with the melt growth case following Burton and Slichter [6], the effective equilibrium constant may be defined:

$$K_{2eff} = \frac{[H^+ \cdot M^{3+}]_{s-a}}{[M^{3+}]_{l-a}[H]_{l-a}}, \quad (7)^*$$

where by a treatment analogous to that of Burton and Slichter,

$$K_{2eff} = \frac{K_2}{K_2+(1-K_2)} e^{-R\delta/D}, \quad (8)$$

where R is the growth rate, $\delta$ is the diffusion layer thickness, and D is the diffusion constant.

Figure 2 shows the dependence of $K_{2eff}$ on R. Concentrations are expressed in atoms/cm$^3$. $[H^+]_l$ is taken as $[OH^-]_l$ since the $H^+$ in the lattice arises from the $(OH)^-$ mineralizer. $[M^{3+}]_l$ was taken as $[Fe^3]_l + [Al^{3+}]_l$ and comes from the nutrient. Typical analysis gives 50 ppm for each in nutrient [7]. Assuming a solubility of $SiO_2$ of 5 g/cm$^3$ [8], we estimated $[M^{3+}]$ as $1.5 \cdot 10^{-4}$ moles/liter. The $K_{2eff}$ asymptote at 48 was calculated on the assumption that $[H^+]_s$ at high rate was limited by $[Fe^{3+}]_s + [Al^{3+}]_s$ and that at high growth rates all of the $Fe^{3+}$ and $Al^{3+}$ in the nutrient would be incorporated in the grown quartz. That is, at high rates the concentration of $M^{3+}$ in the grown quartz would be the same as in the nutrient. This is perfectly analogous to the high-rate case in melt growth, where the concentration of the impurity in the liquid and the solid becomes equal. However, in the coupled-substitution polycomponent growth situation which we have under present consideration, this does not result in $K_{eff} \rightarrow 1.0$.

The plot shows the typical S-shaped dependence of the Burton and Slichter $K_{eff}$, with $K_{eff}$ approaching $K_1$ at low rates [6]. Departures occur only at temperatures above 350°C, as might be expected if $K_1$ had a retrograde temperature dependence.

---

*$[\ ]_a$ represents actual concentrations at rates appreciably removed from equilibrium.

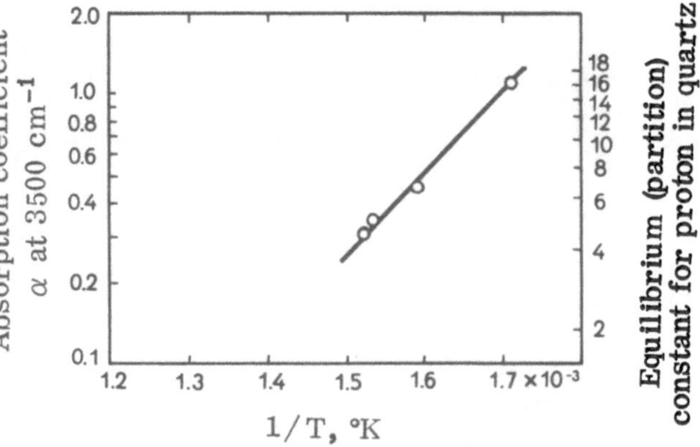


Fig. 3. Dependence of log $\alpha_{3500}$ and log K on 1/T for RbOH-grown quartz for rates below about 0.12 mm/day.

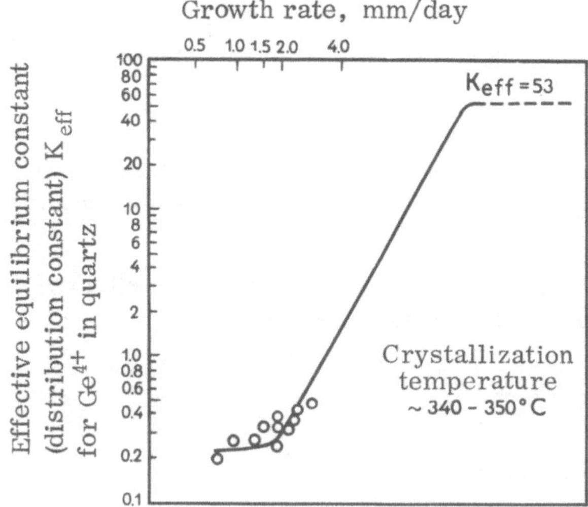


Fig. 4. Logarithm of the effective equilibrium constant for the segregation of Ge vs the logarithm of the growth rate for NaOH-grown quartz.

In Fig. 2 we have also shown values for $K_{2eff}$ for RbOH-grown quartz. As might be expected, $K_{2eff}$ for reaction (2) (where initial proton charge compensates substitutional plus three ions at silicon sites) in RbOH is larger. This can be explained as follows. $[M^{3+}]_s$ compensation will be by means of $M^+$ interstitials. $Rb^+$ is a poor interstitial fit* in comparison to $Na^+$, so that when $Rb^+$ is present, more $H^+$ must enter the lattice to compensate $M^{3+}$ and thus $K_{2eff(RbOH)} > K_{2eff(NaOH)}$.

Figure 3 shows that the dependence of log $\alpha_{3500}$ and $K_{2RbOH}$ are linear with 1/T (with a retrograde slope) at rates below about 0.12 mm/day. Above about 0.12 mm/day $K_2$ becomes rate-dependent so that the Van't Hoff dependence is no longer followed. The heat of reaction (essentially $\Delta H$, the enthalpy, since the pressure was approximately constant) for equation (2) in RbOH as calculated from the slope of Fig. 2 is −13.7 kcal/mole.

$Ge^{4+}$ enters the quartz lattice substitutionally at $Si^{4+}$ sites so that no charge compensation is required. Figure 4 shows a Burton–Slichter dependence of $K_{eff}$ for $Ge^{4+}$ doping where

$$K_{eff} = \frac{[Ge]_s}{[Ge]_l}. \qquad (9)$$

* $Li^+$ is better, and that is why $Li^+$ salts repress $H^+$ uptake and raise acoustic Q [9].

$[Ge]_l$ is determined from the amount of $GeO_2$ added to the hydrothermal system and $[Ge]_s$ by spectrochemical analysis. The asymptote at $K_{eff} = 53$ occurs when all the Ge in a given volume of solution is included in the crystal when the quartz in that volume crystallizes.

## Conclusions

Partition constants for the coupled substitution of $M^{3+}$, $M^{2+}$, and $H^+$ obey the Burton–Slichter relationship as do constants for $Ge^{4+}$. The solubility of $H^+$ in quartz is retrograde with temperature, and in RbOH growth the Van't Hoff relationship is shown to be obeyed. Retrograde impurity solubility in quartz is probably quite common, and the concept of the effective equilibrium constant is probably generally applicable to coupled substitution in polycomponent growth.

## Literature Cited


1. N. C. Lias, E. E. Grudenski, E. D. Kolb, and R. A. Laudise, J. Crystal Growth (1972).
2. E. D. Kolb, D. A. Pinnow, T. C. Rich, N. C. Lias, E. E. Grudenski, and R. A. Laudise, Materials Res. Bull., Vol. 7, No. 5 (1972).
3. O. C. Kopp and P. A. Statts, J. Phys. Chem. Solids, 31:2469 (1970).
4. D. M. Dood and D. B. Fraser, J. Appl. Phys., 37:3911 (1966).
5. T. Bates, Modern Aspects of the Vitreous State, Vol. 2, J. D. MacKenzie, ed., Butterworths, Washington (1962), pp. 195-254.
6. J. Burton and W. P. Slichter, Transistor Technology, H. E. Bridgers and J. N. Shive, eds., Van Nostrand, Princeton (1958), Vol. 1, Chapter 5.
7. R. A. Laudise, A. A. Ballman, and J. C. King, J. Phys. Chem. Solids, 26:1035 (1965).
8. R. A. Laudise and A. A. Ballman, J. Phys. Chem., 65:1396 (1961).
9. J. C. King, A. A. Ballman, and R. A. Laudise, J. Phys. Chem. Solids, 23:1019 (1962).

# DOPE UPTAKE FACTOR IN RELATION TO GROWTH RATE AND SURFACE INCLINATION


## V. V. Voronkov

*State Research and Design Institute for the Rarer-Metal Industries, Moscow*


Crystals of Ge, Si, and other semiconductors grown from the melt along ⟨111⟩ frequently show impurity channels [1-3] that correspond to (111) faces on the convex crystallization front. The concave parts of the front parallel to (111) planes also generate narrower impurity channels [4]. Channels of these two types are shown schematically in Fig. 1 (a case typical of crucible-free zone melting). The width of the tubular channels and the radius of curvature of the concave front allow one to show that the boundary of such a channel corresponds to an angle $\theta \sim 10^{-2}$ (about 0.5°) between the front and (111) planes. The dope concentration C taken up by the crystal is directly proportional to the concentration $C_m$ in the melt directly at the front: $C = KC_m$, where K is the uptake factor. The value for $C_m$ differs for that within the melt if the mixing is not very vigorous [5], so the observed relationship between C and the inclination of the front may in part be due to variation in $C_m$ along the front. However, such variation can hardly be important for the narrow channels on the concave parts. Another case where the main part is certainly played by variation in K rather than $C_m$ is the facetted channels given by Te in InSb [3], so one supposes that such channels are due mainly to the relationship between K and the front inclination. Here we calculate K for a stepped surface consisting of steps of minimal height h. Let the surface be inclined at a small angle $\theta$ to a (111) plane, i.e., be an echelon of steps of mean separation $h/\theta$. The normal speed V of the surface, the step speed $V_2$, and the surface supercooling $\Delta T$ are all related:

$$V_2 = \beta \Delta T \doteq V/\theta, \tag{1}$$

where $\beta$ is the kinetic coefficient for a step, which is 60 cm · sec$^{-1}$ · deg$^{-1}$ for Si [6]. The steps are generated at the center of the face, and this has an average (111) orientation, and therefore we can introduce an effective angle $\theta = V/V_2$, which characterizes the step density. The observed shapes of growth fronts for Si crystals with and without dislocations have been used [6] to determine the supercooling at the center of the face; for our subsequent numerical estimates we used a supercooling of $\Delta T = 0.3$ deg for an Si face bearing dislocations (for a typical growth rate $V = 3 \cdot 10^{-3}$ cm/sec). The corresponding angle is $\theta \approx 10^{-4}$, while $\theta$ for a curved part of the front is usually several degrees ($\Delta T \sim 10^{-3}$ deg). Therefore, $\Delta T$, $\theta$, and $V_2$ vary by three orders of magnitude on passing from a curved part of a front to a face, which explains the sharp boundary of the channel. We now discuss why K is dependent on $\theta$, i.e., on $V_2$. A very simple effect is that the dope is rejected by a moving step and accumulates ahead of it, the more so the higher $V_2$. Unit length of step rejects $Q = (1 - K_2)C_m q V_2$ dope atoms, where $K_2$ is the step trapping factor, q is the surface node density in the layer laid down by a step, and $C_m$ is the concentration (atomic proportion) of the dope in the alloy ahead of the step. The dis-

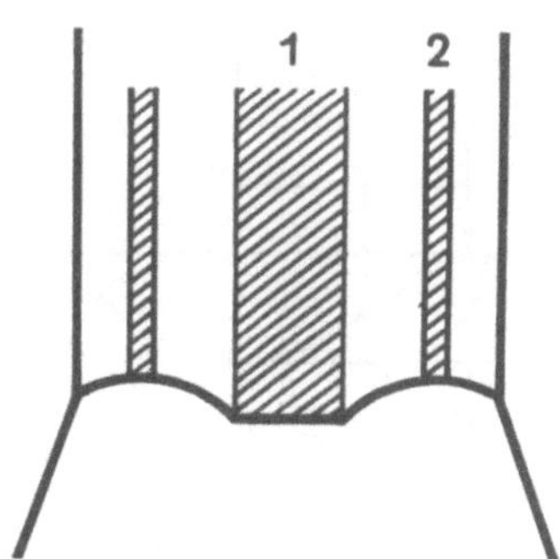


Fig. 1. Impurity channels in a longitudinal cross section of a crystal: (1) central cylindrical channel corresponding to a (111) face; (2) tubular channel corresponding to a concave front parallel to a (111) plane.

tribution of $C'_m$ set up by a moving line source of output Q is [7]

$$\rho_m C'_m = (Q/\pi D_m)\ K_0(V_2 r/2D_m)\ \exp\ (-V_2 r \cos\chi/2D_m), \tag{2}$$

where $D_m$ is the diffusion coefficient for the dope in the melt, $\gamma$ is the angle between the radius vector r and $V_2$, $\rho_m \approx q/h$ is the atomic concentration in the melt, and $K_0$ is a Bessel function. The dope accumulation at the step is characterized by the difference $\Delta C'_m$ along the surface between the points r = h and r = $L_2/2$, where $L_2$ is the distance between adjacent steps. In the case of the fastest steps (on the face), $V_2$ = 20 cm/sec and $L_2 = 2 \cdot 10^{-4}$ cm. We put $D_m = 2 \cdot 10^{-4}$ cm$^2$/sec [8] to get from (2) that $\Delta C'_m/C_m < 0.01$ (the contribution to $\Delta C'_m$ from the other steps in the echelon may be neglected). Therefore, diffusion in the melt can eliminate any accumulation of the dope along the stepped surface, so the $V_2$ dependence of K cannot be explained in terms of simple dope accumulation at steps. It therefore remains to consider the kinetics of step interaction with surface impurity atoms. We show below that step fluctuations play a major part in this, so first of all we examine step motion in the absence of impurities.

## Fluctuations in Moving Steps

By f we denote the reduction in the volume free energy per atom on crystallization; f = $H\Delta T/T$, where H is the latent heat of fusion per atom. This f acts as the driving force of tallization, since it causes step displacement. We first consider displacement of an isolated kink along a low-index step, with the distance $h_1$ between adjacent possible positions for the kink. The kink advances with the mean speed $V_1$ while at the same time wanders in a random fashion with a diffusion coefficient $B_1$. If the probability of finding a kink near point x per unit length is p(x), then the kink flux is $V_1 p - B_1 \partial p/\partial x$, which should become zero for the Gibbs distribution $p = A\exp fx/h_1 kT$, which gives $V_1/B_1 = f/h_1 kT$, which is analogous to Einstein's relation for Brownian motion. During its random walk, a step deviates by a distance $\lambda = \sqrt{2B_1 t} - V_1 t$ from the initial position in time t on average; the maximum deviation is denoted by $\lambda_1$ and is $\sim B_1/V_1 = h_1 kT/f$, and the kink undoubtedly will advance from its initial position within a time $\tau_1 \sim B_1/V_1 \sim \lambda_1/V_1$. It is also possible to derive $\lambda_1$ purely thermodynamically: deviation of the kink causes the free energy to increase by $\delta F = f\lambda/h_1$, and fluctuations for which $\delta F \lesssim kT$ are the most likely, and hence the likely deviation is $\lambda_1 \sim h_1 kT/f$. A distinctive feature of a crystal-melt boundary is the high equilibrium kink density even for a low-index step. For instance, the mean distance $L_1$ between kinks for Si is $\approx 4h_1$ [6], which is much less than $\lambda_1$ even for the maximum f (for a face), where $\lambda_1 \sim 1500h_1$. The time for a kink to diffuse a length $L_1$ is $\sim L_1^2/B_1$, and this is much less than the drift time for this distance, which is $\sim L_1/V_1$, which means that there is rapid fluctuation merging of the kinks, so a step rapidly runs through all possible configurations before advancing substantially. In that case, a step can be considered macroscopically as a line of specific free energy $\alpha$, which is determined by the equilibrium statistics of the kinks.

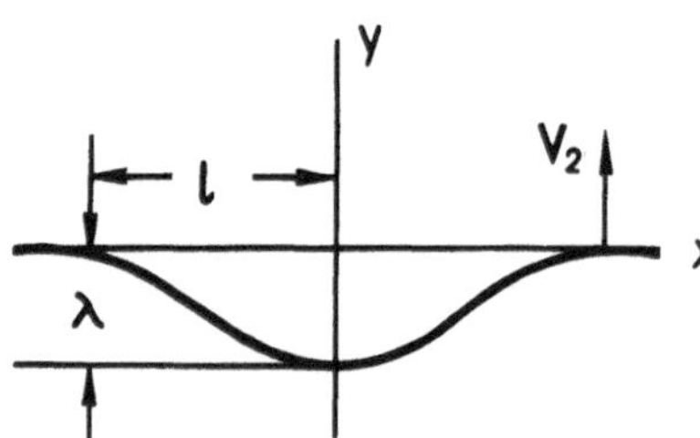


Fig. 2. Fluctuating deviation of part of a step.

During the step motion with mean speed $V_2$, the individual parts fluctuate and may deviate backwards; let a part of length 21 deviate a distance λ (Fig. 2). We estimate the free-energy increment $\delta F$ by describing the step shape via the first harmonic $y = -\lambda \cos \pi x/2l$, in which case $\delta F$ is composed of a bulk part $fq\int y dx$ (where q is the surface density of the layer laid down by the step) and the increase in the edge free energy of the step $(\alpha/2)\int(dy/dx)^2 dx$

$$\delta F = (4/\pi) fq l \lambda + (\pi^2/8)\alpha l^{-1}\lambda^2. \tag{3}$$

The minimum $\delta F$ for a given λ occurs at $l \approx (\alpha\lambda/fq)^{1/2}$ and equals $(2\pi\alpha qf)^{1/2}\lambda^{3/2}$; the probable deviation $\lambda_2$ for a given step element is defined by the conditions $\delta F \sim kT$ and equals

$$\lambda_2 \sim (kT)^{2/3}(2\pi\alpha qf)^{-1/3}. \tag{4}$$

The characteristic length $l$ of a part of a step corresponding to the maximum deviation is

$$l \sim (\alpha kT)^{1/3}(fq)^{-2/3}. \tag{5}$$

When a step element has reached a certain position, it will certainly advance further only after a time $\tau_2 \sim \lambda_2/V_2$, where $\tau_2$ is the characteristic time for which a step fluctuates near a given position.

The free energy α of a step in the presence of a high kink density is almost independent of the step orientation and is $3 \cdot 10^{-6}$ erg/cm for Si [6]. However, even weak anisotropy is important for low-index steps, because α alters in a narrow range of rotation angle φ. In that case, one has to replace α by the effective value $\alpha_{eff} = \alpha + d^2\alpha/d\varphi^2$ [9]. If the mean distance $L_1$ between kinks is much greater than $h_1$ (which is qualitatively so), then we have [10] that

$$\alpha_{eff} = kTL_1h_2^{-2}, \tag{6}$$

where $h_2$ is the distance between adjacent low-index atomic rows. Therefore $\alpha_{eff}/\alpha \approx 5.3$ for Si, but the difference has little effect on $\lambda_2$. We substitute (6) into (4) to get the simple result

$$\lambda_2/h_2 \sim (\lambda_1/L_1)^{1/3}, \tag{7}$$

where $\lambda_1 = h_1kT/f$ is the deviation length for an isolated kink; $\lambda_2$ is minimal for a face and is $\sim 2 \cdot 10^{-7}$ cm, while the corresponding fluctuation time is $\tau_2 \sim 10^{-8}$ sec.

We can derive $\lambda_2$ and $\tau_2$ more rigorously via a kinetic approach, as for a kink; the mean speed of a step of curvature $\partial^2 y/\partial x^2$ becomes zero for an equilibrium radius of curvature $\alpha/qf$ (for a critical two-dimensional nucleus), while $\dot{y} = V_2 = \beta\Delta T$ for a straight step, and therefore

$$\dot{y} = V_2 + (\alpha\beta T/qH)\partial^2 y/\partial x^2. \tag{8}$$

Then the deviation length $\lambda$ (Fig. 2) varies with a mean rate $\dot{\lambda}$ as defined by (8), while it also fluctuates with the diffusion coefficient $B_2$ with $\dot{\lambda}/B_2 = -\partial/\partial\lambda(\delta F/kT)$ (this relation is analogous to that between $V_1$ and $B_1$ for a kink). We substitute for $\delta F$ from (3) and put f = 0 (equilibrium fluctuations) to get

$$B_2 = \beta kT^2/qHl. \tag{9}$$

The displacement due to fluctuation of a length $2l$ in time t is $\sqrt{2B_2t}$ subject to the additional condition that it does not exceed the amplitude $\lambda_e$ of the equilibrium fluctuations; $\lambda_e$ itself is defined by the condition $\delta F \sim kT$ for f = 0 and equals $\lambda_e \sim (kTl/\alpha)^{\frac{1}{2}}$. The maximum fluctuation displacement occurs for the length $l$ (t) such that $\sqrt{2B_2t} \sim \lambda_e$, while the fluctuation time $\tau_2$ is equal to the interval t within which the mean displacement $V_2t$ is comparable with the fluctuations. As a result, we get the previous expressions for $\lambda_2$, $l$, and $\tau_2$. If the length of a part is optimal, as in (5), the step diffusion coefficient is $B_2 \sim V_2\lambda_2/2 \sim 2 \cdot 10^{-6}$ cm$^2$/sec for an Si face.

The steps move about the surface independently provided that the mean distance $L_2$ between them exceeds $\lambda_2$; we introduce the characteristic angle $\theta_0$, which is such that $\lambda_2 = L_2/2$, and from (4) and (1) we have

$$\theta_0 = [qh^3(\alpha/kT)(H/kT)(V/\beta T)]^{1/4}. \tag{10}$$

This $\theta_0$ is only slightly dependent on V and on the constants, being about $2 \cdot 10^{-2}$ (about one degree). The ratio of the time for a step to diffuse a length $L_2$ and the drift time for this distance is $\sim(\theta_0/\theta)^4$, so the steps move independently for $\theta < \theta_0$, while for $\theta > \theta_0$ the surface rapidly runs through all possible configurations. In particular, the effects of curvature of a stepped surface on the speed are determined solely by the configuration part of the statistical sum for the surface, by analogy with (8) for a step [11], namely by the fluctuations between adjacent steps. Therefore, the surface curvature influences the speed only for $\theta > \theta_0$.

## Impurity Capture and Embedding by Fluctuating Steps

A moving step encounters two types of surface impurity atom in its motion (Fig. 3): atoms adsorbed on the surface (type a) and atoms in the surface layer (type s). A step that traps an a atom converts it to the s state. On passing over an s atom (on enclosing it), a step converts such an atom into the bulk state (type b). We commence our analysis with the simpler case of embedding of s atoms. We denote the equilibrium partition coefficient for surface layer of the crystal and the melt by $K_s$, while the same for the bulk of the crystal and the melt is denoted by $K_b$ (the concentration is taken as the atomic fraction in the given subsystem). Most impurities have $K_b < 1$, and in this case $K_s > K_b$, since the s state is intermediate between two phases. The equilibrium probability of finding an impurity atom at a given point is given by the Gibbs distribution as proportional to exp (−F/kT), where F is the free energy of the system (with the impurity at a given point). Therefore, an s atom can be embedded only if a free-energy barrier $kT \ln K_s/K_b$ is overcome, i.e., the s atom hinders step motion. When the step reaches the impurity atom, it fluctuates and repeatedly covers and uncovers the atom, which means that there is an equilibrium ratio $K_b/K_s$ of finding the impurity directly behind

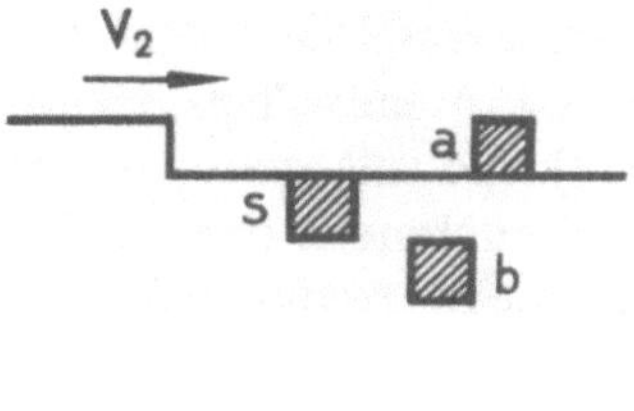

Fig. 3. Impurity atoms ahead of the step.

the step and ahead of it. We denote by $\tau_s$ the time for which the s atom jumps back into the melt and frees the path of the step. A step advances from any given position with a mean fluctuation time $\tau_2$; if $\tau_s \ll \tau_2$, the impurity causes no additional step delay, and in that case there is no correlation between the step and the impurity, so the probability of finding an impurity ahead of a step takes the equilibrium value. Then the embedding coefficient $G_2$ (the ratio of the concentrations behind the step and ahead of the step) is close to the equilibrium value $G_{2e} = K_b/K_s$, where if $\tau_s \gg \tau_2$ the step is hindered at the impurity, and the probability of finding the impurity ahead of the step increases, so $G_2 > G_{2e}$, while if $\tau_s/\tau_2$ is large enough a step will embed completely any encountered impurity and $G_2 = 1$.

To illustrate this principle we consider embedding by an isolated kink, since here the qualitative results can be compared with the exact solution. If $\tau_s$ is much less than the fluctuation time $\tau_1 = B_1/V_1^2$ of the kink, then the embedding factor $G_1$ should be close to $G_{1e} = K_b/K_s$; the contrasting case $G_1 = 1$ occurs when $\tau_s$ is much greater than the embedding time $\tilde{\tau}_1$, i.e., the time taken for a kink to recede from the impurity. If the impurity were not present. the kink would certainly advance in a time $\tau_1$, while the impurity results in a barrier $\tilde{F} = kT\ln K_s/K_b$, so

$$\tilde{\tau}_1 \sim \tau_1 \exp \tilde{F}/kT = \tau_1 G_{1e}^{-1}. \tag{11}$$

An analytic expression for $G_1$ is derived from a simple one-dimensional treatment for the dope distribution with respect to the kink; a solution has been given [12] for an analogous problem for a one-dimensional chain, which is directly applicable to embedding if one assumes [12] that all the frequencies of attachment to a kink are identical. If the supercooling is small, we then get after certain algebraic steps that

$$G_{1e}/G_1 = 1 - 2(1 - G_{1e})[1 + \sqrt{1 + 4\tau_1/\tau_s}]^{-1}. \tag{12}$$

The criteria for the two limiting case ($G_1 \approx G_{1e}$ and $G_1 \approx 1$) that follow from (12) agree with the above general conclusions ($\tau_s \ll \tau_1$ and $\tau_s \gg \tilde{\tau}_1 = \tau_1 G_{1e}^{-1}$).

We now consider the embedding time $\tilde{\tau}_2$ for an impurity atom at a step; if the time t following encounter of the step with the impurity exceeds $\tau_2$, the step is deflected as in Fig. 4, in accordance with the equation of motion (8), which represents diffusion with an effective diffusion coefficient $\tilde{D} = \alpha\beta T/qH$; the solution to the standard diffusion problem [7] gives the deflection angle $\psi$ as

$$\psi = V_2(4t/\pi\tilde{D})^{1/2}. \tag{13}$$

We now determine the detachment time for a given $\psi$; by analogy with (11), this is larger by a factor $G_{2e}^{-1}$ than the time for the step to advance with the obstacle removed. If we neglect the fluctuations, the kink in the step rapidly straightens out after elimination of the obstacle, in accordance with (8). The slope $\partial y/\partial x$ of the step at the initial instant is equal to $\psi$ for $x > 0$ and to $-\psi$ for $x < 0$, and so after a time t' we get the shape

$$\partial y/\partial x = \psi \operatorname{erf}(x/\sqrt{4\tilde{D}t'}). \tag{14}$$

In fact, the fluctuations mean that the step is detached completely only when the fall in the free energy F by comparison with the initial state of Fig. 4 exceeds kT. The main contribution to the reduction in F comes from the edge free energy $(\alpha/2)\int(\partial y/\partial x)^2 dx$ and we substitute from this into (14) and integrate to get that the fall of kT in F occurs in a time $t' = (\pi/8\tilde{D})(kT/\alpha)^2\psi^{-4}$; the detachment time is larger than this by a factor $G_{2e}^{-1}$, and it decreases rapidly as $\psi$ increases. The increase in $\psi$ is in accordance with (13), and in fact the step detaches

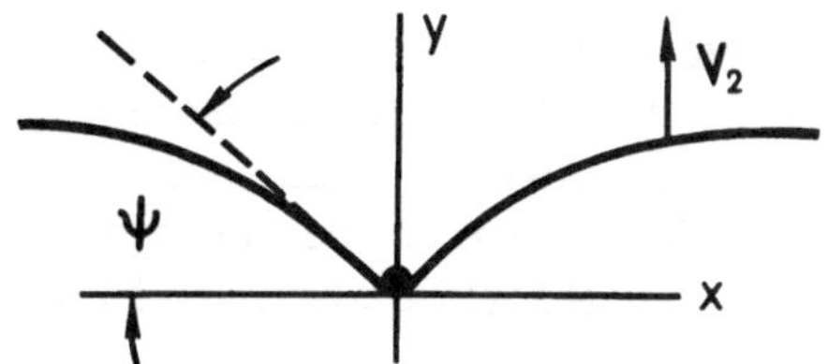


Fig. 4. Step deflection at an impurity obstacle.

from the impurity when $t'G_{2e}^{-1}$ is comparable with the total time elapsed t. Therefore, the embedding time $\tau_2$ is given by (4) with $\lambda_2 = V_2\tau_2$ as

$$\tilde{\tau}_2 \sim \tau_2 G_{2e}^{-1/3}. \qquad (15)$$

This formula applies if $\tilde{\tau}_2$ is less than the time needed for the right-hand and left-hand branches of the kink in Fig. 4 to meet, and the latter time is $\sim\alpha/qfV_2 \sim \tau_2(\lambda_2\alpha/kT)^2$, i.e., (15) is applicable if $G_{2e}$ is not too small. The deflection of the step means that $\tilde{\tau}_2$ is closer to the fluctuation time $\tau_2$ than in the case of a kink, so the transition from equilibrium embedding ($\tau_s \ll \tau_2$) to complete embedding ($\tau_s \gg \tilde{\tau}_2$) is sharper as $\tau_2$ decreases (i.e., as $V_2$ increases).

The results for embedding are applicable directly to trapping of adsorbed atoms a (Fig. 3); the desorption time is $\tau_a \sim (h^2/D_m)$, where U is the adsorption energy (the energy change on exchanging an adsorbed impurity atom for a basic atom in the bulk of the melt). We estimate U by means of a quasichemical model, in which the atoms of both phases are embedded in a single lattice and the system energy is made up of nearest-neighbor pair interactions. We denote these energies for the main atoms as $E_{cc}$, $E_{mm}$, $E_{cm}$, where the subscripts c and m denote the crystal and melt. The excess energy per interphase bond is $W = E_{cm} - (E_{cc} + E_{mm})/2$, while the latent heat of fusion per atom is $H = (E_{mm} - E_{cc})z/2$, where z is the number of neighbors. We have $W > H/z$ for Ge and Si [6], i.e., a basic atom in the melt interacts more strongly with neighboring atoms in the melt than with neighboring atoms in the crystal ($|E_{cm}| < |E_{mm}|$) and one therefore expects that this applies also for an impurity atom, in which case $U < E_{cm} - E_{mm} = W - H/z$. We substitute $W/kT = 1.7$ [6] to get $U/kT < 0.8$ and $\tau_a < 10^{-11}$ sec, i.e., $\tau_a$ is less by three orders of magnitude than $\tau_2$ even for the fastest steps. Therefore, the trapping factor $K_2$ is close to the equilibrium value $K_{2e}$. The relative deviation from the equilibrium value of (12) for the one-dimensional case is equal to the ratio of the mean displacement $V_1\tau_s$ and the fluctuation displacement $\sqrt{B_1\tau_s}$ occurring during $\tau_s$. We utilize this result also in the case of a step. The ratio of the mean and fluctuation displacements in time t is given by the previous section as proportional to $t^{3/4}$ and equal to 1 for $t \sim \tau_2$, which means that the deviation of $K_2$ from $K_{2e}$ is $\sim(\tau_a/\tau_2)^{3/4} < 0.01$. Further, there is surface diffusion of the a atoms, so the difference between $K_2$ and $K_{2e}$ is further reduced.

Therefore, in principle there are two mechanisms that cause K to be dependent on the step speed: 1) the interaction between a step and the adsorbed atoms on the surface, and 2) the interaction between the step and the s atoms in the surface layer of the crystal. However, the first mechanism (which was first proposed in [13]) gives a negligible effect by virtue of the small value of U (of course, if our model estimate of U is correct). We now show that the second mechanism is in quantitative agreement with experiment, which confirms that U is small. For this purpose we estimate $K_s$ and $\tau_s$ within the framework of the quasichemical model. We express $K_s$ in terms of the bulk coefficient $K_b$ [14, 15] on the assumption that the energy of the interaction with an adjacent atom in the melt for a surface atom in the crystal (whether main or impurity) is as though these atoms were in the bulk of the melt. However, a weaker assumption will suffice: let these energies be different, but let the differences be the same for the main and impurity surface atoms. Then when an impurity atom is transferred from the melt into an s state (with a major atom from the surface replacing it), the energy change is due solely to the impurity–crystal interaction and is $(1-\gamma)/\Delta E$, where $\gamma$ is the proportion of the bonds from an s atom directed into the melt and $\Delta E$ is the energy change on transferring the impurity from the melt to the bulk of the crystal. The entropy change $\Delta S$ is assumed to be the same for the s and b states. Then the change in free energy for an s atom is expressed in terms of a corresponding change $\Delta E - T\Delta S$ for a b atom, and therefore $K_s$ is expressed in terms of $K_b$:

$$K_s = K_b^{1-\gamma} \exp \gamma \Delta S/k. \tag{16}$$

Further, $\gamma = 1/4$ for a diamond lattice, while $\Delta S$ is of the order of the difference of the entropies of melting for the main and impurity substances. The entropy of melting for most impurities is substantially less than the values applicable for Ge and Si, so $\Delta S$ is estimated by taking the entropy of melting for the main substance as $H/T$.

A substitutional impurity escaping from the surface layer into the melt leaves a vacant site; then $\tau_s$ can be found from the rate of arrival of the impurity from the melt at the surface vacancies by means of the principle of detailed balancing. The frequency of the jumps from a given vacancy is $\sim D_m h^{-2} C_m$, and the proportion of vacant sites in the volume is denoted by $C_v$. The proportion of vacant sites at the surface is larger, since the energy of formation is lower, and that by an amount $E_{cm} - E_{cc} = W + H/4$. Therefore,

$$\tau_s \sim (h^2/D_m C_v) K_s \exp[-(W+H/4)/kT]. \tag{17}$$

For instance for P and Si ($K_b = 0.35$) we get from (16) that $K_s = 1.1$; we put $C_v = 2 \cdot 10^{-7}$ in (17) [16], which gives $\tau_s \sim 2 \cdot 10^{-6}$ sec. It is of interest to compare $\tau_s$ and the diffusion-jump time $3h^2/8D$ for the solid state [17], where D is the impurity diffusion coefficient in the crystal. We put $D = 10^{-10}$ cm$^2$/sec for P and Si [17] to get $3.5 \cdot 10^{-6}$ sec, which is close to $\tau_s$. The ratio is then $\tau_s/\tau_2 \sim 200$ for a step on a face, i.e., there is complete embedding of the surface layer containing the impurity. From the condition $\tau_s \sim \tau_2 \sim \lambda_2/V_2$ with (4) and (1) we get the angle $\theta_s$ separating the regions of equilibrium embedding ($\theta$ large) and complete embedding ($\theta$ small):

$$\theta_s \sim \theta_0 (\tau_s V/h)^{3/4}, \tag{18}$$

where $\theta_0$ is the angle bounding the region of independent step motion as given by (10). Equation (18) applies only for $\theta_s < \theta_0$; $\theta_s$ and $\theta_0$ are of the same order of magnitude for typical growth rates V.

At large angles $\theta < \theta_0$, the steps fluctuate rapidly between their neighbors, and in that case an s atom may be eliminated from the track of a given step because an adjacent step is absent and releases the s atom (after which it rapidly diffuses into the melt). Therefore, the impurity does not hinder the step motion and K is close to equilibrium $K_c$, no matter what $\tau_s$. Therefore, K starts to rise above the equilibrium $K_b$ at $\theta \sim \theta_0$, i.e., $\theta_0$ defines the boundary of an impurity channel. The above estimate $\theta_0 \sim 1°$ agrees with observations on tubular channels (Fig. 1) for P in Si [4].

## Face Trapping Coefficient

The steps on a face embed the s impurity atoms, which diffuse to the surface, so the final value for K is dependent on the ratio between V and the diffusion rate in the crystal V/h [18]. If V is sufficiently small, equilibrium can be set up, and then $K \approx K_b$, whereas if $V \gg D/h$ all the buried impurity is retained and $K \approx K_s$ (an atomic layer laid down by a step consists of two planes in a diamond lattice, and instead of $K_s$ one needs to use the mean value $\overline{K}_s = (K_s + K_b)/2$).

Consequently, K for a face deviates from $K_s$ and is dependent on the growth rate. The doping level in the channel R (the ratio of the concentration in the channel to that outside it) is $K/K_b$.

The effective trapping coefficient as a function of V has been examined [5, 19], and it has been shown [5] that the result is explicable within the error of measurement in terms of incomplete mixing in the melt. A study on small-diameter billets [14], where mixing variations did not influence the result, showed that the true trapping coefficient K is dependent on V; the above theory indicates that K(V) should occur only for a face, not for a curved front. The billet diameter in [14] was less than the size of the faces at the front found on large-diameter crystals, so the entire front consisted of a single face (a contrary assertion is to be found in [14], which is wrong). The limiting K for large V for P and Al [14] or for B [20] in Si agree closely with $\overline{K}_s = (K_s + K_b)/2$, which is given by (16), while the transition range between $K_b$ and $K_s$ corresponds to rates $V \sim D/h$.

It is simplest to derive K(V) for randomly distributed steps, since then step passage over a given point is a random event; let the frequency of such events be $\omega$, and then the mean speed is $V = \omega h$. We envisage a simple lattice and denote by $C_n$ the impurity concentration in an atomic layer at depth nh from the surface. Here $C_n$ can vary with time on account of impurity diffusion as well as step passage:

$$\dot{C}_n = -\nu_n(C_n - r_n C_{n-1}) + \nu_{n+1}(C_{n+1} - r_{n+1} C_n) - \omega(C_n - C_{n-1}), \qquad (19)$$

where $\nu_n$ is the frequency of diffusion jumps from n to n − 1, while $r_n$ is the equilibrium ratio of $C_n$ and $C_{n-1}$ ($r_n = 1$ for $n \geq 2$, while $r_1 = K_b/K_s$). Equations (19) apply for $n \geq 1$, while for the surface layer (n = 0) we have

$$\dot{C}_0 = \nu_1(C_1 - r_1 C_0) - (\tau_s^{-1} + \omega)(C_0 - C_{0e}), \qquad (20)$$

where $C_{0e}$ is the concentration corresponding to equilibrium with the melt. We are interested in the steady-state distribution ($\dot{C}_n = 0$), and we therefore let all frequencies $\nu_n$ be identical for $n \geq 2$ (although they may differ from $\nu_1$, which is subsequently denoted simply by $\nu$). Then from the string of equations in (19) and the physical condition that $C_n$ cannot increase for $n \to \infty$ we get that $C_n$ = const for $n \geq 1$; we determine the two unknowns $C_0$ and $C_1$ in (19) and (21) to get K from

$$\frac{K - K_b}{K_s - K} = \frac{V}{\nu h}\left[1 + \tau_s\left(\frac{V}{h} + \frac{\nu K_b}{K_s}\right)\right]. \qquad (21)$$

This formula is also qualitatively applicable to a diamond lattice if we replace $K_s$ by the average value $\overline{K}_s$; however, this theory of diffusion after embedding does not incorporate various complicating factors such as the following:

(1) The actual diffusion occurs by impurity uptake by vacancies, with the resulting complex then migrating.

(2) The electrical-image force influences the migration of a charged impurity, and the interaction energy for the first bulk layer is $e^2/4\varepsilon h$ (here e is the charge on the impurity and $\varepsilon$ is the dielectric constant of the crystal), which is 0.65 kT for Ge and Si.

There may be a potential difference $\Delta\varphi$ between the surface and bulk of the crystal over the Debye screening length ($3.5 \cdot 10^{-7}$ cm for Si but half this for Ge). This results in a marked K(V) dependence not only for a face but also for any surface, with K falling for donors in the case $\Delta\varphi > 0$, while it increases for acceptors (the converse applies for $\Delta\varphi < 0$). This asymmetry is not observed for Si [14], since it is clear that $|\Delta\varphi| < kT/e$ for an inherent semiconductor.

The first two points indicate that the jump frequency (from the first bulk layer to the surface layer) should be taken not as $Dh^{-2}$ but as the vacancy trapping frequency. Vacancies may be attached to an impurity node from 12 adjacent nodes in a diamond lattice, and the frequency of an individual jump type is then $8D_v/3h^2$ [17], where $D_v$ is the vacancy diffusion coefficient. Therefore, $\nu = 32D_vC_vh^{-2} \approx 3 \cdot 10^5$ $sec^{-1}$ for Ge and Si [16]. As regards order of magnitude, $\nu$ coincides with $Dh^{-2}$ for $D \sim 3 \cdot 10^{-10}$ $cm^2/sec$. We see from (21) that K(V) varies from $K_b$ to $\overline{K}_s$ and passes through the intermediate value $(K_b + \overline{K}_s)/2$ for $V \approx h\nu \approx 10^{-2}$ cm/sec for all impurities in Si. This value of V agrees with the data for Al and P [14], whose diffusion coefficients differ by a factor 3 [17]. However, the observed transition from $K_b$ to $\overline{K}_s$ is much sharper than (21) would imply, and this discrepancy, the possible errors of experiment, and the above complicating factors are together at least in part responsible for the more nearly equidistant spacing of the steps in the dislocation growth mechanism.

Nevertheless, the result for $R = K/K_b$ given by (21), (16), and (17) describes the observed values for R correctly (as regards order of magnitude); for P [4], Sb [21], and Al [22] in Si one finds channels having R of about 1.5 (if the doping is not too heavy), whereas calculation (for the actual growth rate V) gives 1.5, 1.6, and 1.9 respectively. The corresponding channels in Ge [1] show agreement for As, Sb, and In (theoretical R of 1.4, 1.6, and 1.9), but for P and Ga (theoretical R 1.3) there are considerable discrepancies, probably because the model formula of (16) is not applicable here. One expects R to increase with $K_s/K_b$, i.e., (16) indicates that this should occur at least as $K_b$ falls. However, if $K_b$ is very small (as is $K_s$), then the $\tau_s$ given by (17) becomes less than $\tau_2$, i.e., $G_2$ becomes substantially less than 1. In that case, R is determined by $G_2K_s/K_b$, which approaches 1 as $\tau_s/\tau_2$ falls. Correspondingly, Bi in Ge [1] ($K_b = 5 \cdot 10^{-5}$) gives comparatively weak channels (R = 1.6).

On the whole, the theory is in satisfactory agreement with the observed V dependence and $\theta$ dependence of K, which confirms that adsorption on the faces [13] plays no major part (the adsorption energy is small). The following factors are the main ones: (1) if $\theta > 1°$, step fluctuations cause K to agree with the bulk partition coefficient $K_b$, (2) if $\theta$ is sufficiently small, and particularly on faces, the steps completely embed the impurity in the surface layer (if $K_b$ is not too small), and (3) the embedded impurity is redistributed by diffusion.

## Literature Cited


1. J. A. M. Dikhoff, Solid State Electron., 1:202 (1960).
2. M. G. Mil'vidskii and A. V. Berkova, Fiz. Tverd. Tela, 5:709 (1963).
3. J. B. Mullin and K. F. Hulme, J. Phys. Chem. Solids, 17:1 (1960).
4. A. A. Veselkova, M. I. Osovskii, E. S. Fal'kevich, K. N. Neimark, and G. A. Dobrokhotov, Silicon and Germanium, Vol. 1 [in Russian], Metallurgiya, Moscow (1969), p. 29.
5. J. A. Burton, R. C. Prim, and W. P. Slichter, J. Chem. Phys., 21:1987 (1953).
6. V. V. Voronkov, Kristallografiya, 17:909 (1972).
7. H. S. Carslaw and J. C. Jaeger, Conduction of Heat in Solids [Russian translation], Nauka, Moscow (1961).
8. Yu. M. Shashkov and V. M. Gurevich, Zh. Fiz. Khim., 42:2068 (1968).
9. A. A. Chernov, Usp. Fiz. Nauk 73:277 (1961).
10. V. V. Voronkov, Kristallografiya, 18:32 (1973).
11. V. V. Voronkov, Kristallografiya, 12:831 (1967).
12. D. E. Temkin, Kristallografiya, 17:461 (1972).
13. A. Trainor and B. E. Bartlett, Solid State Electron., 2:106 (1961).
14. V. V. Voronkov, V. P. Grishin, and Yu. M. Shashkov, Neorgan. Mater., 3:139 (1967).
15. V. V. Voronkov and A. A. Chernov, Kristallografiya, 12:222 (1967).
16. V. V. Voronkov, G. I. Voronkova, and M. N. Iglitsyn, Fiz. Tekhn. Poluprovod., 6:20 (1972).
17. B. I. Boltaks, Diffusion and Point Defects in Semiconductors [in Russian], Nauka, Leningrad (1972).

18. A. A. Chernov, In: Growth of Crystals, Vol. 3, Consultants Bureau, New York (1962), p. 35.
19. R. N. Hall, J. Phys. Chem., 57:836 (1953).
20. V. P. Grishin, G. I. Kononov, and Yu. M. Shashkov, Abstracts for the Second Conference on the Physicochemical Principles of Semiconductor Doping [in Russian], Moscow (1972), p. 115.
21. K. E. Benson, Electrochem. Technol., 3:332 (1965).
22. B. M. Turovskii, Neorgan. Mater., 4:307 (1968).

# IMPURITY TRAPPING IN THE MOVEMENT OF A SHORT ELEMENTARY STEP


## S. S. Stoyanov

*Institute of Physical Chemistry, Bulgarian Academy of Sciences, Sofia*


Chernov [1-3] has examined the microscopic processes in impurity trapping during motion of a single step. In his model, a new row is laid down by motion of an isolated kink along a step of infinite length. However, there are many short steps on a surface, especially near the points of emergence of screw dislocations (Fig. 1), and these also move by deposition of fresh series at the ends. The finite sizes of such series affect the growth and impurity-capture kinetics. Here we examine the effects of trapped-impurity concentration on the growing-step length.

Consider the semiinfinite monatomic layer ABCD (Fig. 2) on the assumption that new rows are laid down only at end BC. We assume that new atoms begin to be laid down at the beginning of the next row when a kink reaches the end of one row. The frequency of atomic attachment to a kink is denoted by $\omega_+$, while the detachment frequency for any atom having three nearest neighbors is denoted by $\omega_-$. Further, that frequency for atoms of numbers 1, n + 1, ... having only two nearest neighbors is denoted by $\omega'_-$. The retention probability $U_k$ plays an important part in the kinetic description, since this is the probability that an atom of number k (Fig. 2) will never subsequently leave the crystal. The problem is thus to determine $U_k$ for k = 1, 2, ..., n as a function of the attachment and detachment frequencies.

The elementary theory of probability indicates that the probability of an atom persisting at a kink for a time t is $\exp(-\omega_- t)$, while the probability of its being built in doing the interval from t to t + Δt is $(1/\tau_k)\exp(-t/\tau_k)\,dt$, where $\tau_k$ is the mean containment time. Therefore, we have

$$U = \int_0^\infty \exp(-\omega_- t)\frac{1}{\tau_k}\exp\left(-\frac{t}{\tau_k}\right)dt = \frac{1}{1+\omega_-\tau_k}, \quad k=2,\ 3,\ \ldots,\ n; \tag{1}$$

$$U_1 = \int_0^\infty \exp(-\omega'_- t)\frac{1}{\tau_1}\exp\left(-\frac{t}{\tau_1}\right)dt = \frac{1}{1+\omega'_-\tau_1}.$$

The mean $\tau_k$ is given by

$$\omega_+\tau_k U_{k+1} = 1.$$

We combine this with (1) to get

$$U_1 = U_2[(\omega'_-/\omega_+) + U_2]^{-1},$$
$$U_k = U_{k+1}[\omega_-/\omega_+ + U_{k+1}]^{-1}. \tag{2}$$

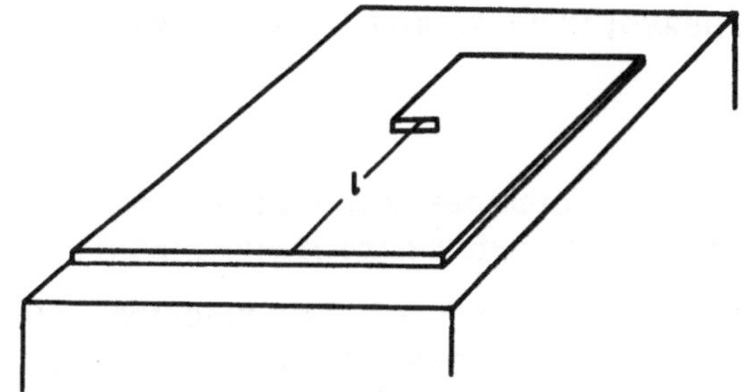


Fig. 1. Dislocation-growth scheme.

We use the periodicity condition

$$U_{n+1}=U_1,$$

and solve the system of equations for the conservation probability as follows:

$$\begin{aligned} U_n &= U_1[(\omega_-/\omega_+)+U_1]^{-1}, \\ U_{n-1} &= U_n[(\omega_-/\omega_+)+U_n]^{-1}=U_1\{(\omega_-/\omega_+)^2+U_1[1+\omega_-/\omega_+]^{-1}, \\ &\cdots\cdots\cdots\cdots\cdots\cdots \\ U_k &= U_1\left\{\left(\frac{\omega_-}{\omega_+}\right)^{n-k+1}+U_1\left[1+\frac{\omega_-}{\omega_+}+\dots+\left(\frac{\omega_-}{\omega_+}\right)^{n-k}\right]\right\}^{-1} \\ &\cdots\cdots\cdots\cdots\cdots\cdots \\ U_2 &= U_1\left\{\left(\frac{\omega_+}{\omega_+}\right)^{n-1}+U_1\left[1+\frac{\omega_-}{\omega_+}+\dots+\left(\frac{\omega_-}{\omega_+}\right)^{n-2}\right]\right\}^{-} \end{aligned} \tag{3}$$

We combine the last equation with the first one from (2) to get

$$U_1=\left[1-\frac{\omega'_-}{\omega_+}\left(\frac{\omega_-}{\omega_+}\right)^{n-1}\right]\left[1+\frac{\omega'_-}{\omega_+}\,\frac{1-(\omega_-/\omega_+)^{n-1}}{1-(\omega_-/\omega_+)}\right]^{-1} \tag{4}$$

The condition for all the $U_k$ to be zero is the condition for the end to be in equilibrium with the

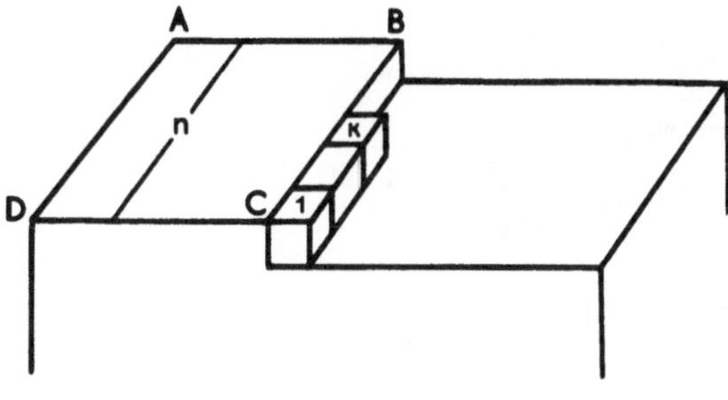


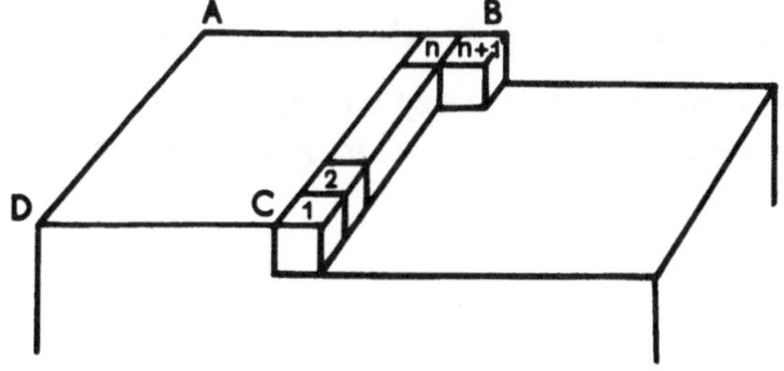


Fig. 2. Scheme for model, with numbers denoting the sequence of atomic attachments.

parent phase; from (3) and (4) it follows that

$$\omega_+^n + \omega'_- \omega_-^{n-1}. \tag{5}$$

This equation coincides with that derived from the method based on mean work of detachment [4]. Equation (5) defines the linear size $n_k$ of a two-dimensional nucleus for any given $\omega_+$.

We now consider impurity trapping. We assume that the main component and the impurity form a substitutional solution. The frequency of impurity-atom attachment to a kink is denoted by $\overline{\omega}_+$, while the detachment frequency is denoted by $\omega_-$, and the detachment frequency for the main atoms is $\overline{\omega}_{-2}$ if one of the nearest neighbors is an impurity atom; the impurity concentration trapped by step motion for a step ending with n atoms is

$$c(n) = \frac{1}{n} \sum_{k=1}^{n} \frac{\overline{\omega}_+ \overline{U}_k}{\omega_+ \overline{U}_k + \omega_+ U_k} \approx \frac{1}{n} \sum_{k=1}^{n} \frac{\overline{\omega}_+ \overline{U}_k}{\omega_+ U_k}, \tag{6}$$

where $U_k$ is the retention probability for an impurity atom. If such an atom is enclosed only by major-component atoms, the values for $U_k$ (k = 2, 3, ..., n − 1) are

$$\overline{U}_k = \frac{\omega_+^2 U_{k+2}}{\overline{\omega}_- \omega_{-2} + (\overline{\omega}_- \omega_+ + \omega_+^2) U_{k+2}}. \tag{7}$$

The expressions for $\overline{U}_1$ and $\overline{U}_n$ are somewhat different, but the difference can be neglected, and we assume in what follows that they are also derived from (7). From (3) we get the ratio $\overline{U}_k/U_k$ and

$$\frac{\overline{U}_k}{U_k} = \frac{\omega_+^2 [1-(U_1/U_\infty)](\omega_-/\omega_+)^{n+1} (\omega_-/\omega_+)^{-k} + \omega_+^2 (U_1/U_\infty)}{\overline{\omega}_- \omega_{-2}\left(1-\frac{U_1}{U_\infty}\right)\left(\frac{\omega_-}{\omega_+}\right)^{n-1}\left(\frac{\omega_-}{\omega_+}\right)^{-k} + U_1(\overline{\omega}_- \omega_+ + \omega_+^2) + \overline{\omega}_- \omega_{-2} \frac{U_1}{U_\infty}}.$$

We substitute this into (6) and replace the sum by an integral to get

$$c(n) = c_\infty + \frac{\mathring{c} - c_\infty}{n \ln(\omega_+/\omega_-)} \ln \frac{M(\omega_+/\omega_-)^n + N}{M+N}, \tag{8}$$

where

$$c_\infty = \frac{\omega_- \omega_+}{(\omega_+ + \overline{\omega}_-)(\omega_+ - \omega_-) + \overline{\omega}_- \omega_{-2}},$$

$$\mathring{c} = \frac{\overline{\omega}_+}{\omega_+} \frac{\omega_-^2}{\overline{\omega}_- \omega_{-2}}, \quad M = \omega_- \omega_{-2}\left(1-\frac{U_1}{U_\infty}\right)\left(\frac{\omega_-}{\omega_+}\right)^{n-1},$$

$$U_\infty = \frac{\omega_+ - \omega_-}{\omega_+}, \quad N = U_1(\overline{\omega}_- \omega_+ + \omega_+^2) + \overline{\omega}_- \omega_{-2} \frac{U_1}{U_\infty}.$$

Here $\mathring{c}$ denotes the equilibrium impurity concentration at the end of a step, while $c_\infty$ denotes the impurity concentration trapped on motion of a step of infinite length.

The dependence of the logarithm in (8) on the step length is very weak, so we take this term as being a constant. We take the value of the logarithm for $n = n_k$ to get the simple approximate formula

$$c(n) = c_\infty + \frac{\mathring{c} - c_\infty}{n/n_k}.$$

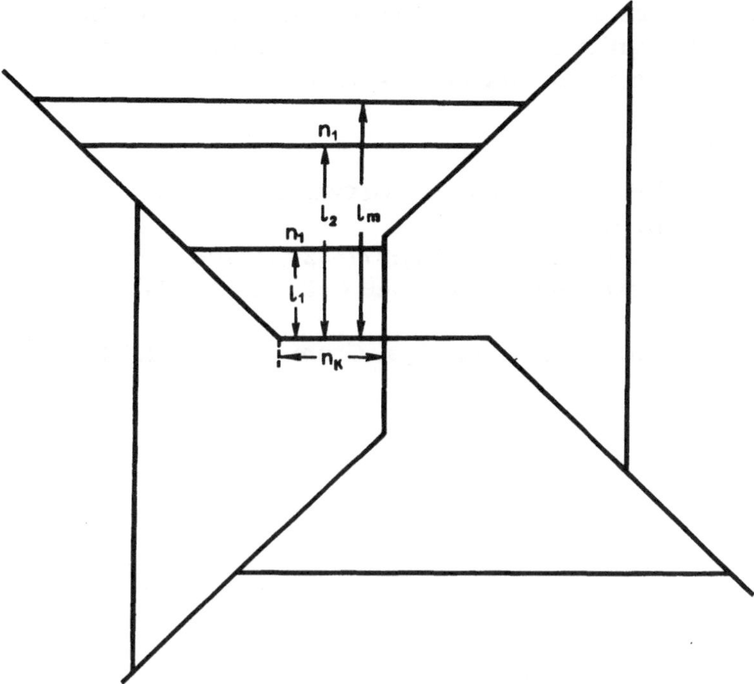


Fig. 3. Scheme for region of integration.

The length of a step is dependent on the distance from the axis of a screw dislocation (Fig. 1), so the trapped-impurity concentration is also dependent on that distance. Near the axis of a dislocation, the step length is nearly equal to the linear dimension of a two-dimensional nucleus, and in that case the value for $\mathring{c}$ is close to the equilibrium one. The impurity concentration falls as the distance to the axis increases if the impurity is strongly adsorbed and vice versa. This means that the mean impurity concentration in the crystal is dependent on the dislocation density at the growing surface:

$$\bar{c} = \rho \int_{S_1} c(x,y)dS \tag{9}$$

Here $\bar{c}$ is the mean impurity concentration in the crystal, $\rho$ is the surface dislocation density, $S_1 = 1/\rho$ is the proportion of the face growing by motion of steps generated by one dislocation, and $c(x, y)$ is local impurity concentration. In order to derive the integral in (9) we assume that the growth rate of a step is independent of the length for $n > n_k$ and is zero for $n \leq n_k$, and then

$$n = n_k + l \quad \text{for} \quad 0 \leqslant l \leqslant n_k,$$
$$n = 2l \quad \text{for} \quad n_k \leqslant l,$$

where $l$ is the distance to the dislocation axis. On this assumption, the region of integration has simple geometry (Fig. 3) and the impurity concentration is dependent only on $l$. Integration gives

$$\bar{c} \approx c_\infty + 2n_k a\sqrt{\rho}(\mathring{c} - c_\infty).$$

In conclusion we note that the isomorphous impurity is unevenly distributed if the trapping is essentially of nonequilibrium type, and the mean impurity concentration is dependent on the surface dislocation density for the growing face.

## Literature Cited

1. A. A. Chernov and B. Ya. Lyubov, In: Growth of Crystals, Vol. 5a, Consultants Bureau, New York (1968), p. 7.
2. A. A. Chernov, Usp. Fiz. Nauk, 100(2):277-328 (1970).
3. A. A. Chernov, Adsorption et Croissance Cristalline. Colloq. Internat. CNRS, Paris (1965), p. 283.
4. I. N. Stranskii and R. Kaishev, Z. Phys. Chem., 26:100 (1934).

# GROWTH CONDITIONS AND STRUCTURED NITROGENOUS INCLUSIONS IN DIAMOND

**Yu. A. Klyuev, N. F. Kirova, V. I. Nepsha, and V. M. Zubkov**

*All-Union Diamond Research Institute, Moscow*

The thermodynamic conditions for diamond production in nature are important not only from the geological viewpoint but also in the production of large artificial single crystals. At present, diamonds grown under the conditions most appropriate for laboratory use (45-90 kbar and 1200-2000°C) differ substantially from natural diamonds, in particular as regards the mode of entry of the most common impurity, namely nitrogen. Although the total nitrogen contents of natural and synthetic diamonds may be comparable ($10^{19}$-$10^{20}$ atoms/cm$^3$) [1, 2], the natural ones contain nitrogen mainly as nonparamagnetic clumps of atoms, which lie at losely adjacent lattice nodes, whereas synthetic diamonds usually contain single paramagnetic nitrogen atoms. As the nitrogen levels in natural and synthetic diamonds are high, and may well be close to the limiting value, it is extremely likely that the factors governing the form taken by the nitrogen are the temperature and pressure of growth rather than the chemical form taken by the nitrogen in the environment.

We have made detailed measurements on nitrogen in synthetic diamonds, including forms similar to those found in natural diamond.

We applied IR methods to synthetic crystals of various habits made during a single synthesis, and we found [3] that a cubic crystal contains the nitrogen only as single atoms, as is clear from the band shape, particularly the strongest one at 8.82 μm [4]. In a cube-octahedron and, to a large extent, an octahedral crystal we find that the single-atom level is somewhat reduced, whereas there is an increase in number of centers responsible for the 7.8 μm absorption (Fig. 1), which is due to the nitrogen clumps characteristic of natural diamonds [1, 5].

The correlation between the form taken by the nitrogen and the habit for synthetic diamonds goes with the P−T diagram for carbon [6, 7] (Fig. 2) to show that the nitrogen clumps most probably form at elevated temperatures near the graphite−diamond equilibrium line. As the habit is determined by the P−T synthesis conditions, the form taken by the nitrogen clearly reflects those conditions.

Synthetic diamonds showing elevated absorption at 7.8 μm, e.g., octahedral diamonds, or diamonds specially doped with nitrogen during synthesis [5], show an additional ESR line, which has been supposed [8] to be due to paramagnetic complexes of substitutional nitrogen atoms that do not lie at adjacent lattice nodes.

The mechanical strength of a synthetic diamond is dependent on the concentration of paramagnetic complexes (Fig. 3) rather than on the concentration of single atoms, which was

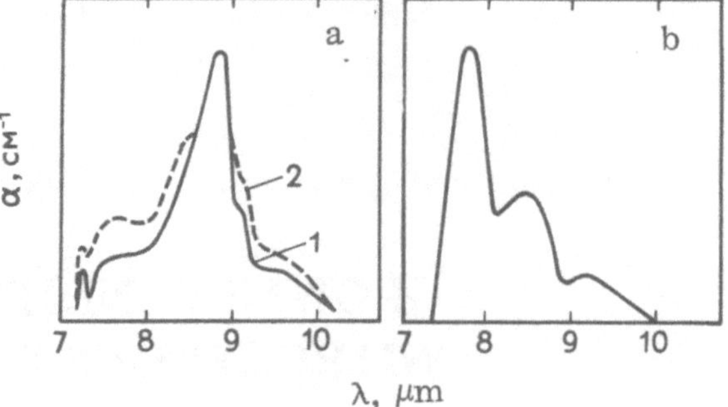


Fig. 1. IR absorption spectra of diamonds in the one-phonon region: (a) synthetic diamonds (1 cube, 2 octahedron); (b) natural diamond of type Ia.

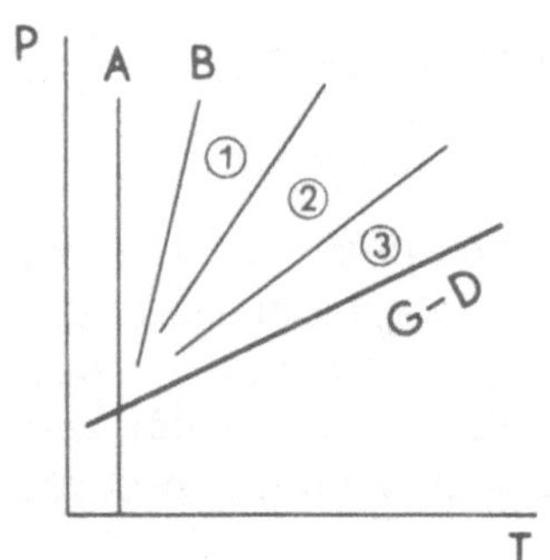


Fig. 2. Regions of formation of various habbit types of diamond in the P – T diagram for carbon: (1) cube; (2) cube-octahedron; (3) octahedron; A melting line for eutectic compositions; B boundary of unsaturated solution region; G – D, graphite – diamond equilibrium line.

the earlier assumption [9]. The relation of strength to single-atom concentration is purely statistical and arises mainly because the probability of complexes increases with the concentration of the single atoms.

Although the paramagnetic complexes in synthetic diamonds differ from nitrogen complexes in natural diamonds, in that some intermediate form is taken such as

$$\left(-\overset{|}{\underset{|}{N}}-\overset{|}{\underset{|}{C}}\ldots\overset{|}{\underset{|}{C}}-\overset{|}{\underset{|}{N}}-\right)$$

which lies between the form typical of synthetic diamonds $\left(-\overset{|}{\underset{|}{N}}-\right)$ and the form found in natural diamonds $\left(-\overset{|}{\underset{|}{N}}-\overset{|}{\underset{|}{N}}-\overset{|}{\underset{|}{N}}-\right)$, nevertheless such complexes strengthen the crystals in the way that nitrogen complexes in natural diamonds do. Consequently, synthetic diamonds of ele-

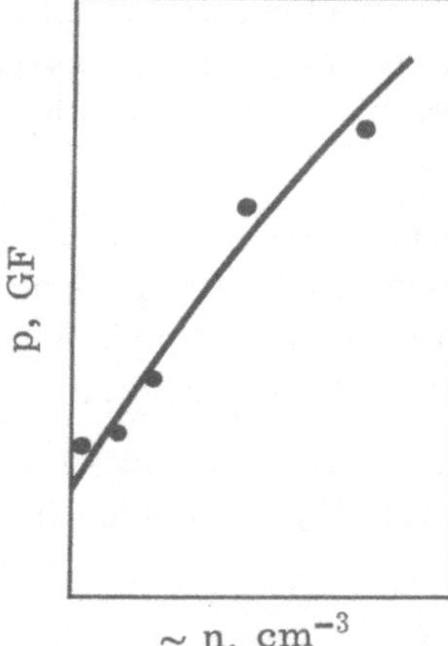


Fig. 3. Strength P of synthetic diamonds in relation to concentration n of paramagnetic nitrogen complexes.

vated absorption at 7.8 μm have features in common with natural ones not only as regards form taken by the nitrogen but also as regards mechanical parameters.

The following conclusions are therefore drawn:

1. The balance between the various forms of nitrogen serves to indicate the P–T growth conditions;

2. The various forms taken by nitrogen in synthetic diamonds go with the growth conditions to define new ranges in the thermodynamic synthesis parameters in which the properties are close to those of natural specimens.

## Literature Cited

1. W. Kaiser and W. L. Bond, Phys. Rev., 115:857 (1959).
2. R. M. Chrenko, H. M. Strong and R. E. Tuft, Phil. Mag., 23:313 (1971).
3. Yu. A. Klyuev, V. I. Nepsha, and G. N. Bezrukov, Almazy, No. 5, 5 (1972).
4. H. B. Dyer, F. A. Roal, L. Du Preez, and J. H. M. Loubser, Phil. Mag., 11:763 (1960).
5. R. T. Elliott, Proc. Phys. Soc., 76:787 (1960).
6. A. A. Diardini and S. E. Tydings, Amer. Mineral., 47:1393 (1962).
7. Yu. A. Litvin, Neorg. Mater., 4:175 (1968).
8. L. A. Shul'man, I. M. Zaritskii, and K. A. Tikhonenko, Fiz. Tverd. Tela, 9:1964 (1967).
9. V. P. Butuzov, M. I. Samoilovich, G. N. Bezrukov, A. I. Novozhilov, and N. F. Kirova, Almazy, No. 3, 5 (1968).

# EFFECTS OF ISOMORPHOUS REPLACEMENT ON SOME PROPERTIES OF SYNTHETIC DIAMONDS

**G. N. Bezrukov and V. P. Butuzov**

*All-Union Synthetic Raw Materials Research Institute, Aleksandrov*

Although there have recently been many studies on the physical properties of diamonds, there has been no study on the effects of isomorphous replacement of carbon. Various workers, in particular Varshavskii and Orlov (natural diamonds) and Butuzov, Vishnevskii, Kirova, Klyuev, and Bezrukov (synthetic diamonds) have indicated individual variations in properties due to the presence of nonstructural impurities. The available methods do not always allow one to establish whether a particular impurity is structural or purely mechanical. On the other hand, a growing diamond crystal tends to trap impurities in amounts such that the mechanical inclusions may exceed the level of isomorphous components of the same composition by a substantial factor. Accumulated experience with diamond growing shows that doping elements affect various characteristics substantially, e.g., the color, shape, electrical parameters, and optical behavior. However, one can speak reliably of isomorphism only in a very restricted number of instances.

Kukharenko [1] has indicated the main factors that result in isomorphous substitution: (1) structural ones (lattice type, packing density in the structural units, coordination numbers of ions, etc.), (2) crystallochemical (chemical symmetry in the immediate environment of an ion, compositions of crystalline phases, valency and volume compensation, etc.), (3) concentration ones (the balance of activity between the mineral-forming and impurity components in the system, the form taken by such components, etc.), and (4) thermodynamic (temperature and pressure in the mineralizing system). As a synthetic diamond is produced in a system containing only a few components (carbon in a form other than diamond plus metal), the scope for isomorphous substitution is very limited, in spite of the above numerous possible factors.

By definition [2], isomorphous substitution is imperfect if the isomorphism is subject to certain limits but is perfect if one type of atom can replace another completely. This implies that high impurity levels result in minerals of variable composition, whereas trace components characterize minerals of constant composition. Although diamond almost always contains traces of nitrogen, boron, nickel, aluminum, and so on, it can be termed a mineral of constant composition. Firstly, the impurity contents are extremely low, and in particular the nitrogen content does not exceed 0.25%, while the usual level is 0.01–0.001%. Secondly, it has been shown that carbon–nickel, carbon–iron, and other such systems form eutectic mixtures, in which the mutual solubility is extremely restricted. As the replacement occurs on only a restricted scale, i.e., the contents of the foreign elements do not exceed the 0.1 or 0.01% level, it is correct to say that one is dealing with isomorphous impurities.

The temperature distribution, pressure, and types of reacting components are the major factors that determine the ordering of isomorphous impurities.

As a rule, isomorphous substitution is accompanied by reduced free energy (reduced chemical potential, elevated entropy) when the mixture is formed from the pure components [3]. It has also been emphasized that entropy increase on isomorphous substitution is a general feature of any multicomponent system. However, the rigid structure of diamond and related substances imposes restrictions on the scope for substitution, which are dependent on various detailed features of the interacting particles (size, shell structure, and bond type), as well as on details of the crystal structure (density, symmetry, coordination number, etc.) [4].

It is extremely difficult to determine the contents of structural impurities in synthetic diamond crystals because such crystals contain numerous mechanical inclusions, which tend to mask the effects due to the structural impurities.

For this reason, isomorphism can be examined only fom the changes in certain physical characteristics. Natural diamonds [5] and synthetic ones [6, 7] contain an extremely wide range of trace elements, but only certain of them (nitrogen, nickel, boron, and aluminum) have definitely been shown to be isomorphous with carbon.

The substitutional atom in such an instance may differ in size from the atom it replaces, and also in valency, etc.; this distorts the lattice, and the changes in parameters are accompanied by stresses and so on. The isomorphous miscibility becomes restricted if the difference in crystallochemical characteristics between the two atoms becomes too large [8].

Grigor'ev [9] indicates that structural anisotropy in a mineral predetermines the entry of various components into different parts of the crystal: face growth pyramids, growth surfaces of edges, growth lines from vertices, and so on. To this we may add that the concentrations of the doping substances tend to vary during the crystal growth, as does the supersaturation of the source solution, the detailed crystallization conditions, and so on. All of these changes are reflected in the zoned structure of the diamond crystal. In some instances, this zoning can be seen by eye, but as a rule this is due to nonstructural impurities, and more often the zoning has to be determined by detailed physical examination, which is the proper basis for speaking of isomorphism.

Some of the specific growth-mechanism features result in nonuniformity in impurity distribution, which results in defects; on the other hand, defects not associated with impuities may cause impurities to localize around themselves.

Isomorphism arises by entry of foreign atoms, which replace the basic ones, so one can say that any isomorphous substitution constitutes a special type of defect that may influence typical physical parameters.

ESR and IR studies on diamonds show that even levels of $10^{-4}$ to $10^{-6}$ wt.% of impurities affect the optical and magnetic characteristics; these trace impurities are very characteristic geochemical indicators for minerals generally and for synthetic diamonds in particular.

A primary classification of diamonds into nitrogen-bearing and nitrogen-free has long been accepted.

Recent researches on artificial diamonds have shown that the form taken by the nitrogen is influenced by the growth conditions; in particular, high growth rates are common with synthetic diamonds, which cause the nitrogen atoms trapped in dispersed form to produce deep donor levels. Most synthetic diamonds are of type Ib, i.e., contain paramagnetic nitrogen. In that case, the difference in ESR spectrum from natural varieties lies solely in the greater line width. Crystals grown in the presence of B or Al result in a different subvariety, where the nitrogen forms or AlN complexes. The nitrogen in that case is detected from the IR absorption or else indirectly from the thermoluminescence [10].

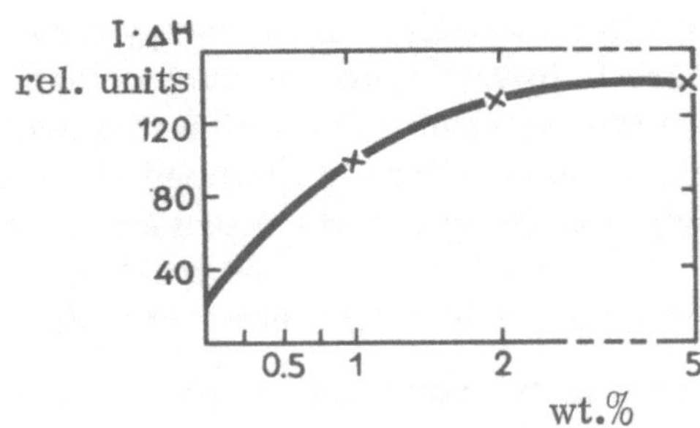


Fig. 1. Contents of dispersed paramagnetic nitrogen in diamond crystals in relation to amount of nitride in initial mixture.

The next variety consists of crystals grown in the presence of excess nitrogen, which is introduced into the crystallization medium as various nitrides. Such a crystal contains not only the normal paramagnetic form of nitrogen but also exchange-interacting nitrogen pairs [11]. Our evidence indicates that synthesis in the presence of manganese nitride increases the nitrogen content by 1.5-2 orders of magnitude, while it also clearly increases the concentration of paramagnetic nitrogen (Fig. 1).

IR spectroscopy has been used to examine the lattice defects; few such studies have been performed for synthetic diamonds, probably because the crystals are extremely small. Klyuev and Nepsha have devised a system enabling one to use crystals weighing only $10^{-4}$-$10^{-3}$ carat. The limit of detection of the method is such that nitrogen defects representing not less than $10^{18}$ atoms/cm$^3$ can be detected.

Measurements have been made on routine-production diamonds of SAM grade, as well as on ones doped with nitrogen and silicon, in addition to measurements on crystals grown under conditions such as to prevent nitrogen trapping; it has been found that the physical characteristics vary considerably. In particular, crystals with elevated nitrogen contents contain nonparamagnetic nitrogen structures similar to those observed in natural diamonds. The natural material has been compared with routine products and nitrogen-doped diamonds; there is a marked difference in absorption strength at about 7.0 μm (Fig. 2), which indicates an increased level of nonparamagnetic nitrogen in natural crystals of type Ib and in crystals made with elevated nitrogen contents. A similar result was obtained on examining routine-production and nitrogen-doped crystals by ESR. A single broad line having g around 2 appears in the spectrum on increasing the level of nonparamagnetic nitrogen centers, this being superimposed on the ordinary nitrogen spectrum.

The shape relations between crystals made in a single run but found in different temperature zones are correlated with the IR absorption spectra; the proportion of the nitrogen in the aggregate form increases on going from cubes to octahedra. On the other hand, the proportion of paramagnetic nitrogen shows the reverse trend, since cubic crystals have the highest levels and octahedral ones the lowest. A similar relationship has been established for natural diamonds [12].

Fig. 2. Intensity ratios in various type Ib diamonds: ——— natural; – – – diamond synthesized under ordinary conditions (SAM grade); · · · · synthesized in the presence of excess nitrogen.

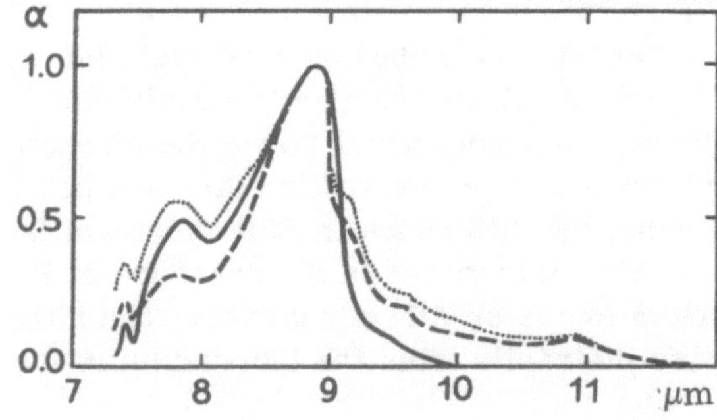

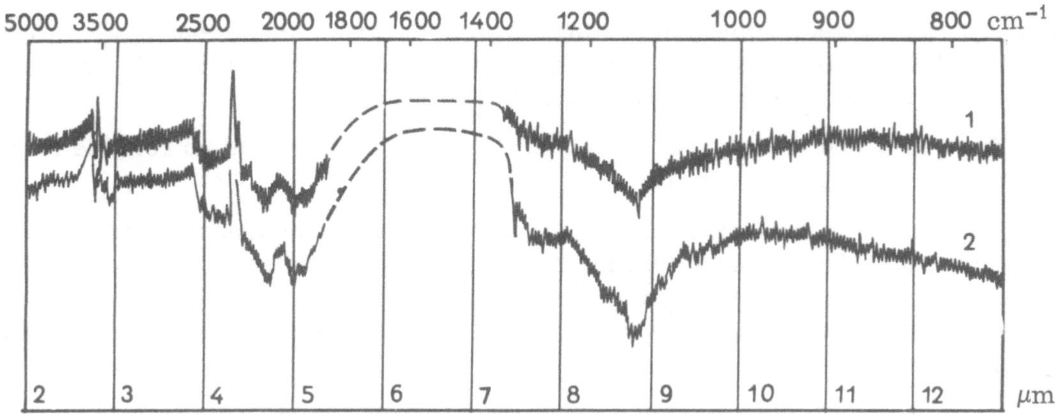


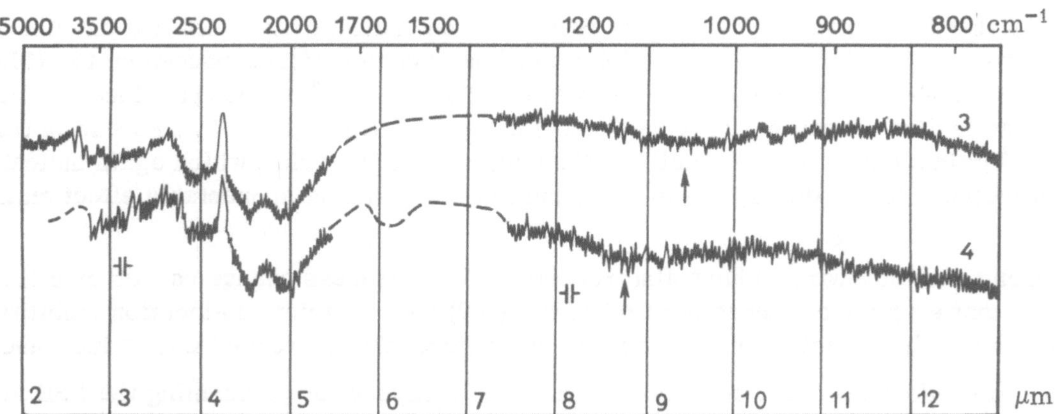


Fig. 3. Spectrograms of diamonds: (1) SDA grade (De Beers); (2) routine-production diamonds of SAM grade; (3) diamonds made under conditions preventing nitrogen capture; (4) silicon-doped diamonds.

Figure 3 shows spectra for routine-production diamonds, ones doped with nitrogen and silicon, nitrogen-free specimens, and ones with low nitrogen contents (SAM), as well as SDA crystals from De Beers. The silicon-doped and nitrogen-free crystals contain no nitrogen in any of the standard forms. A weak paramagetic-nitrogen spectrum is seen for the silicon-doped material, which appears to be due to competition between silicon and nitrogen during growth.

Synthetic diamonds give not only the nitrogen spectrum but also an ESR line associated with paramagnetic nickel; if the medium contains phosphorus and nitrogen, the strength of this line is reduced, because nickel forms two acceptor levels, and therefore the ESR signal strength is directly proportional to the filling of the lower level and inversely proportional to the filling of the upper level. The $^{61}$Ni content in natural material is 1.2%, so these results were indirect until we made diamonds in a system containing an elevated level of $^{61}$Ni. The ESR spectrum then consisted of four lines, as against one in the presence of ordinary nickel. The isotopic hyperfine structure indicates that this ESR spectrum is due to nickel [13].

Boron is the third element that can enter diamond isomorphously; it can appear for example as BN complexes. However, it can also occur in dispersed form, and boron-doped synthetic diamonds represent p-type semiconductors having an acceptor level $E_A = 3.5 \mp 0.1$ eV. The luminescence spectrum of such a crystal consists of a single broad band over the range 3000-

10,000 Å, and it has various features indicating that it arises from donor–acceptor pairs. It is likely that such a pair is provided by the B and N atoms, with the B present in substitutional form. It would seem that aluminum can act in a similar way.

Apart from the above isomorphous components, there is at present no reliable evidence that any other element can behave in the same way.

These differences in the optical characteristics of synthetic diamonds containing N, B, and Al are accompanied by certain difference in mechanical parameters. The load needed to crush a diamond crystal increases with the nitrogen content over a certain range. It would seem that the isomorphous impurity blocks certain parts of the dislocations [14]. Single impurity atoms appear particularly effective here.

There appears to be no correlation between the nitrogen and nickel contents; the ESR spectra also indicate that crystals with imperfect facets contain low nickel levels, while the nitrogen levels vary substantially. The highest nickel levels occur in green and yellow crystals.

Microhardness measurements have been made on ordinary diamonds and on ones containing isomorphous boron; routine-production diamonds have the microhardness on the (100) face in the range 6300-700 $kgf/mm^2$, as against 9400-10,200 $kgf/mm^2$ on the (111) face. Boron-doped crystals give correspondingly 6600 and 7500 and 10,800-11,300. The spread in the microhardness within a single face is small for the boron-doped material, which again indicates that this element enters the crystal in isomorphous form without appreciable effect on the perfection.

Strong and Chrenko [15] have also reported microhardness change caused by nitrogen; they state that sorption on planes parallel to the (100) nets restricts dislocation mobility, which in turn reduces the capacity for plastic flow. Therefore, the microhardness is increased.

We are indebted to M. I. Samoilovich and Yu. A. Klyuev for performing the ESR and optical measurements on the synthetic diamonds.

## Literature Cited

1. A. A. Kukharenko, Abstracts for the 2nd All-Union Symposium on Isomorphism [in Russian], Moscow (1969), p. 84.
2. N. V. Belov and N. L. Smirnova, Abstracts for the 2nd All-Union Symposium on Isomorphism [in Russian], Moscow (1969), p. 2.
3. V. A. Kirkinskii and A. A. Yaroshevskii, Zap. Vses. Min. Obshch., Vol. 96, No. 5 (1967).
4. A. A. Yaroshevskii, Thermodynamic Interpretation of Major Concepts in Isomorphism: Isomorphous Atomic Substitution in Crystals [in Russian], Nauka (1971), p. 48.
5. Yu. L. Orlov, Mineralogy of Diamond, Doctoral Thesis, Moscow (1972).
6. V. P. Butuzov, M. I. Samoilovich, G. N. Bezrukov, A. I. Novozhilov, and N. F. Kirova, Almazy, No. 3, 5 (1968).
7. S. I. Futergendler and V. I. Shemanin, Tr. VNIIASh, Leningrad, No. 7, 15 (1968).
8. V. V. Shcherbina, Geokhimiya, No. 3, 259 (1965).
9. D. P. Grigor'ev, Abstracts for the 2nd All-Union Symposium on Isomorphism [in Russian], Moscow (1969), p. 82.
10. M. I. Samoilovich, V. P. Butuzov, and G. N. Bezrukov, Sinteticheskie Almazy, No. 2, 30 (1970).
11. L. A. Shul'man, I. M. Zaritskii, and K. A. Tikhonenko, Fiz. Tverd. Tela, 9:1964 (1967).
12. Yu. A. Klyuev, Yu. A. Dudenkov, V. I. Nepsha, and Yu. L. Orlov, Almazy, No. 6, 1 (1972).
13. M. I. Samoilovich, G. N. Bezrukov, and V. P. Butuzov, Pis'ma Zh. Eksp. Teor. Fiz., 14:551, (1971).
14. H. G. van Bueren, Imperfections in Crystals, North Holland, Amsterdam (1960).
15. H. M. Strong and R. M. Chrenko, J. Phys. Chem., 75:1838 (1971).